LECTURES ON THE PHYSICS OF HIGHLY CORRELATED ELECTRON SYSTEMS VII

Previous Proceedings in the Series of Lectures on the Physics of Highly Correlated Electron Systems

Year	Title	Publisher	ISBN
2001	Sixth Training Course	AIP Conf. Proceedings Vol. 629	0-7354-0083-0
2000	Fifth Training Course	AIP Conf. Proceedings Vol. 580	0-7354-0019-9
1999	Fourth Training Course	AIP Conf. Proceedings Vol. 527	1-56396-950-5
1998	Third Training Course	*not published*	
1997	Second Training Course	AIP Conf. Proceedings Vol. 438	1-56396-789-8
1996	First Training Course	*not published*	

Other Related Titles from AIP Conference Proceedings

677 Fundamental Physics of Ferroelectrics 2003
Edited by P. K. Davies and D. J. Singh, August 2003, 0-7354-0146-2

633 Structural and Electronic Properties of Molecular Nanostructures: XVI International Winterschool on Electronic Properties of Novel Materials
Edited by Hans Kuzmany, Jörg Fink, Michael Mehring, and Siegmar Roth, November 2002, 0-7354-0088-1

614 Advances in Cryogenic Engineering: Proceedings of the International Cryogenic Materials Conference – ICMC Volume 48
Edited by Balu Balachandran, Donald Gubser, and K. Ted Hartwig, May 2002, 2 vol. hard cover set, CD-ROM included, 0-7354-0060-1

554 Physics in Local Lattice Distortions: Fundamentals and Novel Concepts; LLD2K
Edited by Hiroyuki Oyanagi and Antonio Bianconi, February 2001, 1-56396-984-X

483 High Temperature Superconductivity
Edited by Stewart E. Barnes, Joseph Ashkenazi, Joshua L. Cohn, and Fulin Zuo, August 1999, 1-56396-880-0

427 Lectures on Superconductivity in Networks and Mesoscopic Systems
Edited by Carlo Giovannella and Colin J. Lambert, April 1998, 1-56396-750-2

To learn more about these titles, or the AIP Conference Proceedings Series, please visit the webpage **http://proceedings.aip.org**

LECTURES ON THE PHYSICS OF HIGHLY CORRELATED ELECTRON SYSTEMS VII

Seventh Training Course in the Physics of Correlated Electron Systems and High-Tc Superconductors

Salerno, Italy   14-25 October 2002

EDITORS
Adolfo Avella
Ferdinando Mancini
Università degli Studi di Salerno, Salerno, Italy

SPONSORING ORGANIZATIONS
European Commission (under the TMR Program)
University of Salerno, Italy

AMERICAN INSTITUTE OF PHYSICS

Melville, New York, 2003
AIP CONFERENCE PROCEEDINGS ■ VOLUME 678

Editors:

Adolfo Avella
Ferdinando Mancini

Dipartimento di Scienze Fisiche "E. R. Caianiello"
Unità INFM di Salerno
Università degli Studi di Salerno
Via S. Allende
84081 Baronissi (SA)
ITALY

E-mail: avella@sa.infn.it
mancini@sa.infn.it

L.C. Catalog Card No. 2003109651
ISBN 0-7354-0147-0
ISSN 0094-243X
Printed in the United States of America

CONTENTS

PART II. PARTICIPANT CONTRIBUTIONS

Preface

The ensuing volume contains the lectures delivered at the "Seventh Training Course in the Physics of Correlated Electron Systems and High-Tc Superconductors," held in Vietri sul Mare (Salerno) Italy in October 2002. This volume also contains specific contributions, delivered as seminars by some of the younger researchers in the afternoon sessions.

Following the tradition of previous years, the meeting was devoted to promoting the formation of young scientists by means of training through research. The intent is to bring together for two weeks senior and young researchers in a close and informal atmosphere, paying special attention and effort to an active participation of the young researchers, and to introduce them to specific problems.

The course consisted of four lectures every morning, held by Professors Yuri Izyumov, Natan Andrei, Fakher Assaad, Jozef T. Devreese, and by Dr. Jacques Tempere, and of afternoon activities aimed principally to increase discussions between the students and lecturers, at solving particular exercises, and at oral exposition by the students. The outcome of this type of course was an active participation of the students and a significant interchange of ideas due to the long afternoons devoted to discussions.

It is hoped that both the meeting, which brought together leaders in the field as well as bright and eager beginners, and the present volume may be useful as an up-to-date book for researchers interested in the field.

We wish to acknowledge and thank those whose support made the course and the proceedings possible. The main sponsor of the event is the European Commission, under the TMR Programme. For additional support we gratefully acknowledge the University of Salerno. Finally, we wish to thank Professor Maria Marinaro, President of the International Institute of Advanced Scientific Studies, who hosted the event in the wonderful and warm venue of Vietri sul Mare.

Salerno, 13 May 2003

Adolfo Avella
Ferdinando Mancini

PART I

LECTURES

Polarons

J. T. Devreese

Theoretische Fysica van de Vaste Stoffen (TFVS),
Universiteit Antwerpen (UIA), Universiteitsplein 1, B-2610 Antwerpen, Belgium;
also at: Universiteit Antwerpen (RUCA), Groenenborgerlaan 171, B-2020 Antwerpen, Belgium
and Technische Universiteit Eindhoven, P. O. Box 513, 5600 MB Eindhoven, The Netherlands.

Abstract.
An overview is presented of the theory of continuum ("Fröhlich-Pekar"-) polarons. Experimental manifestations of polarons are analysed and compared to theory. The emphasis is on the response characteristics of one- and many polarons: optical absorption, cyclotron resonance, photoluminescence, Raman scattering. I consider free polarons and polarons confined in space, polarons in 3D, 2D, 0D (quantum dots).

1. THE POLARON CONCEPT

A conduction electron (or hole) together with its self-induced polarization in an ionic crystal or in a polar semiconductor forms a quasiparticle, which is called a polaron (see Fig. 1). The polaron concept was introduced by Landau in 1933 [1].

The physical properties of a polaron differ from those of a band-carrier. A polaron is characterized by its *binding (or self-) energy* E_0, an *effective mass* m^* and by its characteristic *response* to external electric and magnetic fields (e. g. dc mobility and optical absorption coefficient).

If the spatial extension of a polaron is large compared to the lattice parameter of the solid, the latter can be treated as a polarizable continuum. This is the case of a *"large polaron"*, to which most of this paper, except for a few remarks, is devoted.

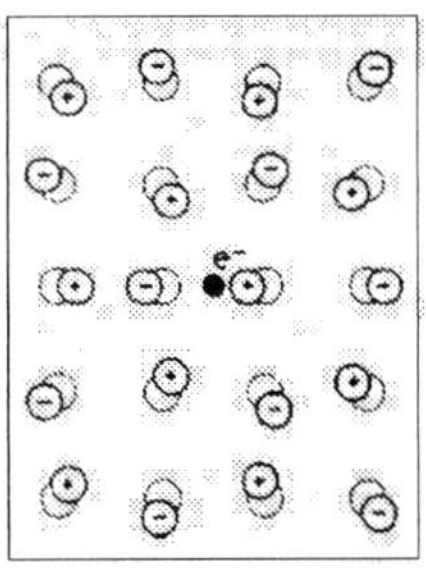

FIGURE 1. Artist view of a polaron. A conduction electron in an ionic crystal or a polar semiconductor repels the negative ions and attracts the positive ions. A self-induced potential arises, which acts back on the electron and modifies its physical properties. (From [2].)

CP678, *Lectures on the Physics of Highly Correlated Electron Systems VII: Seventh Training Course*, edited by A. Avella and F. Mancini

TABLE 1. *Electron-phonon coupling constants*

Material	α	Material	α
InSb	0.022^a	KI	2.5^a
InAs	0.052^a	TlBr	2.55^a
GaAs	0.068^a	KBr	3.05^a
GaP	0.201^a	RbI	3.16^a
CdTe	0.286^b	CdF_2	3.2^a
ZnSe	0.432^a	KCl	3.44^a
CdSe	0.46^a	CsI	3.67^a
CdS	0.527^a	$SrTiO_3$	3.77^e
CdO	0.735^c	RbCl	3.81^a
AgBr	1.53^d	MgO	4.83^f
AgCl	1.84^d	Al_2O_3	5.41^f

[a]From Ref. [7]
[b]From Ref. [8]
[c]From Ref. [9]
[d]From Ref. [10]
[e]After Ref. [11]
[f]From Ref. [12] (for holes)

The polaron is characterised by a coupling constant introduced by Fröhlich, which is defined as:

$$\alpha = \frac{e^2}{\hbar}\sqrt{\frac{m_b}{2\hbar\omega_{\rm LO}}}\left(\frac{1}{\varepsilon_\infty} - \frac{1}{\varepsilon_0}\right). \tag{1}$$

In this definition, $\omega_{\rm LO}$ is the long-wavelength frequency of a longitudinal optical (LO) phonon, ε_∞ and ε_0 are, respectively, the electronic and the static dielectric constant of the polar crystal.

In order to determine the coupling constant α, the electron (hole) band mass m_b must be known. In ionic crystals and polar semiconductors experiments like e. g. cyclotron resonance measurements tend to determine the polaron effective mass m^* rather than the band mass of the carrier — just *because of the polaron effect.* To obtain information about m_b, one can e. g. combine two experiments as will be discussed further on (see Subsection 4.2). In Table 1 polaron coupling constants for selected materials are given.

The first studies on polarons were devoted to the calculation of the self-energy and the effective mass of polarons in the limit of large α, or "strong coupling" (Refs. [3, 4, 5]).

The "weak-coupling" limit, first explored by H. Fröhlich [6], is obtained from the leading terms for $\alpha \to 0$. Well-known weak-coupling results for polarons are:

$$E_0 = -\alpha\hbar\omega_{\rm LO}, \tag{2}$$

$$m^* = \frac{m_b}{1-\alpha/6}. \tag{3}$$

The "weak-coupling" theory is useful and adequate to study polaron effects in materials like InSb ($\alpha = 0.022$) etc.

However, there is nothing magic about $\alpha = 1$, and higher orders in perturbation-theory expansions in powers of α can become significant (see, e. g., Ref. [13]) even if $\alpha < 1$, for example, in GaAs ($\alpha = 0.068$). Also note that, in the "weak-coupling" limit (see, e.

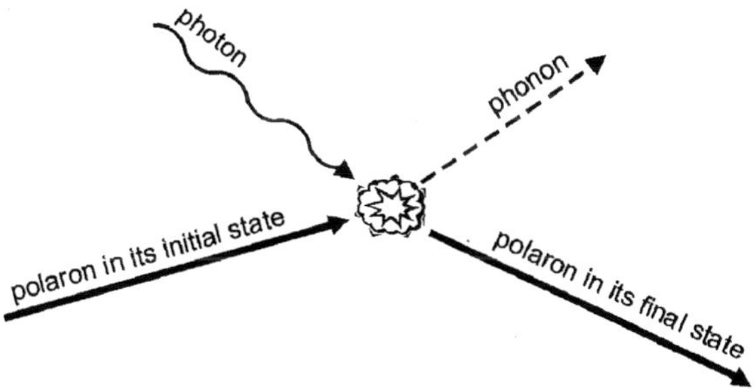

FIGURE 2. Elementary polaron scattering process describing the absorption of an incoming photon and the generation of an outgoing phonon.

g., Ref. [14]), the polaron mobility is dominated by α:

$$\mu = \frac{e}{2m_b\omega_{\rm LO}\alpha}\exp\left(\frac{\hbar\omega_{\rm LO}}{kT}\right) \tag{4}$$

and all physical quantities have series expansion in powers of α.

For many materials it is useful to have an adequate all-coupling polaron theory at one's disposal. Such an all-coupling theory was developed by Feynman using his path-integral formalism [15]. He studied first the self-energy E_0 and the effective mass m^* and later the mobility of polarons [16, 17]. Subsequently the path-integral approach to the polaron problem was generalised and developed to become a tool to study optical absorption, magnetophonon resonance, cyclotron resonance etc.

For some monographs and reviews on the polaron physics, see Refs. [18, 19, 20, 21].

2. OPTICAL PROPERTIES AND INTERNAL STRUCTURE OF POLARONS

2.1. Optical absorption of polarons at weak coupling

At zero temperature and in the weak-coupling limit, the optical absorption is due to the elementary polaron scattering process, schematically shown in Fig. 2.

In the weak-coupling limit ($\alpha \ll 1$) the polaron absorption coefficient was first obtained by Gurevich, Lang and Firsov [22]. At zero temperature, the absorption coefficient for absorption of light with frequency Ω can be expressed in terms of elementary functions in two limiting cases: in the region of comparatively low frequencies $(\hbar(\Omega-\omega_{\rm LO})/\zeta \ll 1)$

$$\Gamma(\omega) = \frac{1}{\varepsilon_0 nc}\frac{2^{1/2}N^{2/3}\alpha}{(3\pi^2)^{1/3}}\frac{e^2}{(\hbar m_b\omega_{\rm LO})^{1/2}}\frac{\omega-1}{\omega^3}\Theta(\omega-1), \tag{5}$$

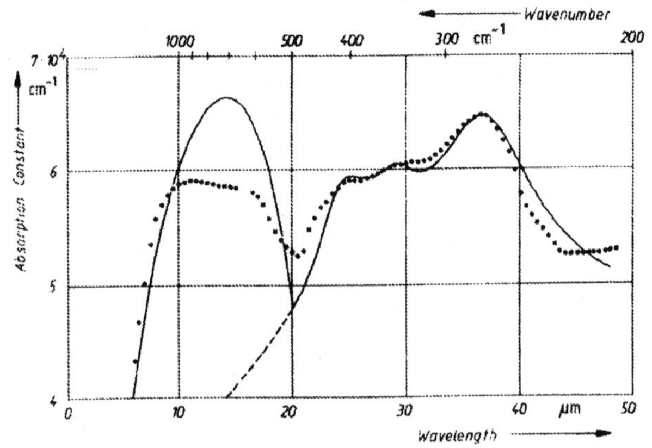

FIGURE 3. Optical absorption spectrum of a CdO-film with the carrier concentration $N = 5.9 \times 10^{19}$ cm^{-3} at $T = 300$ K. The theoretical results are shown with (solid curves) and without (dashed curves) polaron contribution of Eq. (5), and compared to the experimental data (solid dots). (Reprinted from Ref. [23], ©(1969), with permission from Elsevier.)

and in the high-frequency region $(\hbar(\Omega - \omega_{LO})/\zeta \gg 1)$

$$\Gamma(\omega) = \frac{1}{\varepsilon_0 nc} \frac{2Ne^2\alpha}{3m_b\omega_{LO}} \frac{(\omega - 1)^{1/2}}{\omega^3} \Theta(\omega - 1). \tag{6}$$

Here $\omega = \Omega/\omega_{LO}$, ε_0 is the dielectric permittivity of the vacuum, n is the refractive index of the medium, N is the concentration of polarons and ζ is the Fermi level for the electrons. A step function

$$\Theta(\omega - 1) = \begin{cases} 1 & \text{if} \quad \omega > 1, \\ 0 & \text{if} \quad \omega < 1 \end{cases}$$

reflects the fact that at zero temperature the absorption of light accompanied by the emission of a phonon can occur only if the energy of the incident photon is larger than that of a phonon ($\omega > 1$). In the weak-coupling limit, according to Eq. (5), the absorption spectrum consists of a "one-phonon line".

Experimentally, this one-phonon line has been observed for free polarons in the infrared absorption spectra of CdO-films, see Fig. 3. In CdO, which is a weakly polar material with $\alpha \approx 0.74$, the polaron absorption band is observed in the spectral region between 6 and 20 μm (above the LO phonon frequency). The difference between theory and experiment in the wavelength region where polaron absorption dominates the spectrum due to many-polaron effects.

At nonzero temperature, the absorption of a photon can be accompanied not only by emission, but also by absorption of one or more phonons.

A simple derivation in Ref. [24] using a canonical transformation method gives the absorption coefficient of free polarons, which coincides with the perturbation result obtained by Gurevich, Lang and Firsov [22]. As shown in Ref. [24], it is possible to extract an approximation for the optical absorption coefficient of free polarons at $T = 0$ from the impedance calculations of Ref. [16]. On that basis, an analytical expression of the optical absorption is proposed in Ref. [24] for small coupling constants.

2.2. Optical absorption of polarons at strong coupling

The absorption of light by free large polarons was treated in Ref. [25] using the adiabatic strong-coupling approximation of Landau and Pekar. The polaron ground state in this scheme has the energy

$$E_0 = -\frac{\alpha^2}{3\pi}\hbar\omega_{\mathrm{LO}} = -0.106\alpha^2\hbar\omega_{\mathrm{LO}}. \tag{7}$$

If the lattice polarization is allowed to relax or adapt to the electronic distribution of the excited electron (which itself then adapts its wave function to the new potential etc. leading to a self-consistent final state), the so-called *relaxed excited state* (RES) results (see Ref. [4]). Its energy is

$$E_{\mathrm{RES}} = -0.041\alpha^2\hbar\omega_{\mathrm{LO}}. \tag{8}$$

It was argued in Ref. [25], that for sufficiently large α ($\alpha > 3$), the (first) RES of a polaron is a relatively stable state, which can participate in optical absorption transitions. This idea was necessary to understand the polaron optical absorption spectrum in the strong-coupling regime. The following scenario of a transition, which leads to a *"zero-phonon" peak* in the absorption by a strong-coupling polaron, can then be suggested. If the frequency of the incoming photon is equal to

$$\Omega_{\mathrm{RES}} \equiv \frac{E_{\mathrm{RES}} - E_0}{\hbar} = 0.065\alpha^2\omega_{\mathrm{LO}}, \tag{9}$$

then the electron jumps from the ground state (which, at large coupling, is well-characterized by "s"-symmetry for the electron) to an excited state ("$2p$"), while the lattice polarization in the final state is adapted to the "$2p$" electronic state of the polaron.

If one or more real phonons are excited the polaron is said to be in a *scattering state* (see Fig. 4). For photon energies larger than

$$\Omega_{\mathrm{RES}} + \omega_{\mathrm{LO}}, \tag{10}$$

a transition of the polaron towards the first scattering state, belonging to the RES, becomes possible. The final state of the optical absorption process then consists of a polaron in its lowest RES plus a free phonon. A "one-phonon sideband" then appears in the polaron absorption spectrum. This process is called *one-phonon sideband absorption*.

The one-, two-, ... K-, ... phonon sidebands of the zero-phonon peak give rise to a broad structure in the absorption spectrum. The (unstable) polaron state, in which the lattice polarization corresponds to the electron ground state, while the electron is

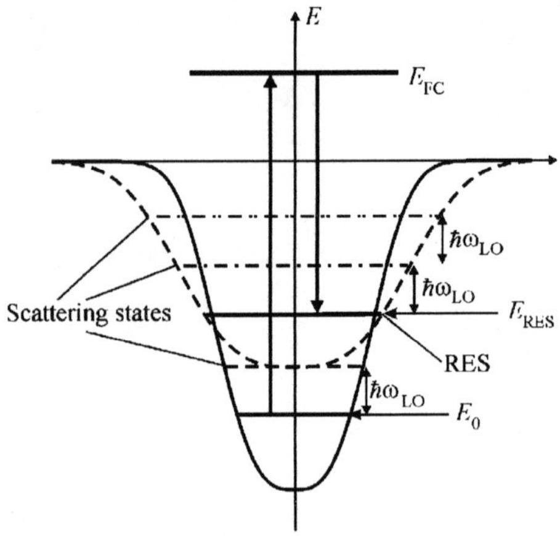

FIGURE 4. Internal excitations of a polaron at strong coupling: E_0 — the ground state, E_{RES} — the (first) relaxed excited state; the Franck-Condon states (E_{FC}) are shown with a set of horizontal lines above the zero energy. In fact, both the Franck-Condon states and the relaxed excited states lie in the continuum and, strictly speaking, are resonances.

excited, is referred to as a *Franck-Condon* (FC) state of the polaron. Within the adiabatic approach, the energy of the lowest FC state is

$$E_{FC} = \frac{\alpha^2}{9\pi}\hbar\omega_{LO} = 0.035\alpha^2\hbar\omega_{LO}. \tag{11}$$

It turns that the *first moment* of the phonon sidebands corresponds to the FC frequency Ω_{FC}:

$$\Omega_{FC} \equiv \frac{E_{FC} - E_0}{\hbar} = 0.141\alpha^2\omega_{LO}. \tag{12}$$

To summarize, the polaron optical absorption spectrum at strong coupling is characterized by the following features ($T = 0$):

a) An intense absorption peak ("zero-phonon line") appears, which corresponds to a transition from the ground state to the first RES at Ω_{RES}.
b) For $\omega > \Omega_{RES} + \omega_{LO}$, a phonon sideband structure arises. This sideband structure peaks around Ω_{FC}.

The qualitative behaviour predicted in Ref. [25], namely, an intense zero-phonon (RES) line with a broader sideband at the high-frequency side, was confirmed after an all-coupling expression for the polaron optical absorption coefficient (13) at $\alpha = 5, 6, 7$ had been studied [26].

In what precedes, the low-frequency end of the polaron absorption spectrum was discussed; at higher frequencies, transitions to higher RES and their scattering states can appear.

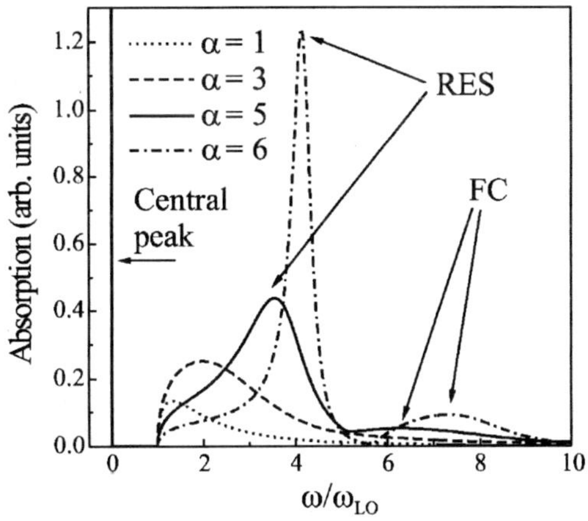

FIGURE 5. Polaron optical absorption spectra at zero temperature, calculated within the path integral method according to Eqs. (11a), (12), (13) of Ref. [26] for different values of the electron-phonon coupling constant α. The peaks are labelled in the picture according to their physical origin as discussed in the text. A δ-like central peak (at $\omega = 0$) is schematically shown by a vertical line.

The study of the optical absorption of polarons at large coupling is mainly of formal interest because all reported coupling constants of polar semiconductors and ionic crystals are smaller than 6.

2.3. Optical absorption of polarons at arbitrary coupling. Internal structure of polarons

The theoretical optical absorption spectrum of a single large polaron (at *all electron-phonon coupling strengths*) was first obtained in Ref. [26]:

$$\Gamma(\omega) = -\frac{1}{n\varepsilon_0 c}\frac{\mathrm{Im}\Sigma(\omega)}{[\omega - \mathrm{Re}\Sigma(\omega)]^2 + [\mathrm{Im}\Sigma(\omega)]^2}. \tag{13}$$

The so-called memory function $\Sigma(\omega)$ contains the dynamics of the polaron and depends on α and ω [26]. This function is related to the impedance function $Z(\omega)$ of the polaron through $-iZ(\omega) = \omega - \Sigma(\omega)$. The starting point in Ref. [26] was the general Eq.(13) together with the $[\mathrm{Im}\Sigma(\omega)]$ obtained in [16] for the Feynman polaron. A crucial ingredient in Ref. [26] is the calculation of $\mathrm{Re}\Sigma(\omega)]$ (which is a Kramers-Kronig-type transform of $[\mathrm{Im}\Sigma(\omega)]$).

The optical absorption spectra calculated according the theory of Ref. [26] are shown in Fig. 5 for values of α in the range from 1 to 6.

The absorption spectrum for $\alpha = 1$ consists of a "one-phonon line", similar to the weak-coupling result discussed in Subsection 2.1. The absorption peak for $\alpha = 3$ contains components with more than one free phonon in the final state. It was suggested [26] that part of the oscillator strength in the polaron absorption for $\alpha = 3$ is due to lattice relaxation in the final states, be it to a much smaller extent than for $\alpha = 6$.

It is remarkable that from (13), as applied in [26] to the Feynman model, the three different kinds of polaron excitations proposed in Ref. [25] appear in the calculated optical absorption spectra for polarons at $\alpha \geq 5$:

a) scattering states where e. g. one real phonon is excited (the structure starting at $\omega = \omega_{\rm LO}$);
b) relaxed excited states (RES);
c) Franck–Condon (FC) states.

Furthermore, at zero temperature, the optical absorption spectrum for one polaron contains a "central peak" [$\sim \delta(\omega)$] at the origin. For non-zero temperature, this "central peak" smears out and the optical absorption spectrum consists of both a broad envelope and an "anomalous" Drude-type low-frequency component.

For example, in Fig. 5 the main peak of the polaron optical absorption for $\alpha = 5$ at $\omega = 3.51\omega_{\rm LO}$ is interpreted as due to transitions to a RES. A "shoulder" at the low-frequency side of the main peak is attributed to one-phonon transitions to "scattering states". The structure centered at about $\omega = 6.3\omega_{\rm LO}$ is interpreted as a FC band. As seen from Fig. 5, when increasing the electron-phonon coupling constant to α=6, the RES peak at $\omega = 4.3\omega_{\rm LO}$ becomes very intense compared with the FC peak centered at $\omega = 7.5\omega_{\rm LO}$.

Free-carrier polaron effects were revealed through measurements of the optical constants, of cyclotron resonance and of carriers' mobility in polar semiconductors and insulating photoconductors. Optical absorption experiments revealed spectra, consistent with the theoretical predictions in Fig. 5 up to $\alpha \sim 3$ [23, 27]. The more dramatic structure of $\Gamma(\omega)$, Eq.(13), characteristic of transitions to the RES for $\alpha \geq 5$ formally arises in cyclotron resonance experiments (see Section 4).

2.4. Optical conductivity for the 2D weak-coupling polaron gas

The electron-electron interaction influences the spectrum of the elementary excitations of the electron gas. For the 2D weak-coupling polaron gas, the optical conductivity taking into account dynamic screening has been investigated in Ref. [28]. The calculation has been performed within the memory-function approach (see Refs. [26, 29] and the previous subsection). Within the aforesaid approach, the real part of the optical conductivity is expressed through the complex memory function $\chi(\omega) \equiv \omega\Sigma(\omega)$:

$$\mathrm{Re}\sigma(\omega) = -\frac{n_0 e^2}{m}\frac{\omega \mathrm{Im}\chi(\omega)}{\left[\omega^2 - \mathrm{Re}\chi(\omega)\right]^2 + \left[\mathrm{Im}\chi(\omega)\right]^2}. \tag{14}$$

The memory function $\chi(\omega)$ can be expressed [28] in terms of the dielectric function of the interacting electron gas. The electron-electron interaction is taken into account within the random-phase approximation. The optical conductivity spectra for a 2D polaron gas in a GaAs heterostructure at zero temperature are plotted in Fig. 6 in units of $\left(n_0 e^2/m\right)$ for several values of the electron density.

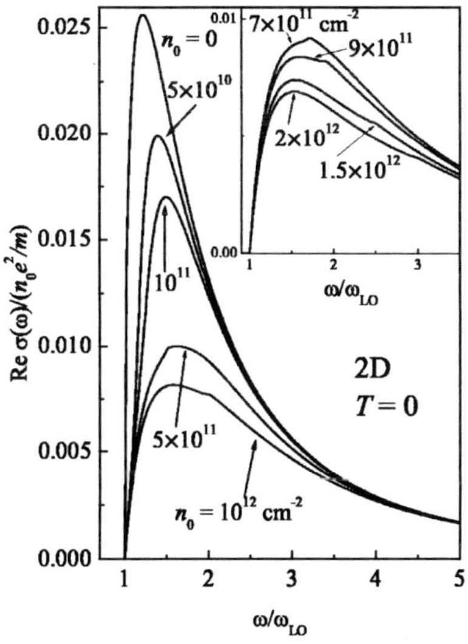

FIGURE 6. Optical conductivity spectra of a 2D polaron gas at zero temperature for different values of the electron density. (From Ref. [28].)

At zero temperature there are no real phonons in the crystal, and no phonon absorption processes are possible. This is the reason why the absorption is zero for frequencies below the LO-phonon frequency, except for a delta-like peak [30] at the origin with weight π/m^*, where m^* is the polaron mass. For frequencies above the LO-phonon frequency the optical conductivity first increases with increasing frequency, it reaches a maximum and then it decreases [cf. Eqs. (5) and (6)]. In the limit of high frequencies, the optical conductivity behaves as $\omega^{-5/2}$ and is independent of the electron density, in the same way as in Ref. [22]. For electron densities $n_0 \sim 10^{12}$ cm^{-2}, the optical conductivity spectrum shows besides the 1-phonon peak also a shoulder structure. The shoulder occurs around $\omega = \omega_{\rm LO} + E_F/\hbar$, where E_F is the Fermi energy of the 2D polaron gas. At lower electron densities this shoulder structure is not present. For higher electron densities, the shoulder moves to the tail of the spectrum and becomes hard to resolve. These shoulders are related to the undamped-plasmon contribution to the memory function. Comparing optical conductivity spectra for different densities, we see that screening *reduces* the optical conductivity.

2.5. Scaling relations for polarons in 2D and in 3D

It is worth mentioning, that several scaling relations connecting the polaron self-energy, the effective mass, the impedance Z and the mobility μ in 2D to the same quantities in 3D exist [31]. Those relations were obtained on the level of the Feynman approximation and are listed here:

$$E_{2D}(\alpha) = \frac{2}{3}E_{3D}\left(\frac{3\pi}{4}\alpha\right), \tag{15}$$

$$\frac{m^*_{2D}(\alpha)}{m_{2D}} = \frac{m^*_{3D}\left(\frac{3\pi}{4}\alpha\right)}{m_{3D}}, \tag{16}$$

$$Z_{2D}(\alpha,\nu) = Z_{3D}\left(\frac{3\pi}{4}\alpha,\nu\right), \tag{17}$$

where ν is the frequency of the external electromagnetic field, and

$$\mu_{2D}(\alpha) = \mu_{3D}\left(\frac{3\pi}{4}\alpha\right). \tag{18}$$

Expressions (15) to (18) provide a straightforward link between polaron characteristics in 3D with those in 2D.

3. MANY-POLARON SYSTEMS

3.1. A gas of interacting polarons: weak-coupling approach

In most polar semiconductors, the weak-coupling regime is realized. Perturbation theory is then adequate to describe the polaron gas in such semiconductors. In early approaches to the problem of a high-density polaron gas, the interaction between polarons was supposed to be completely screened (see, e. g., [32]). The ground-state energy of a gas of interacting polarons has been first derived in Ref. [33]. In Ref. [34], a variational method was formulated for the calculation of the ground-state energy of a single weak-coupling polaron. In Ref. [33], a canonical transformation, inspired by Ref. [34], was extended to the case of a many-polaron system (see also Subsection 3.3). After elimination of the phonons, the reduced Hamiltonian describes a system of charged particles interacting through an effective interparticle potential.

The Hartree-Fock ground-state energy (per particle) of the interacting polaron gas was then derived in the following analytical form [33]:

$$\frac{E_0}{N\hbar\omega_{\mathrm{LO}}} = \frac{3}{5}x_F^2 - \gamma x_F - \alpha\left[1 - \frac{2}{\pi}\arctan(2x_F) + F(x_F)\right], \tag{19}$$

where N is the number of polarons in the gas. The energy per particle (19) as a function of the electron density $n_0 = N/V$ is discussed in terms of a dimensionless parameter

$$r_p = \left(\frac{2m_b\omega_{\mathrm{LO}}}{\hbar}\right)^{1/2} r_0. \tag{20}$$

This parameter compares the interparticle distance r_0, which is defined by the expression: $V = (4\pi r_0^3/3)N$, with the "polaron radius" $\left[\hbar/\left(2m_b\omega_{\mathrm{LO}}\right)\right]^{1/2}$. Using the relation between the Fermi momentum p_F and the density n_0,

$$p_F = \hbar \left(3\pi^2 n_0\right)^{1/3}, \tag{21}$$

the dimensionless Fermi wave number $x_F \equiv (p_F/\hbar)\left[\hbar/\left(2m_b\omega_{\mathrm{LO}}\right)\right]^{1/2}$ can be represented as $x_F = (9\pi/4)^{1/3}(1/r_p)$. In Eq. (19), the dimensionless constant

$$\gamma = \frac{3e^2}{4\pi}\left(\frac{2m_b\omega_{\mathrm{LO}}}{\hbar}\right)^{1/2}\frac{1}{\varepsilon_\infty}\frac{1}{\hbar\omega_{\mathrm{LO}}} \tag{22}$$

measures the strength of the exchange interaction; $F(x_F)$ denotes

$$F(x_F) = \frac{2x_F}{\pi}\left\{\frac{5}{4} - \frac{8}{5}x_F^2 + \frac{(8x_F^2)^2}{2[(2x)^4+3]^{1/2}}\left[g_-(x_F) - g_+(x_F)\right]\right\} \tag{23}$$

and $g_\pm(x_F)$ is given by

$$g_\pm(x_F) = \frac{1}{2}a_\pm(x_F) + a_\pm^2(x_F) + a_\pm^3(x_F)\ln\left|\frac{a_\pm(x_F)-1}{a_\pm(x_F)}\right|, \tag{24}$$

where

$$a_\pm(x_F) = (2x_F)^2 \pm \left[(2x_F)^4+3\right]^{1/2}. \tag{25}$$

In Eq. (19), the ground-state energy is a sum of the following contributions: (i) the average kinetic energy of the electrons; (ii) the exchange energy; (iii) the polaron contribution to the ground-state energy. In Ref. [33], analytical expressions were derived also for those contributions.

The theory was applied to ZnS and ZnO. The material parameters m_b (in units of the electron mass m_e in the vacuum), ε_0, ε_∞, α, the "polaron radius" $[\hbar/(2m_b\omega_{\mathrm{LO}})]^{1/2}$ and the LO-phonon energy $\hbar\omega_{\mathrm{LO}}$, used to calculate the ground-state energy, are listed in Table 2.

TABLE 2. *Material parameters of ZnS and ZnO (from Ref. [33])*

	m_b/m_e	ε_0	ε_∞	α	$\left[\hbar/\left(2m_b\omega_{\mathrm{LO}}\right)\right]^{1/2}$ (nm)	$\hbar\omega_{\mathrm{LO}}$ (meV)
ZnS	0.28	8.6	5.2	0.649	1.767	43.64
ZnO	0.24	8.15	4.0	0.849	1.472	73.27

In Fig. 7, the density dependence of the ground-state energy of a gas of interacting polarons obtained with the approach of Ref. [33] for the case of ZnS and in ZnO is presented. The resulting ground-state energy and different contributions to it are plotted in Fig. 7 as a function of the interparticle distance (in units of the "polaron radius"). The average kinetic energy of the electrons at zero temperature

$$\varepsilon_{\mathrm{kin}} = \frac{3}{5}\frac{p_F^2}{2m_b} \propto n_0^{2/3} \tag{26}$$

is an increasing function of density (resp. a decreasing function of r_p). As is well known from the theory of the electron gas, the sum of the kinetic and the exchange energies (which is just the ground-state energy for the electron gas and is shown by dashed curves in Fig. 7) has a minimum as a function of r_p, because at high densities, the kinetic energy dominates, while as $n_0 \to 0$, both kinetic and exchange energies tend to zero. The polaron contribution to the ground-state energy, ε_{p}, is a monotonically decreasing function of r_p. It is seen from Fig. 7, that, in the approximation of Ref. [33], the minimum of the total ground-state energy per particle for a polaron gas lies at lower density (higher r_p) than that for the electron gas.

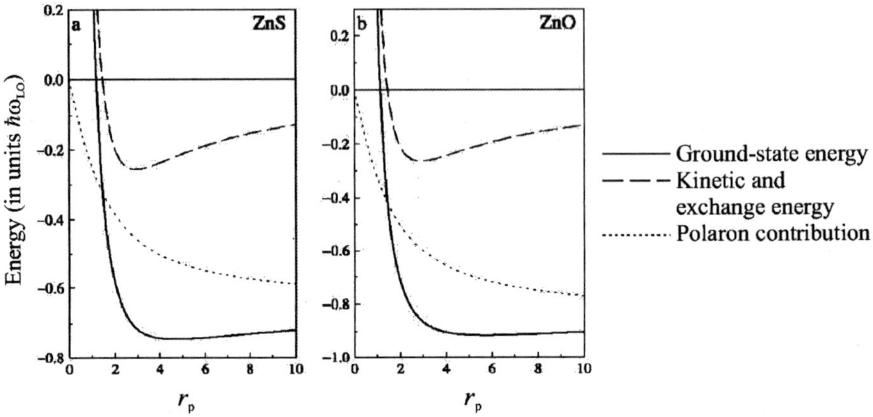

FIGURE 7. Density dependence of the ground-state energy for a gas of interacting polarons in ZnS (a) and in ZnO (b). The solid curve is the total ground-state energy per particle. The dashed curve represents a sum of the kinetic and exchange energies. The dotted curve shows the polaron contribution to the ground-state energy. The energies are measured in units of the LO-phonon energy $\hbar\omega_{\mathrm{LO}}$. The conversion of r_p to a density is given by $n_0 = \left(4.32/r_p^3\right) \times 10^{19}$ cm^{-3} for ZnS, and $n_0 = \left(7.48/r_p^3\right) \times 10^{19}$ cm^{-3}for ZnO. (After Ref. [33].)

3.2. Interacting polarons at arbitrary coupling

The non-degenerate system of interacting polarons in polar doped insulators was analysed by Quemerais and Fratini based on a Feynman-type model for the polaron [35, 36, 37].

Following Ref. [37], as the polaron density increases, at some critical density, there occurs ionisation of electrons from their polarization potential wells. The latter must be screened out by the liberated carriers. The following qualitative criterion has been proposed in Refs. [36, 37] for the dissociation of strong-coupling polarons:

$$\left(n_c^P\right)^{1/3} \frac{\varepsilon^*}{\varepsilon_\infty} R_p \approx \text{const}, \tag{27}$$

where n_c^P is the critical density of the polaron dissociation. In Ref. [36], the idea is used that the crystallized state is unstable when the amplitude of quantum fluctuations of the electron positions exceeds some critical value (the Lindemann criterion). In other words, the crystal melts when $\langle \delta \mathbf{x}^2 \rangle^{1/2} / R_s > \delta$, where R_S is related to the density n_0 according to $n_0 = 3/(4\pi R_S^3)$; δ is a phenomenological constant, which is usually taken as $\delta \approx 1/4$.

At nonzero electron-phonon coupling, following Ref. [37], *two* scenarios are possible for increasing density: (i) polarons may dissociate as described above; (ii) polarons may delocalise to form a polaron liquid. In order to describe both these processes, in Refs. [35, 36, 37], the following model for the crystallized state of interacting polarons has been proposed. First, it is assumed that there is a *jellium* background, which exactly compensates the negative charges. The jellium is constituted of spheres of radius R_S with a uniform positive charge density. Each sphere, which carries a total charge $+e$, is supposed to be centered on the sites of a Bravais lattice. Second, the electrons are supposed to be *localized* around the sites of the same Bravais lattice and to interact with the LO-phonon field.

The ground-state energy of this system is calculated in Refs. [35, 36, 37] using a variational principle. Following Ref. [15], in Ref. [37], a quadratic trial action is introduced, which corresponds to the two-body Lagrangian

$$L_0 = \frac{m}{2}\dot{\mathbf{r}}^2 + \frac{M}{2}\dot{\mathbf{X}}^2 - \frac{K}{2}(\mathbf{r}-\mathbf{X})^2 - \frac{m\omega_W^2}{2\varepsilon_s}\mathbf{r}^2, \tag{28}$$

where ε_s is the static dielectric constant, and $\omega_W = \omega_P/\sqrt{3}$ with the plasma frequency ω_P. The model describes an electron subject to an external harmonic potential $\left(\frac{m\omega_W^2}{2\varepsilon_s}\mathbf{r}^2\right)$ and bound with a spring to a fictitious particle of mass M and coordinate $\mathbf{X}$. The latter represents the polarization cloud associated to the electron. The constant K characterizes the interaction strength for the model system.

In order to delocalise the polarons to create a liquid many-polaron state, the quantum fluctuations of a polaron moving as a whole must be large in comparison with R_s. On the contrary, when the relative fluctuations of the electron with respect to the polarization cloud become large, polaron dissociation is favoured. A simplified Feynman-type model allows one to calculate both the quantum fluctuations of the center of mass $\mathbf{R} = (m^*\mathbf{x} + M\mathbf{X})/(m^* + M)$ (where m^* is the electron mass) and those of the relative coordinate vector $\mathbf{r} = \mathbf{x} - \mathbf{X}$. Consequently, in Refs. [35, 36, 37], two different Lindemann criteria are formulated: (i) $\langle \delta \mathbf{R}^2 \rangle^{1/2} / R_s > 1/4$ for melting of the Wigner crystal; (ii) $\langle \delta \mathbf{r}^2 \rangle^{1/2} / R_s > 1/4$ for the polaron dissociation.

The phase diagram in Fig. 8 presents the critical radii for Wigner crystal melting and polaron dissociation as a function of the electron-phonon coupling constant α. In the low-density limit, the ground state of the many-polaron system is a Wigner lattice of polarons. With increasing density, there is a competition between the melting of the polaron Wigner lattice and the dissociation of the polarons themselves. For fixed α, $\langle \delta \mathbf{r}^2 \rangle^{1/2} / R_s$ increases with increasing density, while $\langle \delta \mathbf{R}^2 \rangle^{1/2} / R_s$ is bounded from above. Moreover, the maximal value of $\langle \delta \mathbf{R}^2 \rangle^{1/2} / R_s$ decreases with increasing α. As a result, a critical value α^* exists, so that for sufficiently large values of the electron-phonon coupling constant $\alpha > \alpha^*$, the formation of a polaron liquid is impossible.

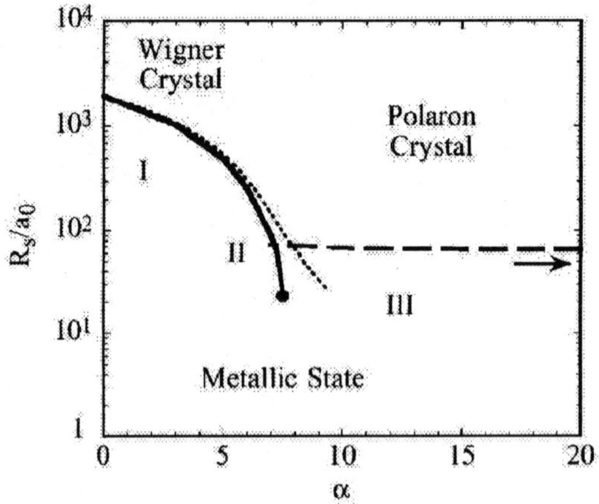

FIGURE 8. The phase diagram of a polaron Wigner lattice for $\varepsilon_\infty = 5$, $\varepsilon_0 = 30$ and $m^* = m_e$. Crystal melting (bold line) and polaron dissociation (broken line). The dotted line is the critical radius corresponding to the low-density approximation. The arrow marks the critical density given by the criterion (27). The melting driven by the fluctuations of the localized particles (criterion (i)) is impossible beyond α^* (full circle). For $\alpha < \alpha^*$, the metallic state beyond the critical density is either a renormalized electron liquid (region I) or a degenerate polaron liquid (region II). For $\alpha > \alpha^*$, the insulator-to-metal transition is described by the dissociation mechanism (ii). In this range, although the polarons do not survive beyond the transition, the resulting electron liquid should present both strong electron-phonon and electron-electron interactions (region III). (From Ref. [37], Fig. 7, ©EDP Sciences, SIF, Springer-Verlag, 2000.)

3.3. Optical absorption coefficient of a many-polaron gas

As is the case for polar semiconductors and ionic crystals [26], insight into the nature of polarons in high-temperature superconductors can be gained i. a. by studying the optical properties of these materials. For an excellent review see Calvani [38].

In Refs. [39, 40], starting from the many-polaron canonical transformations and the variational many-polaron wave function introduced in Ref. [33], the optical absorption coefficient of a many-polaron gas has been derived. This approach to the many-polaron optical absorption [41, 42, 43] allows one to include the many-body effects in terms of the dynamical structure factor of the electron (or hole) system. It is then possible within weak (intermediate) coupling polaron theory to select the level of approximation used in the treatment of the many-polaron gas by choosing the appropriate expression of the dynamical structure factor for the electron (or hole) gas.

The Hamiltonian of a system of N interacting continuum polarons is

$$\begin{aligned} H_0 &= \sum_{j=1}^{N} \frac{p_j^2}{2m_b} + \sum_{\mathbf{k}} \hbar\omega_{\mathrm{LO}} a_{\mathbf{k}}^+ a_{\mathbf{k}} \\ &+ \sum_{\mathbf{k}} \sum_{j=1}^{N} \left[e^{i\mathbf{k}.\mathbf{r}_j} a_{\mathbf{k}} V_{\mathbf{k}} + e^{-i\mathbf{k}.\mathbf{r}_j} a_{\mathbf{k}}^+ V_{\mathbf{k}}^* \right] + \frac{e^2}{2\varepsilon_\infty} \sum_{j=1}^{N} \sum_{\ell(\neq j)=1}^{N} \frac{1}{|\mathbf{r}_i - \mathbf{r}_j|}, \end{aligned} \tag{29}$$

where $\mathbf{r}_j, \mathbf{p}_j$ represent the position and momentum of the N constituent electrons (or holes) with band mass m_b; $a^+_{\mathbf{k}}, a_{\mathbf{k}}$ denote the creation and annihilation operators for longitudinal optical (LO) phonons with wave vector $\mathbf{k}$ and frequency ω_{LO}; $V_{\mathbf{k}}$ describes the amplitude of the interaction between the electrons and the LO phonons; and e is the electron charge. The ground state energy of this many-polaron Hamiltonian has been studied in Ref. [33], for weak and intermediate strengths of the electron-phonon coupling, by introducing a variational wave function:

$$\left|\psi_{\mathrm{LDB}}\right\rangle = U\left|\phi\right\rangle\left|\varphi_{\mathrm{el}}\right\rangle, \tag{30}$$

where $\left|\varphi_{\mathrm{el}}\right\rangle$ represents the ground-state many-body wave function for the electron (or hole) system,$\left|\phi\right\rangle$ is the phonon vacuum and U is a many-body unitary operator which determines a canonical transformation for a fermion gas interacting with a boson field:

$$U = \exp\left\{\sum_{j=1}^{N}\sum_{\mathbf{k}}\left(f_{\mathbf{k}}a_{\mathbf{k}}e^{i\mathbf{k}.\mathbf{r}_j} - f^*_{\mathbf{k}}a^+_{\mathbf{k}}e^{-i\mathbf{k}.\mathbf{r}_j}\right)\right\}. \tag{31}$$

In the one fermion limit, U reduces to a canonical transformation inspired by Tomonaga's work [44] and later applied by several workers, after Lee, Low and Pines [34], to one-polaron theories. In Ref. [33], this canonical transformation was extended and used to formulate a many-polaron theory. The $f_{\mathbf{k}}$ were determined variationally [33] resulting in

$$f_{\mathbf{k}} = \frac{V_{\mathbf{k}}}{\hbar\omega_{LO} + \dfrac{\hbar^2 k^2}{2m_b S(\mathbf{k})}} \tag{32}$$

for a system with total momentum $\mathbf{P} = \sum_j \mathbf{p}_j = 0$. In this expression, $S(\mathbf{k})$ is the static structure factor of the constituent interacting many electron or hole system:

$$NS(\mathbf{k}) = \left\langle \sum_{j=1}^{N}\sum_{j'=1}^{N} e^{i\mathbf{k}.(\mathbf{r}_j - \mathbf{r}_{j'})}\right\rangle. \tag{33}$$

The angular brackets $\langle ... \rangle$ denote the expectation value with respect to the ground state. It may be emphasized that (31), although it appears like a straightforward generalization of the one-particle transformation in [44], constitutes — especially in its implementation — a nontrivial extension of a one-particle approximation to a many-body system.

The many-polaron optical conductivity is governed by the response of the current density to an applied electric field (here along the x-axis) with frequency ω. This applied electric field introduces a perturbation term in the Hamiltonian (29), which couples the vector potential of the incident electromagnetic field to the current-density. As is well known, within linear response theory, the optical conductivity can be expressed through the Kubo formula [45, 46]:

$$\sigma(\omega) = i\frac{Ne^2}{\mathrm{V}m_b\omega} + \frac{1}{\mathrm{V}\hbar\omega}\int_0^\infty e^{i\omega t}\left\langle\left[J_x(t), J_x(0)\right]\right\rangle dt. \tag{34}$$

In this expression, V is the volume of the system, and J_x is the x-component of the current operator $\mathbf{J}$, which is related to the momentum operators of the charge carriers:

$$\mathbf{J} = \frac{q}{m_b} \sum_{j=1}^{N} \mathbf{p}_j = \frac{q}{m_b} \mathbf{P}, \tag{35}$$

with q the charge of the charge carriers ($+e$ for holes, $-e$ for electrons) and $\mathbf{P}$ the total momentum operator of the charge carriers. The real part of $\sigma(\omega)$ at zero temperature, which is proportional to the optical absorption coefficient, can be written as a function of the total momentum operator of the charge carriers as follows:

$$\mathrm{Re}[\sigma(\omega)] = \frac{1}{\mathrm{V}\hbar\omega} \frac{e^2}{m_b^2} \mathrm{Re}\left\{ \int_0^\infty e^{i\omega t} \langle [P_x(t), P_x(0)] \rangle \, dt \right\}. \tag{36}$$

Integrating by parts twice, the real part of the optical conductivity of the many-polaron system can be written in terms of a force-force correlation function:

$$\mathrm{Re}[\sigma(\omega)] = \frac{1}{\mathrm{V}\hbar\omega^3} \frac{e^2}{m_b^2} \mathrm{Re}\left\{ \int_0^\infty e^{i\omega t} \langle [F_x(t), F_x(0)] \rangle \, dt \right\}, \tag{37}$$

with $\mathbf{F} = (i/\hbar) \left[H_0, \mathbf{P} \right]$. The commutator of the Hamiltonian (29) with the total momentum operator of the charge carriers simplifies to

$$\mathbf{F} = -i \sum_{\mathbf{k}} \sum_{j=1}^{N} \mathbf{k} \left(e^{i\mathbf{k}.\mathbf{r}_j} a_{\mathbf{k}} V_{\mathbf{k}} - e^{-i\mathbf{k}.\mathbf{r}_j} a_{\mathbf{k}}^{+} V_{\mathbf{k}}^{*} \right). \tag{38}$$

This result for the force operator illustrates the advantage of using the force-force correlation function rather than the momentum-momentum correlation function. The operator product $F_x(t)F_x(0)$ is proportional to $|V_{\mathbf{k}}|^2$, the charge carrier - phonon interaction strength. This will be a distinct advantage for any expansion of the final result in the charge carrier - phonon interaction strength, since $|V_{\mathbf{k}}|^2$ is factored out beforehand. To calculate $\mathrm{Re}[\sigma(\omega)]$ to order α, e. g., we only need the *α-independent* contributions to $\langle [F_x(t), F_x(0)] \rangle$. Denoting $\rho_{\mathbf{k}} = \sum_{j=1}^{N} e^{i\mathbf{k}\cdot\mathbf{r}_j}$, the real part of the optical conductivity takes the form:

$$\mathrm{Re}[\sigma(\omega)] = \frac{1}{\mathrm{V}\hbar\omega^3} \frac{e^2}{m_b^2} \mathrm{Re}\left\{ \sum_{\mathbf{k},\mathbf{k}'} k_x k'_x \int_0^\infty e^{i\omega t} \right.$$
$$\left. \times \left\langle \left[e^{iH_0 t/\hbar} \left(\rho_{\mathbf{k}} a_{\mathbf{k}} V_{\mathbf{k}} + \rho_{-\mathbf{k}} a_{\mathbf{k}}^{+} V_{\mathbf{k}}^{*} \right) e^{-iH_0 t/\hbar}, \left(\rho_{\mathbf{k}'} a_{\mathbf{k}'} V_{\mathbf{k}'} + \rho_{-\mathbf{k}'} a_{\mathbf{k}'}^{+} V_{\mathbf{k}'}^{*} \right) \right] \right\rangle_0 dt \right\}. \tag{39}$$

Up to this point, no approximations other than *linear response theory* have been made.

Taking the expectation value appearing in the right hand side of the expression (39) with respect to the phonon vacuum, we find

$$\mathrm{Re}[\sigma(\omega)] = -2 \frac{1}{\mathrm{V}\hbar\omega^3} \frac{e^2}{m_b^2} \left\{ \sum_{\mathbf{k}} k_x^2 |V_{\mathbf{k}}|^2 \right.$$
$$\left. \times \int_0^\infty \sin(\omega t) \mathrm{Im} \left[e^{-i\omega_{\mathrm{LO}} t} \left\langle \varphi_{\mathrm{el}} \left| e^{iH' t/\hbar} \rho_{\mathbf{k}} e^{-iH' t/\hbar} \rho_{-\mathbf{k}} \right| \varphi_{\mathrm{el}} \right\rangle \right] dt \right\}. \tag{40}$$

The right hand side of (40) can be written in a more compact form by introducing the dynamical structure factor of the electron (or hole) system.

To find the formula for the real part of the optical conductivity in its final form, we introduce the standard expression for the dynamical structure factor of a system of charge carriers interacting through a Coulomb potential,

$$S(\mathbf{q},\nu) = \int_{-\infty}^{+\infty} \left\langle \varphi_{\mathrm{el}} \left| \frac{1}{2}\sum_{j,\ell} e^{i\mathbf{q}.(\mathbf{r}_j(t)-\mathbf{r}_\ell(0))} \right| \varphi_{\mathrm{el}} \right\rangle e^{i\nu t}dt. \tag{41}$$

Rewriting the expression (40) with the dynamical structure factor of the electron (or hole) gas results in:

$$\mathrm{Re}[\sigma(\omega)] = \frac{n}{\hbar\omega^3}\frac{e^2}{m_b^2}\sum_{\mathbf{k}} k_x^2|V_{\mathbf{k}}|^2 S(\mathbf{k},\omega-\omega_{\mathrm{LO}}), \tag{42}$$

where $n = N/\mathrm{V}$ is the density of charge carriers. As noted before, $V_{\mathbf{k}}$ is the electron-phonon interaction amplitude and k_x is the x-component of the wave vector. Formula (42) for the optical absorption of the many-polaron system has an intuitively appealing form.

In the theory of one-polaron optical absorption for weak coupling constants α, the optical absorption coefficient ($T=0$) as obtained from Fermi's golden rule is [19]:

$$1 \text{ polaron:} \mathrm{Re}[\sigma(\omega)] \propto \omega^{-3}\sum_{\mathbf{k}} k_x^2|V_{\mathbf{k}}|^2\delta[\hbar k^2/(2m_{\mathrm{b}}) - (\omega-\omega_{\mathrm{LO}})]. \tag{43}$$

At low densities, the dynamical structure factor $S(q,\nu)$ is strongly peaked around $q^2/2=\nu$ and is close to zero everywhere else [32]. Substituting a delta-function $\delta(q^2/2-\nu)$ for the dynamical structure factor in Eq. (42) it is easily seen that the one-polaron limit of Ref. [19] is retrieved. The one-polaron result (43) is derived [19] by considering a process in which the initial state consists of a photon of energy $\hbar\omega$ and a polaron in its ground state, and the final state consists of an emitted LO phonon with energy $\hbar\omega_{\mathrm{LO}}$ and the polaron, scattered into a state with momentum $\mathbf{k}$ and kinetic energy $(\hbar k)^2/(2m_{\mathrm{b}}) = \hbar(\omega-\omega_{\mathrm{LO}})$. The many-polaron result, formula (42), is a generalization of this one-polaron picture. The contribution which corresponds to the scattering of a polaron into the momentum state $\mathbf{k}$ and energy $\hbar(\omega-\omega_{\mathrm{LO}})$ is now weighed by the dynamical structure factor $S(k,\omega-\omega_{\mathrm{LO}})$ of the electron (or hole) gas.

Eq. (42) is reminiscent of the Hopfield formula [47] describing the effect of impurities on the optical absorption of metals, which in turn is related to the expression obtained by Ron and Tzoar [48] for the optical absorption by a quantum plasma. Formula (42) also represents a generalization of the results obtained by Gurevich, Lang and Firsov [22]. These authors focused their attention on the many-body effects related to the Fermi statistics, whereas the analysis in Ref. [39] extends the results of [22] to study the influence of plasmons and further many-body effects in the system of many polarons (electrons or holes).

An advantage of the canonical transformation introduced in Ref. [33] for the calculation of *the ground state energy* of a polaron gas is that the many-body effects are

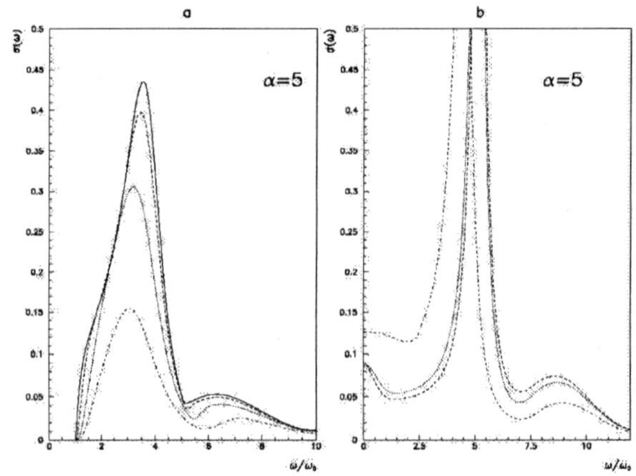

FIGURE 9. Optical conductivity of a polaron gas at $T = 0$ as a function of the frequency for different values of the electron density: $n_0 = 1.4 \times 10^{-5}$ (solid curve), $n_0 = 1.4 \times 10^{-4}$ (dashed curve), $n_0 = 1.4 \times 10^{-3}$ (dotted curve), and $n_0 = 1.4 \times 10^{-2}$ (dotted-dashed curve). The value of $\varepsilon_0/\varepsilon_\infty$ is 3.4. The electron density is measured in units of R_p^{-3}, where R_p is the Fröhlich polaron radius. (From Ref. [42], Fig. 2, ©EDP Sciences, SIF, Springer-Verlag, 1999.)

contained in the *static* structure factor of the electron (or hole) system, which appears in the analytical expression for the energy. The corresponding advantage of the canonical transformation in the present case, for *the optical conductivity*, is that many-body effects are again incorporated through a structure factor, now the *dynamical* structure factor of the electron (or hole) system. The level of approximation made in treating the many-body nature of the polaron system, e. g., to a certain order of perturbation expansion in α, is determined by the choice of the dynamical structure factor of the electron or hole system.

More recently, Cataudella, De Filippis and Iadonisi [42] investigated the optical properties of the many-polaron gas by calculating the correction to the RPA dielectric function of the electron gas due to the electron-phonon interactions. Variational parameters of a single-polaron Feynman model are exploited for the treatment of the interacting many-polaron gas. In order to find the many-polaron optical conductivity, the memory-function approach from Ref. [26, 20] for the optical absorption of non-interacting large polarons is used by these authors as a basis. As a result, the authors of Ref. [42] find a suppression of the optical absorption with increasing density, as seen from Fig. 9.

3.4. Comparison of calculated optical-absorption spectra to the infrared spectrum of $Nd_{2-x}Ce_xCuO_{2-y}$

The experimental optical absorption spectrum (up to 3000 cm^{-1}) of $Nd_2CuO_{2-\delta}$ ($\delta < 0.004$), obtained by Calvani and co-workers [49, 38], is shown in Fig. 10 (shaded area) together with the theoretical curve obtained in Ref. [39] and, for reference, the one-

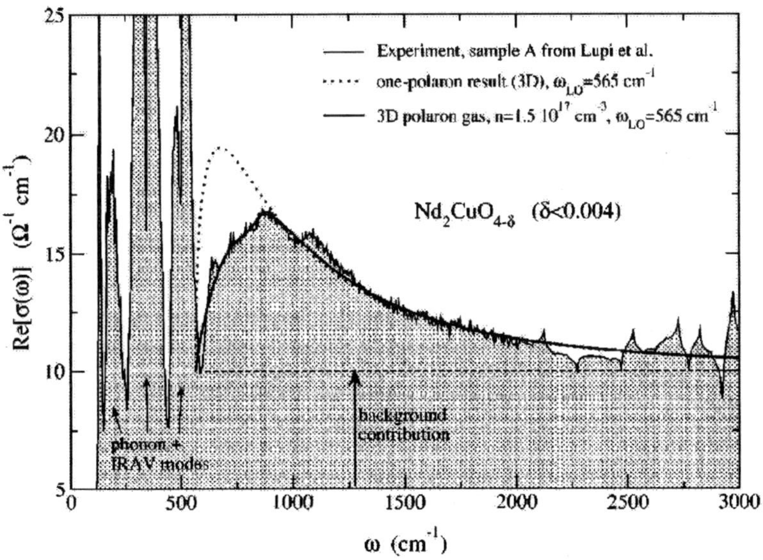

FIGURE 10. The infrared optical absorption of $Nd_2CuO_{2-\delta}$ ($\delta < 0.004$) as a function of frequency. The experimental result of Calvani and co-workers [49, 38] is presented by the thin black curve and by the shaded area. The dotted curve shows the single polaron result calculated according to Ref. [26]. The full black curve presents the theoretical results obtained in Ref. [39] for the interacting many-polaron gas with $n = 1.5 \times 10^{17}$ cm^{-3}, $\alpha = 2.1$ and $m_b = 0.5m_e$. [Recently, a numerical error was found in the fitting procedure, concerning strength of the plasmon branch, so that the parameter values given above may not be the optimal parameter values (see erratum to Ref. [39], in preparation).] (From Ref. [39]).

polaron optical absorption result [26]. At lower frequencies (600-1000 cm^{-1}) a marked difference between the single polaron optical absorption and the many-polaron result is manifest. The experimental 'd-band' can be clearly identified, rising in intensity at about 600 cm^{-1}, peaking around 1000 cm^{-1}, and then decreasing in intensity above that frequency. Good agreement between our theoretical predictions and the experimental results (for a density of $n = 1.5 \times 10^{17}$ cm^{-3}) is found. A background contribution, taken to be constant over the frequency range of the 'd-band', is subtracted in Fig. 10.

The material parameters characterizing polarons $(\varepsilon_0, \varepsilon_\infty, m_b)$ are not known for $Nd_2CuO_{2-\delta}$ ($\delta < 0.004$). For ω_{LO} a value of 7.005×10^{-2} eV was derived from available experimental data. In Ref. [39], the values $\alpha = 2.1$ and $m_b = 0.5m_e$ were suggested. However, these figures need further analysis in view of the limited available experimental data for $\varepsilon_0, \varepsilon_\infty, m_b$. Also in Ref. [39] the "central peak" in the optical absorption ($T = 0$) was not taken into account in the analysis.

The experimental results of Calvani et al. are available in the form of the normalized first frequency moment of the optical absorption spectrum (after substraction of MIR and CT band) $\mathrm{Re}[\sigma_{\exp}(\omega)]$:

$$\langle \omega \rangle = \frac{\int_0^{\omega_{\max}} \omega \mathrm{Re}[\sigma_{\exp}(\omega)] d\omega}{\int_0^{\omega_{\max}} \mathrm{Re}[\sigma_{\exp}(\omega)] d\omega}, \tag{44}$$

where $\omega_{\max} = 10000\ \mathrm{cm}^{-1}$ [49]. Calvani and co-workers [49] determined $\langle\omega\rangle$ for $Nd_{2-x}Ce_xCuO_{2-y}$ (NCCO) samples with a varying cerium doping content x. A comparison of this experimental normalized first frequency moment to the theoretical one presents the advantage that fewer parameters need to be adapted: only the *density* and the electron *band mass* have to be taken from experiment or (if experimental values are lacking) fitted.

Fig. 11 represents the comparison between the present theory and experiment. The squares with error bars show the experimental results for differently doped samples of NCCO, reported in Ref. [49]. The dashed curve shows the normalized first frequency moment of the theoretical optical absorption spectrum, integrated over the entire frequency range ($\omega_{\max} \to \infty$). The tail region of the many-polaron optical absorption still carries a significant weight, just as it does in the one-polaron optical absorption. It is necessary to include a cutoff frequency. The full curve represents the theoretical first frequency moment with a cutoff frequency $\omega_{\max} = 10000\ \mathrm{cm}^{-1}$, which corresponds to the experimental cutoff [49].

There exists a fair agreement between the theoretical and the experimental values of the normalized first frequency moment for the five samples with lowest density, which have a cerium doping content of $x < 0.12$. These correspond to the squares to the left of the dotted vertical line in Fig. 11. For the four remaining samples ($x > 0.12$) a discrepancy between the theoretically predicted first frequency moment for unpaired polarons and the observed first frequency moment appears. It has been observed experimentally that the weight of the low-frequency component in these samples (with $x > 0.12$) is significantly larger that the corresponding weight in samples with $x < 0.12$ [50]. This was interpreted in [50] as a consequence of an *insulator-to-metal* transition taking place around the cerium-doping level of $x = 0.12$. Therefore, it seems reasonable to assume that above this doping level x a change in the nature of the charge carriers takes place. One could hypothesize that, as the formation of bipolarons is stabilized with increasing density of the polaron gas [51], bipolarons start playing a role in the optical absorption spectrum. In a variety of other cuprates and manganates, the presence of bipolarons has also been invoked to interpret a number of response-related properties [52].

4. MAGNETOABSORPTION OF POLARONS

4.1. Polaron cyclotron resonance

Polaron effects in a magnetic field are sometimes referred to as *"magnetopolaron"* effects. Reviews on magnetopolaron effects can be found in Refs. [53, 54].

A powerful technique to study the electron-phonon coupling in semiconductors is cyclotron resonance. Changing the magnetic field B allows one to vary the cyclotron frequency $\omega_c = eB/m_b c$. When ω_c is brought closer to resonance with the phonon frequency ω_{LO}, a strongly renormalized "magnetopolaron" state is formed. This results in a splitting of the cyclotron resonance signal, which is a measure of the electron-phonon (polaron) coupling constant α.

The experimental observation [55] of a discontinuity in the energy spectrum of a

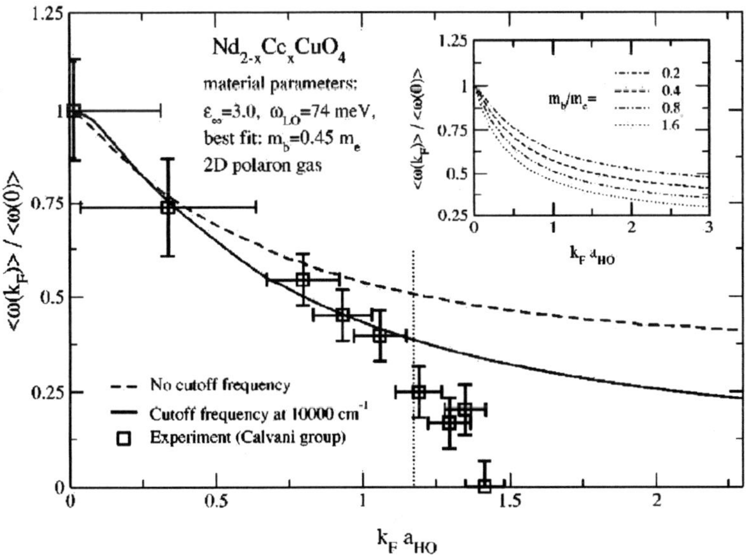

FIGURE 11. The normalized first frequency moment of the optical absorption spectra is shown as a function of the density [expressed through the Fermi wave vector k_F, in units a_{HO}, with $a_{\mathrm{HO}} = \sqrt{\hbar/(m_b \omega_{\mathrm{LO}})}$]. The squares represent the experimental results of Calvani and co-workers [49] in a family of $Nd_{2-x}Ce_xCuO_4$ materials. The dashed curve shows the results from the theoretical two-dimensional many-polaron optical absorption, Ref. [39], obtained by integrating all frequencies in the calculation of the first frequency moment. The full curve shows the theoretical results obtained by integrating up to a cut-off frequency, which is chosen at 10000 cm^{-1} and which corresponds to the maximum frequency in the experiment [49]. The material parameters are listed in the figure, and the effect of choosing a different electron band mass — as a fitting parameter — is illustrated in the inset. The points with $x < 0.12$, to the left of the vertical dotted line, show agreement with the theoretical result from the many-polaron theoretical optical absorption, but it is clear that the experimental cutoff frequency has to be taken into account. For the samples with $x > 0.12$, a discrepancy between the theoretically predicted first frequency moment and the observed first frequency moment is consistent with a possible insulator-to-metal transition at $x = 0.12$ [50]. (From Ref. [39].)

polaron in a magnetic field under the resonance condition $\omega_c = \omega_{\mathrm{LO}}$ has stimulated numerous theoretical works aimed at the description of observed magneto-absorption spectra. In the weak-coupling limit, the energy spectra of a polaron subjected to a magnetic field were studied, e. g., in Refs. [56, 57, 58, 59, 60, 53, 61]. Within the aforementioned approaches, the polaron cyclotron mass is determined from the energy differences between the polaron (Landau-) energy levels.

For an adequate description of the magneto-absorption spectrum itself must be calculated. In the sixties and seventies, several weak-coupling approximations for the magneto-absorption spectra of polarons were suggested (see, e.g., Refs. [62, 63, 30, 64, 65]).

The most direct approach to study the (magneto-) polaron mass is to calculate the magneto-optical absorption spectrum of the polaron (the quantity which is actually measured) and to define the polaron mass, in the same way as the experimentalists do, from the *peak positions* in the spectrum. This approach was chosen in Ref. [73] and elaborated over many years as a basic theoretical tool to analyse experimental data on magnetopolaron spectra (see, e. g., Refs. [10, 66, 67, 68, 69, 70, 71, 72]).

The approach of Ref. [73] is based on the memory-function formalism to study the response of polarons in a magnetic field. Ref. [73] extends the Feynman model [15] for the polaron and its response [16] to the case when an external magnetic field is applied. As a consequence, the treatment of Ref. [73] is expected to be quite accurate at all α, ω_c, temperatures. Other theoretical works (of formal nature) on polarons in magnetic fields are Refs. [74, 75, 76]. In what follows I summarize the method and the results of Ref. [73].

A polaron in a magnetic field is described by the Fröhlich Hamiltonian:

$$H = \frac{1}{2m}\left(\vec{p}+\frac{e}{c}\vec{A}\right)^2 + \sum_{\vec{k}} \hbar\omega_{\vec{k}}\left(a^{\dagger}_{\vec{k}}a_{\vec{k}}+\frac{1}{2}\right)$$
$$+\sum_{\vec{k}}\left(V_{\vec{k}}a_{\vec{k}}e^{i\vec{k}\cdot\vec{r}}+V^{*}_{\vec{k}}a^{\dagger}_{\vec{k}}e^{-i\vec{k}\cdot\vec{r}}\right), \tag{45}$$

where the symmetrical Coulomb gauge is chosen for the vector potential, $\vec{A} = (\mathscr{H}/2)(-y,x,0)$, assuming the magnetic field of strength $\mathscr{H}$ to be parallel to the z-axis. Starting from the Kubo formula [45] and applying the Mori-Zwanzig projection operator technique [77], one obtains [73] that the frequency-dependent magnetooptical absorption in the active-mode Faraday configuration is given by

$$\mu(\omega) = \frac{2e}{m}\mathrm{Im}\frac{1}{\omega-\omega_c-\Sigma_{+}\left(\omega+i0^{+}\right)+i0^{+}}. \tag{46}$$

$\mu(\omega)$ is directly measured in cyclotron resonance experiments. The memory function $\Sigma_{+}(\omega)$ has the form

$$\Sigma_{+}(\omega) = -\frac{1}{m\omega}\sum_{\vec{k}}\left(k_x^2+k_y^2\right)\left|V_{\vec{k}}\right|^2$$
$$\times\int_0^{\infty} dt\left(1-e^{i\omega t}\right)\mathrm{Im}\langle[b_{\vec{k}}(t),b^{\dagger}_{\vec{k}}(0)]+[b_{\vec{k}}(t),b_{\vec{k}}(0)]\rangle, \tag{47}$$

where $b_{\vec{k}} = a_{\vec{k}}e^{i\vec{k}\cdot\vec{r}}$.

The expectation values $\langle\ldots\rangle$ entering Eq. (47) are to be calculated for the electron-phonon system described by the Hamiltonian (45). The exact analytical solution of this problem seems to be impossible. Therefore, one has to look for appropriate approximations. In Ref. [73] the averaging over the exact "eigen"states of the electron-phonon system is replaced by the averaging over states of a model two-particle system with

Hamiltonian, which describes a generalization of the "Feynman model",

$$\begin{aligned} H_m &= \frac{p_z^2}{2m} + \frac{p_z'^2}{2m_\parallel'} + \frac{1}{2}\kappa_\parallel (z-z')^2 \\ &+\frac{1}{2m}\left(p_x - \frac{1}{2}m\omega_c y\right)^2 + \frac{1}{2m}\left(p_y + \frac{1}{2}m\omega_c x\right)^2 \\ &+\sum_{i=x,y}\left[\frac{p_i'^2}{2m_\perp'} + \frac{1}{2}\kappa_\perp \left(r_i - r_i'\right)^2\right] . \end{aligned} \tag{48}$$

The Hamiltonian (48) describes an electron coupled via a harmonic force to a so-called fictitious particle with co-ordinate $\vec{r}'$ and momentum $\vec{p}'$. When applying an external magnetic field, the mass of the fictitious particle and the strength of the effective interaction should depend on the strength as well as on the direction of the applied magnetic field. In order to take into account the magnetic-field-induced anisotropy for the electron-phonon interaction, the authors of Ref. [73] have proposed in Ref. [78] a generalization of the isotropic Hamiltonian (48) to a Hamiltonian with anisotropy in both the fictitious particle mass, given by a diagonal tensor with components $m_{xx}' = m_{yy}' = m_\perp'$, $m_{zz}' = m_\parallel'$, and the coupling strength, described by a diagonal tensor with components $\kappa_{xx} = \kappa_{yy} = \kappa_\perp$, $\kappa_{zz} = \kappa_\parallel$.

The four independent components of the mass and coupling-strength tensors, $m_\perp'$, $m_\parallel'$, $\kappa_\perp$, and $\kappa_\parallel$, are considered as variational parameters in Ref. [78]. In practice, it is more convenient to deal with the following four combinations of these parameters

$$w_i^2 = \frac{\kappa_i}{m_i'}, \qquad v_i^2 = \kappa_i \frac{1+m_i'}{m_i'}, \tag{49}$$

where $i = \perp, \parallel$. The variational procedure used in Ref. [73] exploits the Jensen-Feynman inequality for the Helmholtz free energy F per polaron in an ideal polaron gas,

$$F \leq F_{ph} + F_m - \frac{1}{\beta}\langle S - S_m \rangle_m , \tag{50}$$

where F_{ph} is the free energy of the phonon bath and F_m is the free energy of the above model system. The action S is obtained using the path-integral approach for eliminating the phonon co-ordinates in the action functional of the electron-phonon system. The action S_m is obtained by eliminating the co-ordinates of the fictitious particle in the action of the model system. The angular brackets $\langle \ldots \rangle_m$ in Eq. (50) denote an evaluation of an average with $\mathrm{e}^{-\beta \mathrm{F_m} + \mathrm{S_m}}$ as weight factor.

It should be mentioned that the inequality (50) in the presence of a magnetic field is not valid in all generality (see, e. g., Ref. [20] and references therein). This apriori limitation was recognized in Ref. [73], where (50) was used with a caveat. A generalization to the case of a non-zero magnetic field, was derived in Ref. [79] starting from the ordered-operator calculus. The authors of Ref. [79] have also shown that the Jensen-Feynman inequality can be used — for the present problem — in its *original* form if the following simple constraints on the variational parameters of the trial action S_m are satisfied:

$$\max\left(\omega_{\mathrm{LO}},\omega_c\right)\leq w_\perp\leq v_\perp,\qquad \omega_{\mathrm{LO}}\leq w_\parallel\leq v_\parallel\,, \tag{51}$$

where ω_{LO} is the frequency of LO phonons. From Ref. [79] it follows that the results of Ref. [73] provide upper bounds to the ground state energy of the polaron in a magnetic field except in some extreme limits.

Replacing in Eq. (46) the average $\langle\ldots\rangle$ with the average $\langle\ldots\rangle_m$, we obtain after the calculation of the corresponding path integrals (see Refs. [78, 73] for details) the following expression for the memory function

$$\begin{aligned}\Sigma_+(\omega)=&\frac{\alpha}{2\sqrt{2\pi}\omega}\int_0^\infty \mathrm{d}t\left(1-\mathrm{e}^{\mathrm{i}\omega t}\right)\mathrm{Im}\left\{\left[1+n(\omega_{\mathrm{LO}})\right]\mathrm{e}^{-\mathrm{i}\omega_{\mathrm{LO}}t}+n(\omega_{\mathrm{LO}})\mathrm{e}^{\mathrm{i}\omega_{\mathrm{LO}}t}\right\}\\&\times\left[\frac{1}{\sqrt{D(t)}D_H(t)}+\frac{1}{\sqrt{D(t)}\left[D(t)-D_H(t)\right]}\right.\\&\left.-\frac{1}{2\left[D(t)-D_H(t)\right]^{3/2}}\ln\left(\frac{\sqrt{D(t)}+\sqrt{D(t)-D_H(t)}}{\sqrt{D(t)}-\sqrt{D(t)-D_H(t)}}\right)\right]\end{aligned} \tag{52}$$

with $n(\omega_{\mathrm{LO}})$ the number of LO phonons. The functions $D(t)$ and $D_H(t)$ have the following form[1]:

$$D(t)=\frac{v_\parallel^2}{2w_\parallel^2}\left(-\mathrm{i}t+\frac{t^2}{\beta}\right)+d_0^2\left(1-\mathrm{e}^{\mathrm{i}v_\parallel t}+\frac{4}{\mathrm{e}^{\beta v_\parallel}-1}\sin^2\frac{v_\parallel t}{2}\right), \tag{53}$$

$$D_H(t)=\sum_{j=1}^{3}d_j^2\left(1-\mathrm{e}^{\mathrm{i}s_jt}+\frac{4}{\mathrm{e}^{\beta s_j}-1}\sin^2\frac{s_jt}{2}\right), \tag{54}$$

where

$$d_0^2=\frac{v_\parallel^2-w_\parallel^2}{2v_\parallel^3}, \tag{55}$$

$$d_j^2=\frac{1}{2s_j}\frac{s_j^2-w_\parallel^2}{3s_j^2+2(-1)^j\omega_cs_j-v_\perp^2},\qquad j=1,2,3. \tag{56}$$

The frequencies $s_1\leq s_2\leq s_3$, which are given by positive roots of the equation

$$s^2\left(s^2-v_\perp^2\right)^2-\omega_c^2\left(s^2-w_\perp^2\right)^2=0\,, \tag{57}$$

characterize the internal oscillations in the plane perpendicular to the magnetic field. Equation (52) with Eqs. (53) and (54) from Ref. [73] is a generalization of the result of Ref. [16] to the case when a magnetic field of arbitrary strength is present.

---

[1] In the special case of an isotropic Feynman model and weak magnetic field ($v_\perp=v_\parallel$, $w_\perp=w_\parallel$), these functions were obtained in Ref. [80] in the form of infinite sums.

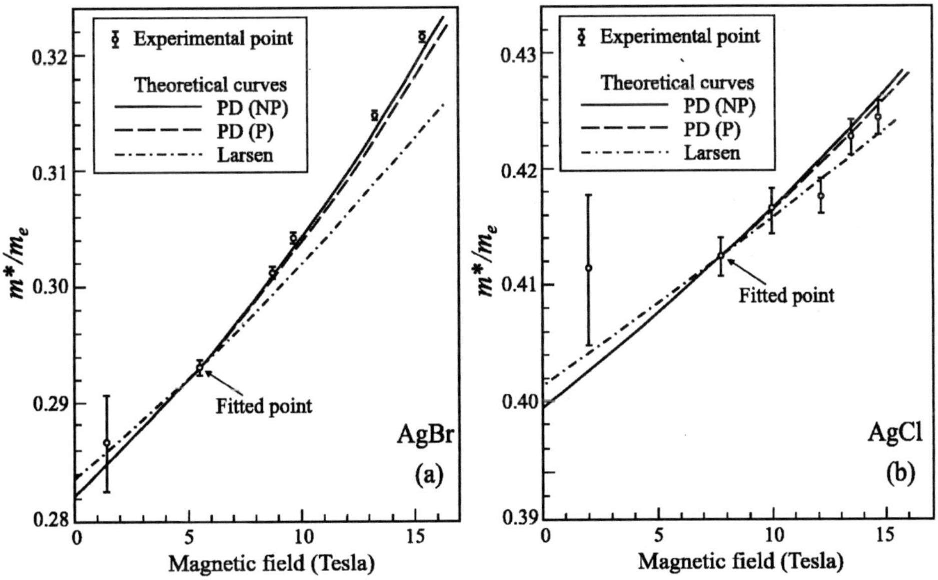

FIGURE 12. The polaron cyclotron mass in AgBr (a) and in AgCl (b): comparison of experiment and theory (Larsen: [82]; PD: [73]); (P) – with parabolic band, (NP) – with corrections of a two-band Kane model. In each case the band mass was adjusted to fit the experimental point at 525 GHz. (From Ref. [10].)

4.2. Cyclotron resonance of polarons in silver halides

In pioneering experimental studies F. C. Brown and co-workers (see Ref. [81]) have combined mobility experiments and cyclotron resonance measurements to demonstrate polaron effects in AgBr. From a theoretical plot of polaron mobility versus band mass compared to experimental Hall mobility data at a given temperature, they estimate the band mass. This allows to estimate α and the polaron mass m^*. This value of m^* can then be compared to the measured cyclotron mass in order to test polaron theory following Ref. [18]. For AgBr, the experimental value $m^*/m_e = 0.27 \pm 0.01$ obtained from the cyclotron resonance at millimetre wave frequencies and at 18 K agrees well with a value $m^*/m_e = 0.27 \pm 0.05$ determined by comparison of mobility experiment and theory.

More direct evidence for the polaron character of charge carriers in AgBr and AgCl was obtained through accurate CR experiments in external magnetic fields up to 16 T (see Fig. 12). Several polaron theories were compared in analysing the CR data. It turns out that the weak-coupling theories (Rayleigh-Schrödinger perturbation theory, Wigner-Brillouin perturbation theory and its improvements) fail (and are all off by at least 20% at 16 T) to describe the experimental data shown in Fig. 12 for the silver halides. The approach of Ref. [82] underestimates the polaron cyclotron mass by 2.5% at 15.3 T. The magneto-absorption calculated in Ref. [73], which is an all-coupling derivation, leads to the best quantitative agreement between theory and experiment for AgBr and AgCl as can be seen from Fig. 12. This analysis provides a clear confirmation of the Fröhlich description of polarons.

The analysis in Ref. [10] on the basis of the theory for polaron magneto-absorption of Ref. [73] leads to the following polaron coupling constants: $\alpha = 1.53$ for AgBr and $\alpha = 1.84$ for AgCl. The corresponding polaron masses are: $m^* = 0.2818$ for AgBr and $m^* = 0.3988$ for AgCl.

For most materials with relatively large Fröhlich coupling constant m_b is not known. The study in Ref. [10] is an example of the detailed analysis of CR data that is necessary to obtain accurate polaron data like α and m_b for a given material. I refer to Ref. [10] for further details like, e. g., the role of the "fitted point" in Fig. 12.

4.3. Cyclotron resonance of polarons in CdTe

Early infrared-transmission studies of hydrogen-like shallow-donor-impurity states in *n*-CdTe were reported in Ref. [83] (see Fig. 13). By studying the Zeeman splitting of the ($1s \rightarrow 2p, m = \pm 1$) transition in the Faraday configuration at magnetic fields up to ~ 160 kOe, a quantitative determination of polaron shifts of the energy levels of a bound electron was performed. The experimental data were shown to be in fair agreement with the weak-coupling theory of the polaron Zeeman effect. In this comparison, however, the value $\alpha = 0.4$ had to be used instead of $\alpha = 0.286$, which would follow from the definition (1). Similarly, the value $\alpha \sim 0.4$ was suggested (see Ref. [60]) for the explanation of the measured variation of the cyclotron mass with magnetic field in CdTe. This discrepancy gave rise to some discussion in the literature (see, e. g., Refs. [84, 85] and references therein).

In Ref. [8], far-infrared photoconductivity techniques were applied to study the energy spectrum of shallow In donors in CdTe layers and experimental data were obtained on the magnetopolaron effect, as shown in Fig. 14. An overall good agreement is found between experiment and a theoretical approach, where the electron-phonon interaction is treated within a second-order improved Wigner-Brillouin perturbation theory and a variational calculation is performed for the lowest-lying donor states ($1s, 2p^{\pm}, 2s, 2p_z, 3d^{-2}, 4f^{-3}$). This agreement is obtained with the coupling constant $\alpha = 0.286$. In an earlier work [86] it was found, that with $\alpha = 0.3$ and $m_b = 0.0920 m_e$ the theoretical results for the magnetic field dependence of the polaron mass in CdTe fall within the error bars of the experimental data.

4.4. Polarons in the layered compound InSe

Nicholas *et al.* demonstrated polaron coupling effects using cyclotron resonance measurements in a 2DEG, which naturally occurs in the polar semiconductor InSe. One clearly sees, over a wide range of magnetic fields ($B = 18$ to 34 T), two distinct magnetopolaron branches separated by as much as 11 meV ($\sim 0.4\omega_{\mathrm{LO}}$) at resonance (Fig. 15). The theoretical curves show the results of calculations for coupling to the LO phonons in bulk (3D), sheet (2D) and after correction for the quasi-2D character of the system with $\alpha = 0.29$ (for motion perpendicular to the *c*-axis). The agreement between theory and experiment is reasonable for the 3D case, but better for the quasi-2D system,

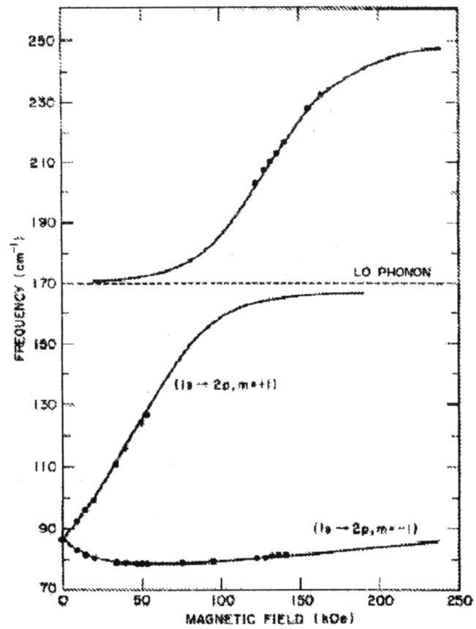

FIGURE 13. Plot of the experimentally determined magnetic field dependence of the ($1s \rightarrow 2p$, $m = \pm 1$) transition frequencies in *n*-CdTe. The solid lines represent the theoretical results for the effects of the electron-LO-phonon interaction upon the shallow-donor-impurity levels. If α is treated as a fitting parameter, the value $\alpha = 0.4$ would result (From Ref. [83].)

if the finite spatial extent of the 2D electron gas in the symmetric planar layer is taken into account.

4.5. Cyclotron resonance of polarons in (quasi-)2D systems

For $GaAs/Al_xGa_{1-x}As$ quantum wells and superlattices, the polaron effect is found to decrease the energy of the shallow donor states at low magnetic fields and leads to a resonant splitting of the energies at higher magnetic fields. The results are in good agreement with available experimental far-infrared optical data in the whole magnetic-field range (see, e. g., Ref. [20] and references therein).

In quasi-2D semiconductor structures the phonon modes are modified by the presence of the interfaces. The study of the cyclotron-resonance spectra in quasi-2D systems provides a means of determining the strength of interaction with specific phonons. In Ref. [69], where the cyclotron resonance of polarons in $GaAs/Al_{0.3}Ga_{0.7}As$ multiple quantum well structures was studied, it appeared possible to observe the anticrossing effects due to different phonon modes. The analysis of the experimental data [69], which is based on the theory [87, 67, 88], has shown that the observed anticrossing effects are caused by the resonant interaction of electrons with the GaAs confined slab phonon modes and the AlAs-like interface optical phonons. The GaAs-like interface phonon modes do not play an important role over the range of magnetic fields and

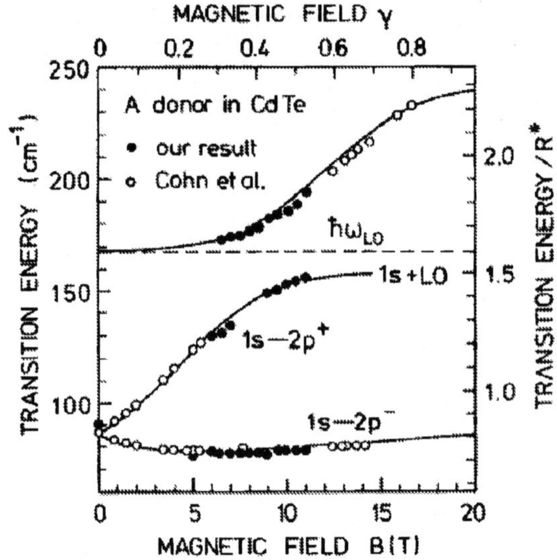

FIGURE 14. Plot of the experimentally determined magnetic field dependence of the $1s \rightarrow 2p^{\pm 1}$ transition energies in CdTe layers grown by molecular-beam epitaxy. The solid lines represent the results of the calculation described in the text without any fitting parameters. The solid dots are the experimental data of Ref. [8] and the open circles represent the data of Ref. [83]. (From Ref. [8].)

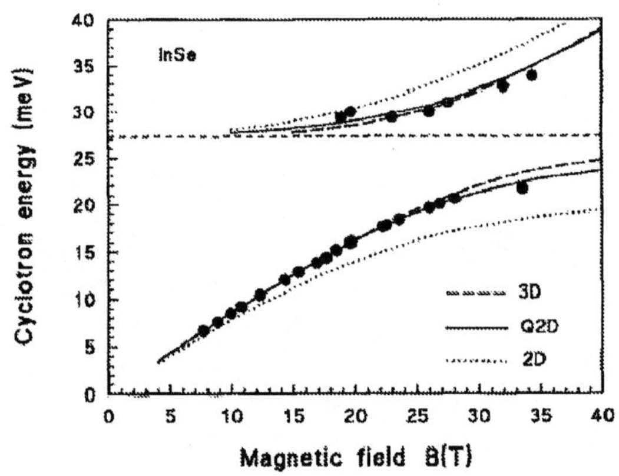

FIGURE 15. The cyclotron resonance position plotted as a function of magnetic field for InSe. (From Ref. [68].)

photon frequencies, for which the polaron cyclotron resonance is observable. This is because (i) the interaction of these modes with electrons is weaker than that of GaAs confined slab modes and (ii) the experiments do not probe into the reststrahlen region where the resonance with the GaAs-like interface phonon modes occurs.

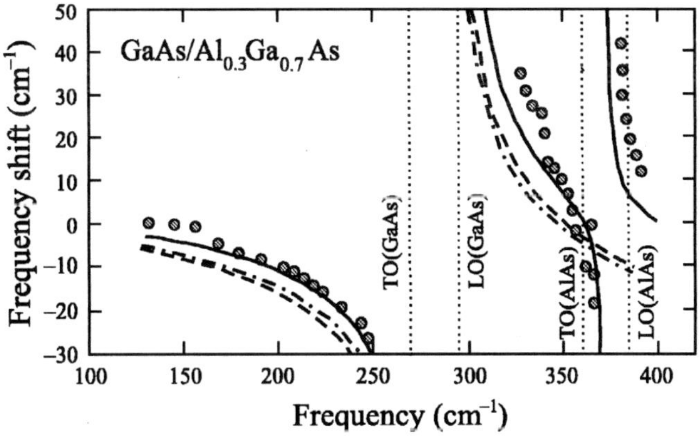

FIGURE 16. The difference between the measured polaron cyclotron-resonance frequency and the unperturbed cyclotron-resonance frequency is shown versus the measured frequency for a $GaAs/Al_{0.3}Ga_{0.7}As$ multiple quantum well structure with the well width 1.2 nm and the nominal doping $1.5 \times 10^{11}\ cm^{-2}$. The dashed lines are calculated results for coupling with only bulk GaAs LO phonons. The dash-dotted lines include screening effects in addition to the coupling with bulk GaAs LO phonons. The solid curves take into account the effects of interface optical phonons and slab LO optical phonons. (After Ref. [69].)

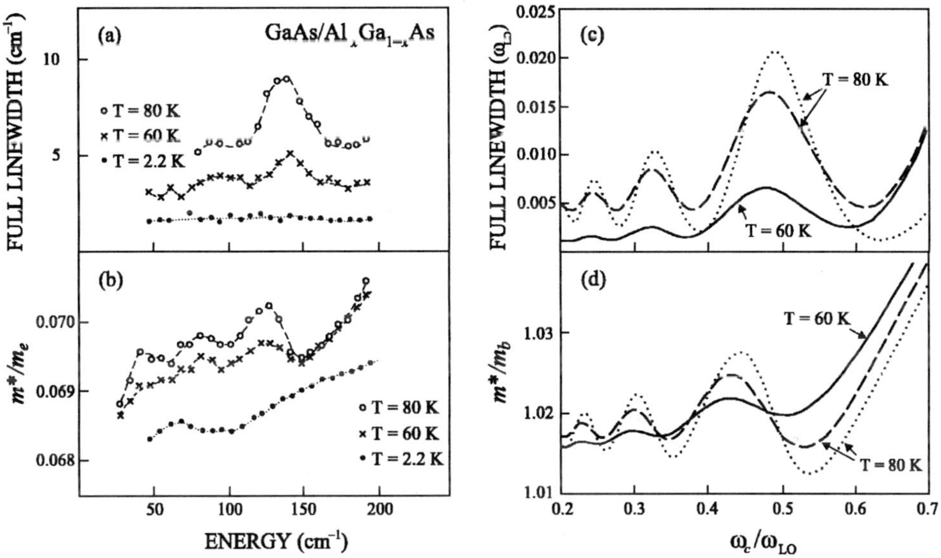

FIGURE 17. Frequency dependence of the resonance full width [panel (a)] and cyclotron mass [panel (b)] in the experiments on the optically detected magnetophonon resonance in $GaAs/Al_xGa_{1-x}As$ heterojunctions at different temperatures. Panels (c) and (d) show the theoretical results for the cyclotron linewidth and effective mass, respectively. In the calculations, the Landau-level broadening was introduced semiempirically in the form $\Gamma/\hbar\omega_{LO} = \Gamma_0(\omega_c/\omega_{LO})^{1/2}$; the dashed and solid curves are for $\Gamma_0 = 0.25$ and the dotted curve is for $\Gamma_0 = 0.2$ (After Ref. [89].)

Under the condition of magnetophonon resonance $N\omega_c = \omega_{LO}$, where N is a positive integer, inelastic scattering of an electron by a LO phonon can induce resonant transitions of this electron between the Landau levels whose indices differ from each other by N. Such a resonant scattering can be revealed by measuring the cyclotron-resonance linewidth. Indeed, a resonant increase of the cyclotron-resonance linewidth at $N\omega_c = \omega_{LO}$ (so-called optically detected magnetophonon resonance) was experimentally observed [89] for GaAs/$Al_xGa_{1-x}As$ heterojunctions (see Fig. 17). A well-pronounced peak in the linewidth at a photon energy of around 140 cm^{-1} (the corresponding resonant magnetic field is about 10.5 T, i. e. close to the $N = 2$ condition) and weaker features close to the $N = 3,4$ conditions are clearly seen in Fig. 17(a). As is seen in Fig. 17(b), the cyclotron-resonance effective mass m^* also shows an oscillatory structure, and it is easier to distinguish the higher resonances in plots of the mass, than in the equivalent linewidth data Ref. [89]. The resonances of the cyclotron-resonance effective mass m^* are, however, shifted in phase from those of linewidth. The theoretical description of the optically detected magnetophonon resonance was given in Ref. [89] on the basis of the approach of Ref. [90] and it appeared to be in a fair agreement with the experimental data [see Figs. 17(c) and 17(d)].

While the resonant *momentum* relaxation at the magnetophonon resonance condition reveals itself though the aforementioned oscillations of cyclotron-resonance linewidth, the resonant changes in the electron *energy* relaxation [91] was observed in the so-called saturated cyclotron resonance [92]. The basic principle of a saturation cyclotron-resonance measurement is to study the bleaching of the cyclotron resonance absorption due to high intensity excitation between the absorbing levels. Such a process can only occur if the ladder of available Landau states is truncated. The mechanism used in Ref. [92] to achieve truncation was band non-parabolicity that is sufficiently high in the narrow-gap material InAs. In the cyclotron-resonance measurements on an InAs/GaSb double heterojunction [92], clear peaks of the saturation intensity were observed at the magnetophonon resonance conditions (see Fig. 18). The Landau-level lifetime, found from the saturation data and displayed in Fig. 18 by filled squares, have well pronounced minima when the conditions of the magnetophonon resonance are satisfied.

4.6. Many-polaron cyclotron resonance in quantum wells

In the presence of a magnetic field, resonant magnetopolaron coupling (i. e., the anticrossing of zero-phonon and one-phonon states of the polaron system) can occur. Numerous experiments on the cyclotron resonance give clear evidence of this resonant magnetopolaron coupling. For low electron densities, the resonant magnetopolaron coupling manifests itself near the LO-phonon frequency (see, e. g., Refs. [93, 94, 95]). Cyclotron-resonance measurements performed on semiconductor quantum wells with high electron density [96] reveal anticrossing near the TO-phonon frequency rather than near the LO-phonon frequency. In Ref. [96], this effect is interpreted by invoking mixing between magnetoplasmons and phonons and in terms of a resonant coupling of the electrons with the mixed magneto-plasmon-phonon modes.

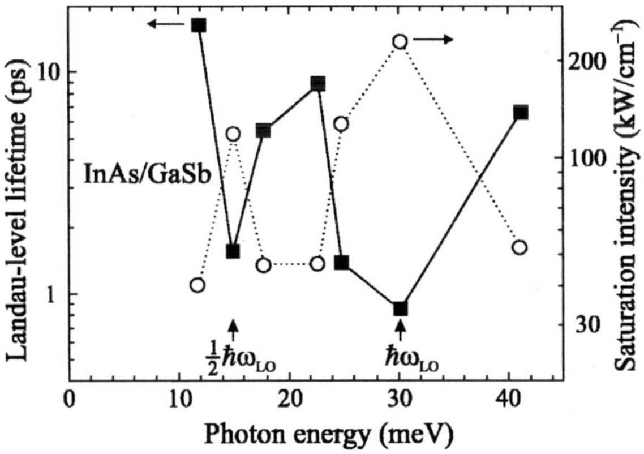

FIGURE 18. Saturation intensity defined as the radiation intensity at which the cyclotron-resonance absorption falls to half its low-intensity limit (open circles) and Landau-level lifetimes (filled squares) as a function of photon energy for an InAs/GaSb double heterojunction. The magnetophonon resonance conditions are indicated by arrows. (After Ref. [92].)

In Ref. [97], CR spectra for a polaron gas in a GaAs/AlAs quantum well are theoretically investigated, taking into account (i) the electron-electron interaction and the screening of the electron-phonon interaction, (ii) the magnetoplasmon-phonon mixing, (iii) the electron-phonon interaction with all the phonon modes specific for the quantum well under investigation. As a result of this mixing, different magnetoplasmon-phonon modes appear in the quantum well, which give contributions to the CR spectra.

In Ref. [97], a finite-barrier quantum well of width d is considered. The quantum well (medium 1) with high-frequency dielectric constant ε_1 is placed into a matrix (medium 2) with high-frequency dielectric constant ε_2. Both these media are supposed to be polar. The external magnetic field $\mathbf{B}$ is applied parallel to the z-axis. The symmetric gauge is chosen for the vector potential of the magnetic field.

The total Hamiltonian of the system of electrons interacting with the phonons and with each other in the quantum well is

$$\begin{aligned} H = & \sum_{nlm\sigma} E_{nl\sigma} a^+_{nlm\sigma} a_{nlm\sigma} + \sum_{\lambda,\mathbf{q}} \hbar\omega_{\lambda,\mathbf{q}} b^+_{\lambda,\mathbf{q}} b_{\lambda,\mathbf{q}} \\ & + \frac{1}{\sqrt{S}} \sum_{\lambda,\mathbf{q}} \sum_{n,k} \left(\gamma_{\lambda,\mathbf{q}}\right)_{nk} \rho_{nk}(\mathbf{q}) \left(b_{\lambda,\mathbf{q}} + b^+_{\lambda,-\mathbf{q}}\right) \\ & + \frac{1}{2S} \sum_{\mathbf{q}} \sum_{n_1 k_1 n_2 k_2} V_C\left(k_1, n_1; n_2, k_2 | \mathbf{q}\right) \mathcal{N}\left[\rho^+_{k_1 n_1}(\mathbf{q}) \rho_{n_2 k_2}(\mathbf{q})\right], \end{aligned} \tag{58}$$

where S is the area of the quantum well in the xy-plane, $\mathcal{N}[\ldots]$ is the normal product symbol, $a^+_{nlm\sigma}$ $\left(a_{nlm\sigma}\right)$ is a creation (annihilation) operator for an electron in the one-particle state

$$\begin{aligned} \Psi_{nlm}(\rho, \varphi, z) &= \psi_n(z)\, \phi_{lm}(\rho, \varphi), \\ \phi_{lm}(\rho, \varphi) &= \frac{1}{\sqrt{2\pi}} e^{im\varphi} \Phi_{lm}(\rho). \end{aligned} \tag{59}$$

The wave function $\psi_n(z)$ corresponds to the n-th size-quantized subband for motion along the z-axis, while $\phi_{lm}(\rho,\varphi)$ is the wave function characterizing the "in-plane" motion of an electron with definite z-projection m of its angular momentum, l is a Landau level quantum number. $\rho_{nk}(\mathbf{q})$ is the electron density operator

$$\rho_{nk}(\mathbf{q}) = \sum_{lml'm'\sigma} \left(e^{i\mathbf{q}\cdot\mathbf{r}_{\parallel}}\right)_{lm,l'm'} \hat{a}^{+}_{nlm\sigma}\hat{a}_{kl'm'\sigma}, \tag{60}$$

with matrix element

$$\left(e^{i\mathbf{q}\cdot\mathbf{r}_{\parallel}}\right)_{lm,l'm'} = \int_0^{\infty} \rho d\rho \int_0^{2\pi} d\varphi e^{i\mathbf{q}\cdot\mathbf{r}_{\parallel}} \phi^{*}_{lm}(\rho,\varphi)\,\phi_{l'm'}(\rho,\varphi). \tag{61}$$

Within the memory-function technique, the real part of the optical conductivity (in the Faraday configuration) in the local-parabolic-band approximation can be written as [88, 98]

$$\mathrm{Re}\sigma(\omega) = -\frac{n_S e^2}{m_b}\mathrm{Im}\frac{1}{\omega-\omega_c-\chi(\omega,\omega_c)/\omega+i\varepsilon} \quad (\varepsilon\to+0), \tag{62}$$

where n_S is the 2D electron density and $\chi(\omega)$ is the memory function. In order to take into account the splitting of the CR peaks due to the nonparabolicity of the conduction band, Eq. (62) is generalized in the following way,

$$\mathrm{Re}\sigma(\omega) = \sum_{n,l,\sigma} \varkappa_{nl\sigma}\mathrm{Re}\sigma_{nl\sigma}(\omega), \tag{63}$$

where each contribution $\mathrm{Re}\sigma_{nl\sigma}(\omega)$ due to the transitions $(l\to l+1)$ is calculated within the local-parabolic-band approximation

$$\mathrm{Re}\sigma_{nl\sigma}(\omega) = -\frac{n_S e^2}{m_b}\mathrm{Im}\frac{1}{\omega-\omega_c^{(nl\sigma)}-\chi\left(\omega,\omega_c^{(nl\sigma)}\right)/\omega+i\varepsilon} \quad (\varepsilon\to+0). \tag{64}$$

The transition frequency

$$\omega_c^{(nl\sigma)} \equiv \frac{E_{n,l+1,\sigma}-E_{nl\sigma}}{\hbar} \tag{65}$$

is used in Eq. (64) instead of the cyclotron frequency ω_c in Eq. (62). The weight $\varkappa_{nl\sigma}$ is proportional to the number of open channels for the transitions $(l\to l+1)$. The normalized weights $\varkappa_{nl\sigma}$ are

$$\varkappa_{nl\sigma} = \frac{f_{nl\sigma}\left(1-f_{n,l+1,\sigma}\right)}{\sum_{n',l',\sigma'} f_{n'l'\sigma'}\left(1-f_{n',l'+1,\sigma'}\right)}, \tag{66}$$

where $f_{nl\sigma}$ is the Fermi occupation number. The memory function $\chi(\omega,\omega_c)$ has the form (cf. Refs. [88, 87])

$$\begin{aligned}\chi(\omega,\omega_c) &= -\sum_{n,k}\sum_{\lambda}\int\frac{d^2\mathbf{q}}{(2\pi)^2}\frac{\left|\left(\gamma_{\lambda,\mathbf{q}}\right)_{nk}\right|^2 q^2}{n_s\hbar m_b\varepsilon^2(q)}\\ &\times\int_{-\infty}^{\infty}dte^{-\varepsilon t}\left(e^{i\omega t}-1\right)\mathrm{Im}\left[\mathscr{D}_{\lambda}(\mathbf{q},t)\,G_{nk}(\mathbf{q},t)\right]\quad(\varepsilon\to+0),\end{aligned}\qquad(67)$$

where $\varepsilon(q)$ is the static screening factor [88], $\mathscr{D}_{\lambda}(\mathbf{q},t)$ is the phonon Green's function

$$\mathscr{D}_{\lambda}(\mathbf{q},t)=-i\Theta(t)\left[\left\langle b_{\lambda,\mathbf{q}}(t)\,b^{+}_{\lambda,\mathbf{q}}(0)\right\rangle+\left\langle b^{+}_{\lambda,-\mathbf{q}}(t)\,b_{\lambda,-\mathbf{q}}(0)\right\rangle\right],\qquad(68)$$

and $G_{nk}(\mathbf{q},t)$ is the electron density-density Green's function,

$$G_{nk}(\mathbf{q},t)=-i\Theta(t)\frac{1}{S}\left\langle\rho_{nk}(\mathbf{q},t)\,\rho^{+}_{nk}(\mathbf{q},0)\right\rangle,\qquad(69)$$

with the Heaviside step function $\Theta(t)$. The averaging in Eqs. (68) and (69) is performed on the equilibrium statistical operator of the electron-phonon system.

In Fig. 19, we plot a set of the calculated magnetoabsorption spectra for a 10-nm GaAs/AlAs quantum well as a density map. This map shows the magnetoabsorption intensity as a function of both the magnetic field and the frequency. For comparison of these spectra with the experimental data of Ref. [96], the peak positions from Fig. 3 of Ref. [96] are shown in the same graph. Splitting of the peaks occurs because the electron Landau levels in a non-parabolic conduction band are non-equidistant. In Fig. 19, the higher-energy part of each split peak corresponds to the transitions between the Landau levels $(0\to 1)$, while the lower-energy part corresponds to the transitions $(1\to 2)$. With increasing magnetic-field strength, the upper filled level with $l=1$ becomes less populated, so that the relative intensity of the lower-energy peak diminishes. For the series of peaks indicated by crosses the splitting due to the band nonparabolicity is not experimentally resolved.

It is clearly seen from Fig. 19, that for the high-density polaron gas, anticrossing of the CR spectra occurs near the GaAs TO-phonon frequency ω_{T1} rather than near the GaAs LO-phonon frequency ω_{L1} for both the experimental and the calculated spectra. This effect is in contrast with the cyclotron resonance of a low-density polaron gas in a quantum well, where anticrossing occurs near the LO-phonon frequency. The appearance of the anticrossing frequency close to ω_{T1} instead of ω_{L1} is due to the screening of the electron-phonon interaction by the plasma vibrations. A similar effect appears also for the magnetophonon resonance: as shown in Ref. [99], the magnetoplasmon-phonon mixing leads to the shift of the resonant frequency of the magnetophonon resonance in quantum wells from $\omega\approx\omega_{L1}$ to $\omega\approx\omega_{T1}$.

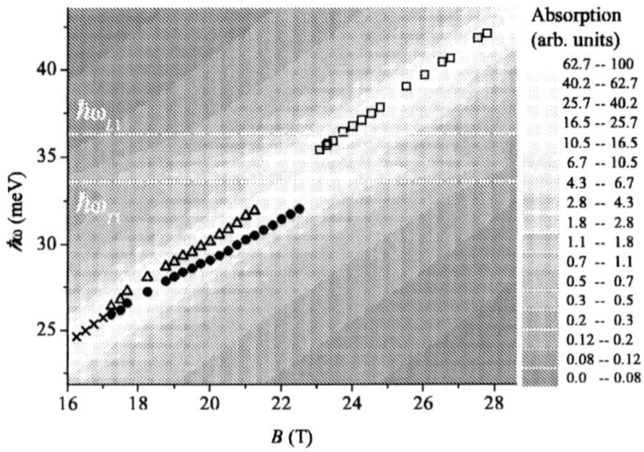

FIGURE 19. Density map of the magnetoabsorption spectra for a 10-nm GaAs/AlAs quantum well as calculated in Ref. [97]. Symbols indicate peak positions of the experimental spectra (which are taken from Fig. 3 of Ref. [96]). Dashed lines show LO- and TO-phonon energies in GaAs. (After Ref. [97].)

4.7. Cyclotron resonance in bulk CdS at ultra-high magnetic fields

The Fröhlich coupling constants in II-VI semiconductors are significantly larger than those for III-V materials and typically lie in the range $\alpha \approx 0.5$, making these materials interesting for the study of polaron effects (see, e. g., Refs. [100, 101]). In particular, the cyclotron mass of charge carriers in II-VI compounds is strongly influenced by the electron-phonon interaction.

The availability of very high magnetic fields allows to achieve the resonant condition $\omega_c = \omega_{\rm LO}$ as well as the condition $\omega_c > \omega_{\rm LO}$ even for II-IV semiconductor compounds, where the effective mass of the conduction electrons can be considerably larger than in II-V semiconductors.

Here we address experiments by Miura et al. on cyclotron resonance in bulk n-type CdS [102] at ultra-high magnetic fields (up to 450 T). Those authors observe a pronounced maximum in the temperature dependence of the cyclotron mass.

The weak-coupling polaron theory predicts a monotonous increase of the cyclotron mass for $\omega_c > \omega_{\rm LO}$ with lowering temperature. However, calculations[71] for $\alpha = 0.527$ based on the all-coupling theory [73] show that in the temperature range from 80 to 120 K the polaron cyclotron mass rapidly decreases as compared to its high-temperature value. Polaron effects beyond the weak-coupling approximation significantly influence the magneto-absorption even for a relatively small Fröhlich coupling constant like 0.527 in n-type CdS.

As the results of Ref. [70] imply, the existing discrepancies between the experimental data and theory can be attributed to the influence of other scattering mechanisms. Extending the approach of Ref. [73] to account for the elastic scattering of electrons by piezoacoustic phonons and lattice defects provides a quantitative interpretation for

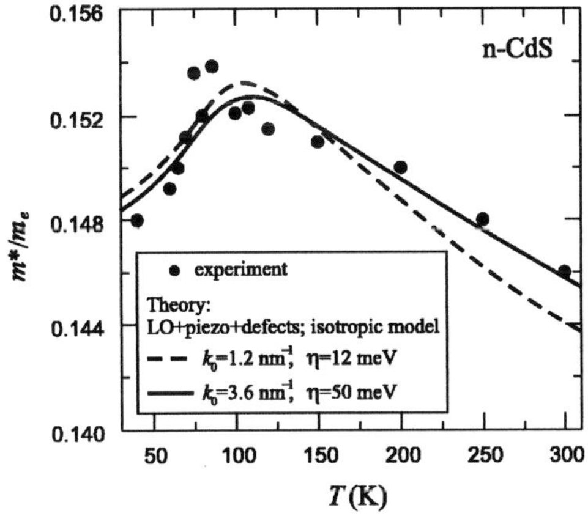

FIGURE 20. Cyclotron mass in n-type CdS, obtained from the calculated magneto-absorption spectra of polarons at the photon wavelength $\lambda = 16.9\,\mu$m ($\omega/\omega_{LO} = 1.92$), in comparison with the experimenal data [102]. The value $m_b = 0.168 m_e$ [102] is used in the calculations for the electron band mass. Along with the electron-LO-phonon interaction, the elastic scattering of electrons by the piezoelectric potential of the acoustic phonons (with piezoelectric-interaction constant $K^2 = 0.0038$ [103]) and by static neutral lattice defects with concentration N is taken into account when calculating the magneto-absorption spectra. A Gaussian potential $V_0 \exp[-k_0^2(\vec{r} - \vec{R}_i)^2]$, where $\vec{r}$ and $\vec{R}_i$ define the positions of the electron and of the i-th defect, respectively, was used to treat the scattering of the electrons by the static lattice defects. Theoretical results [71] are shown for two sets of the parameters k_0 and $\eta = V_0\sqrt{Nk_0^{-3}}$. (After Ref. [71].)

the experimental data [102] (see Fig. 20).

5. OPTICAL PROPERTIES OF QUANTUM DOTS: ROLE OF THE POLARON INTERACTION

5.1. Non-adiabatic approach to the phonon-assisted optical transitions in quantum dots

The interest in the optical properties of quantum dots has been continuously growing i. a. because of the prospects of these structures for optoelectronic applications. Recent measurements [104, 105, 106, 107] of photoluminescence of self-assembled InAs/GaAs quantum dots, reveal high probabilities of the phonon-assisted transitions. Attempts to interpret some of these experiments on the basis of the adiabatic theory meet with difficulties. For spherical CdSe quantum dots, e. g., the Huang-Rhys factor S calculated within the adiabatic approximation takes values which are significantly (by one or two orders of magnitude) smaller than the so-called "experimental Huang-Rhys" factor, determined from the ratio of measured intensities of the phonon satellites. In the framework of the adiabatic approach, various mechanisms, which cause the separation of the electron and hole charges in space [105, 108, 109], are commonly considered as a possible origin for the unexpectedly large Huang-Rhys factor. However, it has been shown [110, 111, 112] that a *non-adiabatic* treatment of the phonon-assisted optical transitions can provide an explanation for the high intensities of the phonon satellites observed in the optical spectra of various quantum-dot structures.

Following the approach used in Ref. [110], the linear light absorption coefficient of a quantum dot is described in the electric dipole approximation by the expression

$$\alpha(\Omega) \propto \mathrm{Re} \sum_{\beta,\beta'} d^*_\beta d_{\beta'} \int_0^\infty dt\, e^{i(\Omega-\Omega_\beta+i\varepsilon)t} \langle \beta | \bar{U}(t) | \beta' \rangle, \tag{70}$$

where Ω is the light frequency, d_β and Ω_β are, respectively, the dipole matrix element and the Franck-Condon frequency of a transition between the exciton vacuum state and the one-exciton state $|\beta\rangle$. The evolution operator averaged over the phonon ensemble, $\bar{U}(t)$, is represented in terms of the chronological ordering operator T:

$$\begin{aligned}\bar{U}(t) = \mathrm{T}\exp\Bigg\{ -\frac{1}{\hbar^2}\sum_\lambda \int_0^t dt_1 \int_0^{t_1} dt_2 \\ \times \Big[(\bar{n}_\lambda + 1)\, \mathrm{e}^{-i\omega_\lambda(t_1-t_2)} \gamma_\lambda(t_1)\gamma^\dagger_\lambda(t_2) \\ + \bar{n}_\lambda \mathrm{e}^{i\omega_\lambda(t_1-t_2)} \gamma^\dagger_\lambda(t_1)\gamma_\lambda(t_2) \Big] \Bigg\}.\end{aligned} \tag{71}$$

In Eq. (71), the index λ labels phonon modes specific for the quantum-dot structure under consideration, ω_λ are phonon frequencies, $\gamma_\lambda(t)$ are the exciton-phonon interaction amplitudes in the interaction representation, and $\bar{n}_\lambda = \left[\exp\left(\hbar\omega_\lambda / k_B T\right) - 1\right]^{-1}$. Within the adiabatic approximation, which has been widely used to treat optical spectra of quantum dots, non-diagonal matrix elements of the exciton-phonon interaction are neglected when calculating the absorption coefficient given by Eq. (70) with Eq. (71). In the adiabatic approach [113, 114] one supposes that (i) both the initial and final states of a quantum transition are non-degenerate, (ii) the energy differences between the exciton states are much larger than the phonon energies. However, these conditions are often violated for optical transitions in quantum dots. In other words, the exciton-phonon system in a quantum dot can be essentially *non-adiabatic*. The polaron interaction of an exciton in a degenerate state results in *internal non-adiabaticity* (the proper Jahn–Teller effect), while the existence of exciton levels separated by an energy comparable with the LO-phonon energy leads to *external non-adiabaticity* (the pseudo Jahn–Teller effect).

5.2. Optical absorption and photoluminescence in quantum dots

In Ref. [110], an approach was developed to calculate the absorption coefficient of a quantum dot given by Eqs. (70) and (71) taking non-adiabaticity into account. Assuming that the exciton-phonon interaction is weak,

$$\eta \equiv \frac{\max |\langle\beta|\gamma_\lambda|\beta'\rangle|}{\hbar\omega_0} \ll 1, \tag{72}$$

and focusing on the intensities of the phonon sidebands, the absorption spectrum can be represented as

$$\begin{aligned}\alpha(\Omega) \propto \quad & \sum_\beta \left[\sum_{K=0}^{\infty} (\bar{n}+1)^K f^{(+)}_{\beta K}\, \delta\left(\Omega_\beta + K\omega_0 - \Omega\right) \right. \\ & \left. + \sum_{K=1}^{\infty} \bar{n}^K f^{(-)}_{\beta K}\, \delta\left(\Omega_\beta - K\omega_0 - \Omega\right) \right],\end{aligned} \tag{73}$$

where for simplicity the frequency dispersion of the optical phonons is neglected: $\omega_\lambda = \omega_0$. The probability of a transition to the state β with emission (absorption) of K phonons is proportional to $f^{(+)}_{\beta K}$ ($f^{(-)}_{\beta K}$) with

$$f^{(\pm)}_{\beta 0} = \left|d_\beta\right|^2, \tag{74}$$

$$\begin{aligned}f^{(\pm)}_{\beta K} = & \sum_{\beta_{-K},\ldots,\beta_K} \delta_{\beta_0\beta}\, \xi_K\left(\beta_{-K},\ldots,\beta_K\right) \\ & \times \prod_{k=1}^{K} \frac{\theta\left(\left|\Omega_{\beta\beta_k} \pm k\omega_0\right| - \Delta\Omega\right)\theta\left(\left|\Omega_{\beta\beta_{-k}} \pm k\omega_0\right| - \Delta\Omega\right)}{\left(\Omega_{\beta\beta_k} \pm k\omega_0\right)\left(\Omega_{\beta\beta_{-k}} \pm k\omega_0\right)}, \quad K \geq 1,\end{aligned} \tag{75}$$

where

$$\begin{aligned}\xi_K\left(\beta_{-K},\ldots,\beta_K\right) = & \frac{d^*_{\beta_{-K}} d_{\beta_K}}{\hbar^{2K}} \sum_{n_1,\ldots n_K=1}^{K} \prod_{i,i'=1}^{K} \left[1 - \delta_{n_i n_{i'}}\left(1-\delta_{ii'}\right)\right] \\ & \times \prod_{k=1}^{K} \sum_\lambda \langle\beta_{-k}|\gamma_\lambda|\beta_{-k+1}\rangle\langle\beta_{n_k-1}|\gamma^\dagger_\lambda|\beta_{n_k}\rangle, \quad K \geq 1,\end{aligned} \tag{76}$$

$$\Omega_{\beta\beta'} = \Omega_\beta - \Omega_{\beta'}, \qquad \theta(x) = \begin{cases} 1, & x \geq 0, \\ 0, & x < 0. \end{cases}$$

The function $\alpha(\Omega)$ given by Eqs. (73) to (76) provides an adequate description (with relative accuracy $\sim \eta^2$) for the absorption coefficient averaged over narrow frequency intervals $(\Omega_\beta \pm K\omega_0 - \Delta\Omega,\ \Omega_\beta \pm K\omega_0 + \Delta\Omega)$, where $\Delta\Omega < \omega_0$.

In Eq. (75), there are three kinds of terms which contain non-diagonal matrix elements $\langle\beta_i|\gamma_\lambda|\beta_k\rangle$ in the coefficients ξ_K and which only occur beyond the adiabatic approximation: (i) terms with the states $|\beta_{-K}\rangle,\ldots,|\beta_K\rangle$ belonging to one and the same

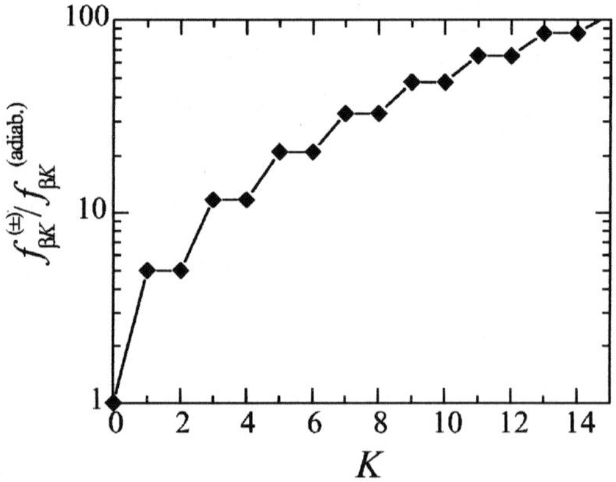

FIGURE 21. Enhancement of the probabilities of optical transitions, participated by d-phonons (phonon modes are classified in analogy with electron states in a hydrogen atom), due to the Jahn-Teller effect in a spherical nanocrystal. $f^{(\pm)}_{\beta K}/f^{(\text{adiab.})}_{\beta K}$ is the ratio of the probabilities, calculated within the non-adiabatic approach ($f^{(\pm)}_{\beta K}$) and the adiabatic approximation ($f^{(\text{adiab.})}_{\beta K}$) for K-phonon optical transitions between the exciton vacuum and the lowest (fourfold degenerate with respect to the z component of the hole angular momentum) one-exciton state. (After Eq. (25) of Ref. [110].)

degenerate energy level which take into account the Jahn-Teller effect; (ii) terms with $|\beta_{-K}\rangle,\ldots,|\beta_K\rangle$ referring to different energy levels that take into account the effects of "external" non-adiabaticity; (iii) terms proportional to $d^*_{\beta_{-K}}d_{\beta_K}$ with $\beta_{-K}\neq\beta_K$ which describe phonon-induced mixing between exciton states with the same symmetry. An example illustrating the enhancement of the probabilities of phonon-assisted optical transitions due to the Jahn-Teller effect is presented in Fig. 21. As follows from Fig. 21, phonon-peak intensities, calculated with non-adiabaticity taken into account, can obey a serial law very different from the Franck-Condon progression, which is predicted by the adiabatic theory and which gives $f^{(\text{adiab.})}_{\beta K}=S^K_\beta/K!$ with $S^K_\beta=\sum_\lambda|\langle\beta|\gamma_\lambda|\beta\rangle/(\hbar\omega_0)|^2$, the Huang-Rhys factor for the exciton state β.

Figure 22 illustrates the crucial role of non-adiabaticity in the light absorption by InAs/GaAs self-assembled quantum dots. The quantum-dot structure is modelled by an InAs cylinder of height h and radius R embedded in GaAs. The absorption spectrum is calculated taking into account the lowest four energy levels of an electron-hole pair. Two of these levels are optically active in the dipole approximation. Within the adiabatic approximation, only the peaks corresponding to the zero-phonon transitions towards the optically active states are clearly seen, while the intensities of the phonon satellites appear to be negligibly small. The effects of non-adiabaticity lead to a mixing of different exciton and phonon states. This non-adiabatic mixing results in a rich structure of the absorption spectrum of the exciton-phonon system [115, 111]. E. g., the probability of phonon-assisted transitions to all the final states significantly rises. These results imply that, due to the effects of non-adiabaticity, a significant enhancement of phonon-satellite intensities arises also in resonant-photoluminescence and photoluminescence excitation

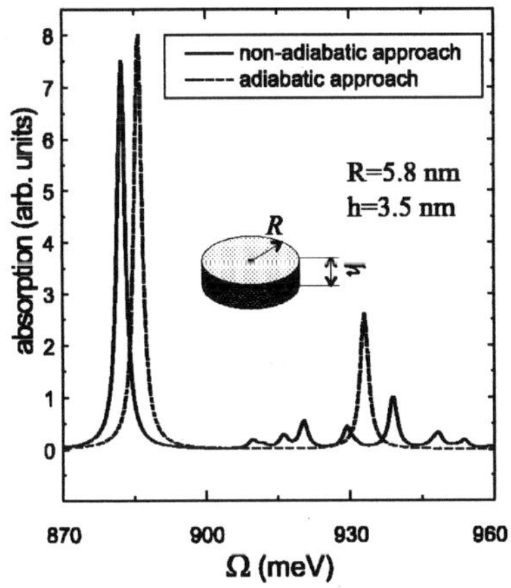

FIGURE 22. Absorption spectra of a cylindrical InAs quantum dot at $T = 4.2$ K calculated within the adiabatic approximation resp. the non-adiabatic approach. (Reprinted from Ref. [111], ©IOP Publishing, 2002.)

(PLE) spectra of quantum dots. Additional experimental evidence for the enhancement of phonon-assisted optical absorption due to the effects of non-adiabaticity has been recently provided by PLE measurements on single self-assembled InAs/GaAs quantum dots [116]. Similar conclusions about the dramatic influence of the exciton-phonon interaction on the optical spectra of quantum dots have been recently formulated in Ref. [117] in terms of a strong coupling regime for the coupling between excitons and LO phonons. Such a strong coupling regime is a particular case of the non-adiabatic mixing related to a (quasi-)resonance when the spacing between exciton levels is close to the LO phonon energy.

The photoluminescence spectra are strongly influenced by relaxation processes during the time interval between excitation and recombination of an exciton. There are two limiting cases where a photoluminescence spectrum can be obtained directly from the corresponding absorption spectra [110]: (i) the *equilibrium luminescence* that takes place, when the time of relaxation between the exciton energy levels τ_0 is much smaller than the radiative lifetime of an exciton τ, and (ii) the opposite case relevant to *slow relaxation*, $\tau_0 \gg \tau$. In Fig. 23, the calculated photoluminescence spectra are compared with the experimental data of Ref. [118] on size-selective photoluminescence of spherical CdSe quantum dots embedded in borosilicate glass. As seen from Fig. 23, taking non-adiabaticity into consideration allows for an adequate description of the observed phonon-peak intensities as well as of their dependence on the size of quantum dots. Another conclusion which follows from Fig. 23 is that probabilities of phonon-assisted optical transitions increase with strengthening confinement (see also Ref. [111]).

As mentioned above, the proposed breakdown [110] of the adiabatic approximation provides an explanation also for discrepancies between the Franck-Condon progression, which is predicted by the adiabatic theory, and the serial law, observed in some multi-phonon photoluminescence spectra, where the measured intensity ratios $S_K^{(\mathrm{exp})}$ for differ-

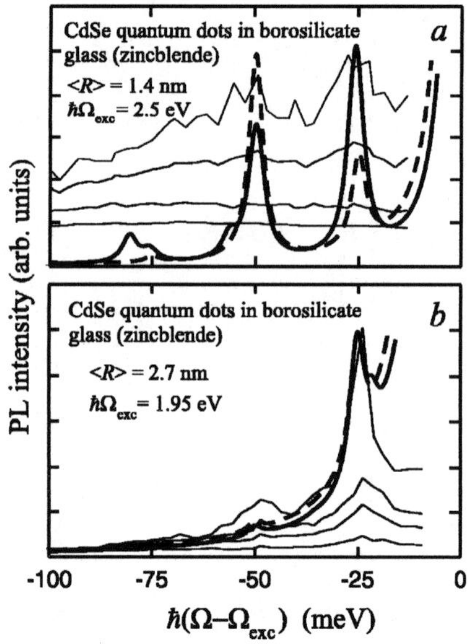

FIGURE 23. Photoluminescence spectra of spherical CdSe quantum dots at different average radii $\langle R \rangle$ and excitation energies $\hbar\Omega_{\text{exc}}$ for $T = 5$ K. Thin solid lines represent the families of experimental time-resolved photoluminescence spectra of Ref. [118] measured at different time intervals between the exciting pulse and the measurement, the upper curve corresponding to the zero time interval. Theoretical curves are displayed for the thermodynamic equilibrium luminescence (heavy solid lines) and in the case of slow relaxation of the exciton energy (dashed lines). (After Ref. [110].)

ent K strongly differ from each other. Figure 24 shows multiphonon photoluminescence spectra calculated by different methods as well as the experimentally observed spectrum [119] for colloidal CdSe quantum dots with wurtzite structure. In these quantum dots, the magnitude of the splitting of the lowest exciton level by the crystal field is close to the LO phonon energy over a wide range of radii of quantum dots. The non-radiative energy relaxation between the exciton energy sublevels, in combination with the pseudo-Jahn–Teller effect, leads to a significant difference between the multiphonon photoluminescence spectra and a Franck-Condon progression.

Recently, the bound-polaron concept was proposed in order to interpret PLE spectra in semiconductor nanostructures [120]. In that work, the multiphonon optical-absorption coefficient for the bound polaron is calculated within a non-adiabatic treatment using the path-integral method. The resulting optical-absorption spectrum is a set of equidistant peaks separated by the LO-phonon frequency, in line with the results of Ref. [121].

5.3. Raman scattering in quantum dots

Experiments on Raman scattering in quantum dots reveal a rich structure of peaks due to optical-phonon-assisted quantum transitions. In Ref. [112], Raman spectra of

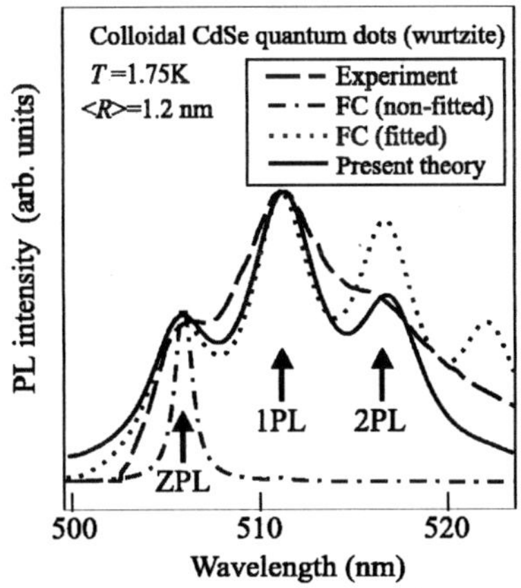

FIGURE 24. Photoluminescence spectra of spherical CdSe quantum dots with average radius $\langle R\rangle = 1.2$ nm. The dashed curve represents the experimental data [119], the dash-dotted curve displays a Franck-Condon progression with Huang-Rhys factor $S = 0.06$ calculated in Ref. [108], the dotted curve shows another Franck-Condon progression with $S = 1.7$, which is obtained by fitting the intensity ratio for two-phonon and one-phonon peaks to the experimental value, and the solid curve results from our theory. (After Ref. [110].)

quantum dots have been theoretically treated taking into account the effects of non-adiabaticity and using the multimode dielectric continuum model to describe phonons specific for the confined systems under consideration. Since the electron-phonon coupling constant α in semiconductor quantum dots is relatively small ($\alpha < 1$), the K-phonon Raman scattering intensity, corresponding to a definite combinatorial frequency $\sum_{j=1}^{K}\omega_{\lambda_j}$, can be analysed to the lowest (K-th) order in α. After averaging over phonon states, the K-phonon scattering intensity is expressed through the squared modulus of the scattering amplitude:

$$F_K^{(\pm)}(\lambda_1,\ldots,\lambda_K) = \sum_{\beta_0,\ldots,\beta_K} \frac{d_{\beta_0}^{I}\left(d_{\beta_K}^{S}\right)^*}{\Omega_{\beta_0} - \Omega_I + i\tilde{\Gamma}_{\beta_0}} \times \prod_{j=1}^{K} \frac{\left\langle \beta_j \middle| \gamma_{\lambda_j} \middle| \beta_{j-1} \right\rangle}{\Omega_{\beta_j} - \Omega_I \pm \sum_{k=1}^{j}\left(\omega_{\lambda_k} \pm i\Gamma_{\lambda_k}\right) + i\tilde{\Gamma}_{\beta_j}}. \quad (77)$$

Here, indices I and S relate to the incoming and scattered light, respectively, Ω_I (Ω_S) is the frequency of the incoming (scattered) light, $\tilde{\Gamma}_\beta$ is the inverse lifetime of an exciton in the state $|\beta\rangle$, Γ_λ is the inverse lifetime of a phonon of the mode λ. The expression (77) takes into account both the Jahn–Teller and the pseudo-Jahn–Teller effect.

For the calculation of Raman spectra in spherical quantum dots in Ref. [112], two fitting parameters are used. (i) A linewidth parameter Γ associated with the finite phonon lifetimes for CdSe we choose $\Gamma = 10\,\mathrm{cm}^{-1}$ [122] and for PbS $\Gamma = 15\,\mathrm{cm}^{-1}$ [123]. (ii) For

PbS nanocrystals, imperfections of the spherical interface are modelled by a potential expanded in spherical harmonics. Random amplitudes of that expansion are assumed to obey a Gaussian distribution with variance $U_0 = 36$ meV. The Raman spectra calculated in Ref. [112] for CdSe and PbS spherical quantum dots show good agreement with the experimental results [122, 123] (see Fig. 25).

The effects of non-adiabaticity result in selection rules for the Raman scattering, which differ from those derived within the adiabatic approximation. For PbS quantum dots, owing to the pronounced LO-phonon dispersion in the bulk material, Raman peaks are easily identified (see Fig. 25*b*). The main contribution into both fundamental and overtone bands is due to $1p$- and $2p$-phonons, while the adiabatic approach would imply the domination of peaks corresponding to s-phonons.

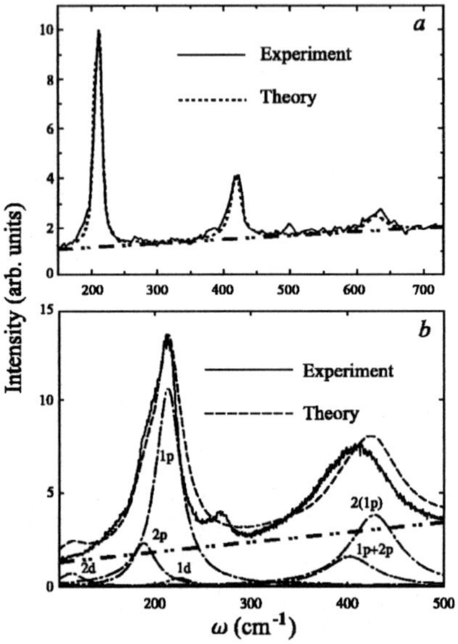

FIGURE 25. Resonant Raman scattering spectra of an ensemble of CdSe quantum dots with average radius 2 nm at $T = 77$ K (panel *a*) and of PbS quantum dots with average radius 1.5 nm at $T = 4.2$ K (panel *b*). The dash-dot-dot curves show the luminescence background. The dash-dot curves in panel b indicate contributions of phonon modes (classified in analogy with electron states in a hydrogen atom) into the Raman spectrum. The dashed curves represent experimental Raman spectra of Refs. [122] (panel *a*) and [123] (panel *b*). (After Ref. [112].)

Non-adiabaticity is an inherent property of exciton-phonon systems in various quantum-dot structures, which *drastically enhances the efficiency of the exciton-phonon interaction*, especially when the exciton levels are separated with energies close to the phonon energies. The effects of non-adiabaticity are important to interpret the surprisingly high intensities of the phonon 'sidebands' observed in the optical absorption, the photoluminescence and the Raman spectra of quantum dots. The observed deviations of the phonon-peak sidebands in some experimental optical spectra from the Franck-Condon progression, which is prescribed by the commonly used adiabatic approximation, find a natural explanation within the non-adiabatic approach.

6. INTERACTING POLARONS IN A QUANTUM DOT

6.1. Ground-state properties of interacting polarons in a quantum dot

In this section, a system of N electrons with mutual Coulomb repulsion and interacting with the lattice vibrations is considered. A parabolic confinement potential, characterised by the frequency parameter Ω_0, is assumed. The total number of electrons is $N = \sum_\sigma N_\sigma$, where N_σ is the number of electrons with spin projection $\sigma = \pm 1/2$. The bulk phonons (characterized by wave vectors $\mathbf{q}$ and frequencies $\omega_\mathbf{q}$) are described by the complex coordinates $Q_\mathbf{q}$. The full set of electron and phonon coordinates is denoted by $\bar{\mathbf{x}} \equiv \left\{\mathbf{x}_{j,\sigma}\right\}$ and $\bar{Q} \equiv \{Q_\mathbf{q}\}$.

The partition function $Z(\{N_\sigma\},\beta)$ of the system can be expressed as a path integral over all electron and phonon coordinates:

$$Z(\{N_\sigma\},\beta) = \sum_P \frac{(-1)^{\xi_P}}{N_{1/2}!N_{-1/2}!} \int d\bar{\mathbf{x}} \int_{\bar{\mathbf{x}}}^{P\bar{\mathbf{x}}} D\bar{\mathbf{x}}(\tau) \int d\bar{Q} \int_{\bar{Q}}^{\bar{Q}} D\bar{Q}(\tau)\, e^{-S[\bar{\mathbf{x}}(\tau),\bar{Q}(\tau)]}, \quad (78)$$

where $S\left[\bar{\mathbf{x}}(\tau),\bar{Q}(\tau)\right]$ is the "action" functional:

$$S\left[\bar{\mathbf{x}}(\tau),\bar{Q}(\tau)\right] = -\frac{1}{\hbar}\int_0^{\hbar\beta} L\left(\dot{\bar{\mathbf{x}}},\dot{\bar{Q}};\bar{\mathbf{x}},\bar{Q}\right) d\tau \quad (79)$$

with the Lagrangian of the electron-phonon system $L\left(\dot{\bar{\mathbf{x}}},\dot{\bar{Q}};\bar{\mathbf{x}},\bar{Q}\right)$. The parameter $\beta \equiv 1/(k_B T)$ is inversely proportional to temperature T. In order to take the Fermi-Dirac statistics into account, the integral over the electron paths $\{\bar{\mathbf{x}}(\tau)\}$ in (78) contains a sum over all permutations P of the electrons with equal spin projections, with ξ_P denoting the parity of a permutation P.

The path integral over the phonon variables in (79) can be calculated analytically [124]. As a result, the partition function of the electron-phonon system (78) factorises into a product of a free-phonon partition function with a partition function $Z_P\left(N_{1/2},N_{-1/2}|\beta\right)$ of interacting polarons, which is a path integral over the electron coordinates only:

$$Z(\{N_\sigma\},\beta) = Z_P(\{N_\sigma\},\beta)\prod_\mathbf{q}\frac{1}{2\sinh\left(\beta\hbar\omega_{\mathrm{LO}}/2\right)}. \quad (80)$$

$$Z_P(\{N_\sigma\},\beta) = \sum_P \frac{(-1)^{\xi_P}}{N_{1/2}!N_{-1/2}!}\int d\bar{\mathbf{x}} \int_{\bar{\mathbf{x}}}^{P\bar{\mathbf{x}}} D\bar{\mathbf{x}}(\tau)\, e^{-S_P[\bar{\mathbf{x}}(\tau)]}. \quad (81)$$

where $S_P[\bar{\mathbf{x}}(\tau)]$ results from the elimination of the phonon variables and contains the "influence phase" of the phonons. It describes the phonon-induced retarded interaction between the electrons, including the retarded self-interaction of each electron. The free energy of a system of interacting polarons $F_P(\{N_\sigma\},\beta)$ is related to the partition function (81) by the relation:

$$F_P(\{N_\sigma\},\beta) = -\frac{1}{\beta}\ln Z_P(\{N_\sigma\},\beta). \tag{82}$$

At present no method is known to calculate the non-Gaussian path integral (81) analytically. For *distinguishable* particles, the Jensen-Feynman variational principle [124] provides a convenient approximation technique. It yields a lower bound to the partition function, and consequently an upper bound to the free energy.

The formulation of a variational principle for the free energy for a system of *identical particles* is a non-trivial problem. However, it can be shown [125] that the path-integral approach to the many-body problem for a fixed number of identical particles can be formulated as a Feynman-Kac functional defined on a state space for N indistinguishable particles, by imposing an ordering on the configuration space and by the introduction of a set of boundary conditions at the boundaries of this state space. The path integral (in the imaginary-time variable) for identical particles was shown to be *positive* within this state space. This implies that a many-body extension of the Jensen-Feynman inequality was found, which can be used for interacting identical particles (Ref. [125], p. 4476). A more detailed analysis of this variational principle for both local and retarded interactions can be found in Ref. [126]. It is required that the potentials are symmetric with respect to all permutations of the particle positions, and that both the exact propagator and the model propagator are antisymmetric (for fermions) with respect to permutations of any two electrons at any moment in time. This means that those propagators must be defined on the same configuration space. Keeping in mind these requirements, the variational inequality for identical particles has the same form as the Jensen-Feynman variational principle:

$$F_P \leq F_0 + \frac{1}{\beta}\langle S_P - S_0\rangle_{S_0}, \tag{83}$$

where S_0 is a model action with corresponding free energy F_0. The angular brackets mean a weighted average over the paths

$$\langle(\bullet)\rangle_{S_0} = \frac{\sum_P \frac{(-1)^{\xi_P}}{N_{1/2}!N_{-1/2}!}\int d\bar{\mathbf{x}}\int_{\bar{\mathbf{x}}}^{P\bar{\mathbf{x}}} D\bar{\mathbf{x}}(\tau)\,(\bullet)\,e^{-S_0[\bar{\mathbf{x}}(\tau)]}}{\sum_P \frac{(-1)^{\xi_P}}{N_{1/2}!N_{-1/2}!}\int d\bar{\mathbf{x}}\int_{\bar{\mathbf{x}}}^{P\bar{\mathbf{x}}} D\bar{\mathbf{x}}(\tau)\,e^{-S_0[\bar{\mathbf{x}}(\tau)]}}. \tag{84}$$

In Ref. [127], we have chosen a model system consisting of N electrons with coordinates $\bar{\mathbf{x}} \equiv \left\{\mathbf{x}_{j,\sigma}\right\}$ and with N_f fictitious particles with coordinates $\bar{\mathbf{y}} \equiv \left\{\mathbf{y}_j\right\}$ in a harmonic confinement potential with elastic interparticle interactions as studied in Ref. [43]. The Lagrangian of this model system takes the form

$$\begin{aligned} L_M\left(\dot{\bar{\mathbf{x}}},\dot{\bar{\mathbf{y}}};\bar{\mathbf{x}},\bar{\mathbf{y}}\right) &= -\frac{m}{2}\sum_\sigma\sum_{j=1}^{N_\sigma}\left(\dot{\mathbf{x}}_{j,\sigma}^2 + \Omega^2\mathbf{x}_{j,\sigma}^2\right) + \frac{m\omega^2}{4}\sum_{\sigma,\sigma'}\sum_{j=1}^{N_\sigma}\sum_{l=1}^{N_{\sigma'}}\left(\mathbf{x}_{j,\sigma} - \mathbf{x}_{l,\sigma'}\right)^2 \\ &\quad -\frac{m_f}{2}\sum_{j=1}^{N_f}\left(\dot{\mathbf{y}}_j^2 + \Omega_f^2\mathbf{y}_j^2\right) - \frac{k}{2}\sum_\sigma\sum_{j=1}^{N_\sigma}\sum_{l=1}^{N_f}\left(\mathbf{x}_{j,\sigma} - \mathbf{y}_l\right)^2. \end{aligned} \tag{85}$$

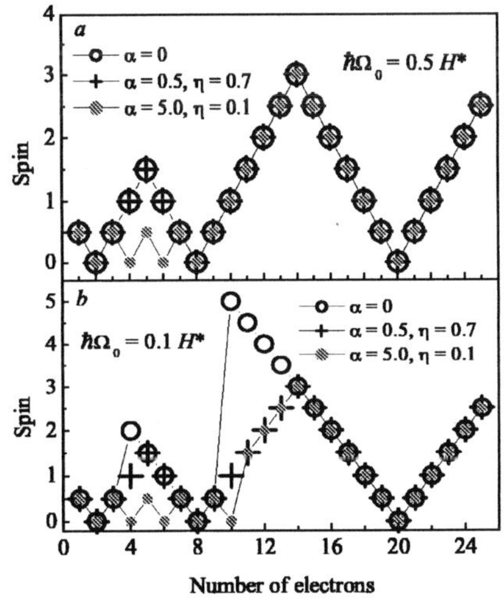

FIGURE 26. Total spin of the system of interacting polarons in a parabolic quantum dot as a function of the number of electrons for $\hbar\Omega_0 = 0.5\ H^*$ (*a*) and for $\hbar\Omega_0 = 0.1\ H^*$ (*b*). (From Ref. [127].)

The frequencies Ω, ω, Ω_f, the mass of the fictitious particle m_f, and the force constant k are variational parameters. Clearly, this Lagrangian is symmetric with respect to electron permutations. Using this model Lagrangian we obtain an upper bound to the free energy F_{var}. Both the free energy and the correlation functions of the model system can be calculated analytically using the generation-function technique [128]. Further on, the zero-temperature case is considered.

In Fig. 26, the total spin S of a system of interacting polarons in their ground state is plotted as a function of the number of electrons in a quantum dot for different values of the confinement frequency Ω_0, of the electron-phonon coupling constant α and of the parameter $\eta = \varepsilon_\infty/\varepsilon_0$. The parameter $\hbar\Omega_0$ is measured in effective Hartrees ($H^* = \left[m/\left(m_0\varepsilon_\infty^2\right)\right] \times 1$ Hartree). Normally, for closed-shell systems $S = 0$, while for open-shell systems S takes its maximal value for a given shell filling (Hund's rule [129]). Hund's rule means, that the electrons in the upper (partly filled) shell are distributed in such a way that the total spin takes its maximal possible value. As seen from Fig. 26 (*a*), for a quantum dot with $\hbar\Omega_0 = 0.5\ H^*$ at $\alpha = 0$ and at $\alpha = 0.5$, the shell filling does obey Hund's rule.

At sufficiently small Ω_0, a spin-polarized state for a system of interacting electrons in a quantum dot can become energetically more favourable than a state satisfying Hund's rule. For a quantum dot with $\hbar\Omega_0 = 0.1\ H^*$, the spin-polarized state at $\alpha = 0$ appears to be energetically favourable for $N = 4$ and $N = 10$, (i.e. for a closed-shell spin-polarized system) as seen from Fig. 26 (*b*).

In the strong-coupling case ($\alpha \gg 1$ and $\eta \ll 1$), the total spin of an open-shell system for the ground state can take its minimal possible value, as seen from Fig. 26(*a*) for

$\alpha = 5, \eta = 0.1$ at $N = 4$ to 6. This trend to minimize the total spin is a consequence of the electron-phonon interaction, presumably due to the fact that the phonon-mediated electron-electron attraction overcomes the Coulomb repulsion, so that a multipolaron state is formed.

In summary, confined few-electron systems without the electron-phonon interaction can exist in one of two phases: the spin-polarized state and a state obeying Hund's rule, depending on the confinement frequency (see, e. g., Ref. [130]). For interacting few-polaron systems, besides the above two phases, there may occur also a third phase — the state with minimal spin — in quantum dots of polar substances with strong electron-phonon coupling (for instance, high-T_c superconductors).

6.2. Optical properties of interacting polarons in a quantum dot

To investigate the optical properties of the many-polaron system, in Ref. [127], the memory-function formalism of Ref. [26] is extended to the case of interacting polarons in a quantum dot. Within this technique, the optical conductivity for a system of interacting polarons in a parabolic confinement potential is given in terms of the memory function $\chi(\omega)$,

$$\mathrm{Re}\sigma(\omega) = -\frac{e^2}{m} \frac{\omega \mathrm{Im}\chi(\omega)}{\left[\omega^2 - \Omega_0^2 - \mathrm{Re}\chi(\omega)\right]^2 + \left[\mathrm{Im}\chi(\omega)\right]^2}, \tag{86}$$

where $\chi(\omega)$ is

$$\chi(\omega) = \sum_{\mathbf{q}} \frac{2|V_{\mathbf{q}}|^2 q^2}{3N\hbar\omega} \int_0^\infty \left(e^{i\omega t} - 1\right) \mathrm{Im}\left[T^*_{\omega_{\mathrm{LO}}}(t) \langle \rho_{\mathbf{q}}(t) \rho_{-\mathbf{q}}(0) \rangle_M\right] dt. \tag{87}$$

Here, $T_\omega(t) = \cos[\omega(t - i\hbar\beta/2)]/\sinh(\beta\hbar\omega/2)$ is the phonon Green's function, while $\langle \rho_{\mathbf{q}}(t) \rho_{-\mathbf{q}}(0) \rangle_M$ is the density-density correlation function calculated using the model system of electrons harmonically interacting with fictitious particles [with the Lagrangian (85)]. It should be noted that the optical conductivity (86) differs from that for a translationally invariant polaron system both by the explicit form of $\chi(\omega)$ and by the presence of Ω_0^2 in the denominator. For $\alpha \to 0$, the optical conductivity (86) tends to a δ-peak at $\omega = \Omega_0$,

$$\lim_{\alpha \to 0} \mathrm{Re}\sigma(\omega) = \frac{\pi e^2}{2m} \delta(\omega - \Omega_0). \tag{88}$$

For a translationally invariant system $\Omega_0 \to 0$ this small-coupling limit (88) reproduces the "central peak" of the polaron optical conductivity [30].

In the zero-temperature limit, the memory function (87) becomes

$$
\begin{aligned}
\chi(\omega) = {} & \lim_{\varepsilon\to+0}\frac{2\alpha}{3\pi N\omega}\left(\frac{\omega_{\mathrm{LO}}}{A}\right)^{3/2} \\
& \times\sum_{p_1=0}^{\infty}\sum_{p_2=0}^{\infty}\sum_{p_3=0}^{\infty}\frac{(-1)^{p_3}}{p_1!p_2!p_3!}\left(\frac{a_1^2}{N\Omega_1 A}\right)^{p_1}\left(\frac{a_2^2}{N\Omega_2 A}\right)^{p_2}\left(\frac{1}{NwA}\right)^{p_3} \\
& \times\Bigg\{\Bigg[\sum_{m=0}^{\infty}\sum_{n=0}^{\infty}\sum_{\sigma}\left[f_1(n,\sigma|\{N_\sigma\},\beta)-f_2(n,\sigma;m,\sigma|\{N_\sigma\},\beta)\right]\Big|_{\beta\to\infty} \\
& \quad\times\Bigg(\frac{1}{\omega-\omega_{\mathrm{LO}}-\left[p_1\Omega_1+p_2\Omega_2+(p_3-m+n)w\right]+i\varepsilon}-\frac{1}{\omega+\omega_{\mathrm{LO}}+p_1\Omega_1+p_2\Omega_2+(p_3-m+n)w+i\varepsilon} \\
& \qquad+\mathscr{P}\left(\frac{2}{\omega_{\mathrm{LO}}+p_1\Omega_1+p_2\Omega_2+(p_3-m+n)w}\right)\Bigg) \\
& \quad\times\sum_{l=0}^{m}\sum_{k=n-m+l}^{n}\frac{(-1)^{n-m+l+k}\Gamma\left(p_1+p_2+p_3+k+l+\frac{3}{2}\right)}{k!l!}\left(\frac{1}{wA}\right)^{l+k}\binom{n+2}{n-k}\binom{2k}{k-l-n+m}\Bigg] \\
& +\Bigg[\Bigg(\frac{1}{\omega-\omega_{\mathrm{LO}}-(p_1\Omega_1+p_2\Omega_2+p_3w)+i\varepsilon}-\frac{1}{\omega+\omega_{\mathrm{LO}}+p_1\Omega_1+p_2\Omega_2+p_3w+i\varepsilon} \\
& \qquad+\mathscr{P}\left(\frac{2}{\omega_{\mathrm{LO}}+p_1\Omega_1+p_2\Omega_2+p_3w}\right)\Bigg) \\
& \quad\times\sum_{m=0}^{\infty}\sum_{n=0}^{\infty}\sum_{\sigma,\sigma'}f_2(n,\sigma;m,\sigma'|\{N_\sigma\},\beta)\Big|_{\beta\to\infty} \\
& \quad\times\sum_{k=0}^{n}\sum_{l=0}^{m}\frac{(-1)^{k+l}\Gamma\left(p_1+p_2+p_3+k+l+\frac{3}{2}\right)}{k!l!}\left(\frac{1}{wA}\right)^{k+l}\binom{n+2}{n-k}\binom{m+2}{m-l}\Bigg]\Bigg\},
\end{aligned}
\tag{89}
$$

where $\mathscr{P}$ denotes the principal value, A is defined as $A\equiv\left[\sum_{i=1}^{2}a_i^2/\Omega_i+(N-1)/w\right]/N$, Ω_1,Ω_2, and w are the eigenfrequencies of the model system, a_1 and a_2 are the coefficients of the canonical transformation which diagonalizes the model Lagrangian (85). $f_1(n,\sigma|\{N_\sigma\},\beta)$ and $f_2(n,\sigma;m,\sigma'|\{N_\sigma\},\beta)$ are, respectively, the one-electron and the two-electron distribution functions in the canonical ensemble.

The transitions between states with different values of the total spin, which occur when varying the confinement frequency, also manifest themselves in the spectra of the optical conductivity. In Fig. 27, optical conductivity spectra for $N=14$ polarons in a quantum dot with $\alpha=5$, $\eta=0.1$ are represented for several values of the confinement energy $\hbar\Omega_0$.

For relatively weak confinement, the ground state is spin-polarized, like for $\hbar\Omega_0^{(1)}=0.00883\ H^*$ (panel *a*). With increasing confinement, the transition from a spin-polarized state (with total spin $S=7$) to the state obeying Hund's rule (with $S=3$) occurs between $\hbar\Omega_0^{(1)}=0.00883\ H^*$ (panel *b*) and $\hbar\Omega_0^{(2)}=0.00884\ H^*$ (panel *c*) of the confinement parameter. At still stronger confinement, like for $\hbar\Omega_0^{(1)}=0.0092\ H^*$ (panel *d*), the ground state obeys Hund's rule.

In the inset to Fig. 27, the first frequency moment of the optical conductivity

$$
\langle\omega\rangle\equiv\frac{\int_0^\infty\omega\mathrm{Re}\sigma(\omega)\,d\omega}{\int_0^\infty\mathrm{Re}\sigma(\omega)\,d\omega},
\tag{90}
$$

as a function of $\hbar\Omega_0$ shows a *discontinuity*, at the value of the confinement energy corresponding to the transition between the spin-polarized ground state and the ground state

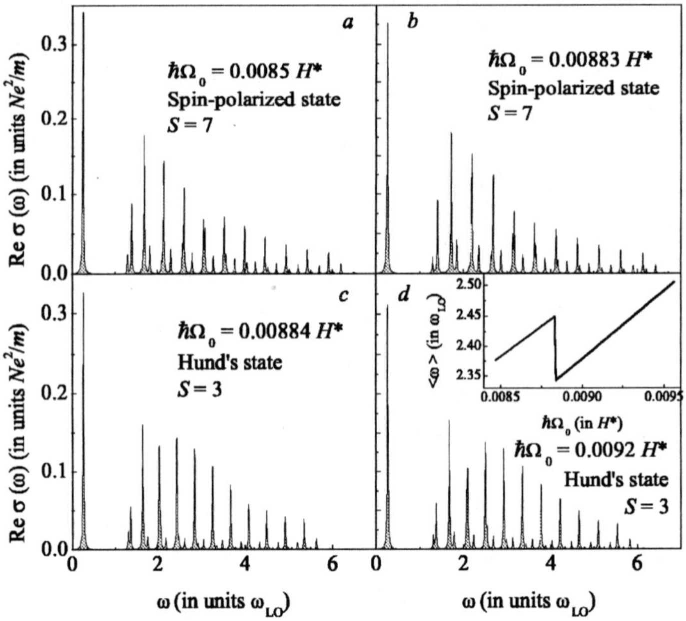

FIGURE 27. Optical conductivity spectra of $N = 14$ interacting polarons in a quantum dot, for different confinement frequencies close to the transition from a spin-polarized ground state to a ground state obeying Hund's rule. *Inset*: the first frequency moment $\langle\omega\rangle$ of the optical conductivity as a function of the confinement energy. (From Ref. [127].)

obeying Hund's rule. This discontinuity should be observable in optical measurements.

The shell structure for a system of interacting polarons in a quantum dot is clearly revealed when analysing both the addition energy and the first frequency moment of the optical conductivity. The addition energy $\Delta(N)$ needed to put an extra electron into a quantum dot containing N electrons is defined as

$$\Delta(N) = E^0(N+1) - 2E^0(N) + E^0(N-1), \tag{91}$$

where $E^0(N)$ is the ground-state energy. In Figs. 28(a) and 28(b), we show both the function

$$\Theta(N) \equiv \langle\omega\rangle|_{N+1} - 2\langle\omega\rangle|_N + \langle\omega\rangle|_{N-1}, \tag{92}$$

and the addition energy $\Delta(N)$. Distinct peaks appear in $\Theta(N)$ and $\Delta(N)$ at the "magic numbers" $N = 10$ and $N = 20$ for closed-shell configurations.

It follows that experimental measurements of the addition energy and the first frequency moment of the optical absorption as a function of the number of polarons in a quantum dot would allow for discrimination between open-shell and closed-shell con-

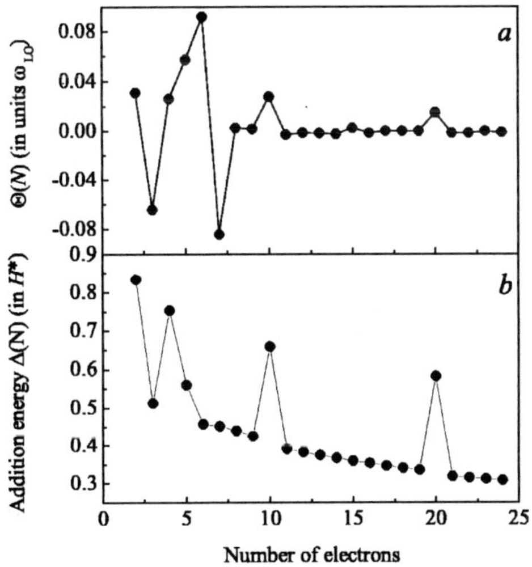

FIGURE 28. The function $\Theta(N)$ (a) and the addition energy $\Delta(N)$ (b) for systems of interacting polarons in a quantum dot with $\alpha = 3$, $\eta = 0.3$ and $\Omega_0 = 0.5\omega_{\rm LO}$ ($\hbar\Omega_0 \approx 0.01361\ H^*$). (From Ref. [127].)

figurations. In particular, the latter configurations may be revealed through peaks in the addition energy and the first frequency moment of the optical absorption.

7. SMALL POLARONS

Small polarons can arise when the self-induced polarisation caused by an electron or hole becomes of the order of a lattice parameter. As distinct from large, or continuum, polarons (discussed in the present paper), small polarons are governed by short-range interactions. Their physical properties can be radically different from those of large polarons. For example, the mobility of small polarons is typically thermally activated, whereas the mobility of large polarons, e. g., at small α, is given by $\mu \sim \exp(\hbar\omega_{\rm LO}/kT)$. Experimentally small-polaron effects have been analysed, e. g., in KCl, LiF, MnO, TiO_2, $BaTiO_3$, $SrTiO_3$, $LaCoO_3$,... For reviews on small polarons I refer, e. g., to Ref. [131] and Section 2 in Ref. [20]. Alexandrov and Mott [21] have surveyed the small-polaron theory mainly in the context of the (bi)polaronic approach in the physics of high-T_c superconductors.

Both large and small polarons have been invoked by Eagles et al. [27] to interpret the infrared optical absorption spectra of $Pr_2NiO_{4.22}$. The infrared reflectivity spectra for the crystal of this material were measured between 30 $\rm cm^{-1}$ and 20000 $\rm cm^{-1}$. Mid-infrared conductivity data for frequencies between 1000 $\rm cm^{-1}$ and 16000 $\rm cm^{-1}$ were obtained by Kramers-Kronig transformation of those reflectivity spectra. The resulting optical

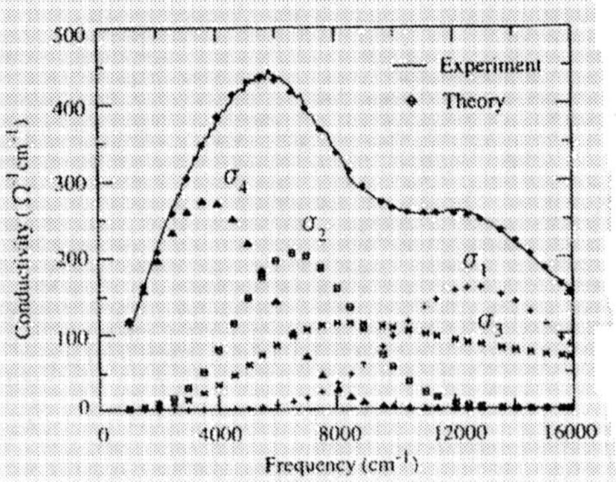

FIGURE 29. Experimental and theoretical conductivity data at 250 K for $Pr_2NiO_{4.22}$ and four contributions to the theoretical curves resulting from the transitions between the following polaron states: σ_1 — small → small; σ_2 — large → small; σ_3 — small → large; σ_4 — large → large. (From Ref. [27].)

absorption spectrum was quantitatively interpreted (see Fig. 29) using the assumption that in $Pr_2NiO_{4.22}$ two types of polarons, small and large, coexist. The contribution due to the transitions between the large polarons states, which dominates the absorption in the low-frequency region, was calculated using the arbitrary coupling theory of Ref. [26] for the coupling constant $\alpha = 3$.

ACKNOWLEDGMENTS

As is manifest from the list of references, I had the pleasure of collaborations on the physics of polarons with several colleagues, post-doctoral scientists and students. I like to thank them all here. I thank V. Fomin, F. Brosens, V. Gladilin, S. Klimin and J. Tempere for discussions during the preparation of this manuscript. This work has been supported by the BOF NOI (UA-UIA), GOA BOF UA 2000, IUAP, FWO-V projects G.0287.95, G.0071.98, G.0274.01, G.0435.03 and the W.O.G. WO.025.99N (Belgium).

REFERENCES

1. Landau, L. D., *Phys. Z. Sowjetunion* **3**, 664 (1933).
2. Devreese, J. T. L., *Moles agitat mentem. Ontwikkelingen in de fysika van de vaste stof.* Rede uitgesproken bij de aanvaarding van het ambt van buitengewoon hoogleraar in de fysica van de vaste stof, in het bijzonder de theorie van de vaste stof, bij de afdeling der technische natuurkunde aan de Technische Hogeschool Eindhoven, March 9, 1979.
3. Landau, L. D., and Pekar, S. I., *Zh. Eksper. Teor. Fiz.* **18**, 419 (1948).
4. Pekar, S. I., *Issledovanija po Ekektronnoj Teorii Kristallov*, Gostekhizdat, Moskva, 1951 (in Russian) [German translation: *Untersuchungen über die Elektronentheorie der Kristalle*, Akademie Verlag, Berlin, 1951].

5. Bogolubov, N. N., *Ukr. Matem. Zh.* **2**, 3 - 24 (1950).
6. Fröhlich, H., *Adv. Phys.* **3**, 325 - 361 (1954).
7. Kartheuser, E., "Dielectric properties of polar crystals" in *Polarons in Ionic Crystals and Polar Semiconductors*, edited by J. T. Devreese, North-Holland, Amsterdam, 1972, pp. 717 - 733.
8. Grynberg, M., Huant, S., Martinez, G., Kossut, J., Wojtowicz, T., Karczewski, G., Shi, J. M., Peeters, F. M., and Devreese, J. T., *Phys. Rev. B* **54**, 1467 - 1470 (1996).
9. *Landolt-Börnstein Numerical Data and Functional Relationships in Science and Technology, New Series*, editor in chief K.-H. Hellwege, Springer, New York, 1982, Group III, Vol. 17, Subvolume b, p. 163.
10. Hodby, J.W., Russell, G., Peeters, F., Devreese, J. T., and Larsen, D. M., *Phys. Rev. Lett.* **58**, 1471 - 1474 (1987).
11. Verbist, G., Peeters, F. M., and Devreese, J. T., *Ferroelectrics*, **130**, 27 - 34 (1992) (for $m_b/m_e = 2.6$).
12. Devreese, J. T., Fomin, V. M., Pokatilov, E. P., Kotomin, E. A., Eglitis, R., Zhukovskii, Yu. F., *Phys. Rev. B* **63** 184304, 6 pages (2001).
13. Peeters, F. M., Wu X., and Devreese, J. T., *Phys. Rev. B*, **34**, 1160 - 1164 (1986).
14. Peeters, F. M., and Devreese, J. T., *Phys. Stat. Sol. (b)*, **115**, 539 - 543 (1983).
15. Feynman, R. P., *Phys. Rev.* **97**, 660 - 665 (1955).
16. Feynman, R. P., Hellwarth, R. W., Iddings, C. K., and Platzman, P. M., *Phys. Rev.* **127**, 1004-1017 (1962).
17. Thornber, K. K., and Feynman, R. P., *Phys. Rev. B* **1**, 4099-4114 (1970).
18. *Polarons and Excitons*, edited by C. G. Kuper and G. D. Whitfield, Oliver and Boyd, Edinburgh, 1963.
19. *Polarons in Ionic Crystals and Polar Semiconductors*, edited by J. T. Devreese, North-Holland, Amsterdam, 1972.
20. Devreese, J. T., "Polarons", in *Encyclopedia of Applied Physics*, edited by G. L. Trigg, VCH, Weinheim, 1996, vol. 14, pp. 383 - 413.
21. Alexandrov, A. S., and Mott, N., *Polarons and Bipolarons*, World Scientific, Singapore, 1996.
22. Gurevich, V. L., Lang, I. G., and Firsov, Yu. A., *Fiz. Tverd. Tela* **4**, 1252-1262 (1962) [*Sov. Phys. Solid State* **4**, 918 - 925 (1962)].
23. Finkenrath, H., Uhle, N., and Waidelich, W., *Solid State Commun.* **7**, 11 - 14 (1969).
24. Devreese, J., Huybrechts, W., and Lemmens, L., *Phys. Stat. Sol. (b)* **48**, 77-86 (1971).
25. Kartheuser, E., Evrard, R., and Devreese, J., *Phys. Rev. Lett.* **22**, 148 - 151 (1969).
26. Devreese, J. T., De Sitter, J., and Goovaerts., M., *Phys. Rev. B*, **5**, 2367 - 2381 (1972).
27. Eagles, D. M., Lobo, R. P. S. M., and Gervais, F., *Phys. Rev. B* **52**, 6440 - 6450 (1995).
28. Wu, X.-G., Peeters, F.M., and Devreese, J. T., *Phys. Rev. B* **34**, 2621 - 2626 (1986).
29. Devreese, J. T., "Internal structure of free Fröhlich polarons optical absorption and cyclotron resonance", in *Polarons in Ionic Crystals and Polar Semiconductors*, North-Holland, Amsterdam, 1972, pp. 83 - 159.
30. Devreese, J. T., Lemmens, L. F., and Van Royen, J., *Phys. Rev. B* **15**, 1212 - 1214 (1977).
31. Peeters, F. M., and Devreese, J. T., *Phys. Rev. B* **36**, 4442 - 4445 (1987).
32. Mahan, G. D., "Polarons in heavily doped semiconductors", in *Polarons in Ionic Crystals and Polar Semiconductors*, edited by J. T. Devreese, North-Holland, Amsterdam, 1972, pp. 553 - 657.
33. Lemmens, L. F., Devreese, J. T., and Brosens, F., *Phys. Stat. Sol. (b)* **82**, 439 - 447 (1977).
34. Lee, T. D., Low, F. E., and Pines, D., *Phys. Rev.* **90**, 297 - 302 (1953).
35. Quemerais, P., and Fratini, S., *Int. Journal of Mod. Phys. B* **12**, 3131 - 3136 (1998).
36. Fratini, S., and Quemerais, P., *Mod. Phys. Lett. B* **12**, 1003 - 1012 (1998).
37. Fratini, P., and Quemerais, P., *Europ. Phys. J. B* **14**, 99 - 113 (2000).
38. Calvani, P., *Optical Properties of Polarons*, Editrice Compositori, Bologna, 2001.
39. Tempere, J., and Devreese, J. T., *Phys. Rev. B* **64**, 104504, 10 pages (2001).
40. Tempere, J., and Devreese, J. T., *Eur. Phys. J. B* **20** 27 - 33 (2001).
41. Iadonisi, G., Chiofalo, M. L., Cataudella, V., and Ninno, D., *Phys. Rev. B* **48**, 12966 - 12978 (1993).
42. Cataudella, V., De Filippis, G., and Iadonisi, G., *Eur. Phys. J. B* **12**, 17 - 22 (1999).
43. Devreese, J. T., Klimin, S. N., Fomin, V. M., and Brosens, F., *Solid State Commun.* **114**, 305 - 310 (2000).
44. Tomonaga, S., *Prog. Theoret. Phys.* **2**, 6 - 24 (1947).
45. Kubo, R., *J. Phys. Soc. Japan* **21**, 570 - 586 (1957).

46. See e. g. Mahan, G. D., *Many-Particle Physics*, Plenum Press, New York, 1993, p. 210.
47. Hopfield, J. J., *Phys. Rev.* **139**, A419 - A424 (1965).
48. Ron, A., and Tzoar, N., *Phys. Rev.* **131**, 12 - 20 (1963).
49. Lupi, S., Maselli, P., Capizzi, M., Calvani, P., Giura P., and Roy, P., *Phys. Rev. Lett.* **83**, 4852 - 4855 (1999).
50. Lupi, S., Calvani, P., Capizzi, M., Maselli, P., Sadowski, W., and Walker, E., *Phys. Rev. B* **45**, 12470 - 12477 (1992).
51. Shanenko, A. A., Smondyrev, M. A., and Devreese, J. T., *Solid State Commun.* **98**, 1091 - 1094 (1996).
52. Alexandrov, A. S., Kabanov, V. V., and Mott, N. F., *Phys. Rev. Lett.* **77**, 4796 - 4799 (1996); Alexandrov, A. S., and Bratkovsky, A. M., *Phys. Rev. B* **60**, 6215 - 6218 (1999); Emin, D., *Phys. Rev. B* **45**, 5525 - 5529 (1992); Alexandrov, A. S., and Ranninger, J., *Phys. Rev. B* **23**, 1796 - 1801 (1981).
53. Larsen, D. M., "Polaron effects in cyclotron resonance", in *Landau Level Spectroscopy* (in 2 volumes), edited by G. Landwehr and E. I. Rashba, North-Holland, Amsterdam, 1991, pp. 109-130.
54. Petrou, A., and McCombe, B. D., "Magnetoscpectroscopy of confined semiconductor systems", in *Landau Level Spectroscopy* (in 2 volumes), edited by G. Landwehr and E. I. Rashba, North-Holland, Amsterdam, 1991, pp. 679 - 775.
55. Johnson, E.J., and Larsen, E. J., *Phys. Rev. Lett.* **16**, 655 - 659 (1966).
56. Larsen, D. M., *Phys. Rev.* **135**, A419 - A426 (1964).
57. Larsen, D. M., *Phys. Rev.* **142**, 428 - 435 (1966).
58. Das Sarma, S., *Phys. Rev. Lett.* **52**, 859 - 862 (1984).
59. Lindemann, G., Lassnig, G., Seidenbusch, W., and Gornik, E.: *Phys. Rev. B* **28**, 4693 - 4703 (1983).
60. Larsen, D. M.,"Polaron energy levels in magnetic and Coulomb fields", in *Polarons in Ionic Crystals and Polar Semiconductors*, edited by J. T. Devreese, North-Holland, Amsterdam, 1972, pp. 237-287.
61. Peeters, F. M., and Devreese, J. T., *Phys. Rev. B* **31**, 3689 - 3695 (1985).
62. Korovin, L. I., and Pavlov, S. T., *Zh. Eksp. Teor. Fiz.* **53**, 1708 - 1716 (1967) [*Sov. Phys. - JETP* **26**, 979 (1968)].
63. Nakayama, M., *J. Phys. Soc. Jpn.* **27**, 636 (1969).
64. Devreese, J. T., De Sitter, J., Johnson, E.J., and Ngai, K. L., *Phys. Rev. B* **17** 3207 - 3220 (1978).
65. Vigneron, J. P., Evrard, R., and Kartheuser, E., *Phys. Rev. B* **18**, 6930 - 6943 (1978).
66. Kobori, H., Ohyama, H., and Otsuka, E., *Solid State Commun.* **84**, 383 -388 (1992)
67. Peeters, F. M., Wu, X.-G., Devreese, J. T., Langerak, C. J. G. M., Singleton, J., Barnes, D. J., and Nicholas, R. J., *Phys. Rev. B* **45**, 4296 - 4300 (1992).
68. Nicholas, R. J., Watts, M., Howell, D. F., Peeters, F. M., Wu, X.-G., Devreese, J. T., van Bockstal, L., Herlach, F., Langerak, C. J. G. M., Singleton, J., and Chery, A., *Phys. Rev. B* **45**, 12144 - 12147 (1992).
69. Wang, Y. J., Nickel, H. A., McCombe, B. D., Peeters, F. M., Shi, J. M., Hai, G. Q., Wu, X.-G., Eustis, T. J., and Schaff, W., *Phys. Rev. Lett.* **79** 3226 - 3229 (1997).
70. Devreese, J. T., Fomin, V. M., Gladilin, V. N., Imanaka, Y., and Miura, N., *J. Crystal Growth* **214-215**, 465 - 468 (2000).
71. Devreese, J. T., Fomin, V. M., Gladilin, V. N., Y. Imanaka, Y., and Miura, N., *Physica B* **298**, 207 - 210 (2001).
72. Wang, Y. J., Leem, Y. A., McCombe, B. D., Wu, X.-G., Peeters, F. M., Jones, E. D., Reno, J. R., Lee, X. Y., and Jiang, H. W., *Phys. Rev. B* **64** 161303(R), 4 pages (2001).
73. Peeters, F. M., and Devreese, J. T., *Phys. Rev. B* **34**, 7246 - 7259 (1986).
74. Thornber, K. K., *Phys. Rev. B* **3**, 1929 - 1941 (1971); **4**, 675 (1971).
75. Saitoh, M., *J. Phys. C* **15**, 6981 - 6989 (1982).
76. Saitoh, M., *J. Phys. A* **16**, 1795 - 1808 (1983).
77. Mori, R., *Prog. Theor. Phys* **33**, 423 - 455 (1965); **34**, 399 -415 (1965).
78. Peeters, F. M., and Devreese, J. T., *Phys. Rev. B* **25**, 7281 - 7301 (1982).
79. Devreese, J. T., and Brosens, F., *Phys. Rev. B* **45**, 6459 - 6478 (1992).
80. Hellwarth, R. W., and Platzman, P. M., *Phys. Rev.* **128**, 1599 - 1604 (1962).
81. Brown, F. C., "Experiments on the polaron", in *Polarons and Excitons*, edited by C. G. Kuper and G. D. Whitfield, Oliver and Boyd, Edinburgh, 1963, pp. 323 - 355.
82. Larsen, D., *Journal of Physics C* **7**, 2877 - 2889 (1974).

83. Cohn, D. R., Larsen, D. M., and Lax, B., *Phys. Rev. B* **6**, 1367 - 1374 (1972).
84. Pfeffer, P., and Zawadzki, W., *Phys. Rev. B* **37**, 2695 - 2698 (1988).
85. Miura, N., Nojiri, H., Pfeffer, P., and Zawadzki, W., *Phys. Rev. B* **55**, 13598 - 13604 (1997).
86. Peeters, F. M., and Devreese, J. T., *Physica* **127B**, 408 - 410 (1984).
87. Hai, G. Q., Peeters, F. M., and Devreese, F. M., *Phys. Rev. B* **47** 10358 - 10374 (1993).
88. Wu, X.-G., Peeters, F. M., and Devreese, J. T., *Phys. Rev. B* **36**, 9760 - 9764 (1987).
89. Barnes, D. J., Nicholas, R. J., Peeters, F. M., Wu, X.-G., Devreese, J. T., Singleton, J., Langerak, C.J.G.M., Harris, J.J., and Foxon, C. T., *Phys. Rev. Lett.* **66**, 794 - 797 (1991).
90. Wu, X.-G., Peeters, F. M., and Devreese, J. T., *Phys. Rev. B* **34**, 8800 - 8809 (1986).
91. Warmenbol, P., Peeters, F. M., Wu, X., and Devreese, J. T., *Phys. Rev. B* **40**, 6258 - 6264 (1989).
92. Vaughan, T. A., Nicholas, R. J., Langerak, C. J. G. M., Murdin, B. N., Pidgeon, C. R., Mason, N. J., and Walker, P. J., *Phys. Rev. B* **53**, 16481 - 16484 (1996).
93. McCombe, B. D., and Kaplan, R., *Phys. Rev. Lett.* **21**, 756 - 759 (1968).
94. Chang, Y.-H., McCombe, B. D., Mercy, J.-M., Reeder, A. A., Ralston, J., and Wicks, G. A., *Phys. Rev. Lett.* **61** 1408 - 1411 (1988).
95. Ziesmann, M., Heitmann, D., and Chang, L. L., *Phys. Rev. B* **35** 4541 - 4544 (1987).
96. Poulter, A. J. L., Zeman, J., Maude, D. K., Potemski, M., Martinez, G., Riedel, A., Hey, R., and Friedland, K. J., *Phys. Rev. Lett.* **86**, 336 - 339 (2001).
97. Klimin, S. N., Fomin, V. M., and Devreese, J. T., to be published.
98. Wu, X., Peeters, F. M., and Devreese, J. T., *Phys. Rev. B* **40**, 4090 - 4094 (1989).
99. Afonin, V. V., Gurevich, V. L.. and Laiho, R., *Phys. Rev. B* **62**, 15913 - 15924 (2000).
100. Nicholas, R. J., Sasaki, S., Miura, N., Peeters, F. M., Shi, J. M., Hai, G. Q., Devreese, J. T., Lawless, M. J., Ashenford, D. E., and Lunn, B., *Phys.Rev. B* **50** 7596 - 7601 (1994).
101. Shi, J. M., Peeters, F. M., Devreese, J. T., Imanaka, Y., and Miura, N., *Phys. Rev. B* **52**, 17205 - 17209 (1995).
102. Imanaka, Y., Miura, N., and Nojiri, H., *Physica B* **246-247**, 328 - 332 (1998).
103. Klyukanov, A. A., and Pokatilov, E. P., *Zh. Exper. Teor. Fiz.* **60**, 1878 - 1889 (1971) [*Sov. Phys. JETP* **33**, 1015 - 1020 (1971)].
104. García-Cristóbal, A., Minnaert, A. W. E., Fomin, V. M., Devreese, J. T., Silov, A. Yu., Haverkort, J. E. M., and Wolter, J. H., *Phys. Stat. Sol. (b)* **215**, 331 - 336 (1999).
105. Heitz, R., Mukhametzhanov, I., Stier, O., Madhukar, A., and Bimberg, D., *Phys. Rev. Letters* **83**, 4654 - 4657 (1999).
106. Bissiri, M., Baldassarri Höger von Högersthal, G., Bhatti, A. S., Capizzi, M., Frova, M., Frigeri, P., and Franchi, S., *Phys. Rev. B* **62**, 4642 - 4646 (2000).
107. Bissiri, M., Capizzi, M., Fomin, V. M., Gladilin, V. N., and Devreese, J. T., *Phys. Stat. Sol (b)* **224**, 639 - 642 (2001).
108. Nomura, S., and Kobayashi, S., *Phys. Rev. B* **45**, 1305 -1316 (1992).
109. Fry, P. W., Itskevich, I. E., Mowbray, D. J., Skolnick, M. S., Finley, J. J., Barker, J. A., O'Reilly, E. P., Wilson, L. R., Larkin, I. A., Maksym, P. A., Al-Knafaji, M., David, J. P. R., Cullis, A. G., Hill, G., and Clark, J. C., *Phys. Rev. Letters* **84**, 733 - 736 (2000).
110. Fomin, V. M., Gladilin, V. N., Devreese, J. T., Pokatilov, E. P., Balaban, S. N., and Klimin, S. N., *Phys. Rev. B* **57**, 2415 - 2425 (1998).
111. Devreese, J. T., Fomin, V. M., Gladilin, V. N., Pokatilov, E. P., and Klimin, S. N., Nanotechnology **13**, 163 - 168 (2002).
112. Pokatilov, E. P., Klimin, S. N., Fomin, V. M., Devreese, J. T., and Wise, F. W., *Phys. Rev. B* **65**, 075316, 8 pages(2002).
113. Pekar, S. I., *Zh. Eksp. Teor. Fiz.* **20**, 267 - 270 (1950).
114. Huang, K., and Rhys, A., *Proc. R. Soc. London, Ser. A* **204** , 406 - 423 (1950).
115. Gladilin, V. N., Balaban, S. N., Fomin, V. M., and Devreese, J. T., in *Proc. 25th Int. Conf. on the Physics of Semiconductors, Osaka*, Springer, Berlin, 2001, pp. 1243 - 1244.
116. Lemaître, A., Ashmore, A. D., Finley, J. J., Mowbray, D. J., Skolnick, M. S., Hopkinson, M., and Krauss, T. F., *Phys. Rev. B* **63**, 161309(R), 4 pages (2001)
117. Verzelen, O., Ferreira, R., and Bastard, G., *Phys. Rev. Lett.* **88** 146803, 4 pages (2002).
118. Jungnickel, V., and Henneberger, F., *J. Lumin.* **70**, 238 - 252 (1996).
119. Nirmal, M., Murray, C. B., Norris, D. J., and Bawendi, M. G., *Z. Phys. D* **26**, 361 - 363 (1993).

120. Woggon, U., Miller, D., Kalina, F., Gerlach, B., Kayser, D., Leonardi, K., and Hommel, D., *Phys. Rev. B* **67**, 045204, 5 pages (2003).
121. Gerlach, B., and Löwen, H., *Rev. Mod. Phys.* **63**, 63 - 90 (1991).
122. Klein, M. C., Hache, F., Ricard, D., and Flytzanis, C., *Phys. Rev. B* **42**, 11123 - 11132 (1990).
123. Krauss, T. D., and Wise, F. W., *Phys. Rev. B* **55**, 9860 - 9865 (1997).
124. Feynman, R. P., *Statistical Mechanics*, Benjamin, Massachusetts, 1972.
125. Lemmens, L. F., Brosens, F., and Devreese, J. T., *Phys. Rev. E* **53**, 4467 - 4476 (1996).
126. Devreese, J. T., "Note of the path-integral variational approach in the many-body theory", in *Fluctuating Paths and Fields*, World Scientific, Singapore, 2001, pp. 289 - 304.
127. Klimin, S. N., Fomin, V. M., Brosens, F., and Devreese, J. T., to be published.
128. Brosens, F., Devreese, J. T., and Lemmens, L. F., *Phys. Rev. E* **55**, 227 - 236 (1997); **55**, 6795 - 6802 (1997); **58**, 1634 - 1643 (1998).
129. See e. g. Messiah, A., *Quantum Mechanics*, North-Holland, Amsterdam, 1966, vol. 2, p. 702.
130. Reimann, S. M., Koskinen, M., and Manninen, M., *Phys. Rev. B* **62**, 8108 - 8113 (2000).
131. Reik, H. G.,"The small polaron concept and its use for the interpretation of d.c. and optical properties of semiconducting transition metal oxides", in *Polarons in Ionic Crystals and Polar Semiconductors*, edited by J. T. Devreese, North-Holland, Amsterdam, 1972, pp. 679 - 714.

Multielectron bubbles in liquid helium: a spherical two-dimensional electron system

Jacques Tempere[*†], Isaac F. Silvera[†] and Jozef T. Devreese[*]

[*]*TFVS, Universiteit Antwerpen, B2610 Antwerpen, Belgium*
[†]*Lyman Laboratory of Physics, Harvard University, Cambridge MA02138, USA*

Abstract. The electron system with its many phases and realizations remains a fascinating subject of study. It has been extensively analysed in three, two, one and zero dimensions, as well as in confined nanoscopic and mesoscopic geometries generated with microfabrication techniques. An obvious next step in this investigation is to study the electron system in curved geometries. In particular we will focus in these lectures on the two-dimensional spherical electron system, realized in for example multielectron bubbles (MEBs), buckyballs, and nanoshells.

INTRODUCTION

Origin and experimental detection of MEBs

Electrons above a free surface of liquid helium cannot penetrate this surface freely: there is a 1 eV barrier preventing them from going into the liquid [1]. Furthermore, liquid helium is a dielectric medium with permittivity $\varepsilon = 1.0572$ different from that of vacuum. This means that a charge q placed at a distance z above this helium surface will generate an image charge $-q(1-\varepsilon)/[2(1+\varepsilon)]$ in the liquid, at a depth z below the surface. So, an electron above a surface of liquid helium will be attracted towards the surface with a Coulomb-like potential. The full potential in the direction perpendicular to the helium surface is thus given by

$$V(z) = \begin{cases} -\dfrac{e^2}{4z}\dfrac{1-\varepsilon}{1+\varepsilon} \text{ for } z > 0 \\ 1 \text{ eV for } z \leq 0 \end{cases} . \tag{1}$$

Here e is the electron charge and z is the distance above the liquid ^{4}He. This potential has a bound ground state with energy -7.5 K, and the point of maximum probability for the ground state wave function lies 7.6 nm above the helium surface [1]. The analogous values for ^{3}He are 4.3 K and 9.0 nm, and for H_2 they are 160 K and 1.7 nm [2]. Along the directions parallel to the surface, the electrons are free to move – the wave function can be separated in the confined perpendicular part and plain waves for the parallel coordinates. What does this mean ? Well, if electrons are injected above a surface of liquid helium, they will be attracted towards this surface and form a thin film 7.6 nm on top of the helium surface and anchored to this helium surface. At temperatures well

CP678, *Lectures on the Physics of Highly Correlated Electron Systems VII: Seventh Training Course*, edited by A. Avella and F. Mancini

below 7.5 K, this electron film is effectively two-dimensional, since in the perpendicular direction only one state (or 'mode') is occupied.

The two-dimensional electron system, realized for electrons on top of liquid helium or in a semiconductor heterostructure, has received a lot of interest, not only because of its potential for technological applications in micro- and nano-electronics, but also because of its intrinsic interest as a testing ground for modern many-body theories. Novel phenomena and quantum mechanical states such as the quantum Hall effect (integer and fractional) have been realized and studied. A particular challenge for theories is the accurate description of the phases of this system and the phase transitions. Electrons can crystallize into a lattice (the Wigner lattice) which melts into the hexatic phase destroying long range order but not long range orientational order, which in its turn melts into an electron fluid where also the long-range orientational order is destroyed.

In the late seventies and early eighties several researchers probed the properties of two-dimensional electron systems as a function of the density of the electrons [2, 3, 4, 5]. They managed to tune the density by putting a positive plate beneath the helium liquid. Increasing the (positive) charge on this anode allowed them to increase the density of electrons. One goal was to observe the so-called quantum melting of the Wigner lattice of electrons: it was predicted that above a certain critical density, when the zero-point motion of the electrons becomes of the order of the lattice parameter, the Wigner lattice will melt. This type of melting occurs also at zero temperature, and is dubbed 'quantum melting'.

Yet, when researchers kept increasing the density of electrons on the helium surface in the hope of reaching the quantum melting regime, they discovered that the helium surface becomes unstable [6, 7]. This can be understood from the competition between the Coulomb repulsion of the electrons, that want a surface as large as possible to spread out, and the surface tension of the helium, that favours a small surface. When the electron density is increased, the Coulomb interaction overcomes the surface tension and the helium surface starts to buckle and deform. These deformations grow and at some point, a bubble can be split off from the surface, carrying charges towards the positive anode.

In the experiments of Refs. [6, 7], the multielectron bubbles (MEBs) move relatively quickly through the helium liquid, so that they remain visible only for a time period of milliseconds, in which they were not yet equilibrated. The bubbles can contain from a few to 10^8 electron charges, and are typically microns ($0.1 - 100\ \mu$m) in radius. Figure 1 illustrates schematically this surface instability on a flat helium surface.

A remarkable aspect of these multielectron bubbles is that the electrons in it are not spread out homogeneously throughout the volume of the bubble, but form a nanometer thin film anchored to the surface of the bubble [8]. The electrons are confined in the radial direction but free to move in the tangential direction. In other words, the electrons in a spherical bubble find themselves always localized at a fixed distance from the center of the bubble, but they can spread out and move freely along the surface of the bubble. They form a two-dimensional electron fluid (or solid). The inside of the bubble, apart from the electron layer at its surface, is vacuum (or contains a gas of helium atoms at the vapour pressure corresponding to the temperature of the liquid). For large bubbles, the situation of the flat surface is retrieved locally: we still find a 2D electron system on a helium surface. The location of the electrons, at a distance d from the bubble surface, is schematically indicated in Fig. 1.

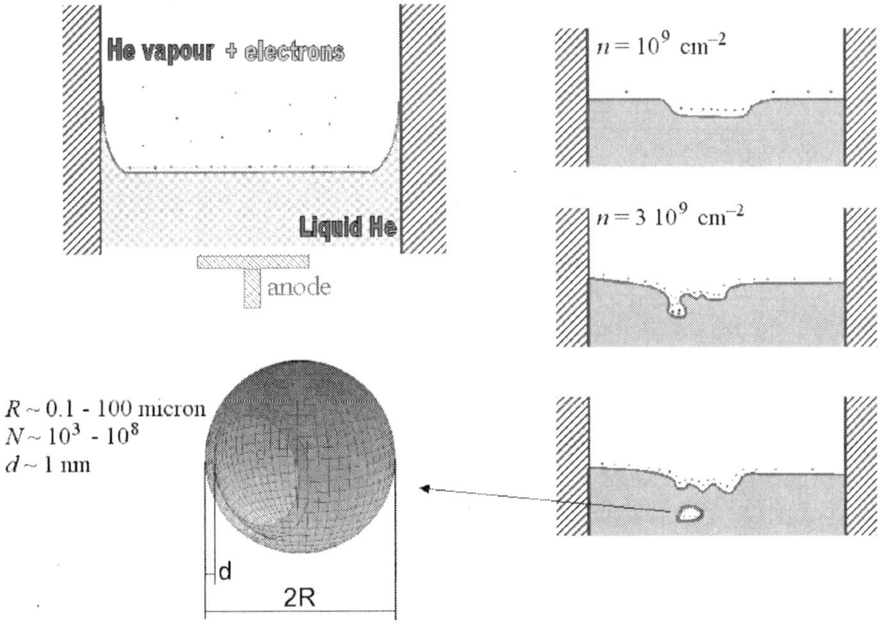

FIGURE 1. When a helium surface is charged and an underlying anode increases the surface density of electrons beyond a critical value, the helium surface becomes unstable and buckles. Bubbles are formed, carrying the excess N electrons to the anode, where these multielectron bubbles are destroyed. The detection of such bubbles is described in refs. [6, 7]. The electrons inside the bubble form a nanometer thin, spherical layer residing at a fixed distance d from the helium surface, to the inside of the bubble.

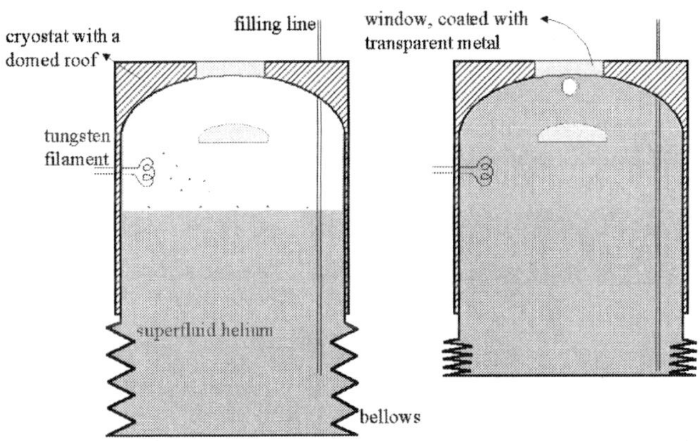

FIGURE 2. A proposed experimental setup [9] that will allow to stabilize multielectron bubbles for study. Because the electrons are confined to the nonsubmerged volume, raising the level of the helium will force the electrons towards the top of the dome-shaped cell. As the helium level reaches the top, the multielectron bubble is formed. It resides between capacitor plates formed by glass pieces coated with a transparent metal, and can be observed through a window at the top of the cell.

Recently, a new experiment have been proposed to study multielectron bubbles in an environment where they are stable and can equilibrate [9]. The setup for such an experiment is illustrated in Fig. 2. The key point of this setup is the use of a domed cell. The cell is first partly filled with liquid helium, which has been cooled below the lambda point so that all non-submerged surfaces of the cell are covered by a superfluid helium film. Then, electrons are injected in the non-submerged volume, using either a tungsten filament or a field-emission tip. Bellows, or a system using a connecting helium reservoir that can be raised and lowered, allow to change the level of the helium. This level is raised until it reaches the top of the dome. The electrons, which cannot penetrate the helium, find themselves restricted to a smaller and smaller nonsubmerged volume as the level rises. Finally, the electrons will find themselves at the top of the dome, where they form the lowest-energy surface, a bubble. This bubble will be detected visually (using a microscope objective looking through a top-window in the cell) and capacitively. For that purpose, capacitor plates will be mounted in such a way that the bubble in its position at the top of the dome is between the capacitor plates. This allows to measure the charge in the bubble, and to perform resonance spectroscopy of the oscillation modes of the bubble by applying an AC field over the capacitor plates. This particular experimental setup is well suited to determine experimentally many of the features that we will discuss in the rest of this lecture.

Radius and surface charge density

A single electron bubble (SEB) has a radius of 34 [10] while, for example, a bubble of 10^4 electrons theoretically should have a radius of around one micron. The surprisingly large diameter of the SEB is the result of a balance between the zero-point energy of the electron and the surface tension energy of the liquid helium. A spherical surface has the minimum energy (in the absence of gravity) and is thus the expected shape of a static bubble or a bubble in its ground state. In larger bubbles electrons repel each other due to Coulomb forces and the zero-point energy of the electrons in a larger bubble is negligible compared to the electrostatic energy of the Coulomb repulsion. Thus, in the large bubbles which are the subject of our lecture, the equilibrium radius will be mainly determined by a balancing of the surface tension energy of the helium and the Coulomb repulsion of the electrons. The total energy, in its simplest approximation, is thus given as a function of the radius R by

$$E_{\text{Coulomb}}(R) = 4\pi\sigma R^2 + \frac{N^2 e^2}{2\varepsilon R} \tag{2}$$

where $\sigma = 3.6 \times 10^{-4}$ is the surface tension of helium, $\varepsilon = 1.0572$ is the dielectric constant of helium, e is the electron charge and N is the total number of electrons. The first term in (2) is the surface tension energy whereas the second term corresponds to the classical electrostatic energy of a spherical surface of charge Ne with helium on the outside and vacuum on the inside. Minimizing this energy with respect to R, we find the coulomb radius:

$$R_C = \sqrt[3]{\frac{N^2e^2}{16\pi\sigma\varepsilon}} = 1.064\ \mu\text{m} \times \left(\frac{N}{10^4}\right)^{2/3}. \tag{3}$$

Plugging this expression for the equilibrium radius back into expression (2) for the energy, we find that the energy increases as $N^{4/3}$, so that the energy of a bubble with N electrons is larger than twice the energy of a bubble with $N/2$ electrons. This raises legitimate concerns about the stability of MEBs, a point to which we shall return in a later section.

Shikin [11] incorporated into expression (2) the confinement effects of the electrons, so crucial for single-electron bubbles. He assumed that the confinement effects were well represented by stating that the electrons in the z-direction are confined in a 'box' of length d. This length is taken equal to the thickness of the spherical electron layer at the bubble surface. The radius R used in the Coulomb term in expression (2) also has to be corrected to $R-d$, so that we get

$$E_{\text{Shikin}}(R,d) = 4\pi\sigma R^2 + \frac{N\hbar^2}{2m_e d^2} + \frac{N^2e^2}{2\varepsilon(R-d)}, \tag{4}$$

with m_e the electron mass (in vacuum). This expression was further refined by Shung and Lin [12], who included quantum mechanical exchange corrections to the electrostatic energy, using density functional theory. They derived a phenomenological expression for the exchange energy which fitted their density functional calculations well. Their expression for the total energy became

$$E_{\text{SL}}(R,d) = 4\pi\sigma R^2 + \frac{N\hbar^2}{2m_e d^2} + \frac{N^2e^2}{2\varepsilon(R-d)} - 0.3176\frac{N^{4/3}e^2}{\varepsilon(R^2d)^{1/3}}. \tag{5}$$

Finally, it is important to note that the helium liquid can be pressurized (up to 25 bar before it solidifies), leading to a substantial pressure difference p between the inside and the outside of the bubble. Also a negative pressure can be applied up to -9 bar before the helium cavitates spontaneously. In the experimental setup of Fig. 2 for example, such a negative pressure could be realized by closing the valve of the filling line after the bubble has been created, and then expanding the bellows. In any case, positive or negative pressure should be taken into account in the determination of the energy of the bubble [13], leading us to the final form of the energy expression:

$$\begin{aligned} E(R,d) &= p\frac{4\pi}{3}R^3 + 4\pi\sigma R^2 + \frac{N^2e^2}{2\varepsilon(R-d)} \\ &\quad + \frac{N\hbar^2}{2m_e d^2} - 0.3176\frac{N^{4/3}e^2}{\varepsilon(R^2d)^{1/3}}. \end{aligned} \tag{6}$$

To find the equilibrium radius of the bubble (and the equilibrium electron shell thickness) as a function of pressure and number of electrons in the bubble, expression (6) needs to be minimized with respect to both R and d. The results of this minimization are shown in Fig. 3. Performing this minimization, it is easily seen that for bubbles with $N > 100$

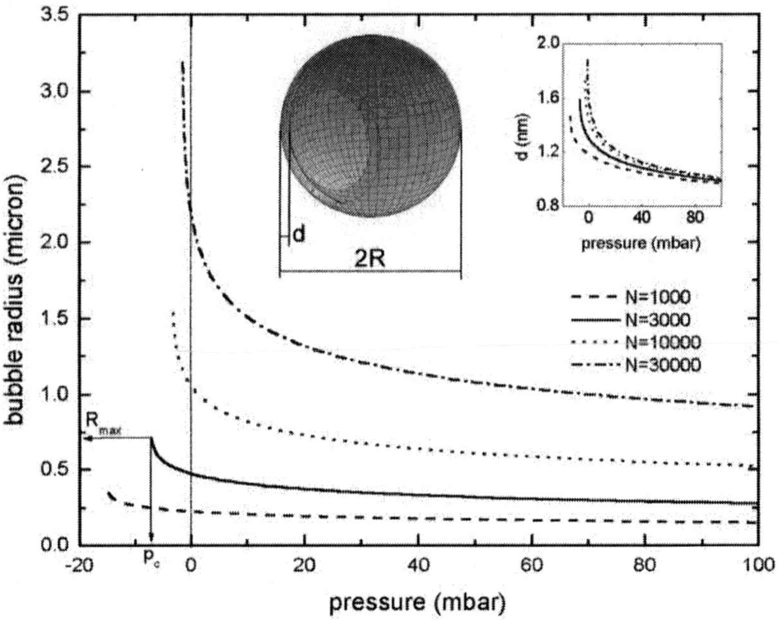

FIGURE 3. The bubble radius, and in the inset the electron layer thickness, are shown as a function of pressure on the bubble for different numbers of electrons in the MEB [13]. Beyond a critical negative pressure, there is no stable solution that minimizes the energy (6).

the coulomb exchange energy term and the confinement term are much smaller than the other terms and can be neglected. For practical purposes, it is sufficient to keep only the first line in expression (6) to determine the radius, putting $R - d \approx R$.

From Fig. 3 it is clear that there is a strong pressure dependence of the radius. A pressure difference of only a few mbar already results in a substantial change for the radius. Multielectron bubbles are very compressible.

Applying a negative pressure on the bubble will expand its radius. But there is a maximum negative pressure beyond which the minimization of the energy (expression (6)) yields no stable (positive real) solution for the radius any more. When the pressure becomes too negative, the bubble will expand until it opens up enough volume to make the pressure less negative than the critical pressure. Using only the first line of expression (6), and thus neglecting the exchange energy and the confinement energy, one finds that the radius should satisfy a fourth degree equation. It's roots can be studied and it is not difficult to derive expressions for the critical negative pressure (and the corresponding maximum bubble radius):

$$R_0^{\max} = \sqrt[3]{\frac{e^2N^2}{4\pi\varepsilon\sigma}} = 4^{1/3}R_C \tag{7}$$

$$p_c = -\frac{3\sigma^{4/3}}{2}\sqrt[3]{\frac{4\pi\varepsilon}{e^2N^2}} \tag{8}$$

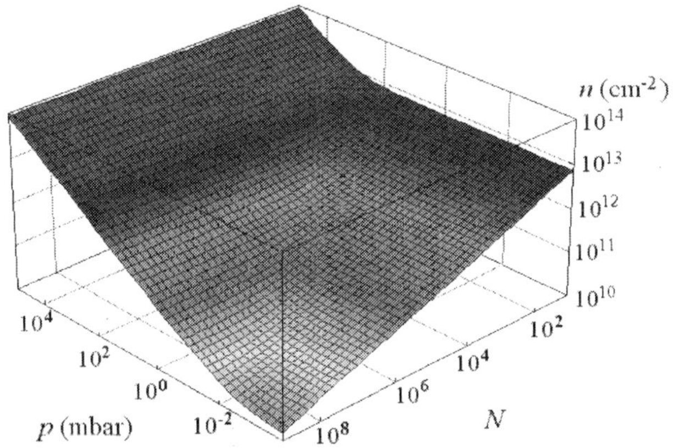

FIGURE 4. The surface density of the spherical two dimensional electron system is shown as a function of the total number of electrons in the bubble and the pressure difference between the inside and the outside of the bubble. Negative pressure can expand the bubble radius up to 1.5 times its zero pressure value, decreasing the lowest achievable density by a further factor of 0.44.

The radius of the bubble also determines the surface density of electrons, $n = N/(4\pi R^2)$. This surface density is shown as function of pressure and total number of electrons in Fig. 4. The pressure is an excellent tuning knob for the radius, and hence for the surface density. Remarkably, the electron surface density in multielectron bubbles can be tuned from around 10^9 to around 10^{14}. Such a large variation (much larger than the 10^{11} – 10^{12} cm^{-2} range available in Si MOSFETs) is very advantageous for the study of Wigner crystallization and electron correlations. Moreover, both in metals and in GaAs/GaAlAs superlattices, it is very difficult to change the density of electrons, whereas here it is tunable by pressure. Part of the interest in MEBs stems from this possibility to explore experimentally the phase diagram of the electron system over such an extended region, and bring the electron system into regimes hitherto inaccessible for probing.

OSCILLATION MODES

(The main results of this section were reported in Refs. [13, 16])

In this section we investigate the modes of oscillation of MEBs. These modes can couple to the electrons and lead to correlations between them. The dispersion relation and the nature of the different modes of oscillation also give insight into the stability of MEBs, and provide a way to detect phase transitions going on in the electronic system inside the bubble, as is the case for electrons on a flat helium surface [4].

In order to derive the dispersion relation for the modes of oscillations, we will assume small amplitude deformations of the bubble, and expand the different terms in the energy, encountered in the previous section, with respect to the amplitude of deformation. The surface of a deformed bubble can be described by a function

$$\begin{aligned}R(\theta,\phi) &= R_b + u(\theta,\varphi) \\ &= R_b + \sum_{\ell=1}^{\infty}\sum_{m=-\ell}^{\ell} Q_{\ell,m} Y_{\ell,m}(\theta,\varphi),\end{aligned} \tag{9}$$

where $Y_{\ell,m}$ is the spherical harmonic function. The function R gives the distance between the center of the bubble and the surface of the bubble along the direction given by the spherical angles θ,ϕ. R_b is the angle-averaged radius - this is the equilibrium radius of a spherical bubble, calculated in the previous section. The displacement out of equilibrium, $u(\theta,\phi)$ is given by the sum over spherical harmonics $Y_{\ell,m}$ multiplied by the amplitude $Q_{\ell,m}$ of that particular spherical harmonic mode. To perform the expansion of the energy terms with respect to the deformation amplitude, we assume that

$$\forall\{\ell > 0, m\} : \sqrt{\ell(\ell+1)} Q_{\ell m}/R_b \ll 1. \tag{10}$$

Geometry of the deformed bubble

The surface of the deformed bubble is

$$\mathrm{S} = \int R^2(\theta,\varphi) \frac{d\Omega}{\cos[\alpha(\theta,\varphi)]} \tag{11}$$

where α is the angle between the normal of the surface and the unit radial vector $\mathbf{e_r}$, so that $R^2 d\Omega/\cos\alpha$ is the surface of an infinitesimal surface element in the direction (θ,φ). This angle is

$$\alpha = \arctan\left[\frac{|\nabla_\Omega u(\theta,\varphi)|}{R(\theta,\varphi)}\right] \tag{12}$$

$$\rightarrow \frac{1}{\cos\alpha} = \sqrt{1 + |\nabla_\Omega u(\theta,\varphi)|^2/R^2} \tag{13}$$

We have up to second order in $|\nabla_\Omega u|/R_b$:

$$\frac{1}{\cos\alpha} = 1 + \frac{1}{2}\frac{|\nabla_\Omega u(\theta,\varphi)|^2}{R_b^2} \tag{14}$$

The deformed surface is, up to second order in u and its derivative

$$\begin{aligned}\mathrm{S} &= \int [R_b + u(\theta,\varphi)]^2 \left[1 + \frac{1}{2}\frac{|\nabla_\Omega u(\theta,\varphi)|^2}{R_b^2}\right] d\Omega \\ &= 4\pi R_b^2 + 2R_b \int u(\theta,\varphi) d\Omega + \int u^2(\theta,\varphi) d\Omega + \frac{1}{2}\int |\nabla_\Omega u(\theta,\varphi)|^2 d\Omega\end{aligned} \tag{15}$$

The term $\int u(\theta,\varphi)d\Omega$ vanishes, and we get:

$$\mathrm{S} = 4\pi R_b^2 + \sum_{\ell=0}^{\infty}\sum_{m=-\ell}^{\ell} |Q_{\ell m}|^2 + \frac{1}{2}\int |\nabla_\Omega u(\theta,\varphi)|^2 d\Omega \tag{16}$$

Finally, we use $\Delta_\Omega Y_{\ell m}(\theta,\varphi) = -\ell(\ell+1)Y_{\ell m}(\theta,\varphi)$ in

$$\begin{aligned}
&\frac{1}{2}\int |\nabla_\Omega u(\theta,\varphi)|^2 d\Omega = \frac{1}{2}\int u(\theta,\varphi)\Delta_\Omega u(\theta,\varphi)d\Omega \\
= &\frac{1}{2}\sum_{\ell=0}^{\infty}\sum_{m=-\ell}^{\ell}\sum_{\ell'=0}^{\infty}\sum_{m'=-\ell'}^{\ell'} Q_{\ell m}Q_{\ell' m'}\int Y_{\ell m}(\theta,\varphi)\Delta_\Omega Y_{\ell m}(\theta,\varphi)d\Omega \\
= &\frac{1}{2}\sum_{\ell=0}^{\infty}\sum_{m=-\ell}^{\ell} \ell(\ell+1)\,|Q_{\ell m}|^2 .
\end{aligned} \tag{17}$$

The surface of the deformed bubble can then be written as

$$\mathrm{S} = 4\pi R_b^2 + \frac{1}{2}\sum_{\ell=0}^{\infty}\sum_{m=-\ell}^{\ell} (\ell^2+\ell+2)\,|Q_{\ell m}|^2 \tag{18}$$

The volume of the bubble is given by

$$\mathrm{V} = \int \frac{1}{3} R^3(\theta,\varphi)d\Omega \tag{19}$$

Expanding this up to second order in the deformation is

$$\mathrm{V} = \frac{1}{3}\int \left[R_b^3 + 3R_b^2 u(\theta,\varphi) + 3R_b u^2(\theta,\varphi)\right] d\Omega \tag{20}$$

Writing the deformation in spherical components gives

$$\mathrm{V} = \frac{4\pi}{3}R_b^3 + R_b \sum_{\ell=0}^{\infty}\sum_{m=-\ell}^{\ell} |Q_{\ell m}|^2 \tag{21}$$

This shows that exciting any mode except the breathing mode will both increase the surface and the volume of the bubble. The surface tension energy (σS) and the pressure related energy term (pV) of the deformed bubble are related straightforwardly to expressions (18) and (21).

Coulomb energy

In the first section, we argued from the findings of ([12]) that the exchange part of the kinetic energy does not play a role for $N > 100$, nor does the confinement energy. Thus, it will be sufficient to derive the direct Coulomb energy based on classical electromagnetism, and take $R_b - d \approx R_b$. The electron layer is at a fixed distance from the helium surface. So, when the helium surface of the bubble deforms according to expression (9), the electron layer deforms with it, in the same manner. It is crucial to take into account that **the electrons can redistribute themselves over the surface of the bubble.** Thus, we need to take into account that the surface charge density becomes a function of the spherical angles:

$$n(\theta,\varphi) = \frac{N}{4\pi R_b^2} + \sum_{\ell=1}^{\infty}\sum_{m=-\ell}^{\ell} n_{\ell m} Y_{\ell m}(\theta,\varphi). \tag{22}$$

This function describes the number of electrons in a solid angle $d\Omega$ the direction given by the spherical angles $\Omega = (\theta,\varphi)$, irrespective of the tilt of the surface in this solid angle. This means that we have chosen a normalization such that

$$N = R_b^2 \int n(\Omega) d\Omega \tag{23}$$

where N is the total number of electrons. The $n_{\ell m}$ represent the amplitudes of the charge redistribution corresponding to the spherical harmonic $Y_{\ell m}(\theta,\varphi)$.

Also the electrostatic potential $V(\mathbf{r})$ can be expanded in spherical harmonics. We use spherical coordinates, and separate the potential into a part $V^{(1)}$ relating to the **inside** of the bubble and $V^{(2)}$ relating to the **outside** of the bubble. We have, taking into account the Laplace equation:

$$V^{(1)}(r,\theta,\varphi) = V_0^{(1)} + \sum_{\ell=1}^{\infty}\sum_{m=-\ell}^{\ell} V_{\ell m}^{(1)} \left(\frac{r}{R_b}\right)^{\ell} Y_{\ell m}(\theta,\varphi), \tag{24}$$

$$V^{(2)}(r,\theta,\varphi) = V_0^{(2)}\frac{R_b}{r} + \sum_{\ell=1}^{\infty}\sum_{m=-\ell}^{\ell} V_{\ell m}^{(2)} \left(\frac{R_b}{r}\right)^{\ell+1} Y_{\ell m}(\theta,\varphi). \tag{25}$$

The coefficients $V_{\ell m}$ can be related to the deformational amplitudes $Q_{\ell m}$ and the charge redistribution amplitudes $n_{\ell m}$ through the use of the electrostatic boundary conditions: (1) the potential should be continuous at the boundary and (2) the normal component of the displacement field should be continuous at the boundary. When these boundary conditions are taken into account, the electrostatic energy is given by

$$U_C = -\frac{eR_b^2}{2}\int_{4\pi} n(\theta,\varphi) V^{(1)}\left(R_b + u(\theta,\varphi),\theta,\varphi\right) d\Omega. \tag{26}$$

Appendix A gives the details of this calculation. The result, up to second order in $Q_{\ell m}$ and $n_{\ell m}$ is given by

$$\begin{aligned} U_C \;=\; & \frac{e^2N^2}{2\varepsilon R_b} - \frac{e^2N^2}{8\pi\varepsilon R_b^3}\sum_{\ell=1}^{\infty}\sum_{m=-\ell}^{\ell} \frac{\ell^2 - \varepsilon(\ell+1)}{\ell + \varepsilon(\ell+1)} \left|Q_{\ell m}\right|^2 \\ & e^2N \sum_{\ell=1}^{\infty}\sum_{m=-\ell}^{\ell} \frac{\ell+1}{\ell+\varepsilon(\ell+1)} n_{\ell m} Q_{\ell m}^{*} + 2\pi e^2 R_b^3 \sum_{\ell=1}^{\infty}\sum_{m=-\ell}^{\ell} \frac{\left|n_{\ell m}\right|^2}{\ell+\varepsilon(\ell+1)}, \end{aligned} \tag{27}$$

where $\varepsilon = 1.0572$ is still the dielectric constant of liquid helium.

Kinetic energy

To find the oscillation frequencies, it is necessary to find also the expansion of the kinetic energy as a function of the amplitude of the deformation. Two terms will be present in the kinetic energy, namely one related to the kinetic energy of the moving helium fluid due to the surface deformation, and one due to the kinetic energy of the motion of the electrons during the charge redistribution along the surface.

Kinetic energy of the moving helium surface

The flow of an ideal fluid is described by a scalar velocity potential $\psi(\mathbf{r})$ as follows:

$$\begin{aligned} \mathbf{v}(\mathbf{r}) &= \nabla\psi(\mathbf{r}), \\ \nabla\psi(\mathbf{r}) &= 0. \end{aligned} \tag{28}$$

It follows from Eq. (28), that $\psi(\mathbf{r})$ obeys the Laplace equation:

$$\Delta\psi(\mathbf{r}) = 0. \tag{29}$$

Let us represent the velocity potential as a series in spherical harmonics $Y_{lm}(\theta,\varphi)$:

$$\psi(\mathbf{r}) = \sum_{\ell=0}^{\infty}\sum_{m=-\ell}^{\ell} \psi_{\ell m}(r)Y_{\ell m}(\theta,\varphi). \tag{30}$$

The equation for radial amplitudes $\psi_{\ell m}(r)$ has the form

$$\frac{d^2\psi_{\ell m}(r)}{dr^2} + \frac{2}{r}\frac{d\psi_{\ell m}(r)}{dr} - \frac{\ell(\ell+1)}{r^2}\psi_{\ell m}(r) = 0. \tag{31}$$

The general solution of this equation is

$$\psi_{\ell m}(r) = Ar^{\ell} + \frac{B}{r^{\ell+1}}. \tag{32}$$

For the fluid outside the sphere, the coefficient $A = 0$, because at $r \to \infty$, $\psi_{\ell m}(r)$ must turn to zero. Hence, we can write down $\psi_{lm}(r)$ as

$$\psi_{\ell m}(r) = \tilde{\psi}_{\ell m}\left(\frac{R_b}{r}\right)^{\ell+1}, \tag{33}$$

where

$$\psi(r,\theta,\varphi) = \sum_{\ell=0}^{\infty}\sum_{m=-\ell}^{\ell} \tilde{\psi}_{\ell m}\left(\frac{R_b}{r}\right)^{\ell+1} Y_{\ell m}(\theta,\varphi). \tag{34}$$

The vector field of velocities is:

$$\mathbf{v}(\mathbf{r}) = \sum_{\ell=0}^{\infty}\sum_{m=-\ell}^{\ell} \tilde{\psi}_{\ell m}\frac{R_b^{\ell+1}}{r^{\ell+2}}\left[-(\ell+1)\mathbf{e}_r Y_{\ell m}(\theta,\varphi) + \nabla_{\Omega}Y_{\ell m}(\theta,\varphi)\right], \tag{35}$$

where $\mathbf{e}_r$ is the radial unit vector: $\mathbf{e}_r \equiv \mathbf{r}/r$, ∇_Ω is the angular gradient operator:

$$\nabla_\Omega = \mathbf{e}_\theta \frac{\partial}{\partial \theta} + \mathbf{e}_\varphi \frac{1}{\sin\theta}\frac{\partial}{\partial \varphi}, \tag{36}$$

$\mathbf{e}_\theta$ and $\mathbf{e}_\varphi$ are unit vectors of the θ- and φ-axes of the spherical coordinate system. The deformation velocity of the bubble surface is

$$\begin{aligned} \dot{R}(\theta,\varphi) &= \mathbf{e}_r \cdot \mathbf{v}(r,\theta,\varphi) && (37)\\ &= -\sum_{\ell=0}^{\infty}\sum_{m=-\ell}^{\ell} \frac{l+1}{R_b}\tilde{\psi}_{\ell m} Y_{\ell m}(\theta,\varphi) && (38)\\ &= \dot{R}_b + \sum_{\ell=1}^{\infty}\sum_{m=-\ell}^{\ell} \dot{Q}_{\ell m} Y_{\ell m}(\theta,\varphi), && (39) \end{aligned}$$

where

$$\dot{Q}_{\ell m} = -\frac{\ell+1}{R_b}\tilde{\psi}_{\ell m}. \tag{40}$$

The kinetic energy is

$$T = \frac{\rho}{2}\int \mathbf{v}^2(\mathbf{r})\,d\mathbf{r} \tag{41}$$

$$\begin{aligned} T &= \frac{\rho}{2}\int [\nabla\psi(\mathbf{r})]^2 d\mathbf{r} = -\frac{\rho}{2}\oint_S \psi(\mathbf{r})\,\mathbf{e}_r \cdot \nabla\psi(\mathbf{r})\,d^2 s \\ &= -\frac{\rho}{2}R_0^2 \int \psi(R_b,\theta,\varphi)\,\dot{u}(\theta,\varphi)\,d\Omega, \end{aligned} \tag{42}$$

The minus sign appears due to the fact, that the vector $\mathbf{e}_r$ is directed outwards of the sphere.

$$\begin{aligned} T &= \frac{\rho R_b^3}{2}\dot{R}_b^2 - \frac{\rho}{2}R_b^2 \int d\Omega \sum_{\ell=1}^{\infty}\sum_{m=-\ell}^{\ell}\left[-\frac{R_b}{l+1}\dot{Q}_{\ell m}Y_{\ell m}(\theta,\varphi)\right]^* \\ &\times \sum_{\ell'=1}^{\infty}\sum_{m'=-\ell'}^{\ell'} \dot{Q}_{\ell' m'} Y_{\ell' m'}(\theta,\varphi). \end{aligned} \tag{43}$$

Taking into account the orthogonality property of the spherical harmonics

$$\int Y_{\ell m}^*(\theta,\varphi) Y_{\ell' m'}(\theta,\varphi)\,d\Omega = \delta_{\ell\ell'}\delta_{mm'}, \tag{44}$$

we find that

$$T = \frac{\rho R_b^3}{2}\dot{R}_b^2 + \frac{\rho R_b^3}{2}\sum_{\ell=1}^{\infty}\sum_{m=-\ell}^{\ell}\frac{|\dot{Q}_{\ell m}|^2}{\ell+1}. \tag{45}$$

This derivation is analogous to that of the kinetic energy of an oscillating droplet by Lord Rayleigh [14], with two differences. Firstly, the volume of the bubble does not need to be kept constant (whereas that of a droplet of incompressible fluid does) so that the term with $\dot{R}_b$ does not occur, and secondly, the expansion in spherical harmonics takes place inside the surface, so that in stead of $\ell + 1$ there is only ℓ in the denominator of (45).

Kinetic energy of the electron motion

To find the kinetic energy associated with the electron motion, we assume that the electrons are localized (as classical point charges, for example in a Wigner lattice) and denote the displacement of electron j from their equilibrium position $\mathbf{r}_j$ by $\mathbf{w}\left(\mathbf{r}_j\right)$. The classical kinetic energy of electrons is

$$T_e = \frac{m_e}{2}\sum_{j=1}^{N}\dot{\mathbf{w}}^2\left(\mathbf{r}_j\right), \tag{46}$$

where $\mathbf{w}\left(\mathbf{r}_j\right)$ is the displacement of an electron from its equilibrium position $\mathbf{r}_j$. The characteristic thickness of the electron layer is ~ 10 nm, what is much less than the average distance between electrons $r_0 = \left(\pi n_0\right)^{-1/2}$, where n_0 is the average surface density of electrons $n_0 = N/\left(4\pi R^2\right)$. Let us express the kinetic energy in the integral form:

$$\begin{aligned} T_e &= \frac{m_e}{2}\sum_{j=1}^{N}\dot{\mathbf{w}}^2\left(\mathbf{r}_j\right)\int_S \delta\left(\mathbf{r}-\mathbf{r}_j\right)d^2\mathbf{r} = \frac{m}{2}\int_S \dot{\mathbf{w}}^2(\mathbf{r})\sum_{j=1}^{N}\delta\left(\mathbf{r}-\mathbf{r}_j\right)d^2\mathbf{r} \\ &= \frac{m_e}{2}\int^2 \dot{\mathbf{w}}^2(\mathbf{r})\,n(\mathbf{r})\,d^2\mathbf{r}, \end{aligned} \tag{47}$$

where the function $n(\mathbf{r}) = \sum_{j=1}^{N}\delta\left(\mathbf{r}-\mathbf{r}_j\right)$ has the sense of the electron density. The two-dimensional integral in Eq. (47) is calculated over the surface of a bubble. When treating the kinetic energy of electrons, we can neglect the deformation of the surface $u(\theta,\varphi)$, since $|u(\theta,\varphi)| \ll R$. So, the radius vector $\mathbf{r} = (R,\theta,\varphi)$, and $\mathbf{w}(\mathbf{r})$, $n(\mathbf{r})$ in fact are the functions of angles θ and φ only. We consider the case when the spatial fluctuations of the density are much less than the average value of the density $\tilde{n}_0$:

$$\frac{\left|n(\mathbf{r}) - n_0\right|}{n_0} \ll 1. \tag{48}$$

In this case, we can use the formula

$$T = \frac{mn_0}{2}\int_S \dot{\mathbf{w}}^2(\mathbf{r})\,d^2\mathbf{r}. \tag{49}$$

The field of the displacements $\mathbf{w}(\mathbf{r})$ can be represented as a sum two fields:

$$\mathbf{w}(\mathbf{r}) = \mathbf{w}_L(\mathbf{r}) + \mathbf{w}_T(\mathbf{r}). \tag{50}$$

Here, $\mathbf{w}_L(\mathbf{r})$ is the field of the tensile/compressive deformation, which can be represented as a gradient of the scalar potential:

$$\mathbf{w}_L(\mathbf{r}) = \nabla\Phi(\mathbf{r}) = \frac{1}{R_b}\nabla_\Omega\Phi(\theta,\varphi), \tag{51}$$

while $\mathbf{w}_T(\mathbf{r})$ is the field of the shear deformation, which satisfies the condition

$$\nabla_\Omega \cdot \mathbf{w}_T(\mathbf{r}) = 0. \tag{52}$$

These two fields are analogous to longitudinal and transverse phonon fields in bulk, or longitudinal and transverse plasmon modes for an electron liquid.

It is easy to show that the integral $\int_S \mathbf{w}_L(\mathbf{r})\,\mathbf{w}_T(\mathbf{r})\,d^2\mathbf{r} = 0$. In other words, potential and solenoidal vector fields are orthogonal to each other. Consequently, the kinetic energy is a sum of two independent terms:

$$T_e = T_L + T_T, \tag{53}$$

where

$$T_{L(T)} = \frac{m_e n_0}{2}\int_S \dot{\mathbf{w}}_{L(T)}^2(\mathbf{r})\,d^2\mathbf{r}. \tag{54}$$

In particular, the contribution to the kinetic energy due to the tensile/compressive deformation can be transformed as follows:

$$\begin{aligned} T_L &= \frac{m_e n_0}{2R_b^2}\int_S \left[\nabla_\Omega \dot{\Phi}(\mathbf{r})\right]^2 d^2\mathbf{r} = \frac{m n_0}{2R_b^2}\int_S \left[\nabla_\Omega \cdot \left(\dot{\Phi}\nabla_\Omega \dot{\Phi}\right) - \dot{\Phi}\Delta_\Omega \dot{\Phi}\right] d^2\mathbf{r} \\ &= \frac{m_e n_0}{2}\int \left[\nabla_\Omega \cdot \left(\dot{\Phi}\nabla_\Omega \dot{\Phi}\right) - \dot{\Phi}\Delta_\Omega \dot{\Phi}\right] d\Omega. \end{aligned} \tag{55}$$

The integral $\int \nabla_\Omega \cdot \left(\dot{\Phi}\nabla\dot{\Phi}\right) d\Omega = 0$. So, we arrive at the formula

$$T_L = -\frac{m_e n_0}{2}\int \dot{\Phi}(\theta,\varphi)\,\Delta_\Omega \dot{\Phi}(\theta,\varphi)\,d\Omega. \tag{56}$$

The reader can check that this approach for the 3D and 2D electron gas yields the known expressions for 3D and 2D plasma frequencies, the latter can be used for a check of the obtained results in the limit $R \to \infty$. The scalar potential $\Phi(\theta,\varphi)$ can be written down as a superposition of spherical harmonics:

$$\Phi(\theta,\varphi) = \sum_{\ell=0}^{\infty}\sum_{m=-\ell}^{\ell} \Phi_{\ell m} Y_{\ell m}(\theta,\varphi). \tag{57}$$

The angular laplacian of this scalar potential is

$$\Delta_\Omega \Phi(\theta,\varphi) = -\sum_{\ell=0}^{\infty}\sum_{m=-\ell}^{\ell} \ell(\ell+1)\,\Phi_{\ell m} Y_{\ell m}(\theta,\varphi). \tag{58}$$

As a result, the kinetic energy T_L takes the form:

$$T_L = \frac{m_e n_0}{2}\sum_{\ell=0}^{\infty}\sum_{m=-\ell}^{\ell} \ell(\ell+1)\left|\dot{\Phi}_{\ell m}\right|^2. \tag{59}$$

We can express T_L in terms of the electron density, using the equation

$$\nabla \cdot \mathbf{w} = \frac{n_0 - n}{n_0}, \tag{60}$$

which is transformed to the equation

$$\frac{1}{R_b^2}\Delta_\Omega \Phi(\theta,\varphi) = 1 - \frac{n(\theta,\varphi)}{n_0}. \tag{61}$$

Using

$$n(\theta,\varphi) = n_0 + \sum_{\ell=1}^{\infty}\sum_{m=-\ell}^{\ell} n_{\ell m} Y_{\ell m}(\theta,\varphi), \tag{62}$$

we get

$$\frac{1}{R_b^2}\sum_{\ell=1}^{\infty}\sum_{m=-\ell}^{\ell} \ell(\ell+1)\Phi_{\ell m} Y_{\ell m}(\theta,\varphi) = \frac{1}{n_0}\sum_{\ell=1}^{\infty}\sum_{m=-\ell}^{\ell} n_{\ell m} Y_{\ell m}(\theta,\varphi). \tag{63}$$

Hence, we arrive at the relation between $\Phi_{\ell m}$ and $n_{\ell m}$:

$$\Phi_{\ell m} = \frac{R_b^2}{n_0 \ell(\ell+1)} n_{\ell m}. \tag{64}$$

In terms of the electron density, the kinetic energy takes the form

$$T_L = \frac{1}{2}\sum_{l=1}^{\infty}\sum_{m=-l}^{l} \frac{4\pi m_e R^6}{N\ell(\ell+1)} \left|\dot{n}_{lm}\right|^2. \tag{65}$$

Lagrangean and Dispersion relation

The full Lagrangian of the bubble, including the ripplon modes and charge redistribution modes, is given by substituting expressions (18), (21), (27), (45), and (65) in $\mathscr{L} = T + T_e - \sigma S - pV - U_C$. The result was the subject of Ref. [16]. The Lagrangean can be brought in the following form

$$\begin{aligned} \mathscr{L} &= \mathscr{L}_R + \sum_{\ell,m} \frac{M_\ell}{2}\left(\dot{Q}_{\ell m}^2 - \omega_r^2(\ell) Q_{\ell m}^2\right) \\ &\quad + \sum_{\ell,m} \frac{m_\ell}{2}\left(\dot{n}_{\ell m}^2 - \omega_p^2(\ell) n_{\ell m}^2\right) + \sum_{\ell,m} \gamma_\ell n_{\ell m} Q_{\ell m}. \end{aligned} \tag{66}$$

We will now proceed to discuss the result term by term. The first term in expression (66) contains the Lagrangian describing the radial motion (the anharmonic breathing mode):

$$\mathscr{L}_R = \frac{\rho R_b^3}{2}\dot{R}_b^2 - \frac{4\pi}{3} p R_b^3 - 4\pi\sigma R_b^2 - \frac{N^2 e^2}{2\varepsilon R_b}. \tag{67}$$

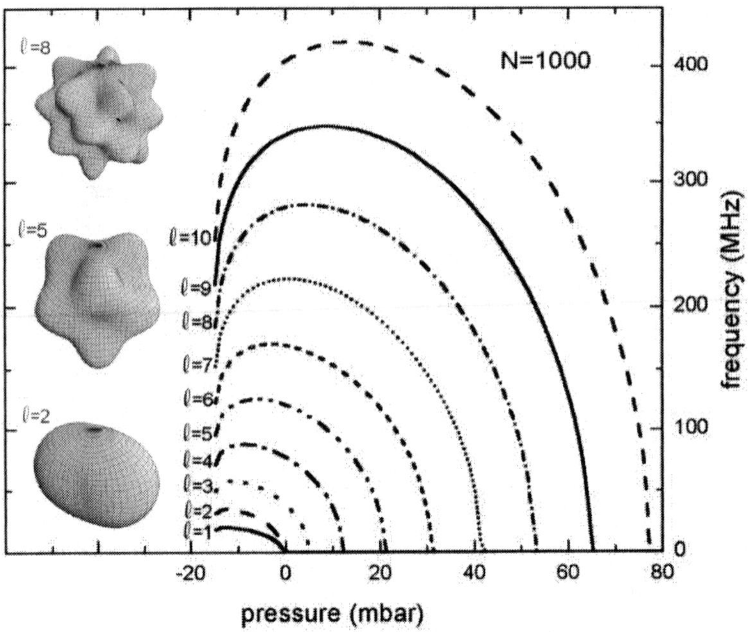

FIGURE 5. The bare ripplon frequency for several low-lying modes of an N=1000 electron bubble, as a function of pressure. Examples of particular deformations are shown on the left, whereby the amplitude has been exaggerated.

In what follows we assume that $\dot{R}_b = 0$ and the bubble radius is given by its equilibrium value.

The Lagrangian $\mathscr{L}$ also contains a part representing the harmonic oscillation of the **ripplon modes**, with oscillator mass term

$$M_\ell = \frac{\rho R_b^3}{\ell+1}, \tag{68}$$

and bare ripplon frequency:

$$\omega_r(\ell) = \sqrt{\frac{1}{M_\ell}\left[\sigma(\ell^2+\ell+1) + pR_b - \frac{N^2e^2}{4\pi\varepsilon R_b^3}\frac{\ell^2-\varepsilon(\ell+1)}{\ell+\varepsilon(\ell+1)}\right]} \tag{69}$$

The bare ripplon frequencies and their dependence on the pressure were the subject of [13]. Note that, in the limit of very large bubbles, $R_b \to \infty$, the dispersion relation for ripplons on a flat helium surface [15] is recovered: if the momentum k is identified with ℓ/R_b, then the typical $k^{3/2}$ is retrieved. For typical multielectron bubbles with radii in the range of 1-100 μm (and numbers of electrons of the order $10^4 - 10^7$) the ripplon frequencies for $\ell < 1000$ lie typically in the MHz-GHz range.

The frequencies of the ripplon modes, as a function of pressure, are shown in Fig. 5 for several low-lying modes. In this figure, R_b has been chosen equal to the equilibrium radius. Note that, upon increasing the pressure, the modes acquire a vanishing frequencies one by at. At zero pressure, the $\ell = 1,2$ modes already have a vanishing frequency, in agreement with the results of Salomaa and Williams [8]. The vanishing of the frequency of certain modes may indicate an instability in the bubble. The bubble stability will be investigated in the next section. However, it is already clear from Fig 5 that applying a small negative pressure (smaller than the critical pressure (8)) will give the modes a nonvanishing frequency.

Next, the Lagrangean (66) contains a sum of terms representing the (small amplitude) harmonic oscillation of the classical charge redistribution around its equilibrium. The oscillator mass term for these oscillations is

$$m_\ell = \frac{4\pi m_e R_b^6}{N\ell(\ell+1)}, \tag{70}$$

and the corresponding frequencies are given by

$$\omega_p(\ell) = \sqrt{\frac{Ne^2}{m_e R_b^3} \frac{\ell(\ell+1)}{\ell+\varepsilon(\ell+1)}}. \tag{71}$$

In the limit of $R_b \to \infty$, this frequency corresponds to that of the **longitudinal plasmon frequency** of a 2D electron system on the helium surface [17]. When the electrons form a Wigner lattice, this corresponds to the longitudinal phonon frequency. If the momentum k is identified with ℓ/R_b, then the typical $k^{1/2}$ dispersion for this longitudinal plasmon is retrieved in the limit of large bubbles. Based on this correspondence, we will refer to these modes in a multielectron bubble as the '(longitudinal) plasmon modes of the MEB' (or 'longitudinal phonon modes of the MEB' if the electrons form a Wigner lattice). Typical multielectron bubbles with radii in the range of 1-100 μm (and numbers of electrons of the order $10^4 - 10^7$) have longitudinal plasmon frequencies with $\ell < 1000$ in the GHz-THz range. For bubbles with $N > 10^5$, the longitudinal plasmon frequencies are larger than the ripplon frequencies, $\omega_p(\ell) \gg \omega_r(\ell)$.

When the plasmon frequencies are much larger than the ripplon frequencies, the electrons can redistribute much faster on the surface of the bubble than the bubble surface can deform. For a given surface deformation (a given set of $Q_{\ell m}$'s, we can then find the surface charge density from the adiabatic approximation. The equations of motion for the electron charge redistribution on a deformed bubble are given by:

$$\begin{aligned} \frac{d}{dt}\frac{\partial \mathscr{L}}{\partial \dot{n}_{\ell m}} &= \frac{\partial \mathscr{L}}{\partial n_{\ell m}} \\ \Leftrightarrow \quad m_\ell \ddot{n}_{\ell m} &= -m_\ell \omega_p^2(\ell) n_{\ell m} + \gamma_\ell Q_{\ell m}. \end{aligned} \tag{72}$$

We find from (72) that the equilibrium charge distribution on the deformed bubble must satisfy the relation

$$\frac{n_{\ell m}}{N} = 2\pi R_b^2 (\ell+1) \frac{Q_{\ell m}}{R_b}. \tag{73}$$

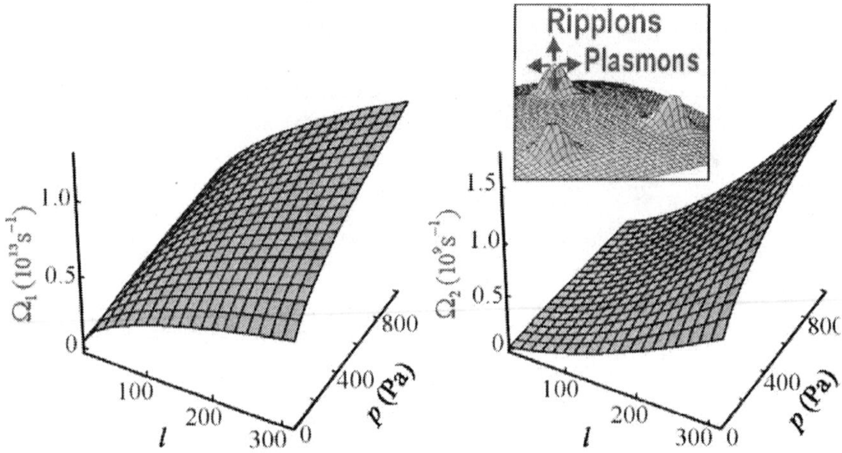

FIGURE 6. The frequency of the coupled ripplon - longitudinal plasmon modes is shown as a function of pressure p and of mode number ℓ. In the inset the nature of the ripplon type of oscillation and the plasmon type of oscillation are illustrated.

Finally, the last part of Langrangean $\mathscr{L}$ represents the coupling between the longitudinal plasmons and the ripplons, with coupling strength given by

$$\gamma_\ell = Ne^2 \frac{(\ell+1)}{\ell+\varepsilon(\ell+1)}. \tag{74}$$

For electrons on a flat helium surface, such a coupling between ripplons and longitudinal plasmons/phonons was derived by Fisher *et al.* [18] and detected experimentally [4]. Here, only ripplon and longitudinal plasmon modes which have the same angular momentum couple to each other. The eigenfrequencies are found after the diagonalisation of the ripplon-plasmon part of the Lagrangian (66),

$$\Omega_{1,2}(\ell) = \sqrt{\frac{1}{2}\left[\omega_p^2(\ell)+\omega_r^2(\ell)\right] \pm \frac{1}{2}\sqrt{\left[\omega_p^2(\ell)-\omega_r^2(\ell)\right]^2+4\gamma_\ell^2}}. \tag{75}$$

Figure 6 shows the dispersion of these coupled modes. One of the modes has a frequency close to that of the bare ripplon frequency (illustrated also in Fig. 5. This mode consists mainly of the ripplon oscillation admixed with a small component of plasmon oscillation. The ripplon oscillation can be seen as an oscillation normal to the spherical surface, whereas the longitudinal plasmon oscillation can be seen as an oscillation of the charge density in a direction tangential to the surface. The second coupled mode is more akin to the longitudinal plasmon oscillation: its frequency lies close to the plasmon frequency, and it's nature consists mostly out of the plasmon type of oscillation, admixed with a small amplitude of ripplon oscillation.

Both of these modes will lead to correlation between the electrons. As we shall see, the longitudinal plasmon frequency corresponds to the confinement frequency of a lattice potential of the Wigner crystal of electrons on the sphere. The electron-ripplon coupling will be essential to localise the electrons so they can form the Wigner lattice where it is present. But before we investigate these effects further, we need to answer concerns about the bubble stability, raised in this section.

BUBBLE STABILITY AND FISSIONING

(The main results of this section were reported in Ref. [19])

In the preceding section, the dispersion relation for the modes of oscillation were calculated, but concerns were raised on the bubble stability. If some modes acquire a vanishing frequency, the amplitude of these modes of deformation can grow until they become of the order of the bubble radius. This deformational instability can then lead to fissioning of the bubble. But since MEBs with N up to 10^8 were observed experimentally [6, 7], albeit in a transient manner (lasting a few msec), there must exist a formation barrier preventing the fissioning of MEBs. Salomaa and Williams [8] have investigated the dynamics of this deformational instability using coupled Rayleigh-Plesset equations for the deformation amplitudes and the bubble radius, and found that when the initial amplitude of the quadrupole mode of deformation is larger than $\sim 6\%$ of the bubble radius, the amplitude of the oscillation keeps growing as a function of time. Because their approach is only valid for small amplitude oscillations and not the large deformations appearing during the fissioning event, a more rigorous method is needed.

The fissioning of a charged bubble can be examined based on a method devised by Bohr [20] to study the fissioning of nuclei. In such models of nuclear fission, and models for the breaking up of homogeneously charged liquid droplets, a certain approximation is used to calculate the properties of the fissioning. This approximation consists of describing the surface of the splitting nucleus (or droplet) in terms of two or three quadratic forms (spheroids and hyperboloids). Such a surface is parametrized in cylindrical coordinates trough

$$\rho = \begin{cases} \sqrt{a_L - b_L(z-f_L)^2} \text{ for } z_0 \leq z \leq z_1 \\ \sqrt{a_M - b_M(z-f_M)^2} \text{ for } z_1 \leq z \leq z_2 \\ \sqrt{a_R - b_R(z-f_R)^2} \text{ for } z_2 \leq z \leq z_3 \end{cases}, \tag{76}$$

where

- a_i is the square of the semi-major axis of the spheroid along the radial direction. For a hyperboloid, a_i is negative;
- b_i is the deformation parameter: the square of radial semi-major axis divided by the square of the longitudinal semi-major axis;
- f_L is the center of the spheroid (hyperboloid);
- $z_0 = f_L - a_L/\sqrt{b_L}$ is the leftmost point of the shape;
- $z_3 = f_R + a_R/\sqrt{b_R}$ is the rightmost point of the shape.

The shape parameters are $\{a_L, a_M, a_R\}, \{b_L, b_M, b_R\}, \{f_L, f_M, f_R\}$ which however are not independent if one imposes that the different surface merge continuously. An example of such a shape, along with the meaning of the different variables is given in Figure 7.

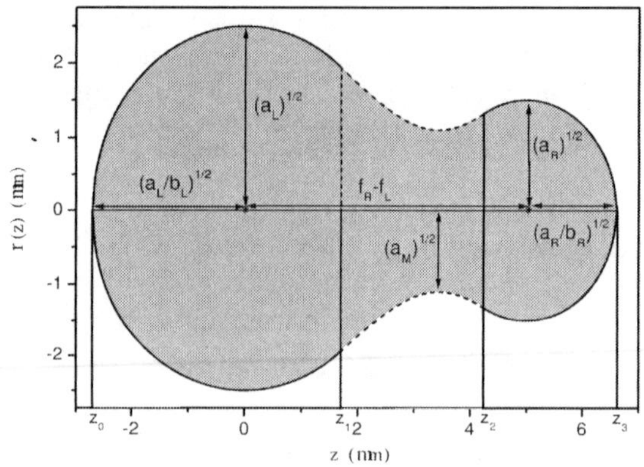

FIGURE 7. The model shape for a fissioning bubble is shown and the shape parameters are indicated.

The points z_1 and z_2 are the points of intersection of the quadratic surfaces - also these can be used as variational parameters (not independent of the other parameters). The left surface has to match the middle surface at an intersection point z_1 and the right surface has to match the middle surface at another intersection point z_2. Imposing continuity and continuous derivatives at these meeting points allow to eliminate six of the nine variables, leaving five independent parameters. These are: $a_L, b_L, a_R, b_R, f_R - f_L$ and z_1. How the other parameters can be calculated given those six is shown in appendix A.

The middle segment determines whether the bubble is fissioned or not. Fig. 8 illustrates how a_M and b_M give rise to different connecting surfaces between the left and the right spheroid. The region with $a_M < 0$ and $b_M > 0$ does not give rise to valid surfaces.

Energy functional

The volume of the bubble can be easily found through

$$V = \int_{f_L-\sqrt{a_L/b_L}}^{f_R+\sqrt{a_R/b_R}} dz\, 2\pi \int_0^{\rho(z)} \rho d\rho = \int_{f_L-\sqrt{a_L/b_L}}^{f_R+\sqrt{a_R/b_R}} dz\, \pi\rho^2(z) \tag{77}$$

This is a sum of integrals of the type

$$\int_{z_a}^{z_b} \pi\left[a+b(z-f)^2\right] dz = \pi a(z_b - z_a) - \frac{b}{3}((z_b - f)^3 - (z_a - f)^3) \tag{78}$$

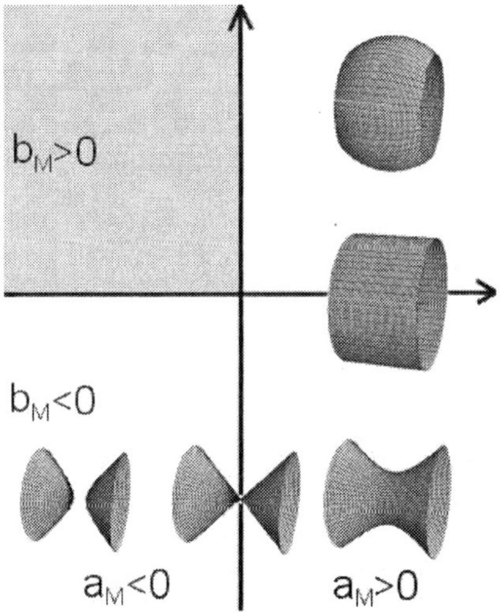

FIGURE 8. Varying the different parameters allow to describe both fissioned bubbles and connected bubbles. Continuous variation of the shape parameters leads to all intermediate shapes.

The surface is obtained by

$$S = \int_{f_L-\sqrt{a_L/b_L}}^{f_R+\sqrt{a_R/b_R}} dz\, \pi\rho(z)\sqrt{1+(d\rho(z)/dz)^2} \tag{79}$$

$$= \pi\sum_{i=1}^{3}\int_{z_{i-1}}^{z_i} \sqrt{a_i+b_i(b_i-1)(z-f_i)^2}dz \tag{80}$$

$$= \pi\sum_{i=1}^{3}\int_{z_{i-1}-f_i}^{z_i-f_i} \sqrt{a_i+b_i(b_i-1)x^2}dx \tag{81}$$

with $z_0 = f_L - \sqrt{a_L/b_L}$ and $z_3 = f_R + \sqrt{a_R/b_R}$; and $a_{1,2,3} = a_{L,M,R}$. The integral is

$$\int_{z_a-f}^{z_b-f} \sqrt{a+b(b-1)x^2}dx$$

$$= \begin{cases} 2\sqrt{a}\left[z\sqrt{1+\frac{b(b-1)}{a}z^2}+\sqrt{\frac{a}{b(b-1)}}\operatorname{arcsinh}\left(\sqrt{\frac{b(b-1)}{a}}z\right)\right]_{z_a-f}^{z_b-f} & \text{for } a>0 \\ 2\sqrt{|a|}\left[\begin{array}{l} z\sqrt{\frac{b(b-1)}{|a|}z^2-1} \\ +\sqrt{\frac{|a|}{b(b-1)}}\log\left(\sqrt{\frac{b(b-1)}{|a|}}z+\sqrt{\frac{b(b-1)}{|a|}z^2-1}\right)\end{array}\right]_{z_a-f}^{z_b-f} & \text{for } a<0 \end{cases}$$

The coulomb energy needs a bit more bookkeeping.

For a sphere, we have a surface density $n_s = N/(4\pi R^2)$. The electrostatic energy is

$$\begin{aligned}
U &= \int d^3\mathbf{r} \int d^3\mathbf{r}' \frac{(n_s e).(n_s e)}{\varepsilon|\mathbf{r}-\mathbf{r}'|} \\
&= \frac{(n_s e)^2}{\varepsilon} \int_0^{\pi} d\theta \int_0^{2\pi} d\phi R^2 \sin\theta \int_0^{\pi} d\theta' \int_0^{2\pi} d\phi' \, R^2 \sin\theta' \frac{1}{2R\sin(\theta/2)} \\
&= \frac{(n_s e)^2}{\varepsilon} \frac{2(2\pi R^2)^2}{R} \int_0^{\pi} d\theta \cos\theta/2 = \frac{(n_s e)^2 (4\pi R^2)^2}{2\varepsilon R} = \frac{(Ne)^2}{2\varepsilon R}
\end{aligned}$$

Analogously, the electrostatic energy of a spherical cap (from 0 till θ) is

$$U_{cap} = \frac{(4\pi R^2 n_s)^2 e^2}{2\varepsilon R} (\sin\theta_{cr}/2) \tag{82}$$

For a general cylindrical symmetrical shape, given by $\rho(z)$, we can use

$$\frac{1}{|\mathbf{r}-\mathbf{r}'|} = \frac{4}{\pi} \int_0^{\infty} dk \cos[k(z-z')] \left\{ \begin{array}{c} \frac{1}{2} I_0(k\rho_<) K_0(k\rho_>) \\ + \sum_{m=1}^{\infty} \cos[m(\phi-\phi')] I_m(k\rho_<) K_m(k\rho_>) \end{array} \right\} \tag{83}$$

where $\rho_<$ is the smallest of the two cylindrical radii.

$$\begin{aligned}
U &= \frac{(n_s e)^2}{\varepsilon} \int_{z_0}^{z_3} dz \int_0^{2\pi} d\phi \int_{z_0}^{z_3} dz' \int_0^{2\pi} d\phi' \, \rho(z)\rho(z') \frac{1}{|\mathbf{r}-\mathbf{r}'|} \\
&= \frac{(2\pi n_s e)^2}{\varepsilon} \int_{z_0}^{z_3} dz \int_{z_0}^{z_3} dz' \frac{2}{\pi} \int_0^{\infty} dk \cos[k(z-z')] I_0(k\rho_<) K_0(k\rho_>) \rho(z)\rho(z')
\end{aligned} \tag{84}$$

Thus, we are left with a triple integral to be performed numerically.

Results and discussion

To investigate fission, we minimize the potential energy of a bubble for a fixed elongation $z_3 - z_0$. The potential energy and the optimum shape is then discussed as a function of the elongation. For this first exploration, the parameters a_L, a_R and z_1 are varied; $b_L = b_R = 1$ is chosen (i.e. bubble fragments are spherical).

Figure 9 shows the results for the optimum potential energy as a function of the elongation **at zero pressure,** for $N = 10000$. The dots show the results of the calculations, the curves are guides for the eye.

The results confirm our previous calculations which show that at zero pressure, the quadrupole frequency $\omega(\ell = 2) = 0$ is zero. Indeed, at elongations smaller than ca.

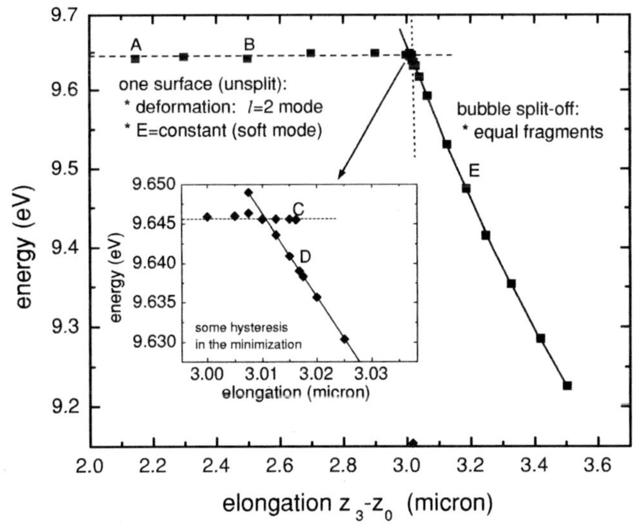

FIGURE 9. The minimum energy obtained by varying the shape parameters while keeping the elongation $z_3 - z_0$ fixed is shown here as a function of this elongation. Below $z_3 - z_0 = 3$ the optimum shape conforms to an $\ell = 2$ type of deformation. In agreement with the results from the previous section, this mode is soft.

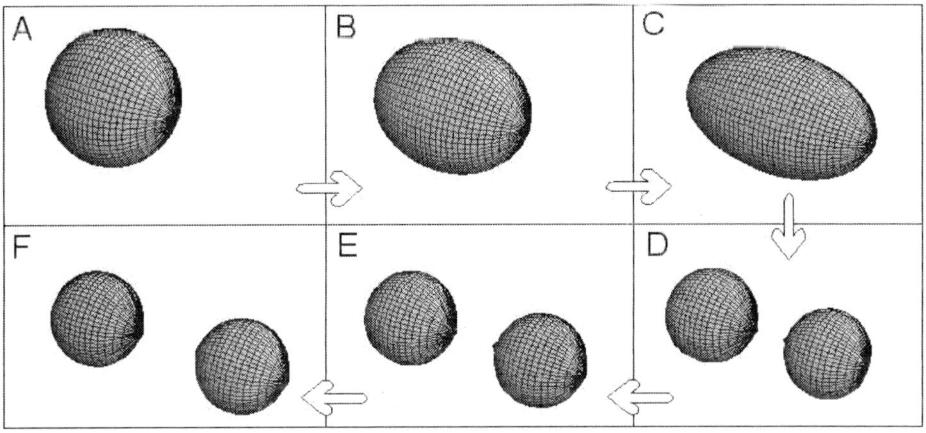

FIGURE 10. Varying the different parameters allow to describe both fissioned bubbles and connected bubbles. Continuous variation of the shape parameters leads to all intermediate shapes.

3, the energy is constant. This is illustrated in Fig. 9 by a dashed line showing the constant value of the energy. The optimum surface for these elongations < 3.01 is an $\ell = 2$ deformed sphere (an ellipsoid). This is shown in Fig. 10 - for the dots in Fig. 9 marked with a capital letter, the optimum shape corresponding to that dot is shown in the panels of Fig. 10. The fact that the optimum shape is an $\ell = 2$ deformed surface further confirms our previous conclusion that the $\ell = 2$ mode is the only mode which does not cost energy to excite. If this were not true, more complicated shapes could be the optimum shape.

Around elongation ≈ 3, the quadrupole-deformation constant energy curve intersects the curve for two separate bubbles with $N/2$ electrons each. The energy of a single spherical bubble with N electrons is

$$E(N) = \frac{3}{4}\left(\frac{16\pi N^4 e^4 \sigma}{\varepsilon^2}\right)^{1/3} \tag{85}$$

When the two separate bubbles are at infinite distance, they have $2E(N/2)$ as energy which is smaller than $E(N)$ since the energy rises with $N^{4/3}$. However, as the two fragments come closer, the total potential energy rises due to the Coulomb repulsion between the fragments. The full line shows the energy of the two-bubble system as a function of the distance between the bubbles (expressed through $z_3 - z_0$). At the distance where the total potential energy of the two fragments equals the energy of the elongated single bubble, fissioning can occur. This is region of the energy landscape is shown more closely in the inset.

As discussed, the potential energy curve in Figure 9 consists of two distinct branches which intersect near $z_3 - z_0 = 3.01$. The two branches correspond to different configuration, namely an elongated ellipsoid on the one hand and two bubbles on the other hand. To go from one state (for example the state indicated by "C" in Figures 9,10) to another state of lower energy in the other branch (for example the state "D" in Figures 9,reffig:10), the bubble shape has to deform continuously between the two geometries. The intermediate geometries have higher energy and thus will constitute an energy barrier.

In Figure 11, some of these intermediate geometries are shown. Also the potential energy is shown as the bubble changes from one shape into the other. The change is characterised by a path in shape space, staring from point 'C' and going to point 'D'. We choose an evolution parameter, equal to 0 in point 'C' and 1 in point 'D'. The intermediate values parametrize the path from 'C' to 'D'. At "evolution parameter" = 0, the bubble is on the horizontal branch of Fig. 9. At "evolution parameter"= 1, the bubble is split. In order to fission, the bubble has to continuously go through the intermediate shapes, which give rise to an energy barrier, in this case of approximately 0.2 eV.

Thus, we conclude that apart from any kinetic barrier found by Salomaa and Williams in solving the time-dependent evolution equations for the quadrupole oscillation, there is a '**shape barrier**' which has to be crossed in order to fission the bubble. The presence of this barrier is also what caused some hysteresis in the minimization routine (otherwise the minimization routine would not get stuck in a local minimum).

The origin of this barrier is the formation of a neck between the fission fragments. This neck involves high-ℓ modes of deformation. These modes are not soft, so that the

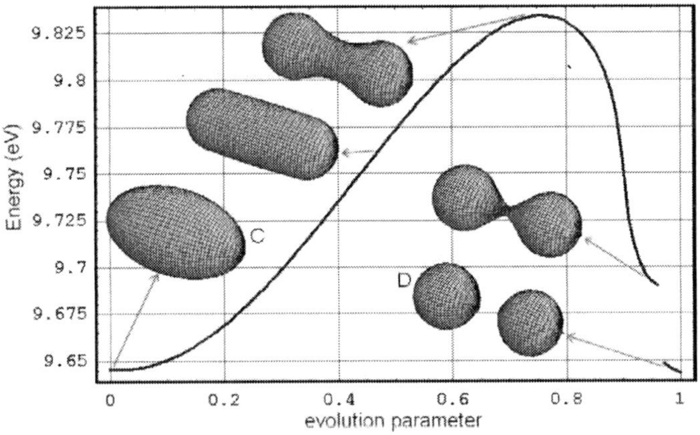

FIGURE 11. During the fissioning process, a neck has to be formed between the two fission fragments. This neck involves high-ℓ modes of deformation and thus costs energy. This leads to an energy barrier stabilizing the MEB against fission.

formation of the neck costs energy. In Ref. [19] we also studied the case of positive and negative pressure. The conclusions from the previous section are reinforced: a small negative pressure will stabilize the bubble further and favour spherical geometry, whereas an increasing positive pressure will make more and more modes soft, lowering the barrier against fissioning and ultimately making the bubble unstable.

THE SPHERICAL TWO-DIMENSIONAL ELECTRON GAS

(The main results of this section were reported in Ref. [21])

Now that we have assured ourselves of the bubble stability, the spherical two-dimensional electron system can be analyzed. In this section we focus on the electron gas (or liquid) state, whereas the next section will focus on the electron solid phase and the phase transition between both.

Fermi energy of the S2DEG

Before including the Coulomb interaction in the description of the S2DEG, let's look at basic quantities for the non-interacting electron gas. In a flat 2DEG, the single-particle electron wave functions are plane waves $\varphi_{\mathbf{k}}(\mathbf{r}) = e^{i\mathbf{k}.\mathbf{r}}$. For the spherical geometry, they are spherical harmonics, characterized by the two indices ℓ, m so that $\varphi_{\ell m}(\Omega) = Y_{\ell m}(\Omega)$. The position of the electron $\mathbf{r}$ is determined by the spherical angle $\Omega = \{\theta, \phi\}$. The single-particle energy levels are given by

$$-\frac{\hbar^2}{2m_eR^2}\Delta_\Omega Y_{\ell m}(\Omega) = E_{\ell m}Y_{\ell m}(\Omega) \tag{86}$$

$$\Leftrightarrow\ E_{\ell m} = \frac{\hbar^2\ell(\ell+1)}{2m_eR^2}, \tag{87}$$

where we have used the laplacian in spherical coordinates,

$$\Delta = \frac{1}{R^2}\frac{\partial}{\partial R}\left(R^2\frac{\partial}{\partial R}\right) + \frac{1}{R^2}\Delta_\Omega$$

$$\text{with } \Delta_\Omega = \frac{1}{\sin\theta}\frac{\partial}{\partial\theta}\left(\sin\theta\frac{\partial}{\partial\theta}\right) + \frac{1}{\sin^2\theta}\frac{\partial^2}{\partial\phi^2}, \tag{88}$$

and the eigenfunction property of the spherical harmonics, $\Delta_\Omega Y_{\ell m}(\Omega) = -\ell(\ell+1)Y_{\ell m}(\Omega)$. The single-particle energy levels are degenerate for $\ell > 0$, and have a degeneracy of $2\ell+1$, since for any ℓ, m can take the integer values $-\ell,..,\ell$. An interesting property of the S2DEG is its Fermi energy E_F as a function of the number of particles. As the number of particles increases, more single-particle energy levels have to be filled up (increasing E_F), but the bubble radius will also increase (decreasing E_F).

The highest filled level L_F is given by

$$N = 2\sum_{\ell=1}^{L_F}(2\ell+1) = 2(L_F+1)^2 \tag{89}$$

$$\rightarrow\ L_F = \left\lceil\sqrt{N/2}-1\right\rceil. \tag{90}$$

where $\lceil x\rceil$ means the smallest integer larger than or equal to x. The factor 2 comes from the spin-up/spin down degeneracy for the electrons. For electron bubbles with $10^6 - 10^8$ electrons, this means $L_F \approx 10^3 - 10^4$. The Fermi-energy is then (for $\sqrt{N/2} \gg 1$),

$$E_F = \frac{\hbar^2 L_F(L_F+1)}{2m_eR^2} \tag{91}$$

$$= \frac{\hbar^2 N}{4m_e}\left(\frac{1}{4}\sqrt[3]{\frac{N^2e^2}{\pi^2\varepsilon\sigma}}\right)^{-2} \tag{92}$$

$$= N^{-1/3}\frac{4\hbar^2}{m_e}\left(\frac{\pi^2\varepsilon\sigma}{e^2}\right)^{2/3}. \tag{93}$$

This rather surprising result shows that the increase of bubble radius with increasing N wins over the level filling, so that the Fermi energy in this system actually decreases with increasing N ! Putting in the parameters, we arrive at

$$E_F = N^{-1/3}\times 2.29\times 10^{-20}\mathrm{J} = N^{-1/3}\times 143\ \mathrm{meV} \tag{94}$$

$$= N^{-1/3}\times 1659.5\ \mathrm{K}. \tag{95}$$

For $N = 10^6$ the Fermi energy is 16 K, for $N = 10^8$ we find $E_F = 3.57$ K. This means that smaller bubbles ($10^6 - 10^7$ electrons) will be in the degenerate Fermi gas regime. Intermediate bubbles (10^8 electrons) will be around the quantum degeneracy point, and for large bubbles (10^9 electrons) the quantum degeneracy will have been partially lifted. This is in contrast with metals at room temperature, which are deep in the quantum degenerate regime.

Many-body theory on a sphere

Second Quantization on a sphere

Many-body theory, and especially the second quantization formalism, commonly treats flat spaces of one, two or three dimensions, using decompositions in plane waves. In this section, we go back to the roots of many-body theory, and build it up for electrons on a spherical shell. First, some notation conventions are listed. Each time, also the familiar plane wave counterpart is listed.

- The position operators $\mathbf{r}_j = \{x_j, y_j\}$ $(j = 1,..,N)$ are replaced by the spherical angle operator $\mathbf{\Omega}_j = \{\theta_j, \phi_j\}$ giving the position on the spherical surface.
- The (non-interacting) single-particle wave functions $\varphi_{\mathbf{k}}(\Omega) = e^{i\mathbf{k}.\mathbf{r}}/\sqrt{L}$ on a flat surface are characterised by the quantum numbers $\mathbf{k} = \{k_x, k_y\}$ with $k_x, k_y = n\pi/L$ where L is the length of the system and n is an integer. The (non-interacting) single-particle wave functions $\varphi_{\ell,m}(\Omega) = Y_{\ell,m}(\Omega)$ for a spherical surface are characterised by the quantum numbers $\{\ell, m\}$ where $\ell \in \aleph$ and m is an integer in the range $\{-\ell, ..., \ell\}$.
- The unitary operator in flat space can be decomposed as $\mathbf{1} = \int d\mathbf{r}\, |\mathbf{r}\rangle \langle \mathbf{r}|$ where $\langle \mathbf{r}|\varphi\rangle = \varphi(\mathbf{r})$. For a sphere, the unitary operator can be decomposed as

$$\mathbf{1} = \int d\Omega\, |\Omega\rangle \langle \Omega| = \int_0^{\pi} d\theta \int_0^{2\pi} d\phi\, \sin\theta\, |\Omega\rangle \langle \Omega|, \tag{96}$$

  where $\langle \Omega|\varphi\rangle = \varphi(\Omega)$.
- The operator $\widehat{c}_{\ell,m}^{+}$ creates an electron in the single-particle state $\varphi_{\ell,m}$ (if none is present) and $\widehat{c}_{\ell,m}$ destroys an electron in the single-particle state $\varphi_{\ell,m}$. These operators satisfy fermionic anticommutation relations $\left[\widehat{c}_{\ell,m}, \widehat{c}_{\ell',m'}^{+}\right]_{+} = \delta_{\ell,\ell'}\delta_{m,m'}$ and are the spherical counterparts of the plane-wave operators $\widehat{c}_{\mathbf{k}}^{+}$ and $\widehat{c}_{\mathbf{k}}$ which respectively create and annihilate an electron in the single-particle state $\varphi_{\mathbf{k}}$. Note that whenever a hat is drawn over a symbol, this will mean that the symbol is an operator and not a complex number.

A one-particle operator on the many-body system is an operator which can be written as:

$$\widehat{A} = \sum_{j=1}^{N} \widehat{A}(\Omega_j). \tag{97}$$

This means they can be expressed as a sum over all particles. In contrast, two-particle operators such as the two-body interaction potential can only be expressed as a sum over all *pairs* of particles. We know from many-body textbooks such as Fetter and Walecka [22] that in second quantisation based on a set of single-particle wave functions $\varphi_{\ell m}(\Omega)$, this operator can be written as

$$\widehat{A} = \sum_{\ell=0}^{\infty} \sum_{m=-\ell}^{\ell} \sum_{\ell'=0}^{\infty} \sum_{m'=-\ell}^{\ell} \left\langle \ell m \middle| \widehat{A} \middle| \ell' m' \right\rangle \widehat{c}^{+}_{\ell m} \widehat{c}_{\ell' m'} \tag{98}$$

Two important one-particle operators are the kinetic energy and the density. Both will be treated next.

The kinetic energy operator $\widehat{T}$ is a one-particle operator

$$\sum_{j=1}^{N} \frac{-\hbar^2}{2m_{\mathrm{e}}R^2} \Delta_{\Omega}. \tag{99}$$

Its expectation value between spherical harmonics is

$$\left\langle \ell m \middle| \widehat{T} \middle| \ell' m' \right\rangle = \int d\Omega\, Y^{*}_{\ell m}(\Omega) \frac{-\hbar^2}{2m_{\mathrm{e}}R^2} \Delta_{\Omega} Y_{\ell' m'}(\Omega) \tag{100}$$

$$= \frac{\hbar^2 \ell(\ell+1)}{2m_e R^2} \int d\Omega\, Y_{\ell m}(\Omega) Y_{\ell' m'}(\Omega) \tag{101}$$

$$= \frac{\hbar^2 \ell(\ell+1)}{2m_e R^2} \delta_{\ell,\ell'} \delta_{m,m'}. \tag{102}$$

Thus, it's second quantized form is

$$\widehat{T} = \sum_{\ell=0}^{\infty} \sum_{m=-\ell}^{\ell} \frac{\hbar^2 \ell(\ell+1)}{2m_e R^2} \widehat{c}^{+}_{\ell m} \widehat{c}_{\ell m}. \tag{103}$$

The density operator is

$$\widehat{n}_{\mathrm{e}}(\Omega) = \sum_{j=1}^{N} \delta(\Omega - \widehat{\Omega}_j), \tag{104}$$

where $\widehat{\Omega}_j$ is the position operator of an electron. Its expectation value gives the density of electrons at location Ω on the sphere. It is a one-body operator (like the kinetic energy) and is expressed in second quantisation accordingly:

$$\widehat{n}_{\mathrm{e}}(\Omega) = \sum_{\ell,m}\sum_{\ell',m'}\left[\int d\Omega'\, Y^*_{\ell',m'}(\Omega')\delta(\Omega-\Omega')Y_{\ell,m}(\Omega')\right]\widehat{c}^+_{\ell'm'}\widehat{c}_{\ell m} \tag{105}$$

$$= \sum_{\ell,m}\sum_{\ell',m'} Y^*_{\ell',m'}(\Omega)Y_{\ell,m}(\Omega)\,\widehat{c}^+_{\ell'm'}\widehat{c}_{\ell m}. \tag{106}$$

The **spherical decomposition** of the density operator is given by

$$\widehat{n}_{\mathrm{e}}(\Omega) = \sum_{\ell=0}^{\infty}\sum_{m=-\ell}^{\ell}\widehat{\rho}_{\mathrm{e}}(\ell,m)\,Y^*_{\ell,m}(\Omega) \tag{107}$$

$$\widehat{\rho}_{\mathrm{e}}(\ell,m) = \int d\Omega\, Y_{\ell,m}(\Omega)\widehat{n}_{\mathrm{e}}(\Omega). \tag{108}$$

Note **the phase convention**, chosen such that $\widehat{\rho}_{\mathrm{e}}(\ell,m) = \sum_{j=1}^{N} Y_{\ell m}(\widehat{\Omega}_j)$. These expressions are analogous to the plane wave (Fourier) decomposition in flat space. We find, substituting (106) into (108),

$$\widehat{\rho}_{\mathrm{e}}(\ell,m) = \sum_{\ell_1,m_1}\sum_{\ell_2,m_2}\left[\int d\Omega\, Y_{\ell,m}(\Omega)Y^*_{\ell_2,m_2}(\Omega)Y_{\ell_1,m_1}(\Omega)\right]\widehat{c}^+_{\ell_2 m_2}\widehat{c}_{\ell_1 m_1}. \tag{109}$$

Its hermitean conjugate is

$$\widehat{\rho}^+_{\mathrm{e}}(\ell,m) = \sum_{\ell_1,m_1}\sum_{\ell_2,m_2}\left[\int d\Omega\, Y^*_{\ell,m}(\Omega)Y_{\ell_2,m_2}(\Omega)Y^*_{\ell_1,m_1}(\Omega)\right]\widehat{c}^+_{\ell_1 m_1}\widehat{c}_{\ell_2 m_2}. \tag{110}$$

which, using $Y_{\ell,m}(\Omega) = (-1)^m Y^*_{\ell,-m}(\Omega)$ can be rewritten to

$$\widehat{\rho}^+_{\mathrm{e}}(\ell,m) = (-1)^m \sum_{\ell_1,m_1}\sum_{\ell_2,m_2}\left[\int d\Omega\, Y_{\ell,-m}(\Omega)Y_{\ell_2,m_2}(\Omega)Y^*_{\ell_1,m_1}(\Omega)\right]\widehat{c}^+_{\ell_1 m_1}\widehat{c}_{\ell_2 m_2}$$

$$= (-1)^m\widehat{\rho}_{\mathrm{e}}(\ell,-m). \tag{111}$$

This relation will be used later on. In the expressions for the spherical components of the density, we have triple products of spherical harmonics to integrate. A double product would be easy to integrate, that's just the orthonormality relation

$$\int d\Omega\, Y_{j,m}(\Omega)Y^*_{j',m'}(\Omega) = \delta_{j,j'}\delta_{m,m'}. \tag{112}$$

Reducing the triple product to a double product relies on the addition of angular momenta. Two spherical harmonics can be combined into a sum of spherical harmonics (because two angular momenta operators can be combined into a sum of angular momenta operators). The addition rules for spherical harmonics (derived in Cohen-Tannoudji's book on Quantum Mechanics [23], p. 1046), is

$$Y_{j_1,m_1}(\Omega)Y_{j_2,m_2}(\Omega) = \sum_{j=|j_1-j_2|}^{j_1+j_2}\sum_{m=-j}^{j}\sqrt{\frac{(2j_1+1)(2j_2+1)}{4\pi(2j+1)}}\langle j_1,0;j_2,0|j,0\rangle \times \langle j_1,m_1;j_2,m_2|j,m\rangle Y_{j,m}(\Omega). \tag{113}$$

From this, it also follows that $Y^*_{j_1,m_1}(\Omega)Y_{j_2,m_2}(\Omega) = (-1)^{m_1}Y_{j_1,-m_1}(\Omega)Y_{j_2,m_2}(\Omega)$. In expression (113), $\langle j_1,m_1;j_2,m_2|j,m\rangle$ is the Clebsch-Gordan coefficient for combining the angular momentum states $|j_1,m_1\rangle$ and $|j_2,m_2\rangle$ into the angular momentum state $|j,m\rangle$, sometimes denoted as $C(j_1,m_1;j_2,m_2|j,m)$. The complex conjugate expression can be found by replacing all Y's by Y^* since the Clebsch-Gordan coefficients are real. Note that the addition rule of spherical harmonics is more restrictive than the addition rule of angular momenta: it contains the extra coefficient $\langle j_1,0;j_2,0|j,0\rangle$ which is different from zero only if j_1+j_2-j is even. The combination of angular momenta of magnitude j_1 and j_2 can lead to all values of angular momentum between $|j_1-j_2|$ and j_1+j_2, but the product of spherical harmonics of order j_1 and j_2 lead to spherical harmonics of order j_1+j_2, j_1+j_2-2, j_1+j_2-4, ..., $|j_1-j_2|$. This insures that the parity of all terms on the right hand side $(-1)^j$ equals the parity of all terms on the left hand side $(-1)^{j_1+j_2}$. The addition rule (113) allows to perform the integral of the triple product of Y's, combining the ℓ and ℓ_1 spherical harmonics.

$$\int d\Omega\, Y_{\ell,m}(\Omega)Y^*_{\ell_2,m_2}(\Omega)Y_{\ell_1,m_1}(\Omega) = \sum_{\ell'=|\ell-\ell_1|}^{\ell+\ell_1}\sum_{m'=-\ell'}^{\ell'}\sqrt{\frac{(2\ell+1)(2\ell_1+1)}{4\pi(2\ell'+1)}} \times \langle \ell,0;\ell_1,0|\ell',0\rangle\langle \ell,m;\ell_1,m_1|\ell',m'\rangle \times \int d\Omega\, Y_{\ell',m'}(\Omega)Y^*_{\ell_2,m_2}(\Omega). \tag{114}$$

The remaining integral over the spherical angle Ω is (112). The result for the spherical components of the density is found by substituting the triple-product integral (114) into expression(109)

$$\widehat{\rho}_e(\ell,m) = \sum_{\ell_1,m_1}\sum_{\ell_2,m_2}\widehat{c}^{+}_{\ell_2 m_2}\widehat{c}_{\ell_1 m_1}\sum_{\ell'=|\ell-\ell_1|}^{\ell+\ell_1}\sum_{m'=-\ell'}^{\ell'}\sqrt{\frac{(2\ell+1)(2\ell_1+1)}{4\pi(2\ell'+1)}} \times \langle \ell,0;\ell_1,0|\ell',0\rangle\langle \ell,m;\ell_1,m_1|\ell',m'\rangle \times \delta_{\ell',\ell_2}\delta_{m',m_2}. \tag{115}$$

The delta-functions allow to perform the summation over the spherical indices ℓ_2, m_2 :

$$\widehat{\rho}_e(\ell,m) = \sum_{\ell_1=0}^{\infty}\sum_{m_1=-\ell_1}^{\ell_1}\sum_{\ell'=|\ell-\ell_1|}^{\ell+\ell_1}\sum_{m'=-\ell'}^{\ell'}\sqrt{\frac{(2\ell+1)(2\ell_1+1)}{4\pi(2\ell'+1)}} \times \langle \ell,0;\ell_1,0|\ell',0\rangle\langle \ell,m;\ell_1,m_1|\ell',m'\rangle \widehat{c}^{+}_{\ell' m'}\widehat{c}_{\ell_1 m_1}. \tag{116}$$

To draw the analogy with the plane wave density operator, we introduce the following notation:

$$\widehat{c}^{+}_{(\ell_1,m_1)\otimes(\ell,m)} = \sum_{\ell'=|\ell-\ell_1|}^{\ell+\ell_1} \sum_{m'=-\ell'}^{\ell'} \sqrt{\frac{(2\ell+1)(2\ell_1+1)}{4\pi(2\ell'+1)}} \langle \ell,0;\ell_1,0|\ell',0\rangle \times \langle \ell,m;\ell_1,m_1|\ell',m'\rangle \, \widehat{c}^{+}_{\ell' m'}. \tag{117}$$

The operator $\widehat{c}^{+}_{(\ell_1,m_1)\otimes(\ell,m)}$ creates an electron in a state which results from the combination of a spherical (ℓ_1,m_1) state and a spherical (ℓ,m) state, as governed by the spherical harmonics addition formula (113). In this notation, we find for the spherical components of the density operator

$$\widehat{\rho}_{\mathrm{e}}(\ell,m) = \sum_{\ell'=0}^{\infty} \sum_{m'=-\ell'}^{\ell'} \widehat{c}^{+}_{(\ell',m')\otimes(\ell,m)} \widehat{c}_{\ell',m'}. \tag{118}$$

This is the spherical counterpart of the second quantisation expression for the plane wave components of the density, $\widehat{\rho}_{\mathrm{e}}(\mathbf{k}) = \sum_{\mathbf{q}} \widehat{c}^{+}_{\mathbf{q}+\mathbf{k}} \widehat{c}_{\mathbf{q}}$. Wave number addition has been replaced by angular momentum addition. Note that the operator $\widehat{c}^{+}_{(\ell_1,m_1)\otimes(\ell,m)}$ originated from

$$\widehat{c}^{+}_{(\ell_1,m_1)\otimes(\ell,m)} = \sum_{\ell_2=0}^{\infty} \sum_{m_2=-\ell_2}^{\ell_2} \int d\Omega \, Y_{\ell,m}(\Omega) Y^{*}_{\ell_2,m_2}(\Omega) Y_{\ell_1,m_1}(\Omega) \widehat{c}^{+}_{\ell_2 m_2}. \tag{119}$$

This representation will come in handy in the next subsection.

Two-particle operators

These are operators which can be expressed as a sum over pairs of particles, but not as a sum over single particles. Examples are the pair correlation function and the interaction part of the Hamiltonian. In this section, we'll focus on the Coulomb interaction between the electrons. The operator which represents the potential energy associated with the Coulomb interaction is

$$\widehat{\mathscr{H}}_{\text{coul}} = \frac{1}{2} \sum_{j=1}^{N} \sum_{j' \neq j=1}^{N} \frac{e^2}{4\pi\varepsilon} \frac{1}{|\widehat{\mathbf{r}}_j - \widehat{\mathbf{r}}_{j'}|}, \tag{120}$$

where $\mathbf{r}_j$ and $\mathbf{r}_{j'}$ are the position operators of electrons j and j' - the electrons are restricted to a spherical shell of radius R and negligible width. The Coulomb interaction is ideally suited to rewrite in spherical harmonics [24]:

$$\widehat{\mathscr{H}}_{\text{coul}} = \frac{1}{2} \sum_{j=1}^{N} \sum_{j' \neq j=1}^{N} \frac{e^2}{4\pi\varepsilon} \frac{1}{R} \sum_{\ell=0}^{\infty} \sum_{m=-\ell}^{\ell} \frac{4\pi}{2\ell+1} Y_{\ell,m}(\widehat{\Omega}_j) Y^{*}_{\ell,m}(\widehat{\Omega}_{j'}). \tag{121}$$

The advantage of the Coulomb interaction in the spherical representation is that the $\widehat{\Omega}_j$ and $\widehat{\Omega}_{j'}$ are factorized. The standard second quantised form of this operator is

$$\widehat{\mathscr{H}}_{\text{coul}} = \sum_{\ell_1,m_1}\sum_{\ell_2,m_2}\sum_{\ell_3,m_3}\sum_{\ell_4,m_4} \langle V_2\rangle \widehat{c}^{+}_{\ell_4,m_4}\widehat{c}^{+}_{\ell_3,m_3}\widehat{c}_{\ell_2,m_2}\widehat{c}_{\ell_1,m_1} \tag{122}$$

$$\begin{aligned}\text{with } \langle V_2\rangle &= \frac{e^2}{2\varepsilon}\frac{1}{R}\sum_{\ell=0}^{\infty}\sum_{m=-\ell}^{\ell}\int d\Omega\int d\Omega'\, Y^*_{\ell_4,m_4}(\Omega)Y^*_{\ell_3,m_3}(\Omega') \\ &\times\frac{Y_{\ell,m}(\Omega)Y^*_{\ell,m}(\Omega')}{2\ell+1}Y_{\ell_2,m_2}(\Omega')Y_{\ell_1,m_1}(\Omega).\end{aligned} \tag{123}$$

The double integration can be factorized:

$$\begin{aligned}\langle V_2\rangle &= \frac{e^2}{2\varepsilon}\frac{1}{R}\sum_{\ell=0}^{\infty}\sum_{m=-\ell}^{\ell}\frac{1}{2\ell+1}\left[\int d\Omega\, Y^*_{\ell_4,m_4}(\Omega)Y_{\ell,m}(\Omega)Y_{\ell_1,m_1}(\Omega)\right] \\ &\times\left[\int d\Omega'\, Y^*_{\ell_3,m_3}(\Omega')Y^*_{\ell,m}(\Omega')Y_{\ell_2,m_2}(\Omega')\right],\end{aligned} \tag{124}$$

so that the Coulomb part of the Hamiltonian can be written as

$$\begin{aligned}\widehat{\mathscr{H}}_{\text{coul}} &= \frac{e^2}{2\varepsilon R}\sum_{\ell_1,m_1}\sum_{\ell_2,m_2}\sum_{\ell=0}^{\infty}\sum_{m=-\ell}^{\ell}\frac{1}{2\ell+1} \\ &\times\sum_{\ell_4,m_4}\int d\Omega\, Y^*_{\ell_4,m_4}(\Omega)Y_{\ell,m}(\Omega)Y_{\ell_1,m_1}(\Omega)\widehat{c}^{+}_{\ell_4,m_4} \\ &\times\sum_{\ell_3,m_3}\int d\Omega'\, Y^*_{\ell_3,m_3}(\Omega')Y^*_{\ell,m}(\Omega')Y_{\ell_2,m_2}(\Omega')\widehat{c}^{+}_{\ell_3,m_3} \\ &\times\widehat{c}_{\ell_2,m_2}\widehat{c}_{\ell_1,m_1}.\end{aligned} \tag{125}$$

The second line is, according to expression (119):

$$\sum_{\ell_4,m_4}\int d\Omega\, Y^*_{\ell_4,m_4}(\Omega)Y_{\ell,m}(\Omega)Y_{\ell_1,m_1}(\Omega)\widehat{c}^{+}_{\ell_4,m_4} = \widehat{c}^{+}_{(\ell,m)\otimes(\ell_1,m_1)}. \tag{126}$$

The third line is

$$\sum_{\ell_3,m_3}\int d\Omega'\, Y^*_{\ell_3,m_3}(\Omega')Y^*_{\ell,m}(\Omega')Y_{\ell_2,m_2}(\Omega')\widehat{c}^{+}_{\ell_3,m_3} = (-1)^m\widehat{c}^{+}_{(\ell,-m)\otimes(\ell_2,m_2)}. \tag{127}$$

Once again, our notation for the creation operator of a combination of spherical states is simplifying the expression. We find for the Coulomb part of the Hamiltonian

$$\widehat{\mathscr{H}}_{\text{coul}} = \frac{e^2}{2\varepsilon R}\sum_{\ell_1,m_1}\sum_{\ell_2,m_2}\sum_{\ell,m}\frac{(-1)^m}{2\ell+1}\widehat{c}^{+}_{(\ell_1,m_1)\otimes(\ell,m)}\widehat{c}^{+}_{(\ell_2,m_2)\otimes(\ell,-m)}\widehat{c}_{\ell_2,m_2}\widehat{c}_{\ell_1,m_1}. \tag{128}$$

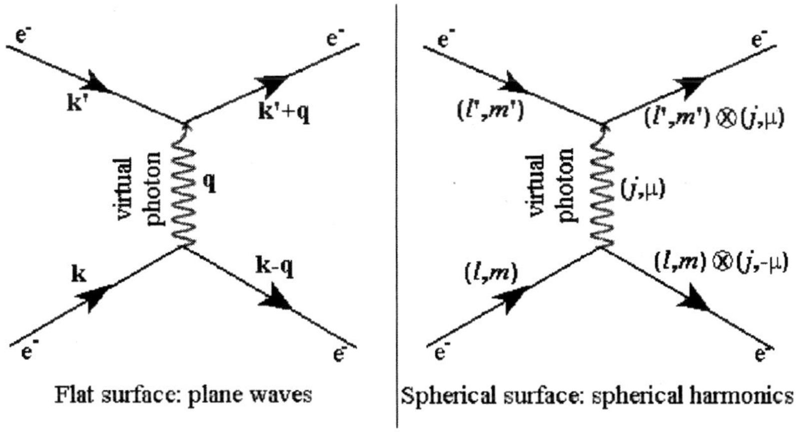

FIGURE 12. The physical process described by the Hamiltonian $\widehat{\mathscr{H}}$ is illustrated in these Feynman diagrams. On a sphere, the momenta have to be replaced by angular momenta and the addition of momenta becomes addition of angular momenta.

Also here, there is a nice analogy with the plane-wave expansion

$$\widehat{\mathscr{H}}_{\text{coul}}^{\text{plane wave}} = \sum_{\mathbf{k}_1}\sum_{\mathbf{k}_2}\sum_{\mathbf{q}} v(\mathbf{q})\widehat{c}^{+}_{\mathbf{k}_1+\mathbf{q}}\widehat{c}^{+}_{\mathbf{k}_2-\mathbf{q}}\widehat{c}_{\mathbf{k}_2}\widehat{c}_{\mathbf{k}_1}, \tag{129}$$

where $v(\mathbf{q})$ is the (ill-behaved) Fourier transform of the Coulomb interaction. The analogy is illustrated in figure 12.

The electrons interact through the exchange of a virtual photon, which for a flat surface transfers momentum $\mathbf{q}$ and on a sphere transfers angular momentum j. A useful form for the Coulomb interaction is its expression as a function of the spherical components of the density. To do this, some creation and annihilation operators should be swapped around first. The first swap is straightforward:

$$\widehat{c}_{\ell_2,m_2}\widehat{c}_{\ell_1,m_1} = -\widehat{c}_{\ell_1,m_1}\widehat{c}_{\ell_2,m_2}. \tag{130}$$

The second swap is more complicated

$$\begin{aligned}\widehat{c}^{+}_{(\ell_2,m_2)\otimes(\ell,-m)}\widehat{c}_{\ell_1,m_1} &= -\widehat{c}_{\ell_1,m_1}\widehat{c}^{+}_{(\ell_2,m_2)\otimes(\ell,-m)} \\ &\quad + \left[\widehat{c}^{+}_{(\ell_2,m_2)\otimes(\ell,-m)},\widehat{c}_{\ell_1,m_1}\right]_{+}.\end{aligned} \tag{131}$$

The anticommutator can be evaluated by expanding $\widehat{c}^{+}_{(\ell_2,m_2)\otimes(\ell,-m)}$ using expression (119):

$$\begin{aligned}\left[\widehat{c}^{+}_{(\ell_2,m_2)\otimes(\ell,-m)},\widehat{c}_{\ell_1,m_1}\right]_{+} &= \sum_{\ell_2=0}^{\infty}\sum_{m_2=-\ell_2}^{\ell_2}\int d\Omega\, Y_{\ell,m}(\Omega)Y^{*}_{\ell_2,m_2}(\Omega)Y_{\ell_1,m_1}(\Omega) \\ &\quad \times \left[\widehat{c}^{+}_{\ell_2 m_2},\widehat{c}_{\ell_1,m_1}\right]_{+}.\end{aligned} \tag{132}$$

The fermionic anticommutation relation leads to

$$\left[\widehat{c}^{+}_{(\ell_2,m_2)\otimes(\ell,-m)}, \widehat{c}_{\ell_1,m_1}\right]_+ = \int d\Omega\, Y_{\ell,-m}(\Omega) Y^*_{\ell_1,m_1}(\Omega) Y_{\ell_1,m_1}(\Omega) \times \delta_{\ell_2,\ell_1}\delta_{m_2,m_1}. \quad (133)$$

Once more, the addition rule for spherical harmonics (113) is used, leading to

$$\begin{aligned}\left[\widehat{c}^{+}_{(\ell_2,m_2)\otimes(\ell,-m)}, \widehat{c}_{\ell_1,m_1}\right]_+ &= \sqrt{\frac{(2\ell+1)(2\ell_1+1)}{4\pi(2\ell_1+1)}}\langle \ell,0;\ell_1,0|\ell_1,0\rangle \\ &\quad \times \langle \ell,0;\ell_1,m_1|\ell_1,m_1\rangle \times \delta_{\ell_2,\ell_1}\delta_{m_2,m_1}. \end{aligned} \quad (134)$$

The result will only differ from zero if $m_1 - m = m_1 \Leftrightarrow m = 0$, and if ℓ is an even number between 0 and $2\ell_1$. Based on these swappings of creation and annihilation operators, the Coulomb part of the Hamiltonian can be rewritten as follows:

$$\begin{aligned}\widehat{\mathscr{H}}_{\text{coul}} &= \frac{e^2}{2\varepsilon R}\sum_{\ell_1,m_1}\sum_{\ell_2,m_2}\sum_{\ell,m}\frac{(-1)^m}{2\ell+1}\widehat{c}^{+}_{(\ell_1,m_1)\otimes(\ell,m)}\widehat{c}_{\ell_1,m_1}\widehat{c}^{+}_{(\ell_2,m_2)\otimes(\ell,-m)}\widehat{c}_{\ell_2,m_2} \\ &\quad -\frac{e^2}{2\varepsilon R}\sum_{\ell_1,m_1}\sum_{\ell=0}^{2\ell_1}A(\ell,\ell_1,m_1)\widehat{c}^{+}_{(\ell_1,m_1)\otimes(\ell,0)}\widehat{c}_{\ell_1,m_1}, \end{aligned} \quad (135)$$

with

$$A(\ell,\ell_1,m_1) = \sqrt{\frac{1}{4\pi(2\ell+1)}}\langle \ell,0;\ell_1,0|\ell_1,0\rangle\langle \ell,0;\ell_1,m_1|\ell_1,m_1\rangle. \quad (136)$$

The four-operator product in the Coulomb part of the Hamiltonian can be simplified using the spherical components of the density operator (118):

$$\begin{aligned}\widehat{\mathscr{H}}_{\text{coul}} &= \frac{e^2}{2\varepsilon R}\sum_{\ell=0}^{\infty}\sum_{m=-\ell}^{\ell}\frac{1}{2\ell+1}\widehat{\rho}_{\text{e}}(\ell,m)\widehat{\rho}^{+}_{\text{e}}(\ell,m) \\ &\quad -\frac{e^2}{2\varepsilon R}\sum_{\ell_1,m_1}\sum_{\ell=0}^{2\ell_1}A(\ell,\ell_1,m_1)\widehat{c}^{+}_{(\ell_1,m_1)\otimes(\ell,0)}\widehat{c}_{\ell_1,m_1}. \end{aligned} \quad (137)$$

The last term is related to the self-interaction of an electron in state (ℓ_1,m_1). The first line of (137) includes this self-interaction (i.e. the interaction of the electron j with itself), and the second line subtracts it again from the first. This annoying self-interaction term also appears in the plane wave case, where it is proportional to $v(0)\widehat{c}^{+}_{\mathbf{k}}\widehat{c}_{\mathbf{k}}$ with $v(0)$ the Fourier transform of the Coulomb interaction, taken in $k=0$. This is a divergent term in the plane wave case. But for the spherical case, $A(\ell,\ell_1,m_1)$ remains finite everywhere, and the self interaction (which gives rise to this term) will hopefully turn out to be less problematic in the spherical case because of this property. To lowest order in the interaction, this term adds a constant ('offset') term to the energy which we denote by $-e^2\mathscr{A}/(2\varepsilon R)$ where

$$\mathscr{A} = \sum_{\ell_1=0}^{L_{\text{F}}}\sum_{m_1=-\ell_1}^{\ell_1}\sum_{\ell=0}^{2\ell_1}\left[\langle \ell,0;\ell_1,0|\ell_1,0\rangle\langle \ell,0;\ell_1,m_1|\ell_1,m_1\rangle\right]^2. \quad (138)$$

(which is proportional to L_F^2). We will investigate

$$\widehat{\mathscr{H}}_{\text{coul}} = \frac{e^2}{2\varepsilon R}\sum_{\ell=0}^{\infty}\sum_{m=-\ell}^{\ell}\frac{1}{2\ell+1}\left[\widehat{\rho}_{\text{e}}(\ell,m)\widehat{\rho}_{\text{e}}^{+}(\ell,m)-\mathscr{A}\delta(\ell=0)\right]. \tag{139}$$

This is to be compared with the plane-wave result

$$\mathscr{H}_{\text{coul}}^{\text{plane wave}} = \sum_{\mathbf{k}_1}\sum_{\mathbf{k}_2}\sum_{\mathbf{q}}v(\mathbf{q})\left[\widehat{\rho}_{\mathbf{q}}\widehat{\rho}_{\mathbf{q}}^{+}-N\delta(\mathbf{q}=0)\right]. \tag{140}$$

The usefulness of rewriting the interaction Hamiltonian as a function of the density will become apparent when we treat the response.

The final result of this section is the second quantised version of the full Hamiltonian of the S2DEG :

$$\begin{aligned}\widehat{\mathscr{H}} &= \sum_{\ell=0}^{\infty}\sum_{m=-\ell}^{\ell}\frac{\hbar^2\ell(\ell+1)}{2m_eR^2}\widehat{c}_{\ell m}^{+}\widehat{c}_{\ell m}\\ &+\frac{e^2}{2\varepsilon R}\sum_{\ell=0}^{\infty}\sum_{m=-\ell}^{\ell}\frac{1}{2\ell+1}\left[\widehat{\rho}_{\text{e}}(\ell,m)\widehat{\rho}_{\text{e}}^{+}(\ell,m)-\mathscr{A}\delta(\ell=0)\right].\end{aligned} \tag{141}$$

Response properties of the S2DEG

An essential quantity from which properties of an electron gas can be deduced is its dielectric function. It allows to calculate both to the static thermodynamics of the electron gas and its response properties. In this section, we will try to evaluate the dielectric function of the S2DEG in the RPA approximation. This approximation is good for high electron density but does not describe Wigner crystallisation.

Let's start by looking at the response of the S2DEG to an external charge, for example an ion, brought into the gas. This probe brings an external charge density in the S2DEG. On the spherical surface, it can be decomposed as

$$\rho_{\text{ext}}(\Omega,t) = \int_{-\infty}^{\infty} d\omega\sum_{\ell,m}\rho_{\text{ext}}(\ell,m;\omega)Y_{\ell m}^{*}(\Omega)e^{-i\omega t}. \tag{142}$$

The Coulomb interaction between the external charge and the S2DEG can be calculated analogously to our previous analysis:

$$\widehat{\mathscr{H}}_{\text{perturb}}(t) = \sum_{\ell=0}^{\infty}\sum_{m=-\ell}^{\ell}V_\ell\,\widehat{\rho}_e(\ell,m;t)\rho_{\text{ext}}^{*}(\ell,m;t)+\text{c.c.} \tag{143}$$

$$\text{with } V_\ell = \frac{e^2}{2\varepsilon R}\frac{1}{2\ell+1}. \tag{144}$$

We need not care about the self-interaction term, since this perturbation Hamiltonian does not include term where a charge can interact with itself. The perturbation Hamiltonian is hermitean by virtue of the complex conjugate term. In linear response theory (valid for a weak perturbation), each spherical component of this problem can be treated separately. To get the general result we can suffice by studying the effect of

$$\widehat{\mathscr{H}}^{\ell m,\omega}_{\text{perturb}}(t) = V_\ell \rho^*_{\text{ext}}(\ell,m;\omega)e^{i\omega t}e^{\delta t} \times \widehat{\rho}_{\text{e}}(\ell,m;t) + c.c. \tag{145}$$

An factor $e^{\delta|t|}$ has been added to switch on the perturbation adiabatically. Note that this form of the perturbation Hamiltonian is generic for perturbations which couple to the density of the electrons in the S2DEG.

The main question of response theory is : how will the electron density $\widehat{\rho}_e$ of evolve in time due to the perturbation ? To find the answer, Kubo's formalism can be used. However, in the spherical case, this poses a problem since the commutator relations for the density operator are complicated (because addition of two angular momenta leads to a set of possible angular momenta, whereas addition of two momenta leads to one total momentum). To avoid this, standard time-dependent perturbation theory can be used. The many-body wave function of the S2DEG can be written as a linear superposition of many-body eigenkets $|\Psi_n\rangle$ of the unperturbed Hamiltonian $\widehat{\mathscr{H}}$ (141):

$$\Psi(t) = \sum_n |\Psi_n\rangle e^{-iE_n t/\hbar} a_n(t). \tag{146}$$

In this expression, E_n denotes the energy of the many-body eigenket $|\Psi_n\rangle$. The corresponding eigenfunction is $\langle \mathbf{r}_1, \mathbf{r}_2, \mathbf{r}_3, .., \mathbf{r}_N | \Psi_n\rangle = \Psi_n(\mathbf{r}_1, \mathbf{r}_2, \mathbf{r}_3, .., \mathbf{r}_N)$. Let $|\Psi_0\rangle$ correspond to the ground state, and E_0 to the ground state energy. The initial condition on the coefficients a_n is that $a_n(t \to -\infty) = 1$ for $n = 0$ and $a_n(t \to -\infty) = 0$ for $n > 0$. This affirms that without the perturbation of the external charge, the S2DEG is in its ground state. To first order in the perturbation, the coefficients $a_n(t)$ are given by

$$a_0(t) = 1 - \frac{i}{\hbar}\int_{-\infty}^{t} \left\langle \Psi_0 \left| \widehat{\mathscr{H}}^{\ell m,\omega}_{\text{perturb}}(t') \right| \Psi_0 \right\rangle dt' \tag{147}$$

$$a_{n\neq 0}(t) = -\frac{i}{\hbar}\int_{-\infty}^{t} \left\langle \Psi_n \left| \widehat{\mathscr{H}}^{\ell m,\omega}_{\text{perturb}}(t') \right| \Psi_0 \right\rangle dt'. \tag{148}$$

The time dependence of the expectation value of the spherical components of the density is given by

$$\begin{aligned}\rho_{\text{e}}(\ell,m;t) &= \langle \Psi(t) | \widehat{\rho}_{\text{e}}(\ell,m) | \Psi(t)\rangle &(149)\\ &= \sum_{n,m} \langle \Psi_n | \widehat{\rho}_e(\ell,m) | \Psi_m\rangle e^{-i(E_m - E_n)t/\hbar} a_m(t) a_n^*(t). &(150)\end{aligned}$$

The shorthand notation $(E_n - E_0)/\hbar = \omega_{n0}$ will be used. To proceed further, the coefficients $a_n(t)$ must be evaluated, and substituted in (150), resulting in

$$\rho_{\text{e}}(\ell,m;t) = \rho_{\text{e}}(\ell,m;-\infty) - \frac{V_\ell \rho_{\text{ext}}(\ell,m;\omega)}{\hbar} e^{-i\omega t + \delta t}$$

$$\times \sum_n \left[\frac{|\langle \Psi_n | \widehat{\rho}_{\rm e}(\ell,m) | \Psi_0 \rangle|^2}{\omega_{n0} - \omega - i\delta} + \frac{|\langle \Psi_0 | \widehat{\rho}_{\rm e}^{+}(\ell,m) | \Psi_n \rangle|^2}{\omega_{n0} + \omega + i\delta} \right]. \quad (151)$$

The term $\rho_{\rm e}(\ell,m;-\infty)$ is the unperturbed charge density. The remaining terms represent the charge density induced by the presence of the external charge. This induced charge density screens the external field. Let's denote this **induced charge density** $\rho_{ind}(\ell,m;t) = \rho_{\rm e}(\ell,m;t) - \rho_{\rm e}(\ell,m;-\infty)$. It is given by

$$\rho_{\rm ind}(\ell,m;t) = \frac{V_\ell}{\hbar} \rho_{\rm ext}(\ell,m;\omega) e^{-i\omega t + \delta t} \, \mathscr{D}_{\rm R}(\ell,m;\omega), \quad (152)$$

with

$$\mathscr{D}_{\rm R}(\ell,m;\omega) = \sum_n \left[\frac{|\langle \Psi_n | \hat{\rho}_{\rm in}(\ell,m) | \Psi_0 \rangle|^2}{\omega - \omega_{n0} + i\delta} - \frac{|\langle \Psi_0 | \hat{\rho}_{\rm in}(\ell,m) | \Psi_n \rangle|^2}{\omega_{n0} + \omega + i\delta} \right]. \quad (153)$$

The reason for introducing the notation $\mathscr{D}_{\rm R}$ is that we will associate $\mathscr{D}_{\rm R}$ with the density-density correlation function, to obtain the same kind of formalism as for a flat surface.

Now that the response of the S2DEG charge density to an external probe has been analysed, we can introduce the dielectric function. The divergence of the electric field **E** caused by the perturbation is :

$$\mathrm{div}(\mathbf{E}) = e\rho_{\rm ext}(\ell,m;\omega) - e\rho_{\rm ind}(\ell,m;\omega). \quad (154)$$

This can be split up into a polarisation field, caused by the induced charge density, and the field **D** created by the external charge density only. The field **D** is the electric veld of the external charge when it is in the electron gas; the effect of this electron gas (like the effect of a dielectric medium) can be described by a factor ε, the dielectric function:

$$\mathrm{div}(\mathbf{D}) = \mathrm{div}\left[\varepsilon(\ell,m;\omega)\mathbf{E}\right] = e\rho_{\rm ext}(\ell,m;\omega). \quad (155)$$

From this we deduce the familiar definition of the dielectric function

$$\varepsilon(\ell,m;\omega) = \frac{\rho_{\rm ext}(\ell,m;\omega)}{\rho_{\rm ext}(\ell,m;\omega) - \rho_{\rm ind}(\ell,m;\omega)}. \quad (156)$$

An expression for the dielectric function can be derived from (152):

$$\frac{1}{\varepsilon(\ell,m;\omega)} = 1 - \frac{\rho_{\rm ind}(\ell,m;\omega)}{\rho_{\rm ext}(\ell,m;\omega)} \quad (157)$$

$$= 1 + \frac{V_\ell}{\hbar} \mathscr{D}_{\rm R}(\ell,m;\omega), \quad (158)$$

so that

$$\frac{1}{\varepsilon(\ell,m;\omega)} = 1 + \frac{e^2}{2\varepsilon_{\rm vac}\hbar R} \frac{1}{2\ell+1} \mathscr{D}_{\rm R}(\ell,m;\omega), \quad (159)$$

with $\varepsilon_{\rm vac}$ the permittivity of vacuum, and $\mathscr{D}_{\rm R}$ given by (153):

First, we show that $\mathscr{D}_{\mathrm{R}}$ is the retarded density-density Green's function. The density-density Green's function is

$$\mathscr{D}(\ell,m;t)=-i\left\langle\Psi_0\left|\mathscr{T}\left\{\widehat{\rho}_{\mathrm{ind}}(\ell,m;t)\widehat{\rho}_{\mathrm{ind}}^{+}(\ell,m;0)\right\}\right|\Psi_0\right\rangle. \tag{160}$$

This is the amplitude that a density fluctuation, created at time $t=0$, will be present at a later time t. $\mathscr{T}$ is the time ordering operator. The retarded Green's function is

$$\mathscr{D}_{\mathrm{R}}(\ell,m;t)=\begin{cases}-i\left\langle\Psi_0\left|\left[\widehat{\rho}_{\mathrm{ind}}(\ell,m;t),\widehat{\rho}_{\mathrm{ind}}^{+}(\ell,m;0)\right]\right|\Psi_0\right\rangle & t>0\\ 0 & t<0\end{cases} \tag{161}$$

Its time Fourier transform is

$$\mathscr{D}_{\mathrm{R}}(\ell,m;\omega) = \int dt\,\mathscr{D}_{\mathrm{R}}(\ell,m;t)e^{i\omega t} \tag{162}$$

$$\Leftrightarrow\quad \mathscr{D}_{\mathrm{R}}(\ell,m;t)=\int\frac{d\omega}{2\pi}\mathscr{D}_{\mathrm{R}}(\ell,m;\omega)e^{-i\omega t} \tag{163}$$

The Lehman spectral representation of $\mathscr{D}_{\mathrm{R}}(\ell,m;t)$ is found by introducing the decomposition of the unitary operator, $\mathbf{1}=\sum_n|\Psi_n\rangle\langle\Psi_n|$, between the density operators. This leads to

$$\begin{aligned}&-i\left\langle\Psi_0\left|\widehat{\rho}_{\mathrm{ind}}(\ell,m;t)\widehat{\rho}_{\mathrm{ind}}^{+}(\ell,m;0)-\widehat{\rho}_{\mathrm{ind}}^{+}(\ell,m;0)\widehat{\rho}_{\mathrm{ind}}(\ell,m;t)\right|\Psi_0\right\rangle\\ =\;&\sum_n -ie^{iE_0t/\hbar}\left\langle\Psi_0\left|\widehat{\rho}_{\mathrm{ind}}(\ell,m;0)e^{-i\widehat{\mathscr{H}}t/\hbar}\right|\Psi_n\right\rangle\left\langle\Psi_n\left|\widehat{\rho}_{\mathrm{ind}}^{+}(\ell,m;0)\right|\Psi_0\right\rangle\\ &+\sum_n ie^{-iE_0t/\hbar}\left\langle\Psi_0\left|\widehat{\rho}_{\mathrm{ind}}^{+}(\ell,m;0)e^{i\widehat{\mathscr{H}}t/\hbar}\right|\Psi_n\right\rangle\left\langle\Psi_n\left|\widehat{\rho}_{\mathrm{ind}}(\ell,m;0)\right|\Psi_0\right\rangle \qquad (164)\\ =\;&-i\sum_n\left|\left\langle\Psi_0\left|\widehat{\rho}_{\mathrm{ind}}(\ell,m;0)\right|\Psi_n\right\rangle\right|^2\left(e^{-i\omega_{n0}t}-e^{i\omega_{n0}t}\right)\ \text{for}\ t>0, \qquad (165)\end{aligned}$$

and 0 elsewhere. Its time Fourier transform, including an $e^{-\delta t}$ convergence factor, is

$$\mathscr{D}_{\mathrm{R}}(\ell,m;\omega)=\sum_n\left|\left\langle\Psi_0\left|\widehat{\rho}_{\mathrm{ind}}(\ell,m;0)\right|\Psi_n\right\rangle\right|^2\left(\frac{1}{\omega-\omega_{n0}+i\delta}-\frac{1}{\omega+\omega_{n0}+i\delta}\right). \tag{166}$$

This is exactly formula (153) obtained in our study of the response of the density to an external probe charge.

The structure factor of the S2DEG is defined as

$$S(\ell,m;\omega)=\sum_n\left|\left\langle\Psi_0\left|\widehat{\rho}_{\mathrm{in}}(\ell,m;0)\right|\Psi_n\right\rangle\right|^2\delta(\omega-\omega_{n0}), \tag{167}$$

so that

$$\mathscr{D}_{\mathrm{R}}(\ell,m;\omega)=\int d\omega'\,S(\ell,m;\omega')\left(\frac{1}{\omega-\omega'+i\delta}-\frac{1}{\omega+\omega'+i\delta}\right). \tag{168}$$

Substituting this result (168) into the expression (152) for the induced charge density yields

$$\rho_{\text{in}}(\ell,m;t) = \frac{V_\ell}{\hbar}\rho_{\text{ext}}(\ell,m;\omega)e^{-i\omega t}\int d\omega' \left(\frac{S(\ell,m;\omega')}{\omega-\omega'+i\delta} - \frac{S(\ell,m;\omega')}{\omega+\omega'+i\delta}\right) \tag{169}$$

Using Plemelj's identity $1/(x \pm i\varepsilon) = \mathscr{P}(1/x) \mp i\pi\delta(x)$ we find the spherical electron gas version of the important **relation between the dielectric function and the structure factor**:

$$\text{Im}\left[\frac{\rho_{\text{in}}(\ell,m;t)}{\rho_{\text{ext}}(\ell,m;\omega)e^{-i\omega t}}\right] = -\frac{V_\ell}{\hbar}\pi S(\ell,m;\omega) \tag{170}$$

$$\Rightarrow \text{Im}\left[\frac{1}{\varepsilon(\mathbf{k},\omega)}\right] = -\frac{\pi e^2}{2\varepsilon_{\text{vac}}\hbar R}\frac{S(\ell,m;\omega)}{2\ell+1} \tag{171}$$

A 'natural' experiment to determine the dynamic structure factor directly is an inelastic scattering experiment. In particular the differential cross section for a probe particle of mass M (such as an incident ion) and initial momentum $\mathbf{p_i}$ to be scattered to a final state $\mathbf{p_f}$ with energy transfer between $\hbar\omega$ and $\hbar(\omega+d\omega)$ to the scattering system is [26]:

$$\frac{d\sigma}{d\theta d\omega} = \frac{M^2}{8\pi^3}\frac{p_f}{p_i}|V_k|^2 S(k,\omega). \tag{172}$$

In this expression, θ is the scattering angle between $\mathbf{p_f}$ and $\mathbf{p_i}$, $|V_\mathbf{k}|$ is the Fourier transform of the interaction potential between the probe particle, and the energy transfer ω is related to the momentum transfer $\mathbf{k}$ by $\hbar\omega = h^2\mathbf{k}.\mathbf{p}_i/m + (\hbar k)^2/(2m)$. The relation between the Fourier decomposition and the spherical decomposition of the density is

$$\begin{aligned}\widehat{\rho}_{\text{e}}(\mathbf{k}) &= \sum_{j=1}^{N}\left\langle e^{i\mathbf{k}.\widehat{\mathbf{r}}_j}\right\rangle = \sum_{j=1}^{N}\sum_{l}\sqrt{4\pi(2l+1)}i^l j_l(kR)\left\langle Y_{l,0}(\widehat{\theta}_j)\right\rangle \\ &= \sum_{l}\sqrt{4\pi(2l+1)}i^l j_l(kR)\widehat{\rho}_{\text{e}}(l,0)\end{aligned} \tag{173}$$

where the z-axis has been taken along the direction of the wave vector $\mathbf{k}$, and $j_l(x)$ is the spherical Bessel function of the first kind, and θ_j is the angle between $\mathbf{k}$ and the position operator of electron j, $\mathbf{r}_j$. The dynamic structure factors are related by

$$S(k,\omega) = \sum_{l}\left|\sqrt{4\pi(2l+1)}j_l(kR)\right|^2 S(l,0;\omega) \tag{174}$$

which can be derived by writing the dynamic structure factor as a density-density correlation function in the Lehmann representation and substituting expression (173) for $\widehat{\rho}_{\text{e}}(\mathbf{k})$.

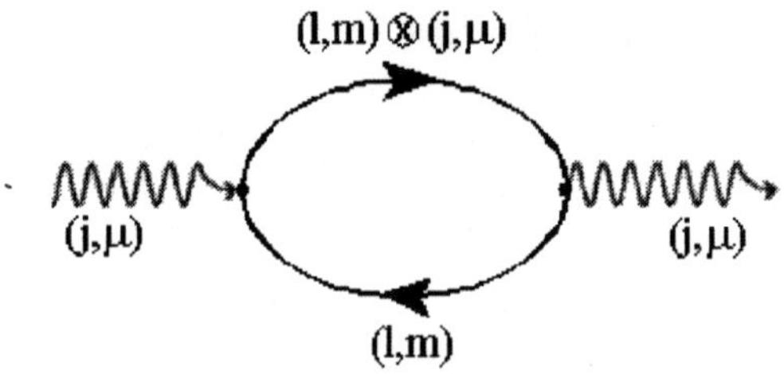

FIGURE 13. The Feynman diagram from which the retarded density-density Greens function, expression (175) is calculated.

The Random-Phase Approximation

The zero-eth order density-density Green's function $\mathscr{D}_{\mathrm{R},0}$ can be calculated using standard many-body techniques. The main diagram to evaluate is shown in Fig. 13 . The calculation proceeds similar to that of the polarization bubble of a flat two-dimensional electron gas, whilst keeping in mind that momenta are now angular momenta and addition of momenta should become addition of angular momenta (leading to the Clebsch-Gordan coefficients in expression (175)). The result is:

$$\begin{aligned}\mathscr{D}_{\mathrm{R},0}(\ell,m;\omega) &= \sum_{\ell' m'} \sum_{L=|\ell-\ell'|}^{\ell+\ell'} \frac{(2\ell+1)(2\ell'+1)}{4\pi(2L+1)} \left|\langle \ell,0;\ell',0|L,0\rangle\right|^2 \\ &\times \left|\langle \ell,m;\ell',m'|L,m+m'\rangle\right|^2 \frac{n(L,m+m')-n(\ell',m')}{\omega-(E_{L,m+m'}-E_{\ell',m'})+i\eta} \end{aligned} \quad (175)$$

where η is an infinitesimal real number and $\langle \ell,0;\ell_1,0|\ell_2,0\rangle, \langle \ell,0;\ell_1,m_1|\ell_2,m_2\rangle$ are Clebsch-Gordan coefficients.

The energy $E_{\ell,m}$ associated with the one-particle state (ℓ,m) was derived in the beginning of this section :

$$E_{\ell m} = \frac{\hbar^2 \ell(\ell+1)}{2m_e R^2}. \quad (176)$$

Finally, $n(\ell,m)$ is the occupation of the state (ℓ,m), given by

$$n(\ell,m) = \frac{1}{1+\exp[(E_{\ell m}-\mu)/(k_B T)]}, \quad (177)$$

where μ is the chemical potential, fixed by the condition $\sum_{\ell,m} n(\ell,m) = N$ where N is the number of electrons in the bubble. This allows for a description at any temperature, but we will here only investigate the zero-temperature limit, so that $n(\ell,m) = 1$ for $\ell \leq L_{\mathrm{F}}$ (where L_{F}, the Fermi level number, is the smallest integer larger than $\sqrt{N/2}-1$), and $n(\ell,m) = 0$ for $\ell > L_{\mathrm{F}}$. Note that this allows to treat the filled-shell problem only, meaning that the $(2L_{\mathrm{F}}+1)$-degenerate Fermi level is fully occupied.

The spherical decomposition of the coulomb interaction can be denoted by

$$U_0(\ell) = \frac{e^2}{2\varepsilon_{\text{vac}}\hbar R}\frac{1}{2\ell+1}. \tag{178}$$

Using the zero-eth order result $\mathscr{D}_{\text{R},0}$ in the dielectric function (159) would give

$$\varepsilon_{\text{Hartree-Fock}}(\ell,m;\omega) = \left[1+U_0(\ell)\mathscr{D}_{\text{R},0}(\ell,m;\omega)\right]^{-1}. \tag{179}$$

This result is known as the Hartree-Fock approximation to the dielectric function.

The Random-Phase Approximation (RPA) is an approximation to the sum over all Feynman diagrams needed to calculate the density-density Green's function $\mathscr{D}_{\text{R}}$. The approximation consists in only summing a subset of those diagrams, which can be summed analytically. This subset leads to

$$\begin{aligned} \mathscr{D}_{\text{R}} &\approx \mathscr{D}_{\text{R},0} + \mathscr{D}_{\text{R},0}U_0\mathscr{D}_{\text{R},0} + \mathscr{D}_{\text{R},0}U_0\mathscr{D}_{\text{R},0}U_0\mathscr{D}_{\text{R},0} + \ldots && (180)\\ &= \mathscr{D}_{\text{R},0}\left[1+U_0\mathscr{D}_{\text{R},0} + \left(U_0\mathscr{D}_{\text{R},0}\right)^2 + \ldots\right] && (181)\\ &= \frac{\mathscr{D}_{\text{R},0}}{1-U_0\mathscr{D}_{\text{R},0}}. && (182) \end{aligned}$$

Using this, we find for the dielectric function

$$\begin{aligned} \frac{1}{\varepsilon_{\text{RPA}}(\ell,m;\omega)} &= 1+U_0\mathscr{D}_{\text{R}} = 1+\frac{U_0\mathscr{D}_{\text{R},0}}{1-U_0\mathscr{D}_{\text{R},0}} = \frac{1}{1-U_0\mathscr{D}_{\text{R},0}} && (183)\\ &\Rightarrow \varepsilon_{\text{RPA}}(\ell,m;\omega) = 1-U_0(\ell)\mathscr{D}_{\text{R},0}(\ell,m;\omega), && (184) \end{aligned}$$

which leads to

$$\varepsilon_{\text{RPA}}(\ell,m;\omega) = 1-\frac{e^2}{2\varepsilon_{\text{vac}}\hbar R}\frac{1}{2\ell+1}\mathscr{D}_{\text{R},0}(\ell,m;\omega). \tag{185}$$

Results and discussion

The interpretation of this result is found by considering Plemelj's rule:

$$\begin{aligned} \text{Im}\frac{1}{\omega-(E_{L,m+m'}-E_{\ell',m'})+i\eta} &= -i\pi\delta[\omega-(E_{L,m+m'}-E_{\ell',m'})] && (186)\\ \text{Re}\frac{1}{\omega-(E_{L,m+m'}-E_{\ell',m'})+i\eta} &= \mathscr{P}\left[\frac{1}{\omega-(E_{L,m+m'}-E_{\ell',m'})}\right] && (187) \end{aligned}$$

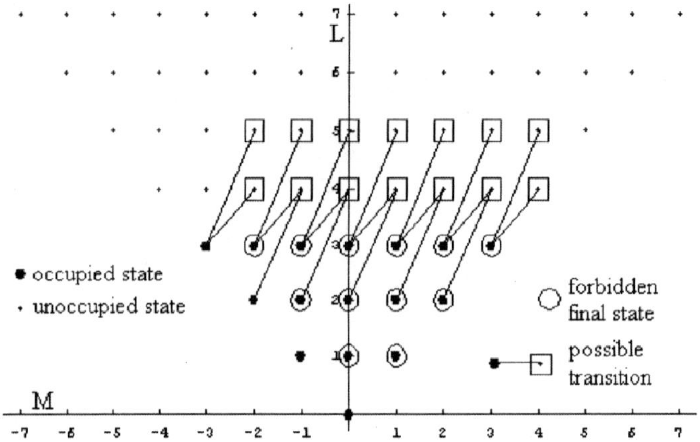

FIGURE 14. The Fermi sea of the S2DEG and single-particle excitations with $(\ell, m) = (2, 1)$.

where $\mathscr{P}$ means the prime value integral. The imaginary part becomes:

$$\mathrm{Im}\varepsilon_{\mathrm{RPA}}(\ell, m; \omega) = \pi U_0(\ell) \sum_{\ell' m'} \sum_{L=|\ell-\ell'|}^{\ell+\ell'} \frac{(2\ell+1)(2\ell'+1)}{4\pi(2L+1)} n(\ell' m')[1 - n(L, m+m')]$$
$$\times |\langle \ell, 0; \ell', 0 | L, 0 \rangle|^2 |\langle \ell, m; \ell', m' | L, m+m' \rangle|^2 \delta[\omega - (E_{L,m+m'} - E_{\ell',m'})] \tag{188}$$

From (171), we know that the structure factor is given by

$$S(\ell, m; \omega) = -\frac{1}{\pi U_0(\ell)} \frac{\mathrm{Im}[\varepsilon]}{(\mathrm{Re}[\varepsilon])^2 + (\mathrm{Im}[\varepsilon])^2}. \tag{189}$$

The structure factor $S(\ell, m; \omega)$ can be interpreted as a probability function. It is the probability that an excitation can be created in the interacting electron system so that this excitation has angular momentum (ℓ, m) and energy $\hbar\omega$ [25]. We find that $S(\ell, m; \omega)$ is zero, unless adding the angular momentum (ℓ, m) can excite a single electron from an occupied state $n(\ell' m')$ into an unoccupied state $1 - n(L, m+m')$. These are the single-particle excitations (called the "Landau Continuum" in the case of the flat electron gas). Figures 14 and 15 illustrate the concept of single particle excitations.

- **Case 1**: $\ell < L_F$. For figure 2, we investigate the case of angular momentum (2,1) which is imparted on the system. The filled disks represent the occupied states. In this example the Fermi-sea is filled up till $L_F = 3$. The open circles and squares show the collection of states which can result from adding the angular momentum (2,1) to any of the occupied states. Not all resulting states can be obtained, since to excite an electron it has to go into a final state which is unoccupied (the open squares). Final states which are already occupied (open circles) are excluded by the Fermi statistics. This means in practice that the electrons which can participate in the absorption of angular momentum (ℓ, m) are the ones close to the Fermi level

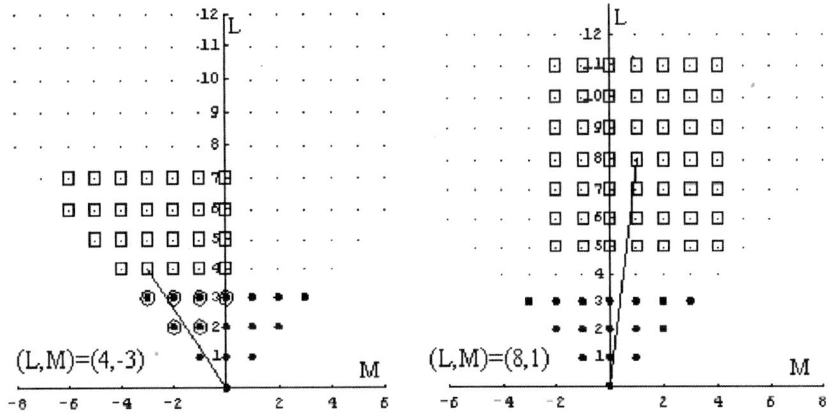

FIGURE 15. Single-particle excitations and the Fermi-sea of a S2DEG

(in fact, those from level $L_F - \ell$ up to L_F). In the picture, these states have a line starting from them.

- **Case 2:** $2L_F > \ell > L_F$. This is shown in the left graph of figure 3 where single-particle excitations of angular moment (4,-3) are investigated. Now all the electrons in the Fermi sea can participate, but not all resulting final states are unoccupied
- **Case 3:** $\ell > 2L_F$. An example is shown in the right graph of figure 3, for single-particle excitations of angular momentum (8,-1). Now not only can all electrons participate in the scattering, but also all possible final states are unoccupied and thus accessible.

The maximum energy difference between two states (connected by a line in figure 14) is

$$\begin{aligned}\omega_{\max} &= \frac{\hbar^2}{2m_eR^2}[(L_F+\ell)(L_F+\ell+1)-(L_F)(L_F+1)] \\ &= \frac{\hbar^2}{2m_eR^2}\left[\ell^2+\ell(2L_F+1)\right].\end{aligned} \tag{190}$$

The smallest energy difference is, in cases 1 and 2:

$$\begin{aligned}\omega_{\min} &= \frac{\hbar^2}{2m_eR^2}[(L_F+1)(L_F+2)-(L_F)(L_F+1)] \\ &= \frac{\hbar^2(L_F+1)}{m_eR^2}.\end{aligned} \tag{191}$$

The smallest energy difference in case 3 ($\ell > 2L_F$) is:

$$\omega_{\min} = \frac{\hbar^2}{2m_eR^2}[(\ell-L_F)(\ell-L_F+1)-(L_F)(L_F+1)]$$

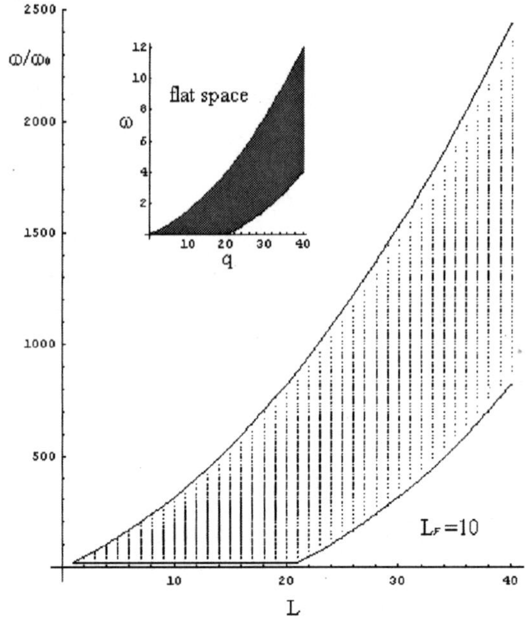

FIGURE 16. Structure factor for single particle excitations; $\omega_0 = \hbar/(m_e R^2)$.

$$= \frac{\hbar^2}{2m_e R^2}\left[\ell^2 - \ell(2L_F - 1) - 2L_F\right]. \tag{192}$$

In Figure 16, the possible ℓ, ω combinations are shown for the single-particle excitations, in the case $L_F = 10$ (in the inset, the similar result for the flat electron gas is shown).The minimum and maximum frequencies ω_{min} and ω_{max} mark off an area in the (ℓ, ω) plane in which single-particle excitations can be created. In the non-curved electron-gas these frequencies are $\omega_{\text{min}} = \max(q^2/2 - q, 0)$ and $\omega_{\text{max}} = q^2/2 + q$ as shown in the inset. For any q, ω in the dark area in the inset, a single-particle excitation can be created with this momentum and energy. This is called the Landau continuum. In the spherical electron-gas, this continuum is discretized. When looking closely, you will see that it consists of a collection of dots. This is a result of the discrete nature of the energy spectrum in the case of the spherical electron gas. It will be averaged out when the temperature is raised $(k_b T \gg \hbar^2/(m_e R^2))$.

Apart from single-particle excitations in the S2DEG, also collective excitations are possible. Technically speaking "Collective excitations" are defined as poles of the density-density Green's function $\mathscr{D}$, whereas single-particle excitations are defined as poles of the single-particle Green's function G. This means collective excitations appear when $\varepsilon = 0$. In the flat electron gas, these collective excitations are called plasmons. The imparted angular momentum is shared between all the particles.

This means that in a collective mode the entire spherical shell of electrons will oscillate. The frequency of the collective modes (the plasma frequency) will depend on the type of oscillation (ℓ,m). The plasma frequency $\omega_{\rm pl}(\ell,m)$ of the S2DEG can be found by solving

$$1-\frac{e^2}{2\varepsilon_{\rm vac}\hbar R}\frac{1}{2\ell+1}\mathrm{Re}\left[\mathscr{D}_{\mathrm{R},0}(\ell,m;\omega)\right]=0 \tag{193}$$

$$\mathrm{Im}\left[\mathscr{D}_{\mathrm{R},0}(\ell,m;\omega)\right]=0 \tag{194}$$

The last condition means for the flat electron gas that the plasma branch does not lie in the Landau continuum. In fact, as soon as the plasma branch reaches the Landau continuum, it will be strongly damped. For the spherical electron gas, the Landau continuum is no longer a continuum but a set of points. Does this mean that the S2DEG plasma branch will be much less damped than in the corresponding flat electron gas case ? At temperature zero, yes, but as the temperature is raised a little this is no longer true; the typical temperature $\hbar^2/(m_{\rm e}R^2)$ (in the case $L_F=10$) corresponding to the separation of the points is of the order of $10^{-2}-10^{-3}$ nK. So for practical purposes, we can continue to consider the region of the single-particle excitations as a continuum.

Figure 17 shows the plasmon branch (or rather, the value of the plasma frequency $\omega_{\rm pl}(\ell,m)$ for every ℓ. The plasma frequency is the same for every m, it seems only to depend on ℓ. This corresponds to the fact that the plasma frequency of the flat electron gas does not depend on the direction of the wave vector but only on its magnitude. The plasma frequency is shown in units $\hbar/(m_{\rm e}R^2)$. The spherical decomposition of the coulomb interaction is given by $U_0/(2\ell+1)$ so that $U_0=e^2/(2\varepsilon_{\rm vac}\hbar R)$. In the units of frequency which we use, the dimensionless value of U_0 is $e^2m_{\rm e}R/(2\varepsilon_{\rm vac}\hbar^2)$ which is of the order of 10^4.

In figure 17 we have used $U_0=0.5\ 10^4$. Note that as R increases, U_0 also increases. The calculation shows that for increasing U_0 the plasma frequency also increases. The plasma frequency shows a striking resemblance to the result of a flat 2DEG (the so-called "acoustic plasmons"). Yet it is no longer a branch but rather a set of separate excitations. Another difference with the 2DEG is that there exists a minimum value larger than zero for the plasmon frequency; the branch starts at a value $\omega_{\rm pl}\neq 0$ which is a characteristic of (optical) plasmons in the three-dimensional electron gas.

In this section, we have presented an overview of the spherical two-dimensional electron gas. A second quantisation formalism in spherical harmonics is introduced and the Hamiltonian is written in this representation. This formalism is used to calculate the dielectric function of the spherical two-dimensional electron gas. Some aspects of the behaviour of the S2DEG were derived: as the bubble grows, the surface density and the Fermi-energy go down, yet the plasma frequency increases. The Landau continuum region seems to be preserved and to be the analog of the flat electron gas single-particle excitation spectrum

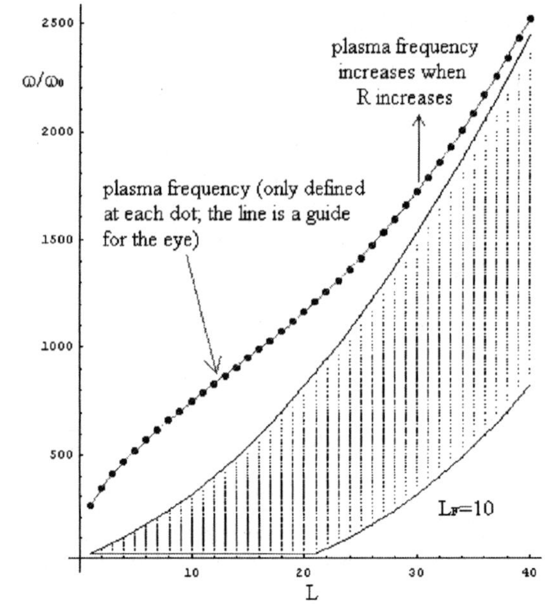

FIGURE 17. Regions in the ℓ, ω plane where $S(\ell, \omega) \neq 0$.

THE WIGNER SOLID OF ELECTRONS IN THE BUBBLE

(The main results of this section were reported in Ref. [27])

In the previous section we described the electrons being a gas or a weakly interacting liquid. This description did not allow to treat the electron solid phase. In this section, we investigate the properties of the electron solid phase in the bubble and show that it is different from the usual Wigner lattice of electrons in that the localization is due to the electron-ripplon interaction rather than to the Coulomb repulsion of the electrons. We conclude this section by discussing the phase transition between the electron liquid and solid phase.

Electron-ripplon interaction

The expression for the interaction energy of electrons with ripplons is derived straightforwardly: the electron is pressed towards the helium surface of the bubble by the electric field **E** generated by the other electrons (this field is directed along the radius of the bubble). The helium surface has surface tension, so the situation is similar to an object pulled down by gravity onto a rubber trampoline sheet. We expect a dimple to form underneath the electron just as we expect a dimple to form in the rubber sheet underneath the mass resting on it. This is illustrated in Figure 18. Later in this section we will denote

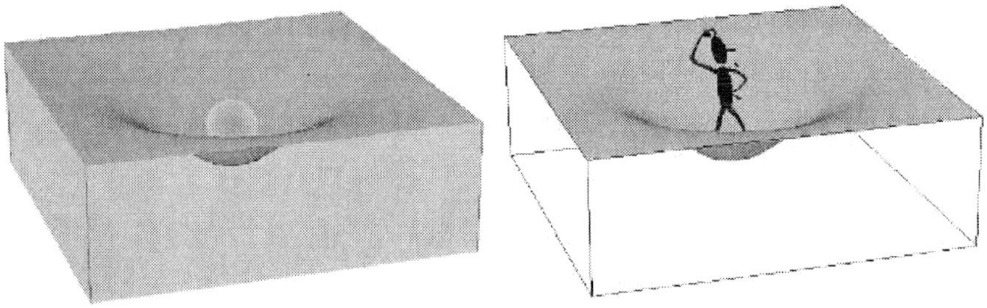

FIGURE 18. Electrons pressed down on a helium surface with surface tension form a dimple in the helium surface, much like a person pressed down on a trampoline.

the combined system of the electron with its surface deformation or dimple by 'ripplonic polaron' or 'ripplopolaron', a name derived from 'polarons' which are electrons together with the crystal deformation they induce in an ionic or polar crystal.

We know furthermore from Sec. 2 that the electrons layer deforms along with the surface, being anchored to it at a fixed distance of about a nanometer. This means that when a ripple on the surface comes by, the electron moves up and down on the crests and troughs of this rippling wave. It is a charge, moving in an electric field - the interaction energy then equals the displacement of the helium (= the displacement of the electron) times the force (=$e\mathbf{E}$) acting on the electron:

$$U_D(\theta,\varphi) = e|\mathbf{E}|u(\theta,\varphi), \tag{195}$$

The surface deformation was introduced in Sec. 2:

$$u(\theta,\varphi) = \sum_{\ell=1}^{\infty}\sum_{m=-\ell}^{\ell} Q_{\ell m} Y_{\ell m}(\theta,\varphi). \tag{196}$$

and the electric field created by the other electrons is

$$|\mathbf{E}| = -\frac{eN}{2\varepsilon R_b^2}, \tag{197}$$

with $\varepsilon = 1.0572$ still the dielectric constant of helium and R_b the bubble equilibrium radius. Hence, Eq. (195) is

$$U_D(\theta,\varphi) = -\frac{e^2 N}{2\varepsilon R_b^2}\sum_{\ell=1}^{\infty}\sum_{m=-\ell}^{\ell} Q_{\ell m} Y_{\ell m}(\theta,\varphi). \tag{198}$$

The transition to the second-quantization formalism is performed using the canonical definitions of the ripplon creation and annihilation operators $a_{\ell m}^{+}$ and $a_{\ell m}$:

$$Q_{\ell m} = (-1)^{\frac{m-|m|}{2}}\left[\frac{\hbar(\ell+1)}{2\rho R^3\omega_\ell}\right]^{1/2}\left(a_{\ell m} + a_{\ell,-m}^{+}\right), \tag{199}$$

$$P_{\ell m} = i(-1)^{\frac{m-|m|}{2}} \left[\frac{\rho R^3 \hbar \omega_\ell}{2(\ell+1)}\right]^{1/2} \left(a^+_{\ell m} - a_{\ell,-m}\right), \tag{200}$$

where $P_{\ell m}$ is the canonical momentum corresponding to the normal coordinate $Q_{\ell m}$ and ω_ℓ is the bare ripplon frequency derived earlier. Within these notations, the Hamiltonian of the deformational electron-ripplon interaction becomes

$$U_D(\theta,\varphi) = -\frac{e^2 N}{2\varepsilon R_b^2} \sum_{\ell=1}^{\infty} \sum_{m=-\ell}^{\ell} (-1)^{\frac{m-|m|}{2}} \left[\frac{\hbar(\ell+1)}{2\rho R_b^3 \omega_\ell}\right]^{1/2} Y_{\ell m}(\theta,\varphi)\left(a_{\ell m} + a^\dagger_{\ell,-m}\right). \tag{201}$$

Let us write down the Hamiltonian of the deformational electron-ripplon interaction in the form

$$U_D(\theta,\varphi) = \sum_{\ell=1}^{\infty} \sum_{m=-\ell}^{\ell} \left(\gamma_{\ell m} a_{\ell m} + \gamma^*_{\ell m} a^\dagger_{\ell m}\right), \tag{202}$$

where γ_{lm} is the interaction amplitude:

$$\gamma_{\ell m} = -U_\ell (-1)^{\frac{m-|m|}{2}} \frac{1}{R_b} Y_{\ell m}(\theta,\varphi), \tag{203}$$

$$U_\ell \equiv \frac{e^2 N}{2\varepsilon R_b^2}\left[\frac{\hbar(\ell+1)}{2\rho R_b \omega_\ell}\right]^{1/2}. \tag{204}$$

In particular, for "large" bubbles (with $N \gg 1$), the ripplon modes with $\ell \gg 1$ give a key contribution into the physical quantities which are the sums over ripplon modes (such as the ripplonic-polaron energy). In this case, we find

$$\frac{\ell}{R_b} \to q, \tag{205}$$

where q is the modulus of the ripplon in-plane wave vector. Since the functions $\frac{1}{R_0} Y_{\ell m}(\theta,\varphi)$ are normalized to unity at the bubble surface,

$$\int \left|\frac{1}{R_0} Y_{\ell m}(\theta,\varphi)\right|^2 ds = \frac{1}{4\pi}\int \left|Y_{\ell m}(\theta,\varphi)\right|^2 d\Omega = 1, \tag{206}$$

the Hamiltonian (202) in the aforesaid limit is equivalent to the following Hamiltonian of a planar system:

$$U_D(\mathbf{r}) = \sum_{\mathbf{q}} \left[V(q)\frac{1}{\sqrt{\mathcal{S}}} e^{i\mathbf{q}\cdot\mathbf{r}} a_{\mathbf{q}} + V^*(q)\frac{1}{\sqrt{\mathcal{S}}} e^{-i\mathbf{q}\cdot\mathbf{r}} a^\dagger_{\mathbf{q}}\right]. \tag{207}$$

According to the correspondence principle (205), the amplitude $Q(q)$ is given by

$$V(q) = \left(\frac{\hbar q}{2\rho\omega(q)}\right)^{1/2} e|\mathbf{E}|, \tag{208}$$

while the ripplon frequency is

$$\omega(q) = \sqrt{\frac{\sigma}{\rho} q^3 + \frac{p}{\rho R_b} q}. \tag{209}$$

Another mechanism of the electron-ripplon interaction was proposed for a planar system by Cole [1]. Within this mechanism, the electron-ripplon interaction is provided by modulation of the electrostatic image potential due to the surface deformation. Since the dielectric constant of helium ($\varepsilon = 1.0572$) is close to unity, the electron-ripplon interaction due to modulations of the image potential appears to be weak with respect to the deformational electron-ripplon interaction, as shown in Ref. [28].

As in Sec. 2, we consider multielectron bubbles with $N > 1000$. We suppose, that when the strong-coupling regime is realized for a ripplonic polaron in such a MEB, the characteristic size of a ripplonic polaron well is much smaller than the radius of a MEB. Indeed, if we are interested in the Wigner lattice of electrons (or ripplonic polarons) then the Lindemann melting criterion [29] requires that the size of the ripplopolaron be smaller than the lattice constant of the solid. This gives us grounds to estimate the polaron parameters using the auxiliary model of a ripplonic polaron in a locally planar system with the electric field $|\mathbf{E}|$ determined by Eq. (197), and with the Hamiltonian of the electron-ripplon interaction given by Eq. (207). This model will allows us to use the Feynman arbitrary-coupling variational method. But first we need to take into account, on some mean-field level, the effect of the other electrons on the investigated electron of the ripplopolaron.

The lattice potential

What is the effect of the other electrons (ripplopolarons) in the lattice on one particular electron (ripplopolaron) ? Fratini and Quémerais [30] in their treatment of a lattice of polarons suggested to describe this influence in a mean-field way, by replacing the other electrons by a homogeneously spread out charge density, out of which a hole with radius equal to the lattice parameter has been punched out. Figure 19 illustrates this. The lattice potential is then the electrostatic energy associated with this 'punched' charge density. Since in a Wigner crystal the electrons are localized to within a fraction of the lattice parameter, exchange and correlation effects will be small. The lattice potential for the two-dimensional case with lattice parameter d, can be expressed in terms of elliptic functions $E(x)$ and $K(x)$,

$$V_{FQ}(\mathbf{r}) = -\frac{2e^2}{\pi d^2}\left[|d-r|\,E\left(-\frac{4rd}{(d-r)^2}\right) + (d+r)\,\mathrm{sgn}(d-r)\,K\left(-\frac{4rd}{(d-r)^2}\right)\right]. \tag{210}$$

Here, $\mathbf{r}$ is the position of the electron under investigation, measured from its lattice site. The expansion of this potential for $r << d$ yields a harmonic lattice confinement potential

$$V_{FQ}(r << d) = -\frac{2e^2}{d} + \frac{m_e\Omega_W^2}{2} r^2 + \mathcal{O}(r^4), \tag{211}$$

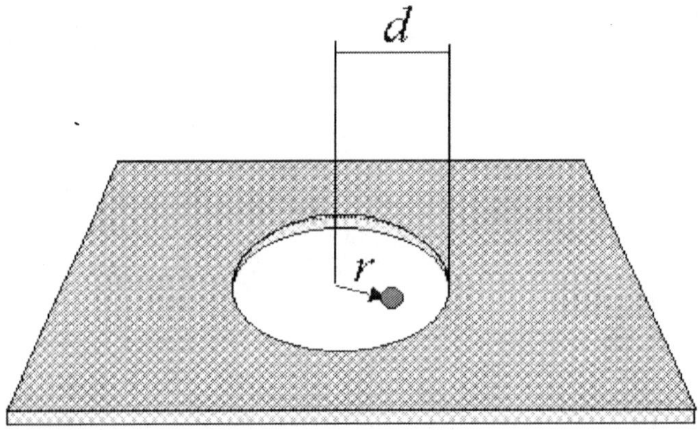

FIGURE 19. The effect, on a particular electron, of the other electrons in the lattice is treated in a mean-field approach replacing those electrons by a homogeneous charge density out of which a circular hole, centered in the lattice site of the electron under investigation and with a radius equal to the lattice distance, is punched out.

with the confinement frequency

$$\Omega_W = \left(\frac{e^2}{m_e d^3}\right)^{1/2}. \tag{212}$$

Now we that both the electron-ripplon interaction and the lattice potential have been derived, we can write down the Hamiltonian of a single ripplopolaron in the lattice:

$$\begin{aligned} \widehat{\mathscr{H}} &= \frac{m_e}{2}\hat{r}^2 + V_{FQ}(\hat{\mathbf{r}}) + \sum_{\ell=1}^{\infty}\sum_{m=-\ell}^{\ell} \hbar\omega_\ell \hat{a}^+_{\ell m}\hat{a}_{\ell m} \\ &+ \sum_{\ell=1}^{\infty}\sum_{m=-\ell}^{\ell} \left(\gamma_{\ell m}\hat{a}_{\ell m} + \gamma^*_{\ell m}\hat{a}^\dagger_{\ell m}\right), \end{aligned} \tag{213}$$

where the interaction amplitude is $\gamma_{\ell m}$ is defined by (203), and $\hat{\mathbf{r}}$ is the position operator of the constituent electron in the ripplopolaron. This Hamiltonian is very reminiscent of the Fröhlich Hamiltonian of a polaron [31]; the role of the phonons is now taken by the ripplons. Methods suitable for the study of single polarons have been used to analyse the single ripplopolaron on a flat surface [28, 32]. Undoubtebly the most versatile and accurate description of polarons is described by the path-integral technique. In this technique, the free energy of the ripplopolaron can be estimated from the Jensen-Feynman variational principle. This variational principle varies parameters of a model system (Lagrangian) rather than parameters of a model wave function. The variational functional

$$F_v = F_0 + \frac{1}{\beta}\langle S - S_0\rangle_{S_0} \tag{214}$$

is an upper bound for the ripplopolaron free energy. F_0 is the free energy of the model system, S is the action functional of the ripplopolaron system under investigation and S_0 is the action functional of the model system. The expectation value $\langle ... \rangle_{S_0}$ is evaluated with respect to the ground state of the model system, at temperature T. The factor $\beta = 1/(k_B T)$ is the inverse temperature.

Path-Integral Treatment

Model system

We use the following model Lagrangian for a ripplopolaron in the Wigner lattice

$$\mathscr{L}_M = \frac{m_e \dot{\mathbf{r}}^2}{2} + \frac{M\dot{\mathbf{R}}^2}{2} + \frac{k}{2}(\mathbf{r}-\mathbf{R})^2 + \frac{m_e \omega_W^2 \mathbf{r}^2}{2}, \tag{215}$$

with three independent variational parameters M, k, and ω_W. This Lagrangian represents a point mass m_e representing the electron attached to its lattice site by a spring with spring constant $m_e\omega_W^2$, and attached to a fictitious mass M representing the effect of the ripplons by another spring, with spring constant k. The variational parameters of this model Lagrangian are the two spring constants and the fictitious mass. Let us write down this "Lagrangian" in the form

$$\mathscr{L}_M = \frac{m_e \dot{\mathbf{r}}^2}{2} + \frac{M\dot{\mathbf{R}}^2}{2} + \frac{1}{2}m_e\omega^2\mathbf{r}^2 + \frac{1}{2}Mw^2\mathbf{R}^2 - \sqrt{m_e M}\gamma\mathbf{r}\cdot\mathbf{R}, \tag{216}$$

where the notations are used:

$$\omega = \sqrt{\frac{k}{m_e} + \omega_W^2}, \quad w = \sqrt{\frac{k}{M}}, \quad \gamma = \frac{k}{\sqrt{m_e M}}. \tag{217}$$

The coefficients for the canonical transformation diagonalizing the model "Lagrangian" are

$$a_1 = \left(\frac{1+\chi}{2}\right)^{1/2}, \quad a_2 = \left(\frac{1-\chi}{2}\right)^{1/2}. \tag{218}$$

The aforesaid canonical transformation has the form

$$\begin{aligned} \mathbf{r} &= \frac{1}{\sqrt{m}}(a_1\mathbf{X} + a_2\mathbf{Y}), \\ \mathbf{R} &= \frac{1}{\sqrt{M}}(-a_2\mathbf{X} + a_1\mathbf{Y}). \end{aligned} \tag{219}$$

The Lagrangian of the model system is transformed as follows,

$$L_M = \frac{1}{2}\dot{\mathbf{X}}^2 + \frac{1}{2}\dot{\mathbf{Y}}^2 + \frac{\Omega_1^2}{2}\mathbf{X}^2 + \frac{\Omega_2^2}{2}\mathbf{Y}^2, \tag{220}$$

with the eigenfrequencies

$$\Omega_1^2 = \frac{\omega^2 + w^2 + \sqrt{(\omega^2 - w^2)^2 + 4\gamma^2}}{2},$$
$$\Omega_2^2 = \frac{\omega^2 + w^2 - \sqrt{(\omega^2 - w^2)^2 + 4\gamma^2}}{2}. \qquad (221)$$

Three frequencies Ω_1, Ω_2 and w can be used as the independent variational parameters. As defined, they obey the following inequalities:

$$\Omega_1 \geq w \geq \Omega_2. \qquad (222)$$

The parameters of the Lagrangian (215) are expressed in terms of (Ω_1, Ω_2, w) as follows:

$$\gamma = \sqrt{(\Omega_1^2 - w^2)(w^2 - \Omega_2^2)},$$
$$\omega = \sqrt{\Omega_1^2 + \Omega_2^2 - w^2}, \quad \omega_W = \frac{\Omega_1 \Omega_2}{w},$$
$$M = m_e \frac{(\Omega_1^2 - w^2)(w^2 - \Omega_2^2)}{w^4}. \qquad (223)$$

The coefficients (218) of the transformation (219) are also expressed in terms of the aforesaid variational frequency parameters,

$$a_1 = \left(\frac{\Omega_1^2 - w^2}{\Omega_1^2 - \Omega_2^2} \right)^{1/2}, \quad a_2 = \left(\frac{w^2 - \Omega_2^2}{\Omega_1^2 - \Omega_2^2} \right)^{1/2}. \qquad (224)$$

As follows from the "Lagrangian" (220), the partition function of the model system is

$$Z_M = \frac{1}{\left(2\sinh \frac{\beta\hbar\Omega_1}{2}\right)^2} \frac{1}{\left(2\sinh \frac{\beta\hbar\Omega_2}{2}\right)^2}, \qquad (225)$$

where $\beta = (k_B T)^{-1}$. Consequently, the free energy of the model system has the form

$$F_M = -\frac{1}{\beta} \ln Z_M$$
$$= \frac{2}{\beta} \ln \left(2\sinh \frac{\beta\hbar\Omega_1}{2} \right) + \frac{2}{\beta} \ln \left(2\sinh \frac{\beta\hbar\Omega_2}{2} \right). \qquad (226)$$

Variational functional

In this subsection, we derive an explicit expression for F_v. The term F_0 has the form

$$F_0 = F_M - \frac{2}{\beta} \ln \left[2\sinh \left(\frac{\beta\hbar w}{2} \right) \right]$$

$$= \frac{2}{\beta}\ln\left(2\sinh\frac{\beta\hbar\Omega_1}{2}\right)+\frac{2}{\beta}\ln\left(2\sinh\frac{\beta\hbar\Omega_2}{2}\right)$$
$$-\frac{2}{\beta}\ln\left[2\sinh\left(\frac{\beta\hbar w}{2}\right)\right]. \tag{227}$$

The model "action" is

$$S_0=\frac{1}{\hbar}\int_0^{\hbar\beta}d\tau\left(\frac{m_e\dot{\mathbf{r}}^2}{2}+\frac{1}{2}m_e\omega^2\mathbf{r}^2\right)-\Phi_0, \tag{228}$$

where Φ_0 is the influence phase of the fictitious particle,

$$\Phi_0=\frac{Mw^3}{4\hbar}\int_0^{\hbar\beta}d\tau\int_0^{\hbar\beta}d\sigma\frac{\cosh\left[w\left(|\tau-\sigma|-\hbar\beta/2\right)\right]}{\sinh\left(\beta\hbar w/2\right)}\mathbf{r}(\tau)\cdot\mathbf{r}(\sigma). \tag{229}$$

The action of the ripplopolaron system described by the Hamiltonian (213) is, after elimination of the ripplon degrees of freedom,

$$S = -\frac{1}{\hbar}\int_0^{\hbar\beta}d\tau\left\{\frac{m_e}{2}\dot{r}^2(\tau)+V_{\text{lat}}[r(\tau)]\right\}+\sum_{\mathbf{q}}|V(q)|^2\int_0^{\hbar\beta}d\tau\int_0^{\hbar\beta}d\sigma$$
$$\times\frac{\cosh[\nu(|\tau-\sigma|-\hbar\beta/2)]}{\sinh(\beta\hbar\nu/2)}\exp\{i\mathbf{q}\cdot[\mathbf{r}(\tau)-\mathbf{r}(\sigma)]\}, \tag{230}$$

So, $\frac{1}{\beta}\left\langle S-S_0\right\rangle_{S_0}$ is a sum of four terms,

$$\frac{1}{\beta}\left\langle S-S_0\right\rangle_{S_0}=\left\langle U_W(\mathbf{r})\right\rangle_{S_0}-\frac{1}{2}m_e\omega^2\left\langle \mathbf{r}^2\right\rangle_{S_0}+\frac{1}{\beta}\left\langle \Phi_0\right\rangle_{S_0}-\frac{1}{\beta}\left\langle \Phi\right\rangle_{S_0}. \tag{231}$$

Let us write down explicit expressions for each of these terms.

1) The average of the mean-field potential takes the form

$$\left\langle U_W(\mathbf{r})\right\rangle_{S_0}=-\frac{\sqrt{\pi}e^2}{\sqrt{D}}\exp\left(-\frac{d^2}{2D}\right)\left[I_0\left(\frac{d^2}{2D}\right)+I_1\left(\frac{d^2}{2D}\right)\right], \tag{232}$$

where the parameter D is

$$D\equiv\frac{\hbar}{m}\sum_{i=1}^{2}\frac{a_i^2}{\Omega_i}\coth\left(\frac{\beta\hbar\Omega_i}{2}\right). \tag{233}$$

2) The term $\frac{1}{2}m\omega^2\left\langle \mathbf{r}^2\right\rangle_{S_0}$ is expressed as follows,

$$\frac{1}{2}m\omega^2\left\langle \mathbf{r}^2\right\rangle_{S_0}=\frac{\hbar\omega^2}{2}\sum_{i=1}^{2}\frac{a_i^2\coth\left(\beta\Omega_i/2\right)}{\Omega_i}. \tag{234}$$

3) The average of the influence phase of a fictitious particle is

$$\frac{1}{\beta}\left\langle \Phi_0\right\rangle_{S_0} = \frac{\hbar\gamma^2}{2}\sum_{i=1}^{2}\frac{a_i^2}{w^2-\Omega_i^2}\left[\frac{\coth\left(\beta\hbar\Omega_i/2\right)}{\Omega_i} - \frac{\coth\left(\beta\hbar w/2\right)}{w}\right]. \tag{235}$$

4) Finally, the average of the ripplon influence phase takes the form

$$\begin{aligned}\frac{1}{\beta}\langle\Phi\rangle_{S_0} &= \frac{1}{2\pi\hbar}\int_{q_0}^{\infty} dq\, q\left|V(q)\right|^2\int_0^{\hbar\beta/2} d\tau\frac{\cosh\left[\omega(q)\left(\tau-\hbar\beta/2\right)\right]}{\sinh\left(\beta\hbar\omega(q)/2\right)}\\ &\times\exp\left[-\frac{\hbar q^2}{2m_e}\sum_{j=1}^{2}a_j^2\frac{\cosh\left(\beta\hbar\Omega_j/2\right)-\cosh\left(\Omega_j\left(\tau-\hbar\beta/2\right)\right)}{\Omega_j\sinh\left(\beta\hbar\Omega_j/2\right)}\right]\end{aligned} \tag{236}$$

with the parameters

$$\omega(q)=\sqrt{\frac{\sigma}{\rho}q^3+\frac{p}{\rho R_b}q}, \quad V(q)=\left(\frac{\hbar q}{2\rho\omega(q)}\right)^{1/2}e|\mathbf{E}|. \tag{237}$$

After collecting all the aforementioned terms together and some identity transformations, the variational functional is substantially simplified:

$$\begin{aligned}F_v &= \frac{2}{\beta}\ln\left(2\sinh\frac{\beta\hbar\Omega_1}{2}\right)+\frac{2}{\beta}\ln\left(2\sinh\frac{\beta\hbar\Omega_2}{2}\right)\\ &-\frac{2}{\beta}\ln\left[2\sinh\left(\frac{\beta\hbar w}{2}\right)\right]-\frac{\hbar}{2}\sum_{i=1}^{2}a_i^2\Omega_i\coth\left(\frac{\beta\hbar\Omega_i}{2}\right)\\ &-\frac{\sqrt{\pi}e^2}{\sqrt{D}}\exp\left(-\frac{d^2}{2D}\right)\left[I_0\left(\frac{d^2}{2D}\right)+I_1\left(\frac{d^2}{2D}\right)\right]-\frac{1}{\beta}\langle\Phi\rangle_{S_0}.\end{aligned} \tag{238}$$

Optimal values of the variational parameters (Ω_1, Ω_2 and w) are determined by the numerical minimization of the variational functional F_v.

Melting criterion for a ripplopolaron Wigner lattice

The most common method of estimating the critical electron density, when the Wigner lattice melts, is by using the Lindemann melting criterion which states that melting will occur when the mean square electron displacement is some fraction δ_0 of the intersite separation [29]. Fratini and Quémerais [30] have extended this criterion in order to describe melting of the Wigner lattice of polarons in ionic crystals. Within the approach [30], the Wigner lattice of polarons melts, when at least 7one of two following conditions (Lindemann criteria) is fulfilled:

$$\delta_R \equiv \frac{\sqrt{\langle\mathbf{R}_c^2\rangle_{S_0}}}{d} > \delta_0, \tag{239}$$

$$\delta_\rho \equiv \frac{\sqrt{\langle \vec{\rho}^2 \rangle_{S_0}}}{d} > \delta_0, \tag{240}$$

where d is the inter-electron distance, $\mathbf{R}_c$ and $\vec{\rho}$ are the center-of-mass and relative coordinate vectors for the model system,

$$\mathbf{R}_c = \frac{m_e \mathbf{r} + M\mathbf{R}}{m_e + M}, \quad \vec{\rho} = \mathbf{r} - \mathbf{R}. \tag{241}$$

The appearance of two different Lindemann criteria takes into account the composite nature of polarons. As follows from the physical sense of coordinates $\mathbf{R}_c$ and $\vec{\rho}$, the first criterion (239) is related to the melting of the polaron Wigner lattice towards a polaron liquid (where each polaron move as a whole). The second criterion (240) is related to the polaron dissociation.

In Ref. [27], we apply this method to the ripplopolaron Wigner lattice in a multi-electron bubble. The averages $\langle \mathbf{R}_c^2 \rangle_{S_0}$ and $\langle \vec{\rho}^2 \rangle_{S_0}$ are expressed through the variational parameters as

$$\langle \mathbf{R}_c^2 \rangle_{S_0} = \frac{\hbar w^4}{m_e} \frac{\frac{\Omega_2^4 (\Omega_1^2 - w^2)}{\Omega_1} \coth\left(\frac{\beta\hbar\Omega_1}{2}\right) + \frac{\Omega_1^4 (w^2 - \Omega_2^2)}{\Omega_2} \coth\left(\frac{\beta\hbar\Omega_2}{2}\right)}{\left(w^2 \left(\Omega_1^2 + \Omega_2^2\right) - \Omega_1^2\Omega_2^2\right)^2 \left(\Omega_1^2 - \Omega_2^2\right)}, \tag{242}$$

$$\langle \vec{\rho}^2 \rangle_{S_0} = \frac{\hbar}{m_e} \frac{\Omega_1^3 \left(w^2 - \Omega_2^2\right) \coth\left(\frac{\beta\hbar\Omega_1}{2}\right) + \Omega_2^3 \left(\Omega_1^2 - w^2\right) \coth\left(\frac{\beta\hbar\Omega_2}{2}\right)}{\left(\Omega_1^2 - \Omega_2^2\right)\left(\Omega_1^2 - w^2\right)\left(w^2 - \Omega_2^2\right)}. \tag{243}$$

The numerical calculation shows, that for ripplopolarons in a MEB with $N < 10^9$, the inequality $\Omega_1 \gg w$ is fulfilled ($w/\Omega_1 \sim 10^{-3}$ to 10^{-2}), what means that the strong-coupling ripplopolaron regime is realized. Owing to this inequality, we find from Eqs. (242) and (243) that

$$\langle \mathbf{R}_c^2 \rangle_{S_0} \ll \langle \vec{\rho}^2 \rangle_{S_0}. \tag{244}$$

So, the destruction of the ripplopolaron Wigner lattice in a MEB occurs through the dissociation of ripplopolarons. This conclusion is in agreement with that made in Ref. [30].

The precise value of the Lindemann parameter δ_0 would be very hard to calculate from first principles. Luckily, for electrons on a flat helium surface, there has been a direct measurement of the melting of a Wigner lattice of electrons by Grimes and Adams [4]. From their measurement, it is possible to extract the value $\delta_0 = 0.13$, so that when electrons move on average more than 13% of the lattice parameter away from their lattice site, the Wigner lattice melts. We have no a priori reason to assume that this value will remain the same for ripplonic polarons in a spherical geometry, but a value of δ_0˜0.1 shows up consistently in the literature on electron Wigner solids, so it seems an adequate first choice. Note furthermore that considerations based on the Mermin-Wagner theorem (putting the validity of the Lindemann criterion in doubt) are not applicable here, since we are dealing with a finite, confined system.

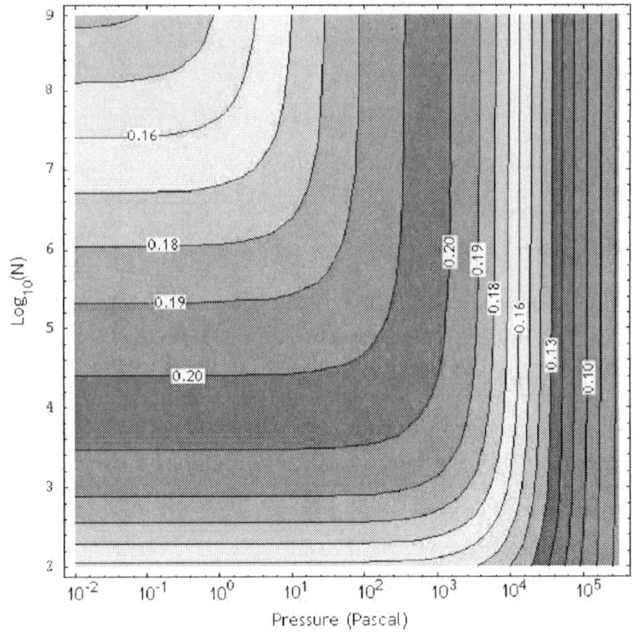

FIGURE 20. The Lindemann parameter δ_ρ as a function of pressure and number of electrons in the bubble is shown in a countourplot. The contour $\delta = 0.13$ separates the region of Wigner crystallization of ripplopolarons (at higher pressure) from the region of the electron liquid (at lower pressure).

In Figure 20, the result for δ_ρ as a function of the pressure and the number of electrons in the bubble are presented as a contour plot. The contour with $\delta_\rho = \delta_0 = 0.13$ separates the region of the Wigner solid of ripplopolarons and the electron fluid. The region at higher pressure is the Wigner solid, the region at lower pressure is the electron fluid. This can be understood as follows: when the pressure is decreased, the radius of the bubble expands and the electric field pressing the electrons against the helium surface decreases. This reduces the coupling between electrons and ripplons: the dimples become more shallow and broader. Below a certain critical pressure, the electron-ripplon coupling is too weak to keep the electrons self-localized by their dimple, and the Wigner lattice of ripplopolarons melts by dissociation of the ripplopolarons as the electrons shed their dimple.

The temperature dependence of this phase diagram, presented here in the limit of temperature $T \to 0$ ($\beta \to \infty$) can be investigated straightforwardly with the path-integral formalism. We find that the $\delta_\rho = 0.13$ curve does not change location significantly for temperatures up to the critical temperature of liquid helium. This is consistent with estimates of the ripplopolaron binding energy of about $T = 10$ K in typical multielectron bubbles.

In Figure 20 we investigate a broad region of phase space, a region that would be accessible in experiments on multielectron bubbles. If we extend the region further, what other phases might appear ? Is there a liquid of ripplopolarons ? This is highly unlikely, since the ripplon frequency and the longitudinal plasmon frequency (or lattice frequency) are orders of magnitude different: the ripplons are much slower than the electrons. To form a liquid, these ripplopolarons should move at a velocity fast enough to prevent crystallization. But at those velocities, the the dimples cannot follow the electrons in the moving ripplopolaron, and the electrons would shed their dimple.

If the number of electrons in the bubble is increased further, and thus the electron density is decreased, the limit of the flat helium surface will slowly be retrieved. We found the electron Wigner lattice (distinct from the ripplopolaron Wigner lattice) for N larger than about 7×10^9, and with a melting temperature of about 0.5 K [27] in agreement with the observation of Grimes and Adams [4]. From Fig. 20 a first hint of this phase can be seen: the value of δ_p reduces again in the upper left hand corner. There are several interesting differences between the electron Wigner lattice and the ripplopolaron Wigner lattice. Firstly, the ripplopolaron solid melts by dissociation of the ripplopolarons, which can take place also at $T = 0$ and can at that temperature be interpreted as quantum melting. The electron solid, by contrast, melts through the amplitude of the motion (thermal or zero-point) of the electrons around their lattice point. Secondly, the driving force behind the formation of a ripplopolaron lattice is the electron-ripplon interaction. Changing the strength of this interaction (by manipulating the bubble radius) can melt the lattice. The driving force for the formation of the electron Wigner lattice is the Coulomb interaction. How could this ripplopolaron Wigner lattice be detected ? A possibility in the current set-up is the resonant spectroscopy discussed in Sec. 1. A measurement of the oscillation frequencies will, as is the case for electrons on a flat helium surface [4] reveal the phase transition between ripplopolaron solid and electron liquid.

CONCLUSIONS

In this review on multielectron bubbles, we started by examining the oscillation modes of multielectron bubbles (Sec. 2) and the bubble stability (Sec. 3). After these necessary preliminary steps, we investigated the electronic system in the bubble. Two phases were analyzed: (1) the electron gas (or liquid), in Sec. 4 and (2) the electron solid, in Sec. 5.

Regarding the electron gas or weakly-interacting liquid, we find that a random phase approximation yields plasmon modes which lie close to the edge of the Landau continuum, and that there is a strictly positive minimum plasmon frequency. The Fermi energy of this system decreases as electrons are added to the bubble.

Regarding the electron solid, we come to the remarkable conclusion that the localization of the electrons is here due to the electron-ripplon interaction and not the Coulomb repulsion: the electrons create dimples in the helium surface, and the induced potential by the dimple localizes the wave function of the electrons. Thus, in multielectron bubbles, a lattice of ripplopolarons forms rather than a Wigner lattice of electrons (except in the limit of very large bubbles).

This also changes the nature of the transition between the solid and fluid: the ripplopolaron lattice does not melt classically, but by dissociation of the ripplopolarons (the electrons shed their dimple). This melting transition can be driven by changing the pressure - since the pressure determines the radius, which in turn influences the electric field and thus the electron-ripplon coupling strength.

ACKNOWLEDGMENTS

The authors are grateful for fruitful discussions and interactions with S. N. Klimin and V. M. Fomin. Discussions with J. Huang are also acknowledged. One of the authors (J.T.) is supported financially by the Fund for Scientific Research - Flanders (Fonds voor Wetenschappelijk Onderzoek - Vlaanderen). This research has been supported by the Department of Energy, grant DE-FG002-85ER45190, and by the GOA BOF UA 2000, IUAP, the FWO-V projects Nos. G.0306.00, G.0274.01, G.0435.03 and the WOG project WO.025.99N (Belgium).

REFERENCES

1. M. W. Cole, Rev. Mod. Phys. **46**, 451 (1974).
2. *Two-Dimensional Electron Systems on Helium and other Cryogenic Substrates* (ed. E. Y. Andrei, Kluwer Acad. Publ, Dordrecht, The Netherlands, 1997).
3. C. C. Grimes, Surf. Sci. 73, 379 (1978).
4. C. C. Grimes and G. Adams, Phys. Rev. Lett. 42, 795 (1979).
5. A. P. Volodin and V. S. Edel'man, Sov. Phys. JETP 54, 198 (1981).
6. A. P. Volodin, M.S. Khaikin, and V.S. Edel'man, Pis'ma Zh. Eksp. Teor. Fiz. **26**, 707 (1977) [JETP Lett. **26**, 543 (1977)]
7. U. Albrecht and P. Leiderer, Europhys. Lett. **3**, 705 (1987).
8. M. M. Salomaa and G. A. Williams, Phys. Rev. Lett. 47, 1730 (1981).
9. I. F. Silvera, Bull. Am. Phys. Soc. **46**, 1016 (2001).
10. C. C. Grimes and G. Adams, Phys. Rev. B **41**, 6366 (1990).
11. V. B. Shikin, JETP Lett. **27**, 39 (1978).
12. K. W. K. Shung and F. L. Lin, Phys. Rev. B **45** , 7491 (1992).
13. J. Tempere, I.F. Silvera, and J.T. Devreese, Phys. Rev. Lett. **87**, 275301 (2001).
14. J. W. Rayleigh, Philosophical Magazine **14**, 184 (1882).
15. K. R. Atkins, *Liquid Helium* (Cambridge University Press, London, 1959).
16. S. N. Klimin, V. M. Fomin, J. Tempere, I. F. Silvera, J. T. Devreese, submitted to Solid State Commun.
17. P. M. Platzman and H. Fukuyama, Phys. Rev. B **10**, 3150 (1974).
18. D. S. Fisher, B. I. Halperin, and P. M. Platzman, Phys. Rev. Lett. **42**, 798 (1979).
19. J. Tempere, I. F. Silvera, and J. T. Devreese, Phys. Rev. B **67**, 035402 (2003).
20. N. Bohr, Nature **137**, 344 and 351 (1936); L. Meitner and O. R. Frisch, Nature **143**, 239 (1939); N. Bohr and J. A. Wheeler, Phys. Rev. **56**, 426 (1939). See also: W. D. Myers, *Droplet Model of Atomic Nuclei* (IFI/Plenum, New York, 1977).
21. J. Tempere, I. F. Silvera, J. T. Devreese, Phys. Rev. B 65, 195418 (2002).
22. A. L. Fetter, J. D. Walecka, *Quantum Theory of Many-Particle Systems* (McGrawHill, New York, 1971).
23. C. Cohen-Tannoudji, B. Diu, F. Laloë, *Quantum Mechanics* (J. Wiley & Sons, New York, 1977), p. 1046.
24. J. D. Jackson, *Classical Electrodynamics* (J. Wiley & Sons, New York, 1975), p. 104.
25. G. D. Mahan, *Many-Body Physics* (Plenum Press, New York, 1981), p. 143.
26. D. Pines, *The Many-Body Problem* (Addison-Wesley, Reading MA, 1997).
27. J. Tempere, S. Klimin, V. M. Fomin, I. F. Silvera, J. T. Devreese, submitted to European Phys. Journ. B.

28. S. A. Jackson and P. M. Platzman, Phys. Rev. B **24**, 499 (1981).
29. F. Lindemann, Z. Phys. **11**, 609 (1910); C.M. Care and N.H. March, Adv. Phys. **24**, 101 (1975).
30. S. Fratini and P. Quémerais, Eur. Phys. J. B. **14**, 99 (2000); Eur. Phys. J. B **29**, 41 (2002).
31. H. Fröhlich, Advances in Physics **3**, 325 (1954).
32. S.A. Jackson and F.M. Peeters, Phys. Rev. B **30**, 4196 (1984).

The numerical approach to the correlated electron problem: quantum Monte Carlo methods.

F. F. Assaad

Institut für Theoretische Physik III,
Universität Stuttgart, Pfaffenwaldring 57, D-70550 Stuttgart, Germany.
Max Planck institute for solid state research, Heisenbergstr. 1, D-70569, Stuttgart. email

Abstract. We present a review of the auxiliary field (i.e. determinantal) Quantum Monte Carlo method applied to various problems of correlated electron systems. The ground state projector method, the finite temperature approach as well as the Hirsch-Fye impurity algorithm are described in details. It is shown how to apply those methods to a variety of models: Hubbard Hamiltonians, periodic Anderson model, Kondo lattice and impurity problems, as well as hard core bosons and the Heisenberg model. An introduction to the world-line method with loop upgrades as well as an appendix on the Monte Carlo method is provided.

INTRODUCTION

The correlated electron problem remains one of the central challenges in solid state physics. Given the complexity of the problem numerical simulations provide an essential source of information to test ideas and develop intuition. In particular for a given model describing a particular material we would ultimately like to be able to carry out efficient numerical simulations so as to provide *exact* results on thermodynamic, dynamical, transport and ground state properties. If the model shows a continuous quantum phase transition we would like to characterize it by computing the critical exponents. Without restriction on the type of model, this is an extremely challenging goal.

The are however a set of problems for which numerical techniques have and will provided invaluable insight. Here we list a few which are *exact*, capable of reaching large system sizes (the computational effort scales as a power of the volume), and provide ground state, dynamical as well as thermodynamic quantities. i) Density matrix renormalization group applied to general one-dimensional systems [1] ii) world-line loop Quantum Monte Carlo (QMC) applied to non-frustrated spin systems in arbitrary dimensions [2] and iii) auxiliary field QMC methods [3]. The latter method is the only algorithm capable of handling a class of models with spin and charge degrees of freedom in dimensions larger than unity. This class contains fermionic lattice models with an attractive interactions (e.g. attractive Hubbard model), models invariant under a particle-hole transformation, as well as impurity problems modeled by Kondo or Anderson Hamiltonians.

Here we will concentrate primarily on the auxiliary field QMC method and introduce briefly the world line method with loop updates. Both algorithms are based on a path

CP678, *Lectures on the Physics of Highly Correlated Electron Systems VII: Seventh Training Course*, edited by A. Avella and F. Mancini

integral formulation of the imaginary time propagator which maps a d-dimensional quantum system on a $d+1$-dimensional classical system. The additional dimension is nothing but the imaginary time. For example, within the World Line QMC algorithm [4], this mapping relates the one-dimensional XYZ quantum spin chain to an eight vertex model [5] or the one-dimensional t-J model to the 15-vertex model [6]. The classical models may then be solved exactly as in the case of the eight vertex model [7] or simulated very efficiently by means of cluster Monte Carlo methods [2]. The latter approach has proved to be extremely efficient for the investigation of non-frustrated quantum spin systems [8] in arbitrary dimensions. The efficiency lies in the fact that i) the computational time scales as the volume of the $d+1$ dimensional classical system so that very large system sizes may be achieved and ii) the autocorrelation times are small. In the next section we will briefly, by way of introduction, review the World Line approach and thereby show how the XXZ chain maps onto the 6-vertex model. The attractive feature of the World Line approach lies in it's simplicity. It will also allow us to aquire some insight into the so called sign problem. This is a major, open, issue in QMC methods applied to correlated systems. When it occurs the computational effort scales exponentially with system size and inverse temperature.

In spacial dimensions larger than unity, the World Line approach often fails (i.e. the occurrence of a sign problem) already at the *mean-field* level. That is: consider the paramagnetic mean-field solution of the two dimensional Hubbard model which boils down to solving a free electron problem in a tight binding approximation. This simple model, already leads to a severe sign problem when formulated within the World Line approach. The auxiliary field QMC method [3] relies on a different formulation which solves the mean-field problem exactly. With the use of a Hubbard Stratonovich transformation the partition function of a Hamiltonian H at temperature $T = 1/\beta$ and chemical potential μ is written as:

$$Z = \mathrm{Tr}\left[e^{-\beta(H-\mu N)}\right] = \int D\Phi e^{-S(\Phi)}. \tag{1}$$

$S(\Phi)$ is the action of a one-body problem in a imaginary time and space dependent field Φ. As we will see for a given field Φ the computational cost required to compute the action scales as the product of the volume to the cubed and inverse temperature. [1] The functional integral is carried by means of Monte Carlo sampling. In this approach the mean-field solution is given by the saddle point approximation: the functional integral over Φ is replaced by a single field Φ^* for which $\left.\frac{\partial S(\Phi)}{\partial \Phi}\right|_{\Phi=\Phi^*} = 0$. The nature of the mean-field solution depends on the choice of the Hubbard Stratonovich decoupling. Thus, in the auxiliary field QMC mean-field Hamiltonians are solved exactly, the price being the above mentioned scaling of the computational effort. In the above framework, the Monte Carlo integration over the field Φ may be seen as a means of taking into account all fluctuations around the mean-field solution. This will in many cases introduce a sign problem. Nevertheless the method has the advantage that symmetries of the

[1] For the Hirsch-Fye impurity algorithm computational effort scales as the cubed of the inverse temperature.

model, such as particle-hole symmetry, may be put to use to avoid the sign problem in many non-trivial cases. Other classes of models where the sign problem does not occur include models with attractive interactions which couple independently to an internal symmetry with an even number of states. The attractive Hubbard model is a member of this class. It is also worth mentioning that when the sign problem occurs in the auxiliary field QMC it is often less severe than in World Line approach so that at least high temperature properties may be investigated.

The auxiliary field quantum Monte-Carlo method is the central topic of this article. We will review in all details both the finite temperature [9, 10, 11] and ground state [12, 13, 14] formulation of the method. The application of the method to various models (Hubbard model, periodic Anderson model, Kondo lattice model, hard core boson systems and Heisenberg models) will be discussed. Since the computational effort scales as the cube of the volume, it is important to control size effects. A simple method to reduce size effects by an order of magnitude in temperature will also be discussed. We equally review a very much related algorithm, the Hirsch-Fye impurity algorithm, which has been used extensively in the context of dynamical mean-field theories [15, 16]. We will apply this algorithm to the single impurity Kondo problem.

Finally, we provide an Appendix for the Monte Carlo method and error analysis.

THE WORLD LINE APPROACH FOR THE XXZ MODEL AND RELATION TO THE 6-VERTEX MODEL

To illustrate with on a simple example the World Line quantum Monte Carlo method, we consider the XXZ quantum spin chain defined as:

$$H = J_X \sum_i \left(S_i^x S_{i+1}^x + S_i^y S_{i+1}^y \right) + J_Z \sum_i S_i^z S_{i+1}^z \tag{2}$$

where $\vec{S}_i$ are the spin $1/2$ operators on site i satisfying the commutation relations:

$$\left[S_i^\eta, S_j^\nu \right] = \delta_{i,j} i \varepsilon^{\eta,\nu,\gamma} S_i^\gamma. \tag{3}$$

In the above, $\varepsilon^{\eta,\nu,\gamma}$ is the antisymmetric tensor and the sum over repeated indices is understood. Our aim is to compute observables:

$$\langle O \rangle = \frac{\mathrm{Tr}\left[e^{-\beta H} O \right]}{\mathrm{Tr}\left[e^{-\beta H} \right]}. \tag{4}$$

The basic idea of the World Line algorithm is to split the above Hamiltonian into a set of independent - in this case - two site problems. The way to achieve this decoupling is with the use of a path integral and the Trotter decomposition. First we write

$$H = \underbrace{\sum_n H^{(2n+1)}}_{H_1} + \underbrace{\sum_n H^{(2n+2)}}_{H_2} \tag{5}$$

with

$$H^{(i)} = J_X \left(S_i^x S_{i+1}^x + S_i^y S_{i+1}^y \right) + J_Z S_i^z S_{i+1}^z$$

One may verify that H_1 and H_2 are sums of commuting (i.e. independent) two site problems. Hence, on their own H_1 and H_2 are trivially solvable problems. However, H is not. To put to use this fact, we wish split the imaginary propagation $e^{-\beta H}$ into successive infinitesimal propagations of H_1 and H_2. This is achieved with the Trotter decomposition [17]:

$$\begin{aligned} \left(e^{-\Delta\tau H_1} e^{-\Delta\tau H_2}\right)^m &= \left(e^{-\Delta\tau H} + \frac{\Delta\tau^2}{2}[H_1,H_2] + \mathcal{O}(\Delta\tau^3)\right)^m \\ = &\left(e^{-\Delta\tau(H-\Delta\tau[H_1,H_2]/2)} + O(\Delta\tau^3)\right)^m = e^{-\beta(H-\Delta\tau[H_1,H_2]/2)} + \mathcal{O}(\Delta\tau^2) \\ = &\ e^{-\beta H} + \frac{\Delta\tau}{2}\int_0^\beta d\tau e^{-(\beta-\tau)H}[H_1,H_2]e^{-\tau H} + \mathcal{O}(\Delta\tau^2) \end{aligned} \tag{6}$$

where $m\Delta\tau = \beta$ and $\mathcal{O}(\Delta\tau^n)$ means that for fixed values of β the error scales as $\Delta\tau^n$. In many cases, we will not take the limit $\Delta\tau \to 0$ and is important to understand the order of the systematic error produced by the above decomposition. [2] A priori, it is of the oder $\Delta\tau$. However, in many non-trivial cases, the prefactor of the error of order $\Delta\tau$ vanishes. In the World line approach we compute:

$$\frac{\mathrm{Tr}\left[\left(e^{-\Delta\tau H_1} e^{-\Delta\tau H_2}\right)^m O\right]}{\mathrm{Tr}\left[\left(e^{-\Delta\tau H_1} e^{-\Delta\tau H_2}\right)^m\right]} = \frac{\mathrm{Tr}\left[e^{-\beta H} O\right] + \frac{\Delta\tau}{2}\mathrm{Tr}\left[AO\right]}{\mathrm{Tr}\left[e^{-\beta H}\right] + \frac{\Delta\tau}{2}\mathrm{Tr}\left[A\right]} + O(\Delta\tau^2). \tag{7}$$

Here $A = \int_0^\beta d\tau e^{-(\beta-\tau)H}[H_1,H_2]e^{-\tau H}$ and $m\Delta\tau = \beta$. Since A is an antihermetian operator, $A^\dagger = -A$, it follows that $\overline{\mathrm{Tr}[A]} = \mathrm{Tr}\left[A^\dagger\right] = -\mathrm{Tr}[A]$ as well as $\overline{\mathrm{Tr}[AO]} = -\mathrm{Tr}[AO]$. Note that the observable O is a hermitian operator. Thus, if O, H_1 and H_2 are simultaneously real representable in a given basis, the systematic error proportional to $\Delta\tau$ vanishes since in this case the trace is real. Hence the systematic error is of order $\Delta\tau^2$.

With the above, the estimation of the partition function reads:

$$\begin{aligned} \mathrm{Tr}\left[e^{-\beta H}\right] &= \mathrm{Tr}\left[(e^{-\Delta\tau H})^m\right] = \mathrm{Tr}\left[(e^{-\Delta\tau H_1} e^{-\Delta\tau H_2})^m\right] + \mathcal{O}(\Delta\tau^2) = \\ \sum_{n_1\cdots n_{2m}} &\langle n_1|e^{-\Delta\tau H_1}|n_{2m}\rangle\cdots\langle n_3|e^{-\Delta\tau H_1}|n_2\rangle\langle n_2|e^{-\Delta\tau H_2}|n_1\rangle + \mathcal{O}(\Delta\tau^2) \end{aligned} \tag{8}$$

where $m\Delta\tau = \beta$ and the states $|n_\tau\rangle$ span the Hilbert space. We choose the states $|n_\tau\rangle$ to be eigenstates of S_i^z. For each set of states $|n_1\rangle\cdots|n_{2m}\rangle$ with non-vanishing contribution to the partition function we have a simple graphical representation in terms of world lines as shown in Fig. 1. Observables are now given by:

---

[2] Within the loop algorithm a continuous time formulation may be achieved [18]

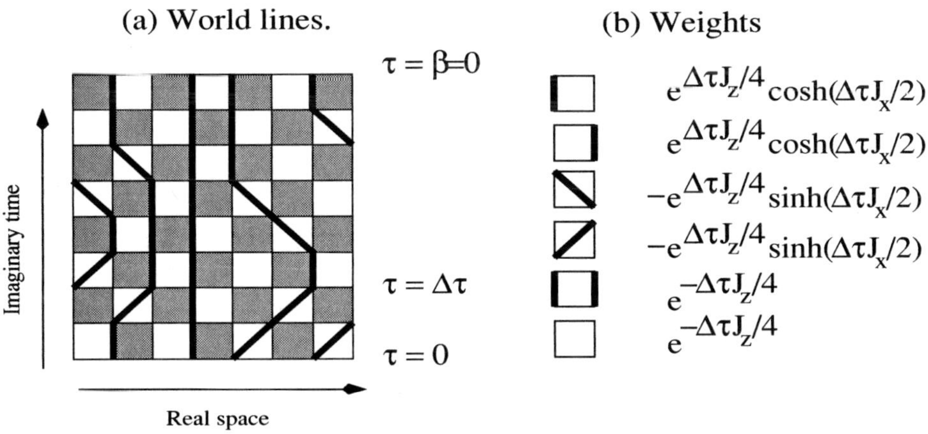

FIGURE 1. (a) World line configuration for the XXZ model of Eq. (2). Here, $m = 4$ and the system size is $L = 8$. The bold lines follow the time evolution of the up spins and empty sites, with respect to the world lines, correspond to the down spins. A full time step $\Delta\tau$ corresponds to the propagation with H_1 followed by H_2. Periodic boundary conditions are chosen in the spacial direction. In the time direction, periodic boundary conditions follow from the fact that we are evaluating a trace. (b) The weights for a given world line configuration is the product of the weights of plaquettes listed in the figure. Note that, although the spin-flip processes come with a minus sign the overall weight for the world line configuration is positive since each world line configuration contains an even number of spin flips.

$$\langle O \rangle = \frac{\sum_w \Omega(w) O(w)}{\sum_w \Omega(w)} \tag{9}$$

where $\Omega(w)$ corresponds to the weight of a given world line configuration as obtained through multiplication of the weights of the individual plaquettes listed in Fig. 1. Note that although spin-flip processes have negative weight, no sign problem occurs. One can for example carry out the transformation: $S_i^x \rightarrow (-1)^i S_i^x, S_i^y \rightarrow (-1)^i S_i^y, S_i^z \rightarrow S_i^z$ which leaves the commutations relation unaltered (i.e. is canonical) but changes the sign of J_x. Observables O which locally conserve the z-component of the spin are easy to compute. If we decide to measure on time slice τ then $O|n_\tau\rangle = O(w)|n_\tau\rangle$. [3]

The problem is now cast into one which may be solved with classical Monte Carlo methods (see Appendix). To generate a Markov chain through the space of World Lines we need to devise an updating mechanism. Local updates where one locally deforms a World Line configuration have been used successfully (see Fig. 2). The local updates conserve the z-component of the total spin (i.e. canonical in the hard core boson notation introduced in Section) and are ergodic only in the case of open boundary conditions. Choosing periodic boundary conditions and starting with a configuration with zero winding one will remain in this sector. That is: the configuration of Fig. 4e will not be generated starting from the configuration of Fig. 4a with local updates. Note however, that this is a boundary problem so that when the thermodynamic limit is taken with the

[3] In practice, one will measure on all time slices so as to reduce statistical fluctuations

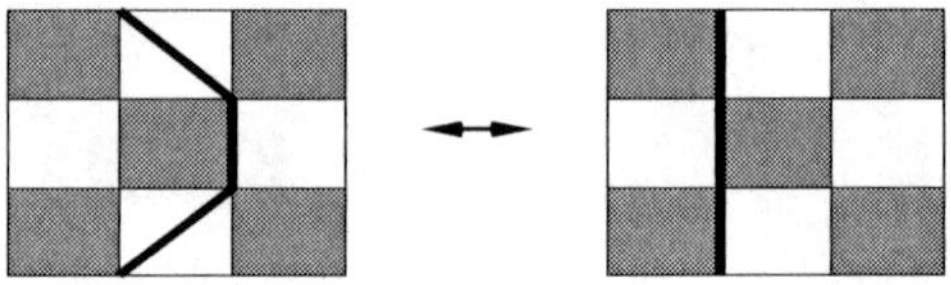

FIGURE 2. Local updates. A shaded plaquette is chosen randomly and a Word Line is shifted from left to right or vice versa across the shaded plaquette.

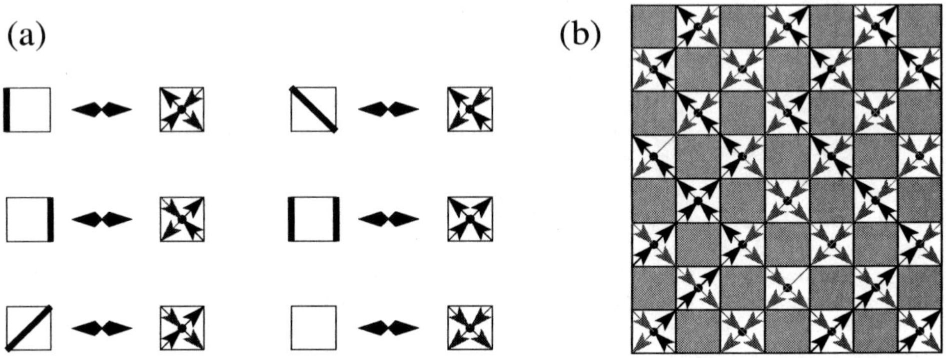

FIGURE 3. (a) Identification of world lines configurations on plaquettes with the vertices of the 6-vertex model. (b) The World Line configuration of Fig. 1 in the language of the 6-vertex model.

above local updates the correct thermodynamic result is obtained [6].

That the XXZ quantum spin chain is equivalent to the classical two-dimensional 6-vertex model follows from a one to one mapping of a World Line configuration to one of the 6-vertex model. The identification of single plaquettes is shown in Fig. 3(a). The world line configuration of Fig. 1 is plotted in the language of the 6-vertex mode in Fig. 3(b). The vertex model lies on a 45 degrees rotated lattice denoted by bullets in Fig. 1(b). At each vertex (bullets in Fig. 1(b)) the number of incoming arrows equals the number of outgoing arrows. In the case of the XYZ chain, source and drain terms have to be added to yield the 8-vertex model.

The identification of the XXZ model to the 6-vertex model gives us an intuitive picture of loop upgrades [2]. Consider the World Line configuration in Fig.4a and it's corresponding vertex formulation (Fig.4b). One can pick a site at random and follow the arrows of the vertex configuration. At each plaquette there are two possible arrow paths to follow. One is chosen, appropriately, and one follows the arrows to arrive to the next plaquette. The procedure is then repeated until one returns to the starting point. Such a loop is shown in Fig. 4c. Along the loop, changing the direction of the arrows generates another valid vertex configuration (see Fig. 4d). The corresponding World Line configuration (after *flipping* the loop) is shown in Fig. 4e. As apparent, this is a global update which changes the winding number and is not achievable with local moves.

To understand the loop algorithm in more details it is convernient to introduce graphs.

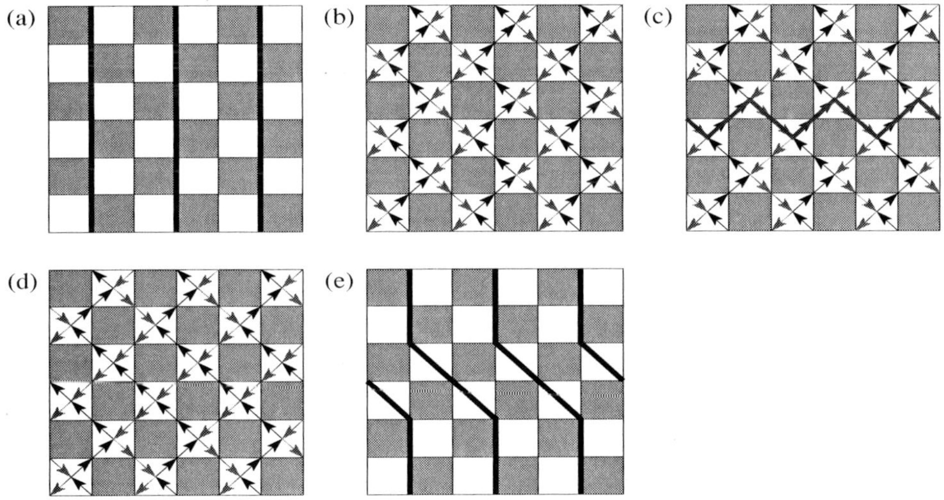

FIGURE 4. Example of a loop update.

For each plaquette and associated vertex there are many possible choices of graphs which merely reflect the fact that different arrow paths may be followed in the construction of the loop. Fig. 5 illustrates this. Given the vertex configurations 1 in Fig. 5 one can follow the arrows vertically (graph 1) or horizontally (graph 2). There is also the possiblilty to flip all the spins of the vertex. This corresponds to graph 3 in Fig. 5. Thus the graph configuration defines the loops along which one will propose to flip the orientation of the arrows of the vertex model. Starting from a vertex configuration S on a plaquette we choose an allowed graph with probability.

$$P(S \to (S,G)) = \frac{W(S,G)}{W(S)}. \tag{10}$$

thereby expanding the space of vertices to the space of vertices and graphs. In the above equation $W(S)$ is the weight of the vertex S and $W(S,G)$ the weight of the combined vertex (S) and graph (G) configuration on the plaquette. We require that:

$$\sum_G W(S,G) = W(S) \tag{11}$$

where the sum runs over all allowed graphs given the vertex configration S. For example graphs $1,4,$ and 3 if $S=3$ (See Fig. 5). To achieve a constant acceptance rate we require that for a given graph G

$$W(S,G) = W(S',G) \tag{12}$$

where S' is obtained from S by flipping the arrows of the vertex configuration according the rules of the graph G. That is for $G=1$ in Fig. 5 we require: $W(S=1,G=1) = W(S=3,G=1)$. Choosing the heat-bath algorithm the probability of flipping the arrows along the loop is given by:

$$P\left((S,G)\rightarrow(S',G)\right)=\frac{\frac{W(S',G)}{W(S,G)}}{1+\frac{W(S',G)}{W(S,G)}}\equiv\frac{1}{2}\tag{13}$$

With Eqn. (12) and (13) the detailed balance in the space of graphs and spins

$$W(S,G)P[(S,G)\rightarrow(S',G)]=W(S',G)P[(S',G)\rightarrow(S,G)]\tag{14}$$

is trivially satisfied. Detailed balance in the space of spins follows from:

$$\begin{aligned}&W(S)P\left(S\rightarrow S'\right)=W(S)\sum_G P[S\rightarrow(S,G)]P[(S,G)\rightarrow(S',G)]=\\&\sum_G W(S)\frac{W(S,G)}{W(S)}P[(S,G)\rightarrow(S',G)]=\\&\sum_G W(S')\frac{W(S',G)}{W(S')}P[(S',G)\rightarrow(S,G)]=W(S')P\left(S'\rightarrow S\right).\end{aligned}\tag{15}$$

This completes the formal discription of the algorithm. We will now illustrate the algorithm in the case of the Heisenberg model ($J=J_x=J_z$) since this turns out to be a particularly simple point. Eqn. (11) and (12) lead to the equations:

$$\begin{aligned}e^{\Delta\tau J/4}\cosh(\Delta\tau J/2)\equiv W_1&=W_{1,1}+W_{1,2}+W_{1,3}\\e^{\Delta\tau J/4}\sinh(\Delta\tau J/2)\equiv W_2&=W_{2,2}+W_{2,4}+W_{2,3}\\e^{-\Delta\tau J/4}\equiv W_3&=W_{3,1}+W_{3,4}+W_{3,3}\text{ with}\\W_{3,1}=W_{1,1},W_{1,2}&=W_{2,2}\text{ and }W_{2,4}=W_{3,4}.\end{aligned}\tag{16}$$

Here we adopt the notation: $W_{i,j}=W(S=i,G=j)$ and $W_i=W(S=i)$. To satisfy the above equations for the sepcial case of the Heisenberg model, we can set $W_{\bullet,3}=W_{\bullet,4}=0$ and thereby consider only the graphs $G=1$ and $G=2$. Te reader will readiliy verify that the equations

$$\begin{aligned}e^{\Delta\tau J/4}\cosh(\Delta\tau J/2)\equiv W_1&=W_{1,1}+W_{1,2}\\e^{\Delta\tau J/4}\sinh(\Delta\tau J/2)\equiv W_2&=W_{2,2}=W_{1,2}\\e^{-\Delta\tau J/4}\equiv W_3&=W_{1,1}=W_{3,1}\end{aligned}\tag{17}$$

are satisfied.

To gain further insight into the loop algorithm the reader is referred to the review article of H.G. Evertz [19] and references therein. Let us however mention that the loop algorithm has been applied with great success to non-frustrated spin systems. Critical exponents of the order-disorder transition for a two dimensional depleted Heisenberg model were pinned down to show that the transition belongs to the universality class of the three dimensional classical $O(3)$ model [8]. Furthermore this algorithm has been used to study single particle dynamics in non-frustrated quantum antiferromagnets on various topologies [20, 21].

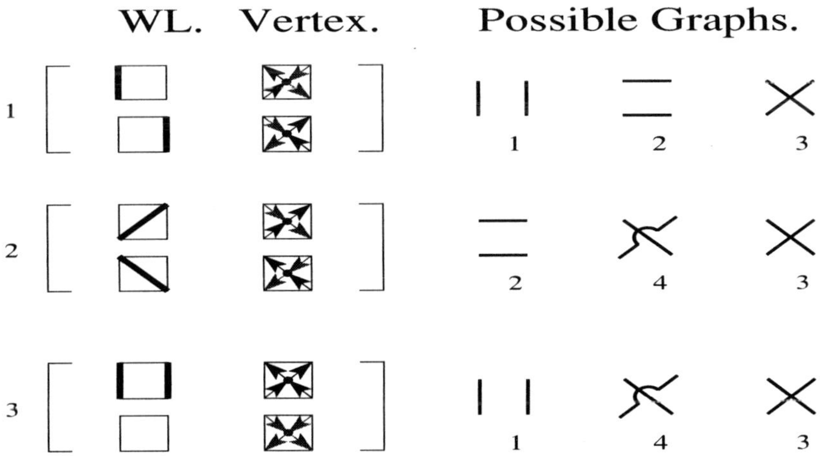

FIGURE 5. Possible graphs for vertex configurations.

The sign problem in the World Line approach.

The Quantum Monte Carlo approach is often plagued by the so-called sign problem. Since the origin of this problem is easily understood in the framework of the World Line algorithm we will briefly discuss it in this section on a specific model. Consider spinless electrons on an L-site linear chain

$$H = -t\sum_i c_i^\dagger \left(c_{i+1} + c_{i+2}\right) + \text{H.c. with } \left\{c_i^\dagger, c_j^\dagger\right\} = \left\{c_i, c_j\right\} = 0, \left\{c_i^\dagger, c_j\right\} = \delta_{i,j}. \quad (18)$$

Here, we consider periodic boundary conditions, $c_{i+L} = c_i$ and $t > 0$. To apply the World Line algorithm to the above Hamiltonian we split it into a set of independent four site problems:

$$H = \underbrace{\sum_{n=0}^{L/4-1} H^{(4n+1)}}_{H_1} + \underbrace{\sum_{n=0}^{L/4-1} H^{(4n+3)}}_{H_2} \quad (19)$$

with

$$H^{(i)} = -tc_i^\dagger\left(\frac{1}{2}c_{i+1} + c_{i+2}\right) - tc_{i+1}^\dagger\left(c_{i+2} + c_{i+3}\right) - \frac{t}{2}c_{i+2}^\dagger c_{i+3} + \text{H.c.}$$

With this decomposition one obtains the graphical representation of Fig. 6. [22]

The sign problem occurs from the fact that the weights $\Omega(w)$ are not necessarily positive. An example is shown in Fig. 18. In this case the origin of negative signs lies in Fermi statistics. To *solve* the problem, one decides to carry out the sampling with an auxiliary probability distribution:

$$\overline{Pr}(\omega) = \frac{|\Omega(w)|}{\sum_w |\Omega(w)|} \tag{20}$$

which in the limit of small values of $\Delta\tau$ corresponds to the partition function of the Hamiltonian of Eq. (18) but with fermions replaced by hard-core bosons. Thus, we can now evaluate Eq. 9 with:

$$\langle O \rangle = \frac{\sum_w \overline{Pr}(\omega)\text{sign}(w)O(w)}{\sum_w \overline{Pr}(\omega)\text{sign}(w)} \tag{21}$$

where both the numerator and denominator are evaluated with MC methods. Let us first consider the denominator:

$$\langle \text{sign} \rangle = \sum_w \overline{Pr}(\omega)\text{sign}(w) = \frac{\sum_w \Omega(w)}{\sum_w |\Omega(w)|} = \frac{\text{Tr}\left[e^{-\beta H}\right]}{\text{Tr}\left[e^{-\beta H_B}\right]} \tag{22}$$

Here, H_B corresponds to the Hamiltonian of Eq. (18) but with fermions replaced by hard-core bosons. In the limit of large inverse temperatures, β, the partition functions is dominated by the ground state. Thus in this limit

$$\langle \text{sign} \rangle \sim e^{-\beta\left(E_0 - E_0^B\right)} = e^{-\beta L \Delta} \tag{23}$$

where $\Delta = \left(E_0 - E_0^B\right)/L$ is an intensive, in general positive, quantity. The above equation corresponds to the sign problem. When the temperature is small or system size large, the average sign becomes exponentially small. Hence, the observable $\langle O \rangle$ is given by the quotient of two exponentially small values which are determined stochastically. When the error-bars become comparable to the average sign, uncontrolled fluctuations in the evaluation of $\langle O \rangle$ will occur. Two comments are in order. (i) In this simple example the sign problem occurs due to Fermi statistics. However, sign problems occur equally in frustrated spin-1/2 systems which are nothing but hard core boson models. Note that replacing the fermions by hard core bosons in Eq. (18) and considering hopping matrix elements of different signs between nearest and next nearest neighbors will generate a sign problem in the above formulation. (ii) The sign problem is formulation dependent. In the World Line algorithm, we decide to work in real space. Had we chosen Fourier space, the Hamiltonian would have been diagonal and hence no sign problem occurs. In the auxiliary field approach discussed in the next section the sign problem would not occur for this non-interacting problem since as mentioned in the introduction one body operators are treated exactly. That is, the sum over all World Lines is carried out exactly.

AUXILIARY FIELD QUANTUM MONTE CARLO ALGORITHMS

In the remaining, we will concentrate on auxiliary field Quantum Monte Carlo algorithms which we will describe in detail. Consider the Hamiltonian

$$H = H_t + H_I \tag{24}$$

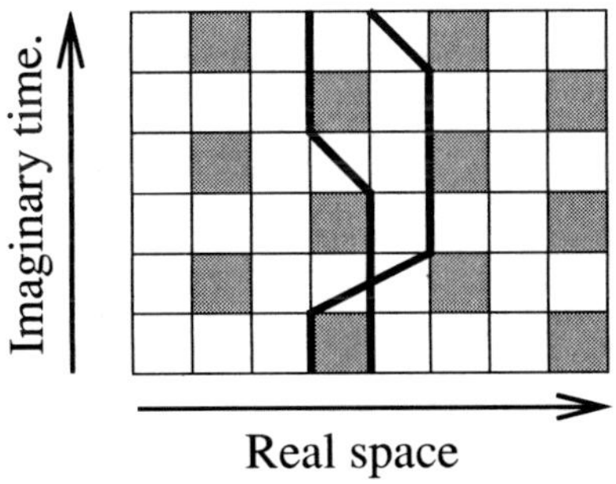

FIGURE 6. World line configuration for the model of Eq. (18). Here, $m = 3$. Since the two electrons exchange their positions during the imaginary time propagation, this world line configuration has a negative weight.

where H_t is the kinetic energy and H_I a two-body interaction term. The ground state expectation value of an observable O is at best obtained by projecting a trial wave function $|\Psi_T\rangle$ along the imaginary time axis:

$$\frac{\langle\Psi_0|O|\Psi_0\rangle}{\langle\Psi_0|\Psi_0\rangle} = \lim_{\Theta\to\infty} \frac{\langle\Psi_T|e^{-\Theta H} O e^{-\Theta H}|\Psi_T\rangle}{\langle\Psi_T|e^{-2\Theta H}|\Psi_T\rangle}. \tag{25}$$

The above equation is readily verified by writing $|\Psi_T\rangle = \sum_n |\Psi_n\rangle\langle\Psi_n|\Psi_0\rangle$ with $H|\Psi_n\rangle = E_n|\Psi_n\rangle$. The assumptions that $\langle\Psi_T|\Psi_0\rangle \neq 0$ and that the ground state is non-degenerate are however required. The algorithm based on Eq. (25) is known as the projector quantum Monte Carlo (PQMC) algorithm [12, 13, 14].

Finite temperature properties in the grand canonical ensemble are obtained by evaluating

$$\langle O\rangle = \frac{\mathrm{Tr}\left[e^{-\beta(H-\mu N)} O\right]}{\mathrm{Tr}\left[e^{-\beta(H-\mu N)}\right]} \tag{26}$$

where the trace runs over the Fock space, $\beta = 1/k_B T$ and μ is the chemical potential. The algorithm based on Eq. (26) will be referred to as finite temperature QMC (FTQMC) method [9, 10]. Comparison of both algorithms are shown in Fig. (7) for the Hubbard model in standard notation.

$$H_U = -t \sum_{\langle\vec{i},\vec{j}\rangle,\sigma} e^{\frac{2\pi i}{\Phi_0}\int_{\vec{i}}^{\vec{j}} \vec{A}\cdot d\vec{l}} c^{\dagger}_{\vec{i},\sigma} c_{\vec{j},\sigma} + U \sum_{\vec{i}} n_{\vec{i},\uparrow} n_{\vec{i},\downarrow}. \tag{27}$$

Here, and for future use, we have included a magnetic field $\vec{B} = \nabla \times \vec{A}$. At half-filling, the ground state is insulating so that charge fluctuations are absent in the low temperature limit on finite lattices. Hence, in this limit both grand canonical and canonical approaches yield identical results. It is however clear that if one is interested solely in ground state properties the PQMC is more efficient. As we will see, this lies in the choice of the trial wave function which is chosen to be a spin singlet.

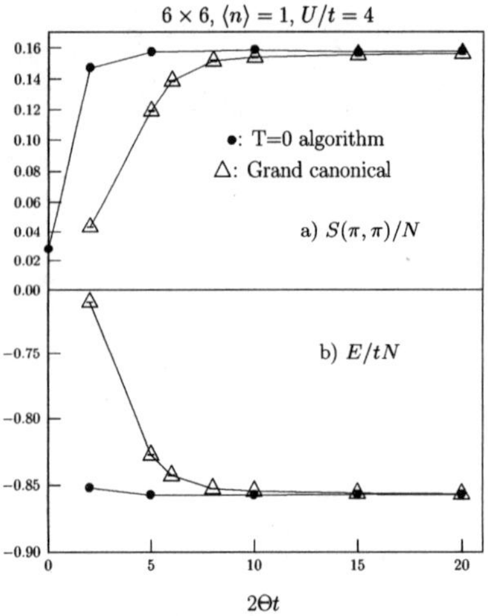

FIGURE 7. Fourier transform of the spin-spin correlation functions at $\vec{Q} = (\pi, \pi)$ (a) and energy (b) for the half-filled Hubbard model in the absence of magnetic field (27). •: PQMC algorithm. Δ: FTQMC algorithm at $\beta = 2\Theta$.

Trotter decomposition and Hubbard Stratonovich transformation

As in the World Line approach we wish to evaluate the imaginary time propagation which we will split into infenitesimal successive propagations with H_t followed by H_I. This is again achieved with the Trotter decomposition

$$\left(e^{-\Delta\tau H_I} e^{-\Delta\tau H_t}\right)^m = e^{-\Theta H} + \frac{\Delta\tau}{2} \int_0^\Theta d\tau e^{-(\Theta-\tau)H} [H_I, H_t] e^{-\tau H} + \mathcal{O}(\Delta\tau^2) \quad (28)$$

or a symmetric variant

$$\left(e^{-\Delta\tau H_t/2} e^{-\Delta\tau H_I} e^{-\Delta\tau H_t/2}\right)^m = e^{-\Theta H} + \mathcal{O}(\Delta\tau^2) \quad (29)$$

where $m\Delta\tau = \Theta$. As mentioned previously and for the FTQMC if O, H_I and H_t are simultaneously real representable in a given basis, the systematic error proportional to $\Delta\tau$ in Eq. (28) vanishes. To achieve this in the PQMC some care has to be taken. To reduce fluctuations one wishes to measure an observable over several time slices. In the QMC evaluation of O based on the Trotter decomposition of Eq. (28) we hence compute:

$$\frac{1}{2N+1} \sum_{n=-N}^{N} \frac{\langle \Psi_T | \left(e^{-\Delta\tau H_I} e^{-\Delta\tau H_t}\right)^{m-n} O \left(e^{-\Delta\tau H_I} e^{-\Delta\tau H_t}\right)^{m+n} |\Psi_T\rangle}{\langle \Psi_T | \left(e^{-\Delta\tau H_I} e^{-\Delta\tau H_t}\right)^{2m} |\Psi_T\rangle}. \quad (30)$$

If $[O,H] = 0$, then one can set $N = m$ and the effective projection parameter is 2Θ. On the other hand, if $[O,H] \neq 0$ one has to choose $N < m$ since the effective projection parameter is $(m-N)\Delta\tau$. It is however crucial to measure symmetrically around the

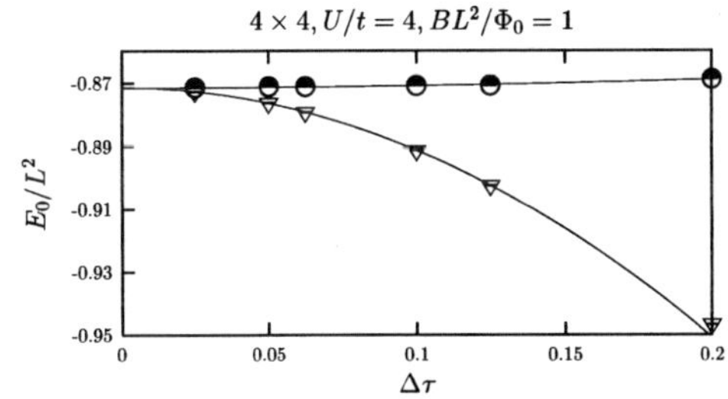

FIGURE 8. Ground state energy of the Half-filled Hubbard model on a 4×4 lattice, $U/t = 4$, $\langle n \rangle = 1$ and $BL^2/\Phi_0 = 1$ as a function of $\Delta\tau$ as obtained with the PQMC. The Trotter decompositions of Eqn. (28) ($\bigtriangledown$) and (29) ($\bigcirc$) are considered. Note that due to the variational principle the the Trotter decomposition of Eq. (29) yields an upper bound to the energy. The solid lines correspond to least square fits to the form $a + b\Delta\tau^2$.

central time slice, since only then can show that the systematic error proportional to $\Delta\tau$ vanishes. This is of course valid provided that O, H_I, H_t as well as $|\Psi_T\rangle$ are simultaneously real representable.

Fig. 8 compares both choices of Trotter decompositions (Eqn. (28) and (29)) for the Hubbard model of Eq. (27). Here we use the PQMC algorithm. As apparent the symmetric decomposition of Eq. (29) is much more accurate and due to the variational principle provides an upper bound to the exact energy. It is however often cumbersome to implement.

Having isolated the interaction term H_I with the Trotter decomposition, we may now proceed with the Hubbard-Stratonovich (HS) transformation. The choice of the HS is important. For Hubbard interactions,

$$H_I = U \sum_{\vec{i}} \left(n_{\vec{i},\uparrow} - 1/2 \right) \left(n_{\vec{i},\downarrow} - 1/2 \right), \tag{31}$$

with $U > 0$ one usually chooses Hirsch's discrete transformation [23]

$$\begin{aligned} \exp\left(-\Delta\tau U \sum_{\vec{i}} \left(n_{\vec{i},\uparrow} - 1/2 \right) \left(n_{\vec{i},\downarrow} - 1/2 \right) \right) & \\ = \tilde{C} \sum_{s_1,\ldots,s_N = \pm 1} & \exp\left(\tilde{\alpha} \sum_{\vec{i}} s_{\vec{i}} \left(n_{\vec{i},\uparrow} - n_{\vec{i},\downarrow} \right) \right). \end{aligned} \tag{32}$$

where $\cosh(\tilde{\alpha}) = \exp(\Delta\tau U/2)$. On an N-site lattice, the constant $\tilde{C} = \exp(\Delta\tau U N/4)/2^N$. As apparent from the above equation, for a fixed set of HS fields, $s_1 \ldots s_N$, $SU(2)$-spin symmetry is broken since the field couples to the z-component of the magnetization. This symmetry is of course restored after summation over the HS fields with the Monte Carlo method. Alternatively, one may consider [23]

$$\exp\left(-\Delta\tau U \sum_{\vec{i}} \left(n_{\vec{i},\uparrow} - 1/2 \right) \left(n_{\vec{i},\downarrow} - 1/2 \right) \right) \tag{33}$$

$$= C \sum_{s_1,\dots,s_N=\pm 1} \exp\left(i\alpha \sum_{\vec{i}} s_{\vec{i}} \left(n_{\vec{i},\uparrow} + n_{\vec{i},\downarrow} - 1 \right) \right).$$

where $\cos(\alpha) = \exp(-\Delta\tau U/2)$ and $C = \exp(\Delta\tau U N/4)/2^N$. With this choice of the HS transformation $SU(2)$ spin invariance is retained for any given HS configuration since the field couples to the density. Even taking into account the overhead of working with complex numbers, it is more convenient to work with this transformation [24] since it is often hard to restore the full $SU(2)$ spin symmetry via Monte Carlo sampling of the HS field. When $U < 0$ the transformation (33) may readily be used and involves only real numbers.

We now consider interaction terms of the form:

$$H_I = -W \sum_{\vec{i}} \left(O^{(\vec{i})} \right)^2 \tag{34}$$

where $O^{(\vec{i})}$ is a one-body operator. In general, $\left[O^{(\vec{i})}, O^{(\vec{j})} \right] \neq 0$ so that the sum in the above equation has to be split into sums of commuting terms: $H_I = \sum_r H_I^r$, $H_I^r = -W \sum_{\vec{i} \varepsilon S_r} \left(O^{(\vec{i})} \right)^2$. For $\vec{i}$ and $\vec{j}$ in the set S_r one requires $\left[O^{(\vec{i})}, O^{(\vec{j})} \right] = 0$. The imaginary time evolution may be written as $e^{-\Delta\tau H_t} \approx \prod_r e^{-\Delta\tau H_t^r}$. Thus we are left with the problem of decoupling $e^{\Delta\tau W O^2}$ where we have omitted the index $\vec{i}$. In principle, one can decouple a perfect square with the canonical HS transformation:

$$e^{\Delta\tau W O^2} = \frac{1}{\sqrt{2\pi}} \int \mathrm{d}\Phi e^{-\frac{\Phi^2}{2} + \sqrt{2\Delta\tau W}\Phi O} \tag{35}$$

However, this involves a continuous field which renders the sampling hard. An alternative formulation is given by [25]:

$$e^{\Delta\tau W O^2} = \sum_{l=\pm 1, \pm 2} \gamma(l) e^{\sqrt{\Delta\tau W}\eta(l) O} + \mathscr{O}(\Delta\tau^4) \tag{36}$$

where the fields η and γ take the values:

$$\gamma(\pm 1) = 1 + \sqrt{6}/3, \quad \gamma(\pm 2) = 1 - \sqrt{6}/3$$
$$\eta(\pm 1) = \pm\sqrt{2\left(3 - \sqrt{6}\right)}, \quad \eta(\pm 2) = \pm\sqrt{2\left(3 + \sqrt{6}\right)}.$$

This transformation is not exact and produces an overall systematic error proportional to $\Delta\tau^3$. However, since we already have a systematic error proportional to $\Delta\tau^2$ from the Trotter decomposition, the transformation is as good as exact. It also has the great advantage of being discrete thus allowing efficient sampling.

Thus, the HS transformation has enabled us to split the two-body interaction term into a one-body operator interacting with an external field. We may now write the imaginary time propagator as:

$$\prod_{n=1}^{m}\left[e^{-\Delta\tau H_I}e^{-\Delta\tau H_t}\right]=\prod_{n=1}^{m}\left[\sum_{\vec{s}}C(\vec{s})e^{H_I(\vec{s})}e^{-\Delta\tau H_t}\right] \tag{37}$$

$$\sum_{\vec{s}_1\cdots\vec{s}_m}C(\vec{s}_1\cdots\vec{s}_m)\prod_{n=1}^{m}\left[e^{H_I(\vec{s}_n)}e^{-\Delta\tau H_t}\right]$$

Here, $H_I(\vec{s}_n)$ is a single body operator. The HS fields have a lattice site, $\vec{i}$ and imaginary time, n, index. We will adopt the notation: $\vec{s}=\left\{s_{\vec{i},n}\right\}\ \vec{i}=1\cdots N$ and $n=1\cdots m$, and $\vec{s}_n$ denotes the HS fields on time slice n. In the special case of the Hubbard model with the HS transformation of Eq. (33),

$$H_I(\vec{s}_n)=i\alpha\sum_{\vec{j},\sigma}s_{\vec{j},n}c^{\dagger}_{\vec{j},\sigma}c_{\vec{j},\sigma},\ C(\vec{s}_1\cdots\vec{s}_m)=\exp\left(\Theta UN/4\right)\exp\left(-i\alpha\sum_{\vec{j},n}s_{\vec{j},n}\right). \tag{38}$$

For an interaction term of the form in Eq. (34) $e^{H_I(\vec{s}_n)}$ has to be replaced by

$$\prod_{r}\exp\left(\sqrt{\Delta\tau W}\sum_{\vec{i}\varepsilon S_r}\eta(l_{\vec{i},n})O^{(\vec{i})}\right) \tag{39}$$

and $C(\left\{s_{\vec{i},n}\right\})$ by $C(\left\{l_{\vec{i},n}\right\})=\prod_{\vec{i},n}\gamma(l_{\vec{i},n})$.

At this stage, the fermionic degrees of freedom may be integrated out. In the framework of the PQMC, we will require the trial wave function to be a Slater determinant:

$$|\Psi_T\rangle=\prod_{y=1}^{N_p}\left(\sum_{x}c^{\dagger}_{x}P_{x,y}\right)|0\rangle=\prod_{y=1}^{N_p}\left(\vec{c}^{\dagger}P\right)_y|0\rangle \tag{40}$$

where we have introduced the indices $\vec{x}=(\vec{i},\sigma)$, which run from $1\cdots 2N$, N_p is the number of particles and $\vec{c}^{\dagger}=\left(c^{\dagger}_1,\cdots,c^{\dagger}_{2N}\right)$. For a Hermitian or anti-hermitian matrix T, one may show that:

$$e^{\vec{c}^{\dagger}T\vec{c}}\prod_{y=1}^{N_p}\left(\vec{c}^{\dagger}P\right)_y|0\rangle=\prod_{y=1}^{N_p}\left(\vec{c}^{\dagger}e^{T}P\right)_y|0\rangle. \tag{41}$$

To derive the above equation, it is useful to go into a basis where T is diagonal: $U^{\dagger}TU=D$. U is a unitary matrix and D a real (purely imaginary) diagonal matrix provided that T is hermetian (anti-hermetian). Thus we can define the fermionic operators $\vec{\gamma}^{\dagger}=\vec{c}^{\dagger}U$ to obtain:

$$e^{\vec{c}^{\dagger}T\vec{c}}\prod_{y-1}^{N_p}\left(\vec{c}^{\dagger}P\right)_y|0\rangle=e^{\vec{\gamma}^{\dagger}D\vec{\gamma}}\prod_{y=1}^{N_p}\left(\vec{\gamma}^{\dagger}UP\right)_y|0\rangle=$$

$$\sum_{y_1,\cdots,y_{N_p}} e^{\sum_x D_{x,x}\vec{\gamma}_x^\dagger \vec{\gamma}_x} \gamma_{y_1}^\dagger \cdots \gamma_{y_{N_p}}^\dagger |0\rangle (UP)_{y_1,1} \cdots (UP)_{y_{N_p},N_p} =$$

$$\sum_{y_1,\cdots,y_{N_p}} e^{D_{y_1,y_1}} \gamma_{y_1}^\dagger \cdots e^{D_{y_{N_p},y_{N_p}}} \gamma_{y_{N_p}}^\dagger |0\rangle (UP)_{y_1,1} \cdots (UP)_{y_{N_p},N_p} =$$

$$\prod_{y=1}^{N_p} \left(\vec{\gamma}^\dagger e^D UP\right)_y |0\rangle = \prod_{y=1}^{N_p} \left(\vec{c}^\dagger U^\dagger e^D UP\right)_y |0\rangle = \prod_{y=1}^{N_p} \left(\vec{c}^\dagger e^T P\right)_y |0\rangle.$$

We can now evaluate the imaginary time propagator. It is convenient to define the notation:

$$H_I = \vec{c}^\dagger h_I(\vec{s}_\tau)\vec{c}, \;\; H_t = \vec{c}^\dagger h_t \vec{c}$$

$$U_{\vec{s}}(\tau_2, \tau_1) = \prod_{n=n_1+1}^{n_2} e^{H_I(\vec{s}_n)} e^{-\Delta\tau H_t} \tag{42}$$

$$\text{and } B_{\vec{s}}(\tau_2, \tau_1) = \prod_{n=n_1+1}^{n_2} e^{h_I(\vec{s}_n)} e^{-\Delta\tau h_t} \tag{43}$$

where, $n_1\Delta\tau = \tau_1$ and $n_2\Delta\tau = \tau_2$, h_t is a $2N \times 2N$ hermetian matrix and $h_I(\vec{s}_n)$ a $2N \times 2N$ hermetian or anti-hermetian matrix depending upon the choice of the HS transformation. One can then derive:

$$\langle \Psi_T | U_{\vec{s}}(2\Theta, 0) | \Psi_T \rangle = \det\left(P^\dagger B(2\Theta, 0) P\right). \tag{44}$$

The above follows from:

$$\langle \Psi_T | U_{\vec{s}}(2\Theta, 0) | \Psi_T \rangle = \langle \Psi_T | \prod_{y=1}^{N_p} \left(\vec{c}^\dagger B_{\vec{s}}(2\Theta, 0) P\right)_y |0\rangle =$$

$$\sum_{\substack{x_1 \cdots x_{N_p} \\ y_1 \cdots y_{N_p}}} P_{1,y_{N_p}}^\dagger \cdots P_{1,y_1}^\dagger \langle 0 | c_{y_1} \cdots c_{y_{N_p}} c_{x_{N_p}}^\dagger \cdots c_{x_1}^\dagger |0\rangle \times$$

$$\left(B_{\vec{s}}(2\Theta, 0)P\right)_{x_1,1} \cdots \left(B_{\vec{s}}(2\Theta, 0)P\right)_{x_{N_p},N_p}.$$

The matrix element $\langle 0 | c_{y_1} \cdots c_{y_{N_p}} c_{x_{N_p}}^\dagger \cdots c_{x_1}^\dagger |0\rangle$ does not vanish provided that i) $x_1 \neq x_2 \neq \cdots \neq x_{N_p}$ and ii) the indices $y_1 \cdots y_{N_p}$ are a permutation (π) of the indices $x_1 \cdots x_{N_p}$: $y_1 = x_{\pi(1)} \cdots y_{N_p} = x_{\pi(N_p)}$. The matrix element is then equal to the sign of permutation: $(-1)^\pi$. Hence,

$$\langle \Psi_T | U_{\vec{s}}(2\Theta, 0) | \Psi_T \rangle =$$

$$\sum_{\substack{x_1 \cdots x_{N_p} \\ \pi \varepsilon \mathscr{S}_{N_p}}} (-1)^\pi P_{1,x_{\pi(1)}}^\dagger \left(B_{\vec{s}}(2\Theta, 0)P\right)_{x_1,1} \cdots P_{N_p,x_{\pi(N_p)}}^\dagger \left(B_{\vec{s}}(2\Theta, 0)P\right)_{x_{N_p},N_p} =$$

$$
\sum_{\pi \varepsilon \mathscr{S}_{N_p}} (-1)^{\pi} \left(P^{\dagger} B_{\vec{s}}(2\Theta,0) P \right)_{\pi(1),1} \cdots \left(P^{\dagger} B_{\vec{s}}(2\Theta,0) P \right)_{\pi(N_p),N_p} =
$$
$$
\det \left(P^{\dagger} B_{\vec{s}}(2\Theta,0) P \right)
$$

where the last sum runs over the space $\mathscr{S}_{N_p}$ consisting of the $N_p!$ permutations of the integers $[1 \cdots N_p]$.

For the FTQMC, we have to evaluate evaluate the trace over the Fock space.

$$
\mathrm{Tr}\left(U_{\vec{s}}(\beta,0) \right) = \det \left(1 + B_{\vec{s}}(\beta,0) \right) \tag{45}
$$

where $m\Delta\tau = \beta$. The above equation is readily verified:

$$
\begin{aligned}
& \det \left(1 + B_{\vec{s}}(\beta,0) \right) = \\
& \sum_{\pi \varepsilon \mathscr{S}_{2N}} (-1)^{\pi} \left(1 + B_{\vec{s}}(\beta,0) \right)_{\pi(1),1} \cdots \left(1 + B_{\vec{s}}(\beta,0) \right)_{\pi(2N),2N} = \\
& \sum_{\pi \varepsilon \mathscr{S}_{2N}} (-1)^{\pi} \delta_{1,\pi(1)} \cdots \delta_{2N,\pi(2N)} + \\
& \sum_{x} \sum_{\pi \varepsilon \mathscr{S}_{2N}} (-1)^{\pi} B_{\vec{s}}(\beta,0)_{\pi(x),x} \delta_{1,\pi(1)} \cdots \widehat{\delta_{x,\pi(x)}} \cdots \delta_{2N,\pi(2N)} + \\
& \sum_{y>x} \sum_{\pi \varepsilon \mathscr{S}_{2N}} (-1)^{\pi} B_{\vec{s}}(\beta,0)_{\pi(x),x} B_{\vec{s}}(\beta,0)_{\pi(y),y} \times \\
& \qquad \delta_{1,\pi(1)} \cdots \widehat{\delta_{x,\pi(x)}} \cdots \widehat{\delta_{y,\pi(y)}} \cdots \delta_{2N,\pi(2N)} + \\
& \sum_{y>x>z} \sum_{\pi \varepsilon \mathscr{S}_{2N}} (-1)^{\pi} B_{\vec{s}}(\beta,0)_{\pi(x),x} B_{\vec{s}}(\beta,0)_{\pi(y),y} B_{\vec{s}}(\beta,0)_{\pi(z),z} \times \\
& \qquad \delta_{1,\pi(1)} \cdots \widehat{\delta_{x,\pi(x)}} \cdots \widehat{\delta_{y,\pi(y)}} \cdots \widehat{\delta_{z,\pi(z)}} \cdots \delta_{2N,\pi(2N)} + \cdots = \\
& 1 + \sum_{x} \langle 0 | c_x U_{\vec{s}}(\beta,0) c_x^{\dagger} | 0 \rangle + \\
& \sum_{y>x} \langle 0 | c_x c_y U_{\vec{s}}(\beta,0) c_y^{\dagger} c_x^{\dagger} | 0 \rangle + \\
& \sum_{y>x>z} \langle 0 | c_x c_y c_z U_{\vec{s}}(\beta,0) c_z^{\dagger} c_y^{\dagger} c_x^{\dagger} | 0 \rangle + \cdots = \\
& \mathrm{Tr}\left(U_{\vec{s}}(\beta,0) \right).
\end{aligned}
$$

Here, $\widehat{\delta_{y,\pi(y)}}$ means that this term is omitted in the product: $\prod_{x=1}^{2N} \delta_{x,\pi(x)}$. We have used Eq. (44) to derive the third equality.

Observables and Wick's theorem

In the last section, we have shown how to carry out the HS transformation and integrate out the fermionic degrees of freedom. Here, we show both for the PQMC and

FTQMC how to compute observables as well as the validity of Wick's theorem for a fixed configuration of HS fields.

PQMC

In the PQMC algorithm we compute:

$$\frac{\langle \Psi_T | e^{-\Theta H} O e^{-\Theta H} | \Psi_T \rangle}{\langle \Psi_T | e^{-2\Theta H} | \Psi_T \rangle} = \sum_{\vec{s}} \mathrm{Pr}_{\vec{s}} \langle O \rangle_{\vec{s}} + O(\Delta\tau^2). \tag{46}$$

For each lattice site, $\vec{i}$, time slice, n, we have introduced an independent HS field, $\vec{s} = \left\{ s_{\vec{i},n} \right\}$ and

$$\mathrm{Pr}_{\vec{s}} = \frac{C_{\vec{s}} \det\left(P^\dagger B_{\vec{s}}(2\Theta, 0) P\right)}{\sum_{\vec{s}} C_{\vec{s}} \det\left(P^\dagger B_{\vec{s}}(2\Theta, 0) P\right)}$$
$$\langle O \rangle_{\vec{s}} = \frac{\langle \Psi_T | U_{\vec{s}}(2\Theta, \Theta) O U_{\vec{s}}(\Theta, 0) | \Psi_T \rangle}{\langle \Psi_T | U_{\vec{s}}(2\Theta, 0) | \Psi_T \rangle}$$

We start by computing the equal time Green function: $O = c_x c_y^\dagger = \delta_{x,y} - \vec{c}^\dagger A^{(y,x)} \vec{c}$ with $A^{(y,x)}_{x_1,x_2} = \delta_{x_1,y} \delta_{x_2,y}$. Inserting a source term, we obtain:

$$\begin{aligned}
&\langle c_x c_y^\dagger \rangle_{\vec{s}} = \delta_{x,y} - \frac{\partial}{\partial \eta} \ln \langle \Psi_T | U_{\vec{s}}(2\Theta, \Theta) e^{\eta \vec{c}^\dagger A^{(y,x)} \vec{c}} U_{\vec{s}}(\Theta, 0) | \Psi_T \rangle \,|_{\eta=0} = \\
&\delta_{x,y} - \frac{\partial}{\partial \eta} \ln\det\left(P^\dagger B_{\vec{s}}(2\Theta, \Theta) e^{\eta A^{(y,x)}} B_{\vec{s}}(\Theta, 0) P \right) |_{\eta=0} = \\
&\delta_{x,y} - \frac{\partial}{\partial \eta} \mathrm{Tr} \ln\left(P^\dagger B_{\vec{s}}(2\Theta, \Theta) e^{\eta A^{(y,x)}} B_{\vec{s}}(\Theta, 0) P \right) |_{\eta=0} = \\
&\delta_{x,y} - \mathrm{Tr}\left[\left(P^\dagger B_{\vec{s}}(2\Theta, 0) P \right)^{-1} P^\dagger B_{\vec{s}}(2\Theta, \Theta) A^{(y,x)} B_{\vec{s}}(\Theta, 0) P \right] \\
&\left(1 - B_{\vec{s}}(\Theta, 0) P \left(P^\dagger B_{\vec{s}}(2\Theta, 0) P \right)^{-1} P^\dagger B_{\vec{s}}(2\Theta, \Theta) \right)_{x,y} \equiv \left(G_{\vec{s}}(\Theta) \right)_{x,y}
\end{aligned} \tag{47}$$

We have used Eq. (44) to go from the second to third equality. The attentive reader will have noticed that Eq. (44) was shown to be valid only in the case of hermetian or anti-hermetian matrices which is certainly not the case of $A^{(y,x)}$. However, since only terms of order η are relevant in the calculation, we may replace $e^{\eta A}$ by $e^{\eta (A + A^\dagger)/2} e^{\eta (A - A^\dagger)/2}$ which is exact up to order η^2. For the latter form, one may use Eq. (44). To obtain the fourth equality we have used the relation: $\det A = \exp \mathrm{Tr} \ln A$.

We now show that any multi-point correlation function decouples into a sum of products of the above defined Green functions. First, we define the cummulants:

$$\langle\langle O_n \cdots O_1 \rangle\rangle_{\vec{s}} = \left. \frac{\partial^n \ln \langle \Psi_T | U_{\vec{s}}(2\Theta,\Theta) e^{\eta_n O_n} \cdots e^{\eta_1 O_1} U_{\vec{s}}(\Theta,0) | \Psi_T \rangle}{\partial \eta_n \cdots \partial \eta_1} \right|_{\eta_1 \cdots \eta_n = 0}$$

$$\text{with } O_i = \vec{c}^{\dagger} A^{(i)} \vec{c}. \tag{48}$$

Differentiating the above definition we obtain:

$$\begin{aligned}
\langle\langle O_1 \rangle\rangle_{\vec{s}} &= \langle O_1 \rangle_{\vec{s}} \\
\langle\langle O_2 O_1 \rangle\rangle_{\vec{s}} &= \langle O_2 O_1 \rangle_{\vec{s}} - \langle O_2 \rangle_{\vec{s}} \langle O_1 \rangle_{\vec{s}} \\
\langle\langle O_3 O_2 O_1 \rangle\rangle_{\vec{s}} &= \langle O_3 O_2 O_1 \rangle_{\vec{s}} - \\
& \langle O_3 \rangle_{\vec{s}} \langle\langle O_2 O_1 \rangle\rangle_{\vec{s}} - \langle O_2 \rangle_{\vec{s}} \langle\langle O_3 O_1 \rangle\rangle_{\vec{s}} - \langle O_1 \rangle_{\vec{s}} \langle\langle O_3 O_2 \rangle\rangle_{\vec{s}} - \\
& \langle O_1 \rangle_{\vec{s}} \langle O_2 \rangle_{\vec{s}} \langle O_3 \rangle_{\vec{s}}.
\end{aligned} \tag{49}$$

The following rule, which may be proven by induction, emerges:

$$\begin{aligned}
& \langle O_n \cdots O_1 \rangle_{\vec{s}} = \langle\langle O_n \cdots O_1 \rangle\rangle_{\vec{s}} + \sum_{j=1}^{n} \langle\langle O_n \cdots \widehat{O_j} \cdots O_1 \rangle\rangle_{\vec{s}} \langle\langle O_j \rangle\rangle_{\vec{s}} + \\
& \sum_{j>i} \langle\langle O_n \cdots \widehat{O_j} \cdots \widehat{O}_i \cdots O_1 \rangle\rangle_{\vec{s}} \langle\langle O_j O_i \rangle\rangle_{\vec{s}} + \cdots + \\
& \langle\langle O_n \rangle\rangle_{\vec{s}} \cdots \langle\langle O_1 \rangle\rangle_{\vec{s}}
\end{aligned} \tag{50}$$

where $\widehat{O_j}$ means that the operator O_j has been omitted from the product.

The cumulants may now be computed order by order. We concentrate on the form $\langle\langle c^{\dagger}_{x_n} c_{y_n} \cdots c^{\dagger}_{x_1} c_{y_1} \rangle\rangle$ so that $A^{(i)}_{x,y} = \delta_{x,x_i} \delta_{y,y_i}$. To simplify notation in the next calculation we introduce:

$$B^{\rangle} = B_{\vec{s}}(\Theta,0) P, \quad \text{and} \quad B^{\langle} = P^{\dagger} B_{\vec{s}}(2\Theta,\Theta) \tag{51}$$

We have already computed $\langle\langle O_1 \rangle\rangle_{\vec{s}}$ (see Eq. (47)):

$$\langle\langle O_1 \rangle\rangle_{\vec{s}} = \langle\langle c^{\dagger}_{x_1} c_{y_1} \rangle\rangle = \mathrm{Tr}\left((1 - G_{\vec{s}}(\Theta)) A^{(1)} \right) = (1 - G_{\vec{s}}(\Theta))_{y_1,x_1} \tag{52}$$

For $n = 2$ we have:

$$\begin{aligned}
\langle\langle O_2 O_1 \rangle\rangle_{\vec{s}} &= \langle\langle c^{\dagger}_{x_2} c_{y_2} c^{\dagger}_{x_1} c_{y_1} \rangle\rangle_{\vec{s}} \\
&= \left. \frac{\partial^2 \mathrm{Tr} \ln \left(P^{\dagger} B_{\vec{s}}(2\Theta,\Theta) e^{\eta_2 A^{(2)}} e^{\eta_1 A^{(1)}} B_{\vec{s}}(\Theta,0) P \right)}{\partial \eta_2 \partial \eta_1} \right|_{\eta_2,\eta_1 - 0} \\
&= \left. \frac{\partial}{\partial \eta_2} \mathrm{Tr} \left[\left(B^{\langle} e^{\eta_2 A^{(2)}} B^{\rangle} \right)^{-1} B^{\langle} e^{\eta_2 A^{(2)}} A^{(1)} B^{\rangle} \right] \right|_{\eta_2 = 0}
\end{aligned}$$

$$
\begin{aligned}
&= -\mathrm{Tr}\left[\left(B^{\langle}B^{\rangle}\right)^{-1}B^{\langle}A^{(2)}B^{\rangle}\left(B^{\langle}B^{\rangle}\right)^{-1}B^{\langle}A^{(1)}B^{\rangle}\right] \\
&\quad +\mathrm{Tr}\left[\left(B^{\langle}B^{\rangle}\right)^{-1}B^{\langle}A^{(2)}A^{(1)}B^{\rangle}\right] \\
&= \mathrm{Tr}\left(\overline{G_{\vec{s}}(\Theta)}A^{(2)}G_{\vec{s}}(\Theta)A^{(1)}\right) \\
&= \langle c^{\dagger}_{x_2}c_{y_1}\rangle_{\vec{s}}\langle c_{y_2}c^{\dagger}_{x_1}\rangle_{\vec{s}}, \text{ with } \overline{G} = 1-G
\end{aligned} \tag{53}
$$

To derive the above, we have used the cyclic properties of the trace as well as the relation $G = 1 - B^{\rangle}\left(B^{\langle}B^{\rangle}\right)^{-1}B^{\langle}$. Note that for a matrix $A(\eta)$, $\frac{\partial}{\partial\eta}A^{-1}(\eta) = -A^{-1}(\eta)\left(\frac{\partial}{\partial\eta}A(\eta)\right)A^{-1}(\eta)$. There is a simple rule to obtain the third cumulant given the second. In the above expression for the second cumulant, one replaces $B^{\rangle}$ with $B^{\langle}e^{\eta_3 A^{(3)}}$. This amounts in redefining the Green function as $G(\eta_3) = 1 - B^{\rangle}\left(B^{\langle}e^{\eta_3 A^{(3)}}B^{\rangle}\right)^{-1}B^{\langle}e^{\eta_3 A^{(3)}}$. Thus,

$$
\begin{aligned}
\langle\langle O_3 O_2 O_1\rangle\rangle_{\vec{s}} &= \langle\langle c^{\dagger}_{x_3}c_{y_3}c^{\dagger}_{x_2}c_{y_2}c^{\dagger}_{x_1}c_{y_1}\rangle\rangle_{\vec{s}} \\
&= \frac{\partial}{\partial\eta_3}\mathrm{Tr}\left(\overline{G_{\vec{s}}(\Theta,\eta_3)}A^{(2)}G_{\vec{s}}(\Theta,\eta_3)A^{(1)}\right)\Big|_{\eta_3=0} \\
&= \mathrm{Tr}\left(\overline{G_{\vec{s}}(\Theta)}A^{(3)}G_{\vec{s}}(\Theta)A^{(2)}G_{\vec{s}}(\Theta)A^{(1)}\right) - \\
&\quad \mathrm{Tr}\left(\overline{G_{\vec{s}}(\Theta)}A^{(3)}G_{\vec{s}}(\Theta)A^{(1)}\overline{G_{\vec{s}}(\Theta)}A^{(2)}\right) \\
&= \langle c^{\dagger}_{x_3}c_{y_1}\rangle_{\vec{s}}\langle c_{y_3}c^{\dagger}_{x_2}\rangle_{\vec{s}}\langle c_{y_2}c^{\dagger}_{x_1}\rangle_{\vec{s}} - \\
&\quad \langle c^{\dagger}_{x_3}c_{y_2}\rangle_{\vec{s}}\langle c_{y_3}c^{\dagger}_{x_1}\rangle_{\vec{s}}\langle c^{\dagger}_{x_2}c_{y_1}\rangle_{\vec{s}}
\end{aligned} \tag{54}
$$

since

$$
\frac{\partial}{\partial\eta_3}G_{\vec{s}}(\Theta,\eta_3)\big|_{\eta_3=0} = -\overline{G_{\vec{s}}(\Theta)}A^{(3)}G_{\vec{s}}(\Theta) = -\frac{\partial}{\partial\eta_3}\overline{G_{\vec{s}}(\Theta,\eta_3)}\Big|_{\eta_3=0}.
$$

Clearly the same procedure may be applied to obtain the nth+1 cumulant given the nth one. It is also clear that the nth cumulant is a sum of products of Green functions. Thus with equation (50) we have shown that any multi-point correlation function may be reduced into a sum of products of Green functions: Wicks theorem. Useful relations include:

$$
\langle c^{\dagger}_{x_2}c_{y_2}c^{\dagger}_{x_1}c_{y_1}\rangle_{\vec{s}} = \langle c^{\dagger}_{x_2}c_{y_1}\rangle_{\vec{s}}\langle c_{y_2}c^{\dagger}_{x_1}\rangle_{\vec{s}} + \langle c^{\dagger}_{x_2}c_{y_2}\rangle_{\vec{s}}\langle c^{\dagger}_{x_1}c_{y_1}\rangle_{\vec{s}}. \tag{55}
$$

For the FTQMC we wish to evaluate:

$$\frac{\mathrm{Tr}\left[e^{-\beta H}O\right]}{\mathrm{Tr}\left[e^{-\beta H}O\right]} = \sum_{\vec{s}} \mathrm{Pr}_{\vec{s}} \langle O \rangle_{\vec{s}} + O(\Delta\tau^2). \tag{56}$$

where

$$\mathrm{Pr}_{\vec{s}} = \frac{C_{\vec{s}} \det\left(1 + B_{\vec{s}}(\beta,0)\right)}{\sum_{\vec{s}} C_{\vec{s}} \det\left(1 + B_{\vec{s}}(\beta,0)\right)}, \quad \langle O \rangle_{\vec{s}} = \frac{\mathrm{Tr}\left[U_{\vec{s}}(\beta,\tau) O U_{\vec{s}}(\tau,0)\right]}{\mathrm{Tr}\left[U_{\vec{s}}(\beta,0)\right]}.$$

Here, we measure the observable on time slice τ. Single body observables, $O = \vec{c}^{\dagger} A \vec{c}$ are evaluated as:

$$\begin{aligned}
\langle O \rangle_{\vec{s}} &= \left. \frac{\partial \ln \mathrm{Tr}\left[U_{\vec{s}}(\beta,\tau) e^{\eta O} U_{\vec{s}}(\tau,0)\right]}{\partial \eta} \right|_{\eta=0} \\
&= \left. \frac{\partial \ln \det\left[1 + B_{\vec{s}}(\beta,\tau) e^{\eta A} B_{\vec{s}}(\tau,0)\right]}{\partial \eta} \right|_{\eta=0} \\
&= \left. \frac{\partial \ln \mathrm{Tr}\left[1 + B_{\vec{s}}(\beta,\tau) e^{\eta A} B_{\vec{s}}(\tau,0)\right]}{\partial \eta} \right|_{\eta=0} \\
&= \mathrm{Tr}\left[B_{\vec{s}}(\tau,0)(1 + B_{\vec{s}}(\beta,0))^{-1} B_{\vec{s}}(\beta,\tau) A\right] \\
&= \mathrm{Tr}\left[\left(1 - \left(1 + B_{\vec{s}}(\tau,0) B_{\vec{s}}(\beta,\tau)\right)^{-1}\right) A\right]
\end{aligned} \tag{57}$$

In particular the Green function is given by:

$$\langle c_x c_y^{\dagger} \rangle_{\vec{s}} = \left(1 + B_{\vec{s}}(\tau,0) B_{\vec{s}}(\beta,\tau)\right)^{-1}_{x,y} \tag{58}$$

Defining the cumulants as

$$\langle\langle O_n \cdots O_1 \rangle\rangle_{\vec{s}} = \left. \frac{\partial^n \ln \mathrm{Tr}\left[U_{\vec{s}}(\beta,\tau) e^{\eta_n O_n} \cdots e^{\eta_1 O_1} U_{\vec{s}}(\tau,0)\right]}{\partial \eta_n \cdots \partial \eta_1} \right|_{\eta_1 \cdots \eta_n = 0}$$

$$\text{with} \quad O_i = \vec{c}^{\dagger} A^{(i)} \vec{c}, \tag{59}$$

one can derive Wicks theorem in precisely the same manner as shown for the PQMC. Thus both in the PQMC and FTQMC, it suffices to compute the equal time Green functions to evaluate any equal time observable.

The Monte Carlo Sampling

We now have to sum over the HS fields. The lowest temperature results in Fig. 7 require summing over 2^{5760} Ising field configurations. An exact summation is thus clearly not possible. We will thus use the Monte Carlo method (see Appendix) and use a single site upgrading method which requires the calculation of the ratio

$$R = \frac{\Pr_{\vec{s}'}}{\Pr_{\vec{s}}} \tag{60}$$

where $\vec{s}$ and $\vec{s}'$ differ only at one point in space and imaginary time, $\vec{i}, n$. For the Ising field required to decouple the Hubbard interaction (Eqn. (32) and (33)):

$$s'_{\vec{i}',n'} = \begin{cases} s_{\vec{i}',n'} & \text{if } \vec{i}' \neq \vec{i} \text{ and } n' \neq n \\ -s_{\vec{i},n} & \text{if } \vec{i}' = \vec{i} \text{ and } n' = n \end{cases} \tag{61}$$

For HS fields $\vec{l}$ required to decouple perfect square terms (Eq. (36)):

$$l'_{\vec{i}',n'} = \begin{cases} l_{\vec{i}',n'} & \text{if } \vec{i}' \neq \vec{i} \text{ and } n' \neq n \\ \mathscr{F}\left(l_{\vec{i},n}\right) & \text{if } \vec{i}' = \vec{i} \text{ and } n' = n \end{cases} \tag{62}$$

where $\mathscr{F}\left(l_{\vec{i},n}\right)$ flips the HS field $l_{\vec{i},n}$ into one of the three other choices with probability $1/3$. The calculation of R boils down to computing the ration of two determinants:

$$\begin{aligned} &\frac{\det\left[1+B_{\vec{s}'}(\beta,0)\right]}{\det\left[1+B_{\vec{s}}(\beta,0)\right]} \quad \text{for the FTQMC} \\ &\frac{\det\left[P^{\dagger}B_{\vec{s}'}(2\Theta,0)P\right]}{\det\left[P^{\dagger}B_{\vec{s}}(2\Theta,0)P\right]} \quad \text{for the PQMC.} \end{aligned} \tag{63}$$

For the Hubbard interaction with HS transformation of Eq. (33) only $h_I(\vec{s}_n)$ will be affected by the move.

$$\begin{aligned} e^{h_I(\vec{s}'_n)} &= e^{i\alpha\sum_{\vec{j}} s'_{\vec{j},n} A^{(\vec{j})}} = e^{i\alpha s'_{\vec{i},n} A^{(\vec{i})}} e^{i\alpha \sum_{\vec{j}\neq\vec{i}} s_{\vec{i},n} A^{(\vec{i})}} \\ &= e^{-i\alpha s_{\vec{i},n} A^{(\vec{i})}} e^{i\alpha \sum_{\vec{j}\neq\vec{i}} s_{\vec{i},n} A^{(\vec{i})}} = e^{-2i\alpha s_{\vec{i},n} A^{(\vec{i})}} e^{i\alpha \sum_{\vec{j}} s_{\vec{i},n} A^{(\vec{i})}} \\ &= (1 + \underbrace{\left(e^{-2i\alpha s_{\vec{i},n} A^{(\vec{i})}} - 1\right)}_{=\Delta^{(\vec{i})}})e^{h_I(\vec{s}_n)}. \end{aligned} \tag{64}$$

Here $A^{(\vec{i})}_{x,y} = \delta_{x,y}\delta_{x,(\vec{i},\sigma)}$ so that $\left[A^{(\vec{i})}, A^{(\vec{j})}\right] = 0\ \forall \vec{i}, \vec{j}$. The matrix $\Delta^{(\vec{i})}$ is diagonal and has only two non-zero entries at $x = (\vec{i}, \sigma)$. Thus, for the Hubbard model:

$$B_{\vec{s}'}(\bullet,0) = B_{\vec{s}}(\bullet,\tau)\left(1+\Delta^{(\vec{i})}\right)B_{\vec{s}}(\tau,0) \tag{65}$$

where the $\bullet$ stands for 2Θ or β and $\tau = n\Delta\tau$. For interaction term in the form of perfect squares isolating the HS filed $l_{\vec{i},n}$ is a bit more cumbersome. First let us assume that $\vec{i}$

is in the set S_r (see paragraph after Eq. (34)) and that $O^{(\vec{i})} = \vec{c}^\dagger A^{(\vec{i})}\vec{c}$. We will work in a basis where $A^{(\vec{i})}$ is diagonal: $U^{(\vec{i}),\dagger}A^{(\vec{i})}U^{(\vec{i})} = D^{(\vec{i})}$.

$$\begin{aligned} e^{h^r_{\vec{l}'}(\vec{l}'_n)} &= e^{\sqrt{\Delta\tau W}\sum_{\vec{j}\varepsilon S_r}\eta(l'_{\vec{j},n})A^{(\vec{j})}} = e^{\sqrt{\Delta\tau W}(\eta(l'_{\vec{i},n})-\eta(l_{\vec{i},n}))A^{(\vec{i})}} e^{h^r_{\vec{l}}(\vec{l}_n)} \\ &= U^{(\vec{i})}\left(1+\underbrace{\left(e^{\sqrt{\Delta\tau W}(\eta(l'_{\vec{i},n})-\eta(l_{\vec{i},n}))D^{(\vec{i})}}-1\right)}_{=\Delta^{(\vec{i})}}\right)U^{(\vec{i}),\dagger}e^{h^r_{\vec{l}}(\vec{l}_n)}. \end{aligned} \quad (66)$$

Thus, $B_{\vec{l}'}(\bullet,0)$ takes the form:

$$\begin{aligned} B_{\vec{l}'}(\bullet,0) = \; & \overbrace{B_{\vec{l}}(\bullet,\tau)\left[\prod_{r'=1}^{r-1} e^{h^{r'}_{\vec{l}}(\vec{l}_n)}\right]U^{(\vec{i})}}^{\tilde{B}_{\vec{l}}(\bullet,\tau)}\left(1+\Delta^{(\vec{i})}\right) \\ & \underbrace{U^{(\vec{i}),\dagger}\left[\prod_{r'\geq r} e^{h^{r'}_{\vec{l}}(\vec{l}_n)}\right]B_{\vec{l}}(\tau,0)}_{\tilde{B}_{\vec{l}}(\tau,0)} \end{aligned} \quad (67)$$

After a redefinition of the B matrices as shown in the above equation, a similar form to that for the Hubbard model is obtained. Note that $\tilde{B}$ matrices satisfy: $\tilde{B}_{\vec{l}}(\bullet,\tau)\tilde{B}_{\vec{l}}(\tau,0) = B_{\vec{l}}(\bullet,0)$. It is often important to work in a basis where the $A^{(\vec{i})}$ matrices are diagonal since computer time is saved if some of the eigenvalues vanish. In the the so-called t-U-W model [26, 25] a factor 6 in CPU time in the upgrading is saved.

Concentrating first on the PQMC, and again introducing the notation $B^{\langle}_{\vec{s}} = P^\dagger B_{\vec{s}}(2\Theta,\tau)$ and $B^{\rangle}_{\vec{s}} = B_{\vec{s}}(\tau,0)P$ we have to evaluate:

$$\begin{aligned} \frac{\det\left[B^{\langle}_{\vec{s}}\left(1+\Delta^{(\vec{i})}\right)B^{\rangle}_{\vec{s}}\right]}{\det\left[B^{\langle}_{\vec{s}}B^{\rangle}_{\vec{s}}\right]} &= \det\left[B^{\langle}_{\vec{s}}\left(1+\Delta^{(\vec{i})}\right)B^{\rangle}_{\vec{s}}\left(B^{\langle}_{\vec{s}}B^{\rangle}_{\vec{s}}\right)^{-1}\right] \\ = \det\left[1+B^{\langle}_{\vec{s}}\Delta^{(\vec{i})}B^{\rangle}_{\vec{s}}\left(B^{\langle}_{\vec{s}}B^{\rangle}_{\vec{s}}\right)^{-1}\right] &= \det\left[1+\Delta^{(\vec{i})}B^{\rangle}_{\vec{s}}\left(B^{\langle}_{\vec{s}}B^{\rangle}_{\vec{s}}\right)^{-1}B^{\langle}_{\vec{s}}\right] \end{aligned} \quad (68)$$

where the last equation follows from the identity $\det[1+AB] = \det[1+BA]$ for arbitrary rectangular matrices. [4] We can recognize the Green function $B^{\rangle}_{\vec{s}}\left(B^{\langle}_{\vec{s}}B^{\rangle}_{\vec{s}}\right)^{-1}B^{\langle}_{\vec{s}} = 1 - G_{\vec{s}}(\tau) \equiv \bar{G}_{\vec{s}}(\tau)$. By construction, $\Delta^{(\vec{i})}$ is diagonal. For a form with Q non vanishing

[4] This identity may be formally shown by using the relation $\det(1+AB) = \exp\mathrm{Tr}\log(1+AB)$, expanding the logarithm and using the cyclic properties of the trace.

entries,

$$(\Delta^{(\vec{i})})_{z,z_1} = \delta_{z,z_1} \sum_{q=1}^{Q} \Delta^{(\vec{i})}_{x_q} \delta_{x_q,z} \tag{69}$$

one has,

$$\begin{aligned} \det & \left[1 + \Delta^{(\vec{i})} \bar{G}_{\vec{s}}(\tau)\right] = \\ & \det \begin{pmatrix} (1 + \Delta^{(\vec{i})} \bar{G}_{\vec{s}}(\tau))_{x_1,x_1} & . & . & (1 + \Delta^{(\vec{i})} \bar{G}_{\vec{s}}(\tau))_{x_1,x_Q} \\ \cdot & & & \cdot \\ \cdot & & & \cdot \\ (1 + \Delta^{(\vec{i})} \bar{G}_{\vec{s}}(\tau))_{x_Q,x_1} & . & . & (1 + \Delta^{(\vec{i})} \bar{G}_{\vec{s}}(\tau))_{x_Q,x_Q} \end{pmatrix}. \end{aligned} \tag{70}$$

For the FTQMC, we have to evaluate:

$$\begin{aligned} & \frac{\det\left[1 + B_{\vec{s}}(\beta,\tau)(1 + \Delta^{(\vec{i})}) B_{\vec{s}}(\tau,0)\right]}{\det\left[1 + B_{\vec{s}}(\beta,0)\right]} = \\ & \det\left[1 + \Delta^{(\vec{i})} B_{\vec{s}}(\tau,0)\left(1 + B_{\vec{s}}(\beta,0)\right)^{-1} B_{\vec{s}}(\beta,\tau)\right] = \\ & \det\left[1 + \Delta^{(\vec{i})}\left(1 - \left(1 + B_{\vec{s}}(\tau,0) B_{\vec{s}}(\beta,\tau)\right)^{-1}\right)\right]. \end{aligned} \tag{71}$$

Since the finite temperature equal time Green function is given by: $G_{\vec{s}}(\tau) = \left(1 + B_{\vec{s}}(\tau,0) B_{\vec{s}}(\beta,\tau)\right)^{-1}$ so that precisely same form as in the PQMC is obtained.

Having calculated the ratio R for a single spin-flip one may now decide stochastically if the move is accepted or not. In case of acceptance, the Green function is to be upgraded since this quantity will be requires for the next spin-flip. The upgrading of the Green function is based on the Sherman-Morrison formula [27]:

$$(A + \vec{u} \otimes \vec{v})^{-1} = A^{-1} - \frac{\left(A^{-1}\vec{u}\right) \otimes \left(\vec{v} A^{-1}\right)}{1 + \vec{v} \bullet A^{-1} \vec{u}} \tag{72}$$

where the outer product is defined by $(\vec{u} \otimes \vec{v})_{x,y} = \vec{u}_x \vec{v}_y$. In the case of the FTQMC and $\Delta^{(\vec{i})}$ given by Eq. (69),

$$\begin{aligned} G_{\vec{s}}(\tau) &= \left(1 + B_{\vec{s}}(\tau,0)(1 + \Delta^{(\vec{i})}) B_{\vec{s}}(\beta,\tau)\right)^{-1} \\ &= \left(1 + B_{\vec{s}}(\tau,0) B_{\vec{s}}(\beta,\tau) + \sum_{q} \vec{u}^{(q)} \otimes \vec{v}^{(q)}\right)^{-1} \end{aligned} \tag{73}$$

where

$$(\vec{u}^{(q)})_x = (B_{\vec{s}}(\tau,0))_{x,x_q} \Delta^{(\vec{i})}_{x_q} \text{ and } (\vec{v}^{(q)})_x = (B_{\vec{s}}(\beta,\tau))_{x_q,x}.$$

Identifying A to $\left(G_{\vec{s}}(\tau)\right)^{-1} = 1 + B_{\vec{s}}(\tau,0) B_{\vec{s}}(\beta,\tau)$, Eq. (72) may now be applied Q times to upgrade the Green function.

For the PQMC, the upgrading of the Green function is equivalent to the upgrading of $(B^{\langle}_{\vec{s}} B^{\rangle}_{\vec{s}})^{-1}$, which is achieved with the Sherman-Morrison formula:

$$(B^{\langle}_{\vec{s}} B^{\rangle}_{\vec{s}})^{-1} = \left(B^{\langle}_{\vec{s}}(1+\Delta^{(\vec{i})})B^{\rangle}_{\vec{s}}\right)^{-1} = \left(B^{\langle}_{\vec{s}} B^{\rangle}_{\vec{s}} + \sum_q \vec{u}^{(q)} \otimes \vec{v}^{(q)}\right)^{-1} \tag{74}$$

with $(\vec{u}^{(q)})_x = (B^{\langle}_{\vec{s}})_{x,x_q} \Delta^{(\vec{i})}_{x_q}$ and $(\vec{v}^{(q)})_x = (B^{\rangle}_{\vec{s}})_{x_q,x}$. Here x runs from $1 \cdots N_p$ where N_p corresponds to the number of particles contained in the trial wave function.

In principle, we now have all elements required to carry out the simulations. The equal time Green function is the central quantity. On one hand it is used to compute any observables. On the other hand, it determines the Monte Carlo dynamics. It is convenient to adopt a sequential upgrading scheme. Given the Green function at imaginary time $\tau = \Delta\tau$, one upgrades the HS fields on this time slice deterministically or randomly. In case of acceptance, the Green function is upgraded after each single spin flip. To reach the next time slice, the relation:

$$G_{\vec{s}}(\tau+1) = B_{\vec{s}}(\tau+1,\tau) G_{\vec{s}}(\tau) \left(B_{\vec{s}}(\tau+1,\tau)\right)^{-1} \tag{75}$$

is used and the procedure is repeated till $\tau = \beta$ (FTQMC) or $\tau = 2\Theta$ (PQMC). Having reached $\tau = \beta$ or $\tau = 2\Theta$ we propagate the Green function to back $\tau = 1$ and on the way upgrade the HS fields. The whole procedure may then be repeated. We note that for interactions of the form (34) the propagation of the Green function from time slice τ to time slice $\tau+1$ is split into intermediate steps according to (67) so as to upgrade the HS fields in the the sets S_r successively.

The above corresponds precisely to the procedure adopted in the case of the FTQMC. For the PQMC, it is more efficient to keep track of $(P^{\dagger} B_{\vec{s}}(2\Theta,0)P)^{-1}$ since (i) it is of dimension $N_p \times N_p$ in contrast to the Green function which is a $N \times N$ matrix and (ii) it is τ independent. When Green functions are required - to compute R, or observables - they are computed from scratch.

Numerical stabilization.

In the previous section, we have assumed that we are able to compute the Green functions. On finite precision machines this is unfortunately not the case. To understand the sources of numerical instabilities, it is convenient to consider the PQMC. The rectangular matrix P is just a set of column orthonormal vectors. Typically for a Hubbard model, at weak couplings, the extremal scales in the matrix $B_{\vec{s}}(2\Theta,0)$ are determined by the kinetic energy and range from $e^{8t\Theta}$ to $e^{-8t\Theta}$ in the two-dimensional case. When the set of orthonormal vectors in P are propagated, the large scales will wash out the small scales yielding an numerically ill defined inversion of the matrix $P^{\dagger} B_{\vec{s}}(2\Theta,0)P$. To be more precise consider a two electron problem. The matrix P then consists of two column orthonormal vectors, $\vec{v}(0)_1$ and $\vec{v}(0)_2$. After propagation along the imaginary time axis, will be dominated by the largest scales in $B_{\vec{s}}(2\Theta,0)$ so that $\vec{v}(2\Theta)_1 = \vec{v}(2\Theta)_2 + \vec{\varepsilon}$, where $\vec{v}(2\Theta)_1 = B_{\vec{s}}(2\Theta,0)\vec{v}_1$. It is the information contained in $\vec{\varepsilon}$ which renders the matrix

$P^\dagger B_{\vec{s}}(2\Theta,0)P$ non-singular. For large values of Θ this information is lost in round-off errors.

To circumvent this problem a set of matrix decomposition techniques were developed [13, 14, 10]. Those matrix decomposition techniques are best introduced with the Gram-Schmidt orthonormalization method of N_p linearly independent vectors. At imaginary time τ, $B_{\vec{s}}(\tau,0)P \equiv B^{\rangle}$ is given by the N_p vectors $\vec{v}_1 \cdots \vec{v}_{N_p}$. Orthogonalizing those vectors yields:

$$\begin{aligned}
\vec{v}'_1 &= \vec{v}_1 \\
\vec{v}'_2 &= \vec{v}_2 - \frac{\vec{v}_2 \cdot \vec{v}'_1}{\vec{v}'_1 \cdot \vec{v}'_1}\vec{v}'_1 \\
&\cdot \\
&\cdot \\
\vec{v}'_{N_p} &= \vec{v}_{N_p} - \sum_{i=1}^{N_p-1} \frac{\vec{v}_{N_p} \cdot \vec{v}'_i}{\vec{v}'_i \cdot \vec{v}'_i}\vec{v}'_i.
\end{aligned} \tag{76}$$

Since $\vec{v}'_n$ depends only on the vectors $\vec{v}_n \cdots \vec{v}_1$ we can write,

$$\left(\vec{v}'_1, \cdots, \vec{v}'_{N_p}\right) = \left(\vec{v}_1, \cdots, \vec{v}_{N_p}\right) V_R^{-1} \tag{77}$$

where V_R is an upper unit triangular $N_p \times N_p$ matrix, that is the diagonal matrix elements are equal to unity. One can furthermore normalize the vectors $\vec{v}'_1, \cdots, \vec{v}'_{N_p}$ to obtain:

$$B^{\rangle} \equiv \left(\vec{v}_1, \cdots, \vec{v}_{N_p}\right) = \underbrace{\left(\frac{\vec{v}'_1}{|\vec{v}'_1|}, \cdots, \frac{\vec{v}'_{N_p}}{|\vec{v}'_{N_p}|}\right)}_{\equiv U^{\rangle}} D_R V_R \tag{78}$$

where D is a diagonal matrix containing the scales. One can repeat the procedure to obtain: $B^{\langle} \equiv P^\dagger B_{\vec{s}}(2\Theta,\tau) = V_L D_L U^{\langle}$. The Green function for the PQMC is now particularly easy to compute:

$$\begin{aligned}
1 - G_{\vec{s}}(\tau) &= B^{\rangle}\left(B^{\langle}B^{\rangle}\right)^{-1}B^{\langle} \\
&= U^{\rangle}D_R V_R\left(V_L D_L U^{\langle}U^{\rangle}D_R V_R\right)^{-1}V_L D_L U^{\langle} \\
&= U^{\rangle}D_R V_R\left(D_R V_R\right)^{-1}\left(U^{\langle}U^{\rangle}\right)^{-1}\left(V_L D_L\right)^{-1}V_L D_L U^{\langle} \\
&= U^{\rangle}\left(U^{\langle}U^{\rangle}\right)^{-1}U^{\langle}
\end{aligned} \tag{79}$$

Thus, in the PQMC, all scales which are at the origin of the numerical instabilities disappear from the problem when computing Green functions. Since the entire algorithm relies solely on the knowledge of the Green function, the above stabilization procedure leaves the physical results invariant. Note that although appealing, the Gram-Schmidt orthonormalization is itself unstable, and hence is is more appropriate to use singular value decompositions based on Housholder's method to obtain the above UDV form for the B matrices [27]. In practice the frequency at which the stabilization is carried out is problem dependent. Typically, for the Hubbard model with $\Delta\tau t = 0.125$ stabilization at every 10th time slice produces excellent accuracy.

The stabilization procedure for the finite temperature algorithm is more subtle since scales may not be doped in the calculation of the Green function. Below, we provide two ways of computing the Green function.

The first approach relies on the identity:

$$\begin{pmatrix} A & B \\ C & D \end{pmatrix}^{-1} = \begin{pmatrix} \left(A - BD^{-1}C\right)^{-1} & \left(C - DB^{-1}A\right)^{-1} \\ \left(B - AC^{-1}D\right)^{-1} & \left(D - CA^{-1}B\right)^{-1} \end{pmatrix} \tag{80}$$

where A, B, C and D are matrices. Using the above, we obtain:

$$\begin{pmatrix} 1 & B_{\vec{s}}(\tau,0) \\ -B_{\vec{s}}(\beta,\tau) & 1 \end{pmatrix}^{-1} = \begin{pmatrix} G_{\vec{s}}(0) & -(1-G_{\vec{s}}(0))B_{\vec{s}}^{-1}(\tau,0) \\ B_{\vec{s}}(\tau,0)G_{\vec{s}}(0) & G_{\vec{s}}(\tau) \end{pmatrix} \tag{81}$$

The diagonal terms on the right hand side of the above equation correspond to the desired equal time Green functions. The off-diagonal terms are nothing but the time displaced Green functions which will be discussed in the next section. To evaluate the left hand side of the above equation, we first have to bring $B_{\vec{s}}(\tau,0)$ and $B_{\vec{s}}(\beta,\tau)$ in UDV forms. This has to be done step by step so as to avoid mixing large and small scales. Consider the propagation $B_{\vec{s}}(\tau,0)$, and a time interval τ_1, with $n\tau_1 = \tau$, for which the different scales in $B_{\vec{s}}(n\tau_1,(n-1)\tau_1)$ do not exceed machine precision. Since $B_{\vec{s}}(\tau,0) = B_{\vec{s}}(n\tau_1,(n-1)\tau_1)\cdots B_{\vec{s}}(\tau_1,0)$ we can evaluate $B_{\vec{s}}(\tau,0)$ for $n=2$ with:

$$B_{\vec{s}}(2\tau_1,\tau_1)\underbrace{B_{\vec{s}}(\tau_1,0)}_{U_1D_1V_1} = \underbrace{((B_{\vec{s}}(2\tau_1,\tau_1)U_1)D_1)}_{U_2D_2V}V_1 = U_2D_2V_2 \tag{82}$$

where $V_2 = VV_1$. The parenthesis determine the order in which the matrix multiplication are to be done. In all operations, mixing of scales are avoided. After the multiplication with diagonal matrix D_1 scales are again separated with the used of the singular value decomposition.

Thus, for $B_{\vec{s}}(\tau,0) = U_RD_RV_R$ and $B_{\vec{s}}(\beta,\tau) = V_LD_LU_L$ we have to invert:

$$\begin{aligned} &\begin{pmatrix} I & V_LD_LU_L \\ -U_RD_RV_R & I \end{pmatrix}^{-1} = \\ &\left[\begin{pmatrix} V_L & 0 \\ 0 & U_R \end{pmatrix}\underbrace{\begin{pmatrix} (V_RV_L)^{-1} & D_L \\ -D_R & (U_LU_R)^{-1} \end{pmatrix}}_{UDV}\begin{pmatrix} V_R & 0 \\ 0 & U_L \end{pmatrix}\right]^{-1} = \\ &\left[\begin{pmatrix} (V_R)^{-1} & 0 \\ 0 & (U_L)^{-1} \end{pmatrix}V^{-1}\right]D^{-1}\left[U^{-1}\begin{pmatrix} (V_L)^{-1} & 0 \\ 0 & (U_R)^{-1} \end{pmatrix}\right] \end{aligned} \tag{83}$$

In the above, all matrix multiplications are well defined. In particular, the matrix D contains only large scales since the matrices $(V_R V_L)^{-1}$ and $(U_L U_R)^{-1}$ act as a cutoff to the exponentially small scales in D_L and D_R. This method to compute Green functions is very stable and has the advantage of producing time displaced Green functions. However, it is numerically expensive since the matrices involved are twice as big as the B matrices.

Alternative methods to compute $G_{\vec{s}}(\tau)$ which involve matrix manipulations only of the size of B include:

$$
\begin{aligned}
(1+B_{\vec{s}}(\tau,0)B_{\vec{s}}(\beta,\tau))^{-1} &= (1+U_R D_R V_R V_L D_L U_L)^{-1} \\
&= (U_L)^{-1}\underbrace{((U_L U_R)^{-1}+D_R(V_R V_L)D_L)}_{UDV}{}^{-1}U_R^{-1} \\
&= (VU_L)^{-1}D^{-1}(U_R U^{-1}).
\end{aligned} \tag{84}
$$

Again, $(U_L U_R)^{-1}$ acts as a cutoff to the small scales in $D_R(V_R V_L)D_L$ so that D contains only large scales.

The accuracy of both presented methods may be tested by in the following way. Given the Green function at time τ we can can upgrade and wrap (see Eq. (75)) this Green function to time slice $\tau+\tau_1$. Of course, for the time interval τ_1 the involved scales should lie within the accuracy of the the computer, $\sim 1^{-12}$ for double precision numbers. The thus obtained Green function at time $\tau+\tau_1$ may be compared to the one computed from scratch using Eq. (83) or Eq. (84). For a 4×4 half-filled Hubbard model at $U/t=4$, $\beta t = 20$, $\Delta\tau t = 0.1$ and $\tau_1 = 10\Delta\tau$ we obtain an average (maximal) difference between the matrix elements of both Green functions of 1^{-10} (1^{-6}) which is orders of magnitude smaller than the statistical uncertainty. Had we chosen $\tau_1 = 50\Delta\tau$ the accuracy drops to 0.01 and 100.0 for the average and maximal differences.

Imaginary time displaced Green functions.

Imaginary time displaced correlation yield important information. On one hand they may be used to obtain spin and charge gaps [28, 29], as well quasiparticle weights [20]. On the other hand, with the use of the Maximum Entropy method [30, 31] dynamical properties such as spin and charge dynamical structure factors, optical conductivity, and single particle spectral functions may be computed. Those quantities offer the possibility of direct comparison with experiments, such as photoemission, neutron scattering and optical measurements.

Since there is again a Wick's theorem for time displaced correlation functions. it suffices to compute the single particle Green function for a given HS configuration. We will first start with the FTQMC and then concentrate on the PQMC.

For a given HS field, we wish to evaluate:

$$G_{\vec{s}}(\tau_1,\tau_2)_{x,y} = \langle T c_x(\tau_1) c_y^\dagger(\tau_2)\rangle_{\vec{s}} \tag{85}$$

where T corresponds to the time ordering. Thus for $\tau_1 > \tau_2$ $G_{\vec{s}}(\tau_1,\tau_2)_{x,y}$ reduces to

$$\begin{aligned} \langle c_x(\tau_1) c_y^\dagger(\tau_2)\rangle_{\vec{s}} &= \frac{\mathrm{Tr}\left[U_{\vec{s}}(\beta,\tau_1) c_x U_{\vec{s}}(\tau_1,\tau_2) c_y^\dagger U_{\vec{s}}(\tau_2,0)\right]}{\mathrm{Tr}\left[U_{\vec{s}}(\beta,0)\right]} \\ &= \frac{\mathrm{Tr}\left[U_{\vec{s}}(\beta,\tau_2) U_{\vec{s}}^{-1}(\tau_1,\tau_2) c_x U_{\vec{s}}(\tau_1,\tau_2) c_y^\dagger U_{\vec{s}}(\tau_2,0)\right]}{\mathrm{Tr}\left[U_{\vec{s}}(\beta,0)\right]} \end{aligned} \tag{86}$$

Evaluating $U^{-1}(\tau_1,\tau_2) c_x U_{\vec{s}}(\tau_1,\tau_2)$ boils down to the calculation of

$$c_x(\tau) = e^{\tau \vec{c}^\dagger A \vec{c}} c_x e^{-\tau \vec{c}^\dagger A \vec{c}}$$

where A is an arbitrary matrix. Differentiating the above with respect to τ yields

$$\frac{\partial c_x(\tau)}{\partial \tau} = e^{\tau \vec{c}^\dagger A \vec{c}} \left[\vec{c}^\dagger A \vec{c}, c_x\right] e^{-\tau \vec{c}^\dagger A \vec{c}} = -\sum_z A_{x,z} c_z(\tau).$$

Thus,

$$c_x(\tau) = \left(e^{-A}\vec{c}\right)_x \text{ and similarly } c_x^\dagger(\tau) = \left(\vec{c}^\dagger e^{A}\right)_x. \tag{87}$$

We can use the above equation successively to obtain:

$$\begin{aligned} U_{\vec{s}}^{-1}(\tau_1,\tau_2) c_x U_{\vec{s}}(\tau_1,\tau_2) &= \left(B_{\vec{s}}(\tau_1,\tau_2)\vec{c}\right)_x \\ U_{\vec{s}}^{-1}(\tau_1,\tau_2) c_x^\dagger U_{\vec{s}}(\tau_1,\tau_2) &= \left(\vec{c}^\dagger B_{\vec{s}}^{-1}(\tau_1,\tau_2)\right)_x \end{aligned} \tag{88}$$

Since B is a matrix and not an operator, we can pull it out of the trace in Eq. (86). Note that the above equation automatically leads to a Wick's theorem for time displaced Green function since the theorem holds for equal time quantities. Thus, we obtain the result:

$$G_{\vec{s}}(\tau_1,\tau_2)_{x,y} = \langle c_x(\tau_1) c_y^\dagger(\tau_2)\rangle_{\vec{s}} = B_{\vec{s}}(\tau_1,\tau_2) G_{\vec{s}}(\tau_2) \quad \tau_1 > \tau_2 \tag{89}$$

where $G_{\vec{s}}(\tau_1)$ is the equal time Green function computed previously. A similar calculation will yield for $\tau_2 > \tau_1$

$$G_{\vec{s}}(\tau_1,\tau_2)_{x,y} = -\langle c_y^\dagger(\tau_2) c_x(\tau_1)\rangle_{\vec{s}} = -\left(1 - G_{\vec{s}}(\tau_1)\right) B_{\vec{s}}^{-1}(\tau_2,\tau_1). \tag{90}$$

Returning to Eq. (81) we see that we have already computed the time displaced Green functions $G_{\vec{s}}(0,\tau)$ and $G_{\vec{s}}(\tau,0)$ when discussing a stabilization scheme for the equal time Green functions. However, this calculation is expensive and is done only at times

$\tau = n\tau_1$ where τ_1 is time scale on which all energy scales fit well on finite precision machines. To obtain the Green functions for arbitrary values of τ one uses the relations:

$$\begin{aligned} G_{\vec{s}}(0,\tau+\tau_2) &= G_{\vec{s}}(0,\tau)B_{\vec{s}}^{-1}(\tau_2,\tau) \\ G_{\vec{s}}(\tau+\tau_2,0) &= B_{\vec{s}}(\tau_2,\tau)G_{\vec{s}}(\tau,0) \end{aligned} \tag{91}$$

where $\tau_2 < \tau_1$.

With the above method, we have access to all time displaced Green functions $G_{\vec{s}}(0,\tau)$ and $G_{\vec{s}}(\tau,0)$. However, we do not use translation invariance in imaginary time. Clearly, using this symmetry in the calculation of time displaced quantities will reduce the fluctuations which may sometimes be desirable. A numerically expensive but elegant way of producing all time displaced Green functions has been proposed by Hirsch [32]. Let β be a multiple of τ_1, $l\tau_1 = \beta$ and τ_1 small enough so that the scales involved in $B_s(\tau+\tau_1,\tau)$ fit on finite precision machines. Using the definition $\tau_i = i\tau_1$ with $i = 1\cdots l$ one can show:

$$\begin{pmatrix} 1 & 0 & . & 0 & B_{\vec{s}}(\tau_1,0) \\ -B_{\vec{s}}(\tau_2,\tau_1) & 1 & 0 & . & 0 \\ 0 & -B_{\vec{s}}(\tau_3,\tau_2) & 1 & . & 0 \\ . & 0 & -B_{\vec{s}}(\tau_4,\tau_3) & . & . \\ . & . & 0 & . & . \\ . & . & . & . & \\ 0 & . & 0 & -B_{\vec{s}}(\tau_l,\tau_{l-1}) & 1 \end{pmatrix}^{-1} = $$

$$\begin{pmatrix} G_{\vec{s}}(\tau_1,\tau_1) & G_{\vec{s}}(\tau_1,\tau_2) & . & . & G_{\vec{s}}(\tau_1,\tau_l) \\ G_{\vec{s}}(\tau_2,\tau_1) & G_{\vec{s}}(\tau_2,\tau_2) & . & . & G_{\vec{s}}(\tau_2,\tau_l) \\ . & . & . & . & . \\ G_{\vec{s}}(\tau_l,\tau_1) & G_{\vec{s}}(\tau_l,\tau_2) & . & . & G_{\vec{s}}(\tau_l,\tau_l) \end{pmatrix} \tag{92}$$

The matrix to inverse is l times the size of the B matrices, and hence expensing to compute. It is worth noting that on vector machines the performance grows with growing vector size so that the above method can become attractive. Having computed the Green functions $G_{\vec{s}}(\tau_i,\tau_j)$ we can obtain Green functions on any two time slices by using equations of the type (91).

PQMC

Zero temperature time displaced Green functions are given by:

$$\begin{aligned} & G_s\left(\Theta+\frac{\tau}{2},\Theta-\frac{\tau}{2}\right)_{x,y} \\ &= \frac{\langle\Psi_T|U_{\vec{s}}(2\Theta,\Theta+\frac{\tau}{2})\, c_x U_{\vec{s}}(\Theta+\frac{\tau}{2},\Theta-\frac{\tau}{2})\, c_y^\dagger U_{\vec{s}}(\Theta-\frac{\tau}{2},0)|\Psi_T\rangle}{\langle\Psi_T|U_{\vec{s}}(2\Theta,0)|\Psi_T\rangle} \\ &= \left[B_{\vec{s}}\left(\Theta+\frac{\tau}{2},\Theta-\frac{\tau}{2}\right)G_{\vec{s}}\left(\Theta-\frac{\tau}{2}\right)\right]_{x,y} \end{aligned} \tag{93}$$

and

$$G_s\left(\Theta-\frac{\tau}{2},\Theta+\frac{\tau}{2}\right)_{x,y} \tag{94}$$
$$= -\frac{\langle\Psi_T|U_{\vec{s}}(2\Theta,\Theta+\frac{\tau}{2})c_y^\dagger U_{\vec{s}}(\Theta+\frac{\tau}{2},\Theta-\frac{\tau}{2})c_x U_{\vec{s}}(\Theta-\frac{\tau}{2},0)|\Psi_T\rangle}{\langle\Psi_T|U_{\vec{s}}(2\Theta,0)|\Psi_T\rangle}$$
$$= -\left[\left(1-G_{\vec{s}}\left(\Theta-\frac{\tau}{2}\right)\right)B_{\vec{s}}^{-1}\left(\Theta+\frac{\tau}{2},\Theta-\frac{\tau}{2}\right)\right]_{x,y}.$$

Here $\tau > 0$ and we have used Eq. (88), as well as the equal time Green function (Eq. (47)). Two comments are in order. (i) For a given value of τ the effective projection parameter is $\Theta-\tau$. Thus, before starting a simulation, one has to set the maximal value of τ which will be considered, τ_M and the effective projection parameter $\Theta-\tau_M$ should be large enough to yield the ground state within the desired precision. (ii) In a canonical ensemble, the chemical potential is meaningless. However, when single particle Green functions are computed it is required to set the reference energy with regards to which a particle will be added or removed. In other words, it is the chemical potential which delimits photoemission from inverse photoemission. Thus, it is useful to have an estimate of this quantity if single particle or pairing correlations are under investigation. For observable such as spin-spin or charge- charge time displaced correlations this complication does not come into play since they are in the particle-hole channel.

We are now left with the problem of computing the Green functions. A direct multiplication of the equal time Green function with B matrices is unstable for larger values of τ. This can be understood in the framework of free electrons on a two-dimensional square lattice:

$$H = -t\sum_{<\vec{i},\vec{j}>} c_{\vec{i}}^\dagger c_{\vec{j}}, \tag{95}$$

where the sum runs over nearest-neighbors. For this Hamiltonian one has:

$$\langle\Psi_0|c_{\vec{k}}^\dagger(\tau)c_{\vec{k}}|\Psi_0\rangle = \exp\left(\tau(\varepsilon_{\vec{k}}-\mu)\right)\langle\Psi_0|c_{\vec{k}}^\dagger c_{\vec{k}}|\Psi_0\rangle, \tag{96}$$

where $\varepsilon_{\vec{k}} = -2t(\cos(\vec{k}\vec{a}_x)+\cos(\vec{k}\vec{a}_y))$, $\vec{a}_x$, $\vec{a}_y$ being the lattice constants. We will assume $|\Psi_0\rangle$ to be non-degenerate. In a numerical calculation the eigenvalues and eigenvectors of the above Hamiltonian will be known up to machine precision, ε. In the case $\varepsilon_{\vec{k}}-\mu > 0$, $\langle\Psi_0|c_{\vec{k}}^\dagger c_{\vec{k}}|\Psi_0\rangle \equiv 0$. However, on a finite precision machine the later quantity will take a value of the order of ε. When calculating $\langle\Psi_0|c_{\vec{k}}^\dagger(\tau)c_{\vec{k}}|\Psi_0\rangle$ this roundoff error will be blown up exponentially and the result for large values of τ will be unreliable. In Eq. (94) the B matrices play the role of the exponential factor $\exp\left(\tau(\varepsilon_{\vec{k}}-\mu)\right)$ and the equal time Green functions correspond to $\langle\Psi_0|c_{\vec{k}}^\dagger c_{\vec{k}}|\Psi_0\rangle$. In the PQMC, the stability problem is much more severe than for free electrons since the presence of the time dependent HS field mixes different scales.

We present two methods to circumvent this problem. The first method is numerically expensive but has been used extensively and hence tested for many models [28]. The second method has recently been developed, is elegant, simple to implement and very cheap in CPU time [33].

The first method lies on the observation that introducing a large but finite temperature $T = 1/\beta$ stabilizes the calculation. For the free electrons:

$$\langle\Psi_0|c^\dagger_{\vec{k}}(\tau)c_{\vec{k}}|\Psi_0\rangle = \lim_{\beta\to\infty} \frac{\exp\left(\tau(\varepsilon_{\vec{k}} - \mu)\right)}{1+\exp\left(\beta(\varepsilon_{\vec{k}} - \mu)\right)}. \tag{97}$$

Even if the eigenvalues are known only up to machine precision, the right hand side of the above equation for large but finite values of β is a numerically stable operation. To implement this idea in the PQMC, we use the fact that the trial wave function is a single Slater determinant. Thus, we can find a single particle Hamiltonian, $H_0 = \sum_{x,y} c^\dagger_x (h_0)_{x,y} c_y$, which has $|\Psi_T\rangle$ as a non-degenerate ground state. Hence the equation

$$\begin{aligned} G_{\vec{s}}\left(\Theta - \frac{\tau}{2}\right) &\equiv \frac{\langle\Psi_T|U_{\vec{s}}(2\Theta,\Theta-\frac{\tau}{2})c_x c^\dagger_y U_{\vec{s}}(\Theta-\frac{\tau}{2},0)|\Psi_T\rangle}{\langle\Psi_T|U_{\vec{s}}(2\Theta,0)|\Psi_T\rangle} \\ &= \lim_{\beta\to\infty} \frac{\mathrm{Tr}\left[e^{-\beta H_O} U_{\vec{s}}(2\Theta,\Theta-\frac{\tau}{2})c_x c^\dagger_y U_{\vec{s}}(\Theta-\frac{\tau}{2},0)\right]}{\mathrm{Tr}\left[e^{-\beta H_O} U_{\vec{s}}(2\Theta,0)\right]} \\ &= \lim_{\beta\to\infty} \left[1 + B_{\vec{s}}\left(\Theta - \frac{\tau}{2},0\right) e^{-\beta h_O} B_{\vec{s}}\left(2\Theta,\Theta-\frac{\tau}{2}\right)\right]_{x,y} \end{aligned} \tag{98}$$

is valid and yields a link between the FTQMC and PQMC, so that at finite but large values of β we can use the methods introduced for the FTQMC to compute time displaced Green functions. In particular, we can use the relation

$$\lim_{\beta\to\infty} \begin{pmatrix} I & B_{\vec{s}}(\Theta-\frac{\tau}{2},0)e^{-\beta h_0}B_{\vec{s}}(2\Theta,\Theta+\frac{\tau}{2}) \\ -B_{\vec{s}}(\Theta+\frac{\tau}{2},\Theta-\frac{\tau}{2}) & I \end{pmatrix}^{-1} = \begin{pmatrix} G_{\vec{s}}(\Theta-\frac{\tau}{2}) & G_{\vec{s}}(\Theta-\frac{\tau}{2},\Theta+\frac{\tau}{2}) \\ G_{\vec{s}}(\Theta+\frac{\tau}{2},\Theta-\frac{\tau}{2}) & G_{\vec{s}}(\Theta+\frac{\tau}{2}) \end{pmatrix} \tag{99}$$

where the Green functions refer to the PQMC Green functions of Eq. (94). To compute in a numerically stable way, similar methods to those applied for the FTQMC are used (see Eq. (83)). A convenient choice of H_0 is obtained in a basis where the trial wave function may be written as:

$$|\Psi_T\rangle = \prod_{n=1}^{N_p} \gamma^\dagger_n|0\rangle. \tag{100}$$

In this basis, we define H_0 through

$$H_0\gamma^\dagger_n|0\rangle = \begin{cases} -\gamma^\dagger_n|0\rangle \text{ if } \gamma^\dagger_n\gamma_n|\Psi_T\rangle = |\Psi_T\rangle \\ +\gamma^\dagger_n|0\rangle \text{ if } \gamma^\dagger_n\gamma_n|\Psi_T\rangle = 0 \end{cases} \tag{101}$$

For the specific case of the Hubbard model and with the above choice of H_0 values of $\beta t \sim 40$ were well sufficient to satisfy Eq. (98) up to an average precision of 10^{-11}.

Clearly this method is ad-hoc since it requires adding a *fictitious* temperature. Furthermore, since the upgrading is done with the PQMC, every time one wishes to compute time displaced correlation function all required quantities have to be computed from scratch.

A more elegant and efficient method rests on the observation that in the PQMC the Green function is a projector. Consider again the free electron case. For a non-degenerate Ground state, $\langle\Psi_0|c^\dagger_{\vec{k}}c_{\vec{k}}|\Psi_0\rangle = 0,1$ so that

$$\langle\Psi_0|c^\dagger_{\vec{k}}(\tau)c_{\vec{k}}|\Psi_0\rangle = \left(\langle\Psi_0|c^\dagger_{\vec{k}}c_{\vec{k}}|\Psi_0\rangle \exp\left((\varepsilon_{\vec{k}}-\mu)\right)\right)^\tau . \tag{102}$$

The above involves only well defined numerical manipulations even in the large τ limit provided that all scales fit onto finite precision machines for a unit time interval.

The implementation of this idea in the QMC algorithm is as follows. First, one has to notice that the Green function $G_{\vec{s}}(\tau)$ is a projector:

$$G_{\vec{s}}(\tau)^2 = G_{\vec{s}}(\tau). \tag{103}$$

We have already seen that for $P^\dagger B_{\vec{s}}(2\Theta,\tau) = V_L D_L U^{\langle}$ and $B_{\vec{s}}(\tau,0) = U^{\rangle} D_R U_R$, $G_{\vec{s}}(\tau) = 1 - U^{\rangle}(U^{\langle}U^{\rangle})^{-1}U^{\langle}$. Since

$$\left[U^{\rangle}(U^{\langle}U^{\rangle})^{-1}U^{\langle}\right]^2 = U^{\rangle}(U^{\langle}U^{\rangle})^{-1}U^{\langle} \tag{104}$$

we have:

$$G^2_{\vec{s}}(\tau) = G_{\vec{s}}(\tau) \text{ and } (1-G_{\vec{s}}(\tau))^2 = 1 - G_{\vec{s}}(\tau). \tag{105}$$

This property implies that $G_{\vec{s}}(\tau_1,\tau_3)$ obeys a simple composition identity

$$G_{\vec{s}}(\tau_1,\tau_3) = G_{\vec{s}}(\tau_1,\tau_2)G_{\vec{s}}(\tau_2,\tau_1). \tag{106}$$

In particular for $\tau_1 > \tau_2 > \tau_3$

$$\begin{aligned} G_{\vec{s}}(\tau_1,\tau_3) &= B_{\vec{s}}(\tau_1,\tau_3)G^2_{\vec{s}}(\tau_3) = G_{\vec{s}}(\tau_1,\tau_3)G_{\vec{s}}(\tau_3) \\ &= \underbrace{G_{\vec{s}}(\tau_1,\tau_3)B^{-1}_{\vec{s}}(\tau_2,\tau_3)}_{G_{\vec{s}}(\tau_1,\tau_2)}\underbrace{B_{\vec{s}}(\tau_2,\tau_3)G_{\vec{s}}(\tau_3)}_{G_{\vec{s}}(\tau_2,\tau_3)} \end{aligned}$$

A similar proof is valid for $\tau_3 > \tau_2 > \tau_1$

Using this composition property (106) we can break up a large τ interval into a set of smaller intervals of length $\tau = N\tau_1$ so that

$$G_{\vec{s}}\left(\Theta+\frac{\tau}{2},\Theta-\frac{\tau}{2}\right) = \prod_{n=0}^{N-1} G_{\vec{s}}\left(\Theta-\frac{\tau}{2}+[n+1]\tau_1,\Theta-\frac{\tau}{2}+n\tau_1\right) \tag{107}$$

The above equation is the generalization of Eq. (102). If τ_1 is *small* enough each Green function in the above product is accurate and has matrix elements bounded by order unity. The matrix multiplication is then numerically well defined.

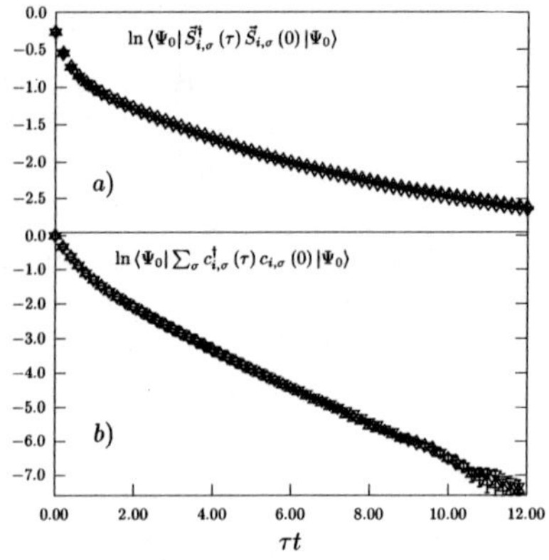

FIGURE 9. Imaginary time displaced on-site spin-spin (a) and Green function (b) correlation function. We consider a 6×6 lattice at half-filling and $J/t = 1.2$. In both (a) and (b) results obtained form Eq. (107) ($\triangle$) and (99) ($\bigtriangledown$) are plotted.

We conclude this section by comparing both presented approaches for the calculation of time displaced correlation functions in the PQMC. We consider the special case of the Kondo lattice model (see Fig. 9). As apparent the results are identical within error-bars. The important point however, is that the method based on Eq. (107) is for the considered case an order of magnitude quicker in CPU time than the method based on Eq. (99).

The sign problem.

The sign problem remains the central challenge in QMC simulations of correlated electron systems. We have shown how it apears in the framework of the World Line algorithm for the case of free fermions with nearest and next nearest neighbor hopping. Since in the PQMC and FTQMC the one body terms are treated exactly the sign problem does not occur in this simple example. However, interactions will in many cases lead to a sign problem. Insight into the origin of the sign problem in the PQMC and FTQMC is obtained through the work of Fahy and Hamann [34, 35] which we will briefly review.

The starting point is to rewrite the imaginary time propagation of the trial wave function as diffusion equation in the space of Slater determinants:

$$\sum_{\vec{s}} U_{\vec{s}}(2\Theta,0)|\Psi_T\rangle = \int d\Psi f(\Psi,2\Theta)|\Psi\rangle \tag{108}$$

Since $U_{\vec{s}}(2\Theta,0)$ is a one-body propagator in a time dependent external field, $U_{\vec{s}}(2\Theta,0)|\Psi_T\rangle$ is a single Slater determinant (See Eq. (41)) so that the sum over the HS fields may be replaced by a weighted average over normalized Slater determinants $|\Psi\rangle$. Fahy and Hamann [34, 35], have shown that $f(\Psi,\tau)$ satisfies a diffusion type equation in the space of Slater determinants which is invariant under parity $|\Psi\rangle \to -|\Psi\rangle$. Thus, the solutions of the diffusion equations may be classified according to this symmetry: $f^+(\Psi,\tau) = f^+(-\Psi,\tau)$ and $f^-(\Psi,\tau) = -f^-(-\Psi,\tau)$. The reader will convince himself that only f^- solutions contribute to the integral over Slater determinants. However, the integrand, $f(\Psi,\tau)$, is exponentially dominated by the even parity solutions $f^+(\Psi,\tau)$ of the diffusion equation [34, 35]. Hence, the relevant odd parity solutions are exponentially damped in comparison to the even parity solutions. This leads to exponential increase of the signal to noise ratio.

Based on the above analysis, Fahy and Hamann have proposed a solution to the sign problem, which is referred to as positive projection [34]. Let us start the diffusion process with the trial wave function $|\Psi_T\rangle$ which we require - without loss of generality - to have positive overlap with the ground state: $\langle\Psi_0|\Psi_T\rangle > 0$. At infinitesimal time $\tau = \varepsilon$ a population of Slater determinants $\{f(\Psi,\varepsilon)|\Psi\rangle\}$ with $f(\Psi,0) = \delta(\Psi - \Psi_T)$ is obtained. If one of those Slater determinants, $|\Psi_1\rangle$ at time $\tau = \varepsilon$, is orthogonal to the ground state, $\langle\Psi_0|\Psi_1\rangle = 0$, it may be discarded since it will provide no information on the ground state at all times $\tau > \varepsilon$. This may easily be seen since starting from $|\Psi_1\rangle$ the population of Slater determinants generated by the diffusion equation has zero overlap with the ground state:

$$\langle\Psi_0| \int \mathrm{d}\Psi f(\Psi,\tau)|\Psi\rangle = \langle\Psi_0|e^{-(\tau-\varepsilon)H}|\Psi_1\rangle = 0, \tag{109}$$

since $f(\Psi,\tau=\varepsilon) = \delta(\Psi - \Psi_1)$ and $\langle\Psi_0|\Psi_1\rangle = 0$. The important point is that computed analytically the ensemble of Slater determinants originating from Ψ_1 cancel. It is the stochastic cancellation of those Slater determinants which lies at the origin of the sign problem. The above procedure is repeated iteratively and at each infinitesimal time step Slater determinants with $\langle\Psi_0|\Psi\rangle = 0$ are discarded. This procedure eliminates the exponential growth of the signal to noise ratio.

The above procedure is exact provided we know a priori the surface $\mathcal{N}$ defined by: $\langle\Psi_0|\Psi\rangle = 0$. This information is in general not available. Thus, the positive projection scheme has to be implemented based on an approximate knowledge of $\mathcal{N}$. An implementation of this scheme has been carried out by [36, 37] in an algorithm which is referred to as the constrained path QMC (CPQMC). However, a major drawback is that the surface $\mathcal{N}$ in the CPQMC is approximated by a single Slater determinant corresponding to a mean-field solution of the model under consideration. In contrast, the Green function method has the ability to incorporate optimized correlated wave functions as starting point for the fixed-node approximation.

There are nevertheless a set of problems where the sign problem may be avoided in FTQMC and PQMC. As we will see in the subsequent section, particle-hole symmetry allows us to avoid the sign problem. Furthermore, models with attractive interactions which couple independently to an internal symmetry with an even number of states lead to weights, for a given HS configuration, which are an even power of a single determinant. If the determinant itself is real (i.e. absence of magnetic fields), the overall weight will be positive. An example is the attractive Hubbard model.

APPLICATION OF THE AUXILIARY FIELD QMC TO SPECIFIC HAMILTONIANS.

In this section, we will concentrate on applications of the FTQMC and PQMC. Both approaches involve a computational effort which scales as the cubed of the volume. This makes it hard to achieve very large lattice sizes. Hence the importance of reducing finite size effects. Below we will describe a method which turns out to be extremely efficient for the case of free electrons where size effects are know to be severe [38]. We then show how to apply the FTQMC and PQMC to the attractive and repulsive Hubbard models, the Kondo-lattice model, hard core bosons and the Heisenberg model. It is beyond the scope of this article to review in detail the physics related to the models. We will concentrate primarily on the technical aspects. In most cases we have the two-dimensional case in mind, the generalization to higher or lower dimensions being straightforward.

Size effects

Size effects become particularly severe when the ground state turns out to be a metallic state with large coherence temperature. On the other hand, insulators are characterized by the localization of the wave function and are hence rather insensitive to boundary conditions on finite sized systems. It thus becomes apparent, that the worst case scenario for severe size effects are just free electrons in a tight binding approximation:

$$H = -t \sum_{\langle \vec{i},\vec{j} \rangle} c^{\dagger}_{\vec{i}} c_{\vec{j}} + \text{H.c.}. \tag{110}$$

In many cases before turning on the interaction which will automatically restrict the size of the lattice under consideration it is important to control size effects for this simple case. We will concentrate on the two dimensional case on a torus geometry which for the above model reduces to imposing periodic boundary conditions: $c^{\dagger}_{\vec{i}+L\vec{e}_x} = c^{\dagger}_{\vec{i}}$, $c^{\dagger}_{\vec{i}+L\vec{e}_y} = c^{\dagger}_{\vec{i}}$ where L is the linear length of the lattice lying in the $\vec{e}_x, \vec{e}_y$ plane.

To reduce size effects on thermodynamic quantities one may in principle consider the Hamiltonian:

$$H(L) = \sum_{\langle \vec{i},\vec{j} \rangle} t_{\vec{i},\vec{j}}(L) c^{\dagger}_{\vec{i}} c_{\vec{j}} + \text{H.c.} \tag{111}$$

where $t_{\vec{i},\vec{j}}(L)$ are arbitrary hopping parameters which have to satisfy

$$\lim_{L\to\infty} t_{\vec{i},\vec{j}}(L) = -t. \tag{112}$$

Clearly this choice of hopping matrix elements on finite lattices will break the lattice symmetry. This is a price we are willing to pay provided that the convergence as a function of system size of thermodynamics quantities is greatly improved. Eq. (112) nevertheless guarantees that in the thermodynamic limit this symmetry is restored. To determine the hopping matrix elements $t_{\vec{i},\vec{j}}(L)$ so as to reduce size effects on say the specific heat, $C_v(L,T) = \frac{\partial E(L)}{\partial T}$, one may minimize

$$\chi^2 = \sum_T [C_v(L,T) - C_v(L=\infty,T)]^2 \tag{113}$$

where the sum extends over a given range of temperatures. Taking into account only amplitude modulations of the hopping matrix elements leads already to a cumbersome minimization problem which does not provide satisfactory results.

Instead of carrying out a complicated minimization problem we can try to guess which matrix elements $t_{\vec{i},\vec{j}}(L)$ will minimize size effects. It turns out that introducing a magnetic field produces remarkable results. The magnetic field is introduced via the Peirls phase factors:

$$H(L) = -t \sum_{\langle \vec{i},\vec{j} \rangle} e^{\frac{2\pi i}{\Phi_0} \int_{\vec{i}}^{\vec{j}} \vec{A}_L(\vec{l}) \cdot d\vec{l}} c^{\dagger}_{\vec{i}} c_{\vec{j}} + \text{H.c.} \tag{114}$$

with $\vec{B}_L(\vec{x}) = \vec{\nabla} \times \vec{A}_L(\vec{x})$ and Φ_0 the flux quanta. The torus geometry imposes restrictions on the $\vec{B}_L$ field. Since, a translation in the argument of the vector potential may be absorbed in a gauge transformation:

$$\begin{aligned} \vec{A}_L(\vec{x}+L\vec{e}_x) &= \vec{A}_L(\vec{x}) + \vec{\nabla}\chi_x(\vec{x}), \\ \vec{A}_L(\vec{x}+L\vec{e}_y) &= \vec{A}_L(\vec{x}) + \vec{\nabla}\chi_y(\vec{x}), \end{aligned} \tag{115}$$

we chose, the boundary condition

$$c^{\dagger}_{\vec{i}+L\vec{e}_x} = e^{\frac{2\pi i}{\Phi_0}\chi_x(\vec{i})} c^{\dagger}_{\vec{i}}, \quad c^{\dagger}_{\vec{i}+L\vec{e}_y} = e^{\frac{2\pi i}{\Phi_0}\chi_y(\vec{i})} c^{\dagger}_{\vec{i}} \tag{116}$$

to satisfy the requirement:

$$[H(L), T_{L\vec{e}_x}] = [H(L), T_{L\vec{e}_y}] = 0. \tag{117}$$

Here, $T_{\vec{x}}$ corresponds to a translation by $\vec{x}$. However, magnetic translation operators belong to the magnetic algebra [39]:

$$T_{L\vec{e}_x} T_{L\vec{e}_y} = e^{-i2\pi \frac{(L\vec{e}_x \times L\vec{e}_y)\cdot\vec{B}}{\Phi_0}} T_{L\vec{e}_y} T_{L\vec{e}_x}. \tag{118}$$

Thus, to obtain a single valued wave function the condition of flux quantization has to be satisfied: $\frac{(L\vec{e}_x \times L\vec{e}_y)\cdot\vec{B}}{\Phi_0} = n$ where n is an integer. Here, we consider a static magnetic field running along the z-axis perpendicular to the x,y plane in which the lattice lies. Hence, the smallest magnetic field which we can consider on a given lattice size satisfies:

$$\frac{BL^2}{\Phi_0} = 1. \tag{119}$$

With this choice of magnetic field and associated vector potential Eq. (112) holds.

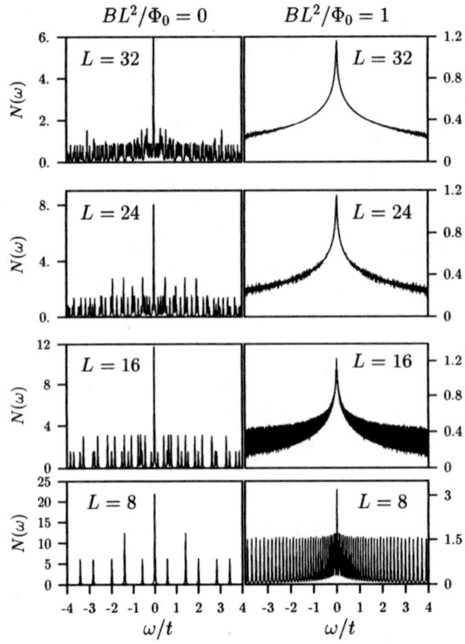

FIGURE 10. Density of states $N(\omega) = \frac{1}{N}\sum_r N(r,\omega)$ with (right column) and without (left column) magnetic field. Here, we consider $\delta = 0.01t$.

To illustrate the reduction of size effects caused by the inclusion of the magnetic field, we first consider the single particle density of states. In a basis where $H(L)$ is diagonal, $H(L) = \sum_{n=1}^{N} \varepsilon_n \gamma_n^\dagger \gamma_n$ with $c_n^\dagger = \sum_m \gamma_m^\dagger U_{m,n}^\dagger$ and $U^\dagger U = I$, the local density of states reads:

$$N(r,\omega) = \mathrm{Im} \sum_{n}^{N} \frac{|U_{n,r}|^2}{\varepsilon_n - \omega - i\delta} \tag{120}$$

where δ is a positive infinitesimal and N the total number of sites. Since the magnetic field breaks translation invariance (it is the site dependent vector potential which enters the Hamiltonian) $N(r,\omega)$ is site dependent. Averaging over sites yields the density of states $N(\omega)$ plotted in Fig. 10. As apparent, without the magnetic field and up to $L = 32$, $N(\omega)$ is dominated by size effects for the considered value of $\delta = 0.01t$. In contrast, the presence of the magnetic field provides remarkable improvements. In particular the van-Hove singularity is well reproduced already on $L = 16$ lattices and at $L = 32$ the result is next to exact for the considered value of δ. It is instructive to look at the $L = 8$ case with and without magnetic fields. When B is turned on, the degeneracy of levels is lifted. Each level - apart from the $\varepsilon_n = 0$ level which is two fold degenerate - is nondegenerate. This is precisely what one expects for Landau levels which have degeneracy L^2B/Φ_0 which is unity in our case. This provides an intuitive understanding of why the method works so well. Since each level becomes singly degenerate, the single particle states cover homogeneously the the energy range of the band-width. Clearly this can only be achieved by breaking the lattice symmetry on finite sized systems.

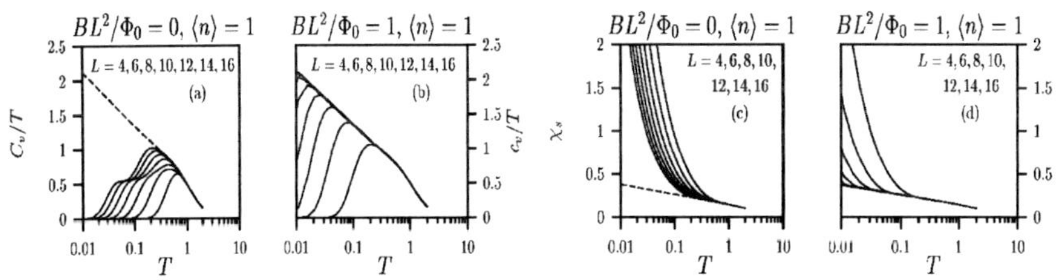

FIGURE 11. Specific and spin susceptibility versus temperature without (a,c) and with (b,d) magnetic field. The curves from right to left correspond to increasingly large lattices as denoted in the figure. The dashed line corresponds to the exact result.

We now turn our attention to the specific heat coefficient $\gamma = C_v/T$ (see Fig. 11a,b). As apparent, for a given system size, the inclusion of the magnetic field provides a gain of more than one order of magnitude in the temperature scale at which size effects set in. In particular the $\ln(1/T)$ behavior of γ due the the van-Hove singularity becomes apparent already on $L = 6$ lattices.

Upon inspection a similar improvement is obtained for the spin susceptibility (see Fig. 11c,d). Note that since we are dealing with free electrons the charge and spin susceptibilities are identical.

One crucial question is whether the magnetic field will introduce a sign problem in the numerical simulations. It turns out that in some non-trivial cases is does not and we can hence benefit from the observed dramatic reduction in size effects. This method has been used successfully in a depleted Kondo lattice model and has opened a whole temperature range where coherence effects may me studied without suffering from size effects [38].

Other schemes have been proposed to reduce size effects. In particular, averaging over boundary conditions has been suggested [40, 41]. This method has the advantage of not breaking translation symmetry. However, the averaging requires several simulations and is hence computationally expensive. In contrast with the presented method the improvement in reduction of size effects is obtained within a single simulation.

The Hubbard model.

The Hubbard model in a magnetic field is given by:

$$H_U = \underbrace{-t \sum_{\langle \vec{i},\vec{j} \rangle,\sigma} e^{\frac{2\pi i}{\Phi_0}\int_{\vec{i}}^{\vec{j}} \vec{A}\cdot d\vec{l}} c^{\dagger}_{\vec{i},\sigma} c_{\vec{j},\sigma}}_{=\sum_{\vec{i},\vec{j},\sigma} c^{\dagger}_{\vec{i},\sigma} h_t(\vec{A})_{\vec{i},\vec{j}} c_{\vec{j},\sigma}} + U \sum_{\vec{i}} \left(n_{\vec{i},\uparrow} - 1/2\right)\left(n_{\vec{i},\downarrow} - 1/2\right) - \mu \sum_{\vec{i}} n_{\vec{i}}. \quad (121)$$

We start by considering the $SU(2)$ invariant HS decoupling of Eq. (33). Since the HS field couples equivalently to up and down spins we obtain for the FTQMC:

$$\mathrm{Tr}\left[e^{-\beta(H-\mu N)}\right] = \sum_{\vec{s}} C_{\vec{s}} \det\left(1 + B_{\vec{s}}(\beta,0)\right)^2 \quad \text{with} \tag{122}$$

$$C_{\vec{s}} = \frac{1}{2^N} e^{\beta U N/4 - i\alpha \sum_{\vec{i},n} s_{\vec{i},n}}, \quad B(\beta,0) = \prod_{n=1}^{m} e^{h_I(\vec{s}_n)} e^{-\Delta\tau(h_t(\vec{A})-\mu)}.$$

Here, $h_I(\vec{s}_n)_{\vec{i},\vec{j}} = \delta_{\vec{i},\vec{j}} i\alpha s_{\vec{i},n}$ and $\cos(\alpha) = \exp(-\Delta\tau U/2)$.

For the PQMC a trial wave function which factorizes in spin up and down sectors is usually chosen:

$$|\Psi_T\rangle = |\Psi_T^{\uparrow}\rangle \otimes |\Psi_T^{\downarrow}\rangle, \quad |\Psi_T^{\sigma}\rangle = \prod_{y=1}^{N_p^{\sigma}} (\vec{c}_{\sigma}^{\dagger} P^{\sigma})_y |0\rangle. \tag{123}$$

Here, $N_p^{\uparrow} = N_p^{\downarrow}$ are the number of particles in the spin up and down sectors. Typically, the trial wave function is chosen to be the solution of the non-interacting system. Hence,

$$P^{\uparrow} = P^{\downarrow} = P \tag{124}$$

so that:

$$\langle \Psi_T | e^{2\Theta H} | \Psi_T \rangle = \sum_{\vec{s}} C_{\vec{s}} \det\left(P^{\dagger} B_{\vec{s}}(2\Theta,0) P\right)^2. \tag{125}$$

Replacing β by 2Θ and setting $\mu = 0$ in the above definitions of $B_{\vec{s}}(2\Theta,0)$ and $C_{\vec{s}}$ yields the form of those quantities for the PQMC.

Before discussing separately the attractive ($U < 0$) and repulsive case ($U > 0$) let us comment on the matrix multiplication $e^{-\Delta\tau h_t(\vec{A})} C$ where C is an $N \times N$ matrix. A straightforward multiplication yields a computational cost scaling as N^3. To reduce this cost the checkerboard decomposition is often used. The matrix h_t is written as $h_t = h_t^{(1)} + h_t^{(2)}$ where $h_t^{(1)}$ and $h_t^{(2)}$ are sums of commuting four site hopping matrices: $h_t^{(1,2)} = \sum_{j=1}^{N/4} (h_t^{(1,2)})_j$ (See Fig. 12). Thus the the multiplication

$$\begin{aligned} e^{-\Delta\tau h_t} A &= e^{-\Delta\tau h_t^{(1)}} e^{-\Delta\tau h_t^{(2)}} A + \mathcal{O}(\Delta\tau^2) \\ &= \prod_{i=1}^{N/4} e^{-\Delta\tau (h_t^{(1)})_i} \prod_{j=1}^{N/4} e^{-\Delta\tau (h_t^{(2)})_j} A + \mathcal{O}(\Delta\tau^2) \end{aligned} \tag{126}$$

Since $(h_t^{(1,2)})_j$ involves hopping on four sites irrespective of the lattice size, the matrix multiplication $e^{-\Delta\tau (h_t^{(1,2)})_j} A$ scales as N. Hence the overall matrix multiplication scales as N_s^2. Since the systematic error involved in the Trotter decomposition is already of order $\Delta\tau^2$ the above checkerboard decomposition does not introduce a new source of errors. However we have gained a power in the computational cost.

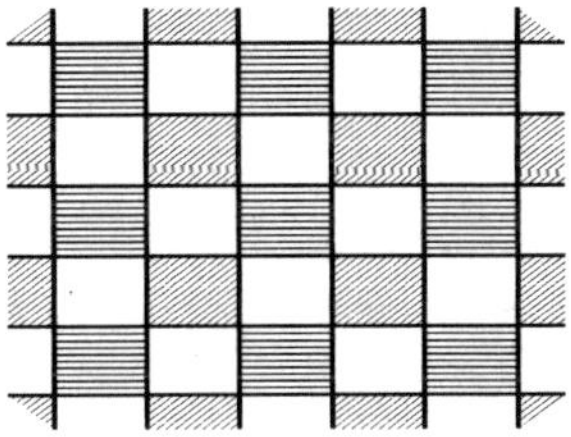

FIGURE 12. Checkerboard decomposition for the two dimensional square lattice with nearest neighbor hopping. $h_t^{(1)}$ ($h_t^{(2)}$) contains hopping processes along the boundaries of the horizontally (diagonally) shaded squares.

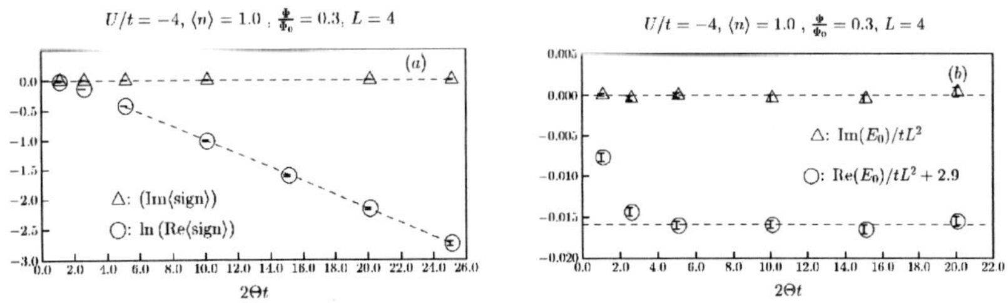

FIGURE 13. (a) Real and imaginary part of the average sign. Within the error-bars the imaginary part vanishes and the real part decays exponentially with the projection parameter Θ. (b) Real and imaginary part of the energy as a function of Θ. For a single HS configuration, the imaginary part of the energy is non-zero. Only after averaging does it vanish. In this case, the sign problem is not severe, since convergence to the ground state is achieved before the average sign becomes prohibitly small.

$U < 0$

The attractive Hubbard model has generated considerable interest since i) in two dimensions it shows a Kosterlitz-Thouless transition to an s-wave superconducting state [42, 43] and ii) in the strong coupling limit pairs form at a temperature scale roughly set by U whereas the Kosterlitz-Thouless transition is expected to scale as t^2/U. Hence, the model offers the possibility of studying the physics of a metallic state whith preformed pairs, a subject of interest in the context of high T_c superconductivity [44].

When $U < 0$, α in Eq. (122) is a pure imaginary number so that in the absence of magnetic fields, the determinant is real. Since the weight is the square of a real number it is positive and hence no sign problem occurs. In the weak coupling limit or BCS regime, the simulations perform well. However, in the strong coupling limit the method suffers from long autocorrelation times.

In the presence of a magnetic field, a sign problem occurs since the determinant becomes complex. We illustrate this point as well as the behavior of the sign problem when the product of determinants is complex by considering a vector potential $\vec{A}(\vec{x}) = \Phi\vec{e}_x/L$ at $\Phi = 0.3\Phi_0$. Such a vector potential introduces a twist in the boundary conditions which corresponds to nothing but threading a magnetic field with flux Φ through the torus on which lies the electronic system. This may be used to probe for flux quantization as well as a Kosterlitz-Thouless transition [43]. Note that insulating states may equally be determined by such a construction [45, 25]. For this choice of $\vec{A}$, the fermionic determinant is complex. Hence in the framework of the PQMC and following Eq. (21), observables are evaluated with:

$$\langle O \rangle = \frac{\sum_{\vec{s}} \overline{\mathrm{Pr}}_{\vec{s}} \mathrm{sign}(\vec{s}) \langle O \rangle_{\vec{s}}}{\sum_{\vec{s}} \overline{\mathrm{Pr}}_{\vec{s}} \mathrm{sign}(\vec{s})} \quad \text{where} \tag{127}$$

$$\mathrm{sign}(\vec{s}) = \frac{\det\left(P^\dagger B_{\vec{s}}(2\Theta,0)P\right)^2}{|\det\left(P^\dagger B_{\vec{s}}(2\Theta,0)P\right)^2|}, \qquad \overline{\mathrm{Pr}}_{\vec{s}} = \frac{|\det\left(P^\dagger B_{\vec{s}}(2\Theta,0)P\right)^2|}{\sum_{\vec{s}} |\det\left(P^\dagger B_{\vec{s}}(2\Theta,0)P\right)^2|}.$$

For a given HS configuration $\mathrm{sign}(\vec{s}) = e^{-i\phi(\vec{s})}$. After summation over the HS fields, the average sign, $\langle \mathrm{sign} \rangle = \sum_{\vec{s}} \overline{\mathrm{Pr}}_{\vec{s}} \mathrm{sign}(\vec{s})$ is a real quantity. This follows from

$$\langle \mathrm{sign} \rangle = \frac{\langle \Psi_T | e^{-2\Theta H} | \Psi_T \rangle}{\sum_{\vec{s}} |\det\left(P^\dagger B_{\vec{s}}(2\Theta,0)P\right)^2|} + \mathscr{O}(\Delta\tau^2) \tag{128}$$

and the fact that H is hermitian so that the numerator is a real quantity. [5] This property gives us a nice internal check for the validity of Monte Carlo sampling. As shown in Fig. (13) the real part of the average sign decays exponentially with the projection parameter Θ.

$U > 0$

The repulsive Hubbard model is given by Eq. (121) with $U > 0$. At half-filling the Hubbard model describes a Mott insulator with long range antiferromagnetic order. The nature of the metallic state in the vicinity of the Mott insulator as well as the metal-insulator transition as a function of doping has attracted considerable recent interest in particular in conjunction with high T_c superconductors [46].

For $U > 0$, α in Eq. (122) is a real number so that the determinant is a complex quantity and hence a sign problem occurs. In the special case of half-filling, $\mu = 0$ in Eq. (121), the model is particle-hole symmetric. This symmetry allows us to avoid the sign-problem. The particle-hole transformation reads.

$$\mathscr{P}^{-1} c^\dagger_{\vec{i},\sigma} \mathscr{P} = (-1)^{i_x+i_y} c^\dagger_{\vec{i},\sigma}. \tag{129}$$

where $\vec{i} = (i_x, i_y)$. Note that the Hamiltonian (121) at $\mu = 0$ is invariant under a combined time reversal and particle-hole transform due to the presence of the vector potential. Using the notation of Eq. (122):

$$\mathscr{P}^{-1} H_t \mathscr{P} \equiv \mathscr{P}^{-1} \vec{c}^{\,\dagger}_\sigma h_t(\vec{A}) \vec{c}_\sigma \mathscr{P} = \vec{c}^{\,\dagger}_\sigma \overline{h_t(\vec{A})} \vec{c}_\sigma \text{ and} \tag{130}$$

[5] This property equally holds for finite values of $\Delta\tau$ since the kinetic and potential terms in H as well as the trial wave function are real representable in Fourier space.

$$\mathscr{P}^{-1}H_I(\vec{s}_n)\mathscr{P} \equiv \mathscr{P}^{-1}\vec{c}_\sigma^\dagger h_I(\vec{s}_n)\vec{c}_\sigma\mathscr{P} = \sum_{\vec{i}} i\alpha s_{\vec{i},n} + \vec{c}_\sigma^\dagger \overline{h_I(\vec{s}_n)}\vec{c}_\sigma.$$

Thus

$$\begin{aligned}
\det(1+B_{\vec{s}}(\beta,0)) &= \mathrm{Tr}\left[\prod_{n=1}^{m} e^{H_I(\vec{s}_n)}e^{-\Delta\tau H_t}\right] \\
&= \mathrm{Tr}\left[\prod_{n=1}^{m} e^{\mathscr{P}^{-1}H_I(\vec{s}_n)\mathscr{P}}e^{-\Delta\tau\mathscr{P}^{-1}H_t\mathscr{P}}\right] \\
&= e^{\sum_{\vec{i},} i\alpha s_{\vec{i},n}}\det(1+\prod_{n=1}^{m} e^{\overline{h_I(\vec{s}_n)}}e^{-\Delta\tau\overline{h_t(\vec{A})}}) \\
&= e^{\sum_{\vec{i}} i\alpha s_{\vec{i},n}}\overline{\det(1+B_{\vec{s}}(\beta,0))} \qquad (131)
\end{aligned}$$

and hence $C_{\vec{s}}\det\left(1+B_{\vec{s}}(\beta,0)\right)^2$ (see Eq. (122)) is a positive quantity even in the presence of a magnetic field. More generally, the above is valid for a half-filled Hubbard model on bipartite lattice [6] with inter-sublattice single-electron hopping. Thus, and for example, simulations on a half-filled honnneycomb lattice are sign free and show Mott semi-metal-insulator transitions [47].

Away from half-filling the sign problem sets in and turns out to be rather severe with the use of the HS transformation of Eq. (33). It is then more efficient to consider the HS which couples to the magnetization (32) to obtain:

$$\mathrm{Tr}e^{-\beta(H-\mu N)} = C\sum_{\vec{s}}\det\left(1+B_{\vec{s}}^\uparrow(\beta,0)\right)\det\left(1+B_{\vec{s}}^\downarrow(\beta,0)\right) \quad \text{with}$$

$$C = e^{\beta UN/4}2^N, \ B_{\vec{s}}^\sigma(\beta,0) = \prod_{n=1}^{m} e^{h_I^\sigma(\vec{s}_n)}e^{-\Delta\tau(h_t(\vec{A})-\mu)} \qquad (132)$$

with $(h_I^\sigma(\vec{s}_n))_{\vec{i},\vec{j}} = \delta_{\vec{i},\vec{j}}\sigma\tilde{\alpha}s_{\vec{i},n}$. In a very similar manner as above, one will show that for the particle-hole symmetric case the sign problem does not occur since: $\det(1+B_{\vec{s}}^\uparrow(\beta,0)) = e^{\sum_{\vec{i},}\sigma\tilde{\alpha}s_{\vec{i},n}}\overline{\det(1+B_{\vec{s}}^\downarrow(\beta,0))}$. Fig. 14 plots the average sign as a function of electronic density for the HS transformation (32). As apparent, it is most severe at low dopings. In general, when the average sign drops below 0.1 accurate simulations become prohibitly expensive. One will notice that the average sign decays more slowly for special band fillings. Those band fillings corresponds to filled shells for which the solution of the non-interacting system is non-degenerate. The dependence of the average sign on different choices of the HS decoupling has been considered in [48].

For the PQMC, we again assume that the trial wave function factorizes in spin up and spin down sectors (123). With the HS transformation (32) we obtain:

$$\langle\Psi_T|e^{-2\Theta H}|\Psi_T\rangle = C\sum_{\vec{s}}\prod_{\sigma}\det\left(P^{\sigma,\dagger}B_{\vec{s}}^\sigma(2\Theta,0)P^\sigma\right) \qquad (133)$$

[6] The lattice sites of a bipartite lattice may be split into two sublattices, A and B such that the nearest neighbors of any site in A belong to B.

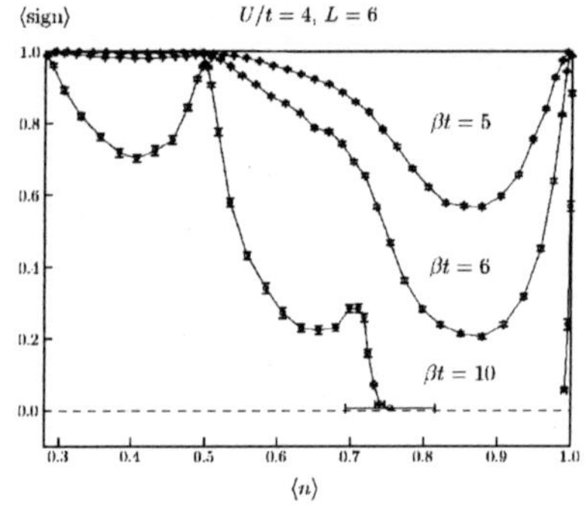

FIGURE 14. Average sign versus band filling $\langle n \rangle$ on a 6×6 lattice at $U/t = 4$. Filled shells are present at $\langle n \rangle = 26/36$ and $\langle n \rangle = 18/36$. As apparent for those band fillings, the decay of the average sign is slow.

with similar definitions for $B^{\sigma}_{\vec{s}}$ as for the FTQMC. At half-filling, the sign problem may again be avoided provided that the trial wave function is appropriately chosen. It is convenient to require the trial wave function to be a non-degenerate ground state of a given single-particle Hamiltonian $H_0 = H_0^{\uparrow} + H_0^{\downarrow}$ with,

$$H_0^{\sigma} = \sum_{\sigma} \vec{c}^{\dagger}_{\sigma} T_0^{\sigma} \vec{c}_{\sigma}. \tag{134}$$

Using Eq. (98) to relate the PQMC to the FTQMC and following Eq. (131) the sign problem is absent provided that:

$$P^{-1} \vec{c}^{\dagger}_{\uparrow} T_0^{\uparrow} c_{\uparrow} P = \vec{c}^{\dagger}_{\uparrow} \overline{T_0^{\downarrow}} \vec{c}_{\uparrow} + C \tag{135}$$

where C is a constant.

The choice of the trial wave function is important. Quick convergence as a function of projection parameter Θ is the measure for the quality of the trial wave function. Away from half-band filling when the sign problem sets in, it is important to reach convergence quickly before the average sign becomes prohibitively small. In the PQMC, different symmetry sectors may be probed independently by judiciously choosing the trial wave function. An example where symmetry considerations are important is the half-filled Hubbard model. The ground state at this band-filling has long-range antiferromagnetic order. This continuous symmetry breaking is accompanied by gapless spin excitations: spin waves. On finite size lattices, however, spin symmetry is not broken, and the ground state is a spin singlet with an energy gap to spin excitations set by v_s/L where v_s is the spin-wave velocity and L the linear size of the lattice. Choosing a spin-singlet trial wave function is a good starting point since it is orthogonal to the low lying spin excitations. Hence, they do not have to be filtered out at great cost (see Fig. 15). To generate a spin singlet trail wave function we consider the non-interacting Hamiltonian:

$$H_0 = \sum_{\langle \vec{i}, \vec{j} \rangle, \sigma} \left(t_{\vec{i},\vec{j}} c^{\dagger}_{\vec{i},\sigma} c_{\vec{j},\sigma} + \text{H.c.} \right) \tag{136}$$

where $\langle \vec{i},\vec{j}\rangle$ is a sum over nearest neighbors and

$$t_{\vec{i},\vec{i}+\vec{a}_x} = \begin{cases} -t(1+\delta) \text{ if } i_x = 2n+1 \\ -t(1-\delta) \text{ if } i_x = 2n \end{cases}, \quad t_{\vec{i},\vec{i}+\vec{a}_y} = -t(1+\delta) \tag{137}$$

with $\delta << t$. The dimerization δ leads to a non-degenerate ground state at half band-filling and hence to a spin singlet ground state which we use as trial wave function $|\Psi_T\rangle$:

$$\vec{S}^2|\psi_T\rangle = 0, \text{ with } \vec{S} = \sum_{\vec{i}} \vec{S}_{\vec{i}} \tag{138}$$

and $\vec{S}_{\vec{i}}$ is the spin operator on site $\vec{i}$. This trial wave function was used to produce the data of Fig. 7.

Unrestricted Hartree Fock solutions may be used as trial wave functions. In contrast to the above, those trial wave functions have overlaps with all symmetry sectors. However, this choice of trial wave function optimizes the overlap with the ground state. At finite dopings, those trial wave functions have been discussed in [49].

Periodic Anderson and Kondo lattice models.

The Kondo lattice model (KLM) as well as the periodic Anderson model (PAM) are the prototype Hamiltonians to describe heavy fermion materials [50] and Kondo insulators [51]. The physics under investigation is that of a lattice of magnetic impurities embedded in a metallic host. The symmetric PAM reads:

$$\begin{aligned} H_{PAM} &= \sum_{\vec{k},\sigma} \varepsilon(\vec{k}) c^{\dagger}_{\vec{k},\sigma} c_{\vec{k},\sigma} - V \sum_{\vec{i},\sigma} \left(c^{\dagger}_{\vec{i},\sigma} f_{\vec{i},\sigma} + f^{\dagger}_{\vec{i},\sigma} c_{\vec{i},\sigma} \right) \\ &+ U_f \sum_{\vec{i}} \left(n^f_{\vec{i},\uparrow} - 1/2 \right) \left(n^f_{\vec{i},\downarrow} - 1/2 \right). \end{aligned} \tag{139}$$

The unit cell, denoted by $\vec{i}$, contains an extended and a localized orbitals. The fermionic operators $c^{\dagger}_{\vec{k},\sigma}$ ($f^{\dagger}_{\vec{k},\sigma}$) create electrons on extended (localized) orbitals with wave vector $\vec{k}$ and $z-$component of spin σ. The overlap between extended orbitals generates a conduction band with dispersion relation $\varepsilon(\vec{k})$. There is a hybridization matrix element, V, between both orbitals in the unit-cell and the Coulomb repulsion- modeled by a Hubbard U_f- is taken into account on the localized orbitals. In the limit of large U_f, charge fluctuations on the localized orbitals are suppressed and the PAM maps onto the KLM [52]:

$$H_{KLM} = \sum_{\vec{k},\sigma} \varepsilon(\vec{k}) c^{\dagger}_{\vec{k},\sigma} c_{\vec{k},\sigma} + J \sum_{\vec{i}} \vec{S}^c_{\vec{i}} \cdot \vec{S}^f_{\vec{i}}. \tag{140}$$

Here $\vec{S}^c_{\vec{i}} = \frac{1}{2}\sum_{s,s'} c^{\dagger}_{\vec{i},s} \vec{\sigma}_{s,s'} c_{\vec{i},s'}$, where $\vec{\sigma}$ are the Pauli $s = 1/2$ matrices. A similar definition holds for $\vec{S}^f_{\vec{i}}$. A magnetic energy scale $J = 8V^2/U$ emerges and there is a constraint of one electron per localized orbital.

It is beyond the scope of this review to discuss the physics of the PAM and KLM. The interested reader is referred to [53]. Here we concentrate only on the technical aspects involved in the simulations. Simulations of the PAM are identical to those for the repulsive Hubbard model [54]. Again, at half-filling, the sign problem is absent due to the underlying particle-hole symmetry.

Simulation of the KLM on the other hand have up to recently been plagued by the sign-problem even in the case of half-filling where the model is invariant under particle-hole transformation [55, 56]. To achieve a sign-free formulation of the problem [29] in the half-filled case we first consider the Hamiltonian:

$$H = \sum_{\vec{k},\sigma} \varepsilon(\vec{k}) c^{\dagger}_{\vec{k},\sigma} c_{\vec{k},\sigma} - \frac{J}{4} \sum_{\vec{i}} \left[\sum_{\sigma} c^{\dagger}_{\vec{i},\sigma} f_{\vec{i},\sigma} + f^{\dagger}_{\vec{i},\sigma} c_{\vec{i},\sigma} \right]^2 . \tag{141}$$

With the use of the HS transformation of Eq. (33) to decouple the perfect square term a QMC algorithm may readily be formulated for the above model. Without constraints on the Hilbert space, the reader will easily persuade himself that the sign problem does not occur since we are dealing with an attractive interaction ($J > 0$) which couples to an internal symmetry – the spin – with an even number of states. We will see that the constraint of single occupancy on f-sites leads to a sign problem away from half-filling.

To see how H relates to H_{KLM} we compute the square in Eq. (141) to obtain:

$$\begin{aligned} H &= \sum_{\vec{k},\sigma} \varepsilon(\vec{k}) c^{\dagger}_{\vec{k},\sigma} c_{\vec{k},\sigma} + J \sum_{\vec{i}} \vec{S}^c_{\vec{i}} \cdot \vec{S}^f_{\vec{i}} \qquad (142) \\ &- \frac{J}{4} \sum_{\vec{i},\sigma} \left(c^{\dagger}_{\vec{i},\sigma} c^{\dagger}_{\vec{i},-\sigma} f_{\vec{i},-\sigma} f_{\vec{i},\sigma} + \mathrm{H.c.} \right) + \frac{J}{4} \sum_{\vec{i}} \left(n^c_{\vec{i}} n^f_{\vec{i}} - n^c_{\vec{i}} - n^f_{\vec{i}} \right). \end{aligned}$$

As apparent, there are only pair-hopping processes between the f- and c-sites. Thus the total number of doubly occupied and empty f-sites is a conserved quantity:

$$[H, \sum_{\vec{i}} (1 - n^f_{\vec{i},\uparrow})(1 - n^f_{\vec{i},\downarrow}) + n^f_{\vec{i},\uparrow} n^f_{\vec{i},\downarrow}] = 0. \tag{143}$$

If we denote by Q_n the projection onto the Hilbert space with $\sum_{\vec{i}} (1 - n^f_{\vec{i},\uparrow})(1 - n^f_{\vec{i},\downarrow}) + n^f_{\vec{i},\uparrow} n^f_{\vec{i},\downarrow} = n$ then:

$$H Q_0 = H_{KLM} + \frac{JN}{4} \tag{144}$$

since in the Q_0 subspace the f-sites are singly occupied and hence the pair-hopping term vanishes. Thus, it suffices to choose

$$Q_0 |\Psi_T\rangle = |\Psi_T\rangle \tag{145}$$

to ensure that

$$\frac{\langle \Psi_T | e^{-\Theta H} O e^{-\Theta H} | \Psi_T \rangle}{\langle \Psi_T | e^{-2\Theta H} | \Psi_T \rangle} = \frac{\langle \Psi_T | e^{-\Theta H_{KLM}} O e^{-\Theta H_{KLM}} | \Psi_T \rangle}{\langle \Psi_T | e^{-2\Theta H_{KLM}} | \Psi_T \rangle}. \tag{146}$$

To obtain a trial wave function which satisfies the requirements $Q_0|\Psi_T\rangle = |\Psi_T\rangle$ we are forced to choose H_0 of the form:

$$H_0 = \sum_{\langle \vec{i},\vec{j}\rangle,\sigma} \left(t_{\vec{i},\vec{j}} c^\dagger_{\vec{i},\sigma} c_{\vec{j},\sigma} + \text{H.c.} \right) + h_z \sum_{\vec{i}} e^{i\vec{Q}\cdot\vec{i}} \left(f^\dagger_{\vec{i},\uparrow} f_{\vec{j},\uparrow} - f^\dagger_{\vec{i},\downarrow} f_{\vec{j},\downarrow} \right) \tag{147}$$

which generates a Néel state ($\vec{Q} = (\pi,\pi)$) on the localized orbitals. and the hopping matrix elements satisfy Eq. (137). With this choice of the trial wave function and the definition of the particle-hole transformation

$$\mathscr{P}^{-1} c^\dagger_{\vec{i},\sigma} \mathscr{P} = (-1)^{i_x+i_y} c^\dagger_{\vec{i},\sigma}$$
$$\mathscr{P}^{-1} f^\dagger_{\vec{i},\sigma} \mathscr{P} = -(-1)^{i_x+i_y} f^\dagger_{\vec{i},\sigma}$$

one may readily show the absence of sign problem at half-filling.

Although attractive, the above approach turns out to be (i) numerically inefficient and (ii) restricted to the PQMC. In the half-filled case, the principle source of inefficiency lies in the coupled constraints, $Q_0|\Psi_T\rangle = |\Psi_T\rangle$ and $|\Psi_T\rangle$ is a Slater determinant factorizable in the spin indices which inhibits the choice of a spin singlet trial wave function. Since the ground state is known to be a spin singlet on finite lattices [57, 58] convergence is bad especially in the small J/t limit for which the ground state has long-range antiferromagnetic order, with small spin-wave velocity.

To alleviate both problems we relax the constraint $Q_0|\Psi_T\rangle = |\Psi_T\rangle$ and add a Hubbard term on the f-sites to the Hamiltonian:

$$\begin{aligned} H = \sum_{\vec{k},\sigma} \varepsilon(\vec{k}) c^\dagger_{\vec{k},\sigma} c_{\vec{k},\sigma} - \frac{J}{4} \sum_{\vec{i}} \left[\sum_{\sigma} c^\dagger_{\vec{i},\sigma} f_{\vec{i},\sigma} + f^\dagger_{\vec{i},\sigma} c_{\vec{i},\sigma} \right]^2 \\ + U_f \sum_{\vec{i}} (n^f_{\vec{i},\uparrow} - 1/2)(n^f_{\vec{i},\downarrow} - 1/2) \end{aligned} \tag{148}$$

This Hamiltonian is again block diagonal in the Q_n subspaces. During the imaginary time propagation, the components $Q_n|\Psi_T\rangle$ of the trial wave function will be suppressed by a factor $e^{-\Theta U_f n/2}$ in comparison to the component $Q_0|\Psi_T\rangle$.

To incorporate the Hubbard term in the QMC simulation the HS transformation of Eq. (33) is recommended at least for the half filled case. Having relaxed the constraint $Q_0|\Psi_T\rangle = |\Psi_T\rangle$ we are now free to choose a spin singlet trial wave function which we generate from:

$$H_0 = \sum_{\vec{k},\sigma} \varepsilon(\vec{k}) c^\dagger_{\vec{k},\sigma} c_{\vec{k},\sigma} - \frac{J}{4} \sum_{\vec{i},\sigma} (c^\dagger_{\vec{i},\sigma} f_{\vec{i},\sigma} + f^\dagger_{\vec{i},\sigma} c_{\vec{i},\sigma}) \tag{149}$$

which is nothing but the non-interacting PAM with hybridization $V = J/4$. The ground state at half-filling is clearly a spin singlet. With this choice of the trial wave function, HS transformations of Eqn. (33), (36) as well as the particle-hole transformation of Eq. (148) the absence of sign problem in the half-filled case is readily shown for $J > 0$.

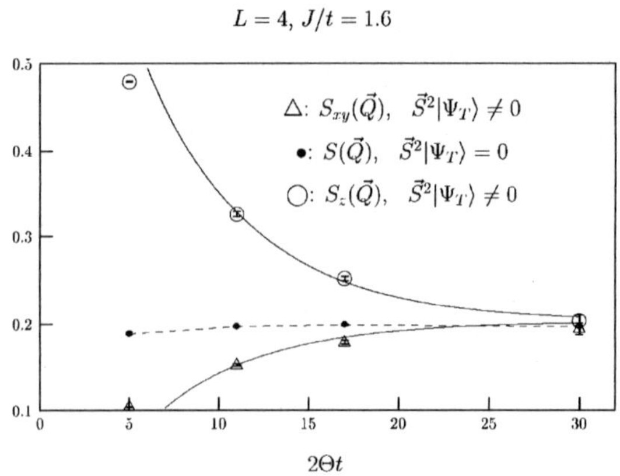

FIGURE 15. Spin-spin correlations as a function of the projection parameter Θ. Here, $S(\vec{Q}) = \frac{4}{3}\langle \vec{S}^f(\vec{Q}) \cdot \vec{S}^f(-\vec{Q})\rangle$, $S^f_z(\vec{Q}) = 4\langle \vec{S}^f_z(\vec{Q}) \cdot \vec{S}^f_z(-\vec{Q})\rangle$, and $S^f_{xy}(\vec{Q}) = 2\left(\langle \vec{S}^f_x(\vec{Q}) \cdot \vec{S}^f_x(-\vec{Q})\rangle + \langle \vec{S}^f_y(\vec{Q}) \cdot \vec{S}^f_y(-\vec{Q})\rangle\right)$. The trial wave function with $\vec{S}^2|\Psi_T\rangle \neq 0$ ($\vec{S}^2|\Psi_T\rangle = 0$) corresponds to the ground state of the Hamiltonian in Eq. (149) (Eq. (147)). In the *large* Θ limit, the results are independent on the choice of the trial wave function. In particular, starting from a broken symmetry state the symmetry is restored at *large* values of Θt. For this system, the spin gap is given by $\Delta_{sp} = 0.169 \pm 0.004$ [29]. Starting with a trial wave function with $\vec{S}^2|\Psi_T\rangle \neq 0$, convergence to the ground state follows approximatively the form: $a + be^{-\Delta_{sp}2\Theta}$. The solid lines correspond to a least square fit to this form.

Fig. 15 demonstrates the importance of using a spin singlet trial wave function. Starting from a Néel order for the f-electrons (See Eq. (147)) convergence to the ground state follows approximatively $e^{-\Delta_{sp}2\Theta}$ where Δ_{sp} corresponds to the spin-gap. When the spin gap is small, convergence is poor and the remedy is to consider a spin singlet trial wave function.

It is equally possible to consider the ferromagnetic exchange $J < 0$. To achieve a sign free simulation at least at half-filling we define the particle hole transformation as.

$$\mathscr{P}^{-1} c^\dagger_{\vec{i},\sigma} \mathscr{P} = (-1)^{i_1+i_2} c^\dagger_{\vec{i},\sigma}$$
$$\mathscr{P}^{-1} f^\dagger_{\vec{i},\sigma} \mathscr{P} = (-1)^{i_1+i_2} f^\dagger_{\vec{i},\sigma}$$

in conjunction with a trial wave function generated by the non-interacting Hamiltonian

$$H_0 = \sum_{\vec{k},\sigma} \varepsilon(\vec{k}) c^\dagger_{\vec{k},\sigma} c_{\vec{k},\sigma} - \frac{J}{4} \sum_{\langle \vec{i},\vec{j} \rangle,\sigma} (c^\dagger_{\vec{i},\sigma} f_{\vec{j},\sigma} + f^\dagger_{\vec{j},\sigma} c_{\vec{i},\sigma}) \tag{150}$$

leads to a sign free simulation for the ferromagnetic half-filled case.

Having introduced the Hubbard term on the f-sites, the FTQMC may be formulated. For a given temperature U_f has to be large enough so as to eliminate double occupancy on the f-sites and hence ensure projection onto the Q_0 subspace. At this point, it is very convenient to choose the $SU(2)$-invariant HS decomposition of Eq. (33) since one can take the limit $U_f \to \infty$ by setting $\alpha = \pi/2$. Hence irrespective of the considered temperature, we are guaranteed to be in the correct Hilbert space. Note that since the Hubbard U_f term commutes with the Hamiltonian of Eq. (141) no uncontrolled Trotter error is involved in taking the limit $U_f \to \infty$.

It is beyond the scope of this review to describe the results obtained with the above algorithm for the Kondo Lattice model and the reader is referred to the Refs. [29, 59, 38].

Hard-core Bosons systems and the Heisenberg Hamiltonian.

The spin $1/2$ Heisenberg model is defined by the Hamiltonian

$$H = J \sum_{\langle \vec{i},\vec{j} \rangle} \vec{S}_{\vec{i}} \vec{S}_{\vec{j}}, \tag{151}$$

where the spin operator, $\vec{S}_{\vec{i}} = (1/2) \sum_{\sigma,\sigma'} c^{\dagger}_{\vec{i},\sigma} \vec{\sigma}_{\sigma,\sigma'} c_{\vec{i},\sigma'}$ act on the states, $|\uparrow\rangle$, $|\downarrow\rangle$ and $\vec{\sigma}$ are the Pauli spin $1/2$ matrices. Defining the raising and lowering spin operators,

$$S^{+}_{\vec{i}} = S^{x}_{\vec{i}} + iS^{y}_{\vec{i}} \quad S^{-}_{\vec{i}} = S^{x}_{\vec{i}} - iS^{y}_{\vec{i}} \tag{152}$$

transforms the Heisenberg model to:

$$H = J \sum_{\langle \vec{i},\vec{j} \rangle} \frac{1}{2} \left(S^{+}_{\vec{i}} S^{-}_{\vec{j}} + S^{-}_{\vec{i}} S^{+}_{\vec{j}} \right) + S^{z}_{\vec{i}} S^{z}_{\vec{j}}. \tag{153}$$

The raising and lowering operators satisfy the commutation relations:

$$\left[S^{+}_{\vec{i}}, S^{-}_{\vec{j}} \right] = \delta_{\vec{i},\vec{j}} 2 S^{z}_{\vec{i}}. \tag{154}$$

The mapping onto hard core bosons follows from the identification

$$|\uparrow\rangle \to |1\rangle, \quad |\downarrow\rangle \to |0\rangle \tag{155}$$

Hard core boson operators, $b^{\dagger}$ acting on the states $|1\rangle$ and $|0\rangle$ satisfy the commutation rules:

$$\left[b^{\dagger}_{\vec{i}}, b_{\vec{j}} \right] = \delta_{\vec{i},\vec{j}} 2 \left(b^{\dagger}_{\vec{i}} b_{\vec{i}} - \frac{1}{2} \right) \tag{156}$$

which follows from the usual bosonic commutation rules with constraint of no double occupancy: $b^{\dagger}_{\vec{i}} b^{\dagger}_{\vec{i}} = 0$. The identification of states (155) leads to:

$$S^{+}_{\vec{i}} \to b^{\dagger}_{\vec{i}}, \; S^{-}_{\vec{i}} \to b_{\vec{i}}, \text{ and } S^{z}_{\vec{i}} \to \left(b^{\dagger}_{\vec{i}} b_{\vec{i}} - \frac{1}{2} \right). \tag{157}$$

Since both the S and b operators satisfy identical commutation rules on their respective Hilbert spaces (Eqn. (154) and (156)), the Heisenberg model may be written in it's hard core boson form:

$$H = J \sum_{\langle \vec{i},\vec{j} \rangle} \frac{1}{2} \left(b^{\dagger}_{\vec{i}} b_{\vec{j}} + b^{\dagger}_{\vec{j}} b_{\vec{i}} \right) + \left(b^{\dagger}_{\vec{i}} b_{\vec{i}} - \frac{1}{2} \right) \left(b^{\dagger}_{\vec{j}} b_{\vec{j}} - \frac{1}{2} \right) \tag{158}$$

To carry out simulations of hard core boson Hamiltonians such as the Heisenberg model within the framework of the FTQMC of PQMC, we have to explicitly build the hard core bosons from fermions ($b^\dagger_{\vec{i}} = c^\dagger_{\vec{i},\uparrow} c^\dagger_{\vec{i},\downarrow}$) and restrict the Hilbert space to doubly or empty occupied sites in terms of fermions. One can achieve this goal by considering the Hamiltonian

$$\begin{aligned} H &= -t \sum_{\langle i,j \rangle} \left(\sum_{\sigma} c^\dagger_{\vec{i},\sigma} c_{\vec{j},\sigma} + H.c. \right)^2 - V \sum_{\langle i,j \rangle} \left(n_{\vec{i}} \pm n_{\vec{j}} \right)^2 \\ & -U \sum_{i} n_{\vec{i},\uparrow} n_{\vec{i},\downarrow}. \end{aligned} \tag{159}$$

Here, $n_{\vec{i}} = n_{\vec{i},\uparrow} + n_{\vec{i},\downarrow}$ and $n_{\vec{i},\sigma} = c^\dagger_{\vec{i},\sigma} c_{\vec{i},\sigma}$. At first glance, we can see that for $t, U, V > 0$ and with the use of the HS transformations (33),(36), the sign is absent since the weight for a given configuration of HS fields is a product of two identical real determinants.

To see how the above Hamiltonian relates to that of hard core boson systems, we expand the squares to obtain (up to a chemical potential term):

$$\begin{aligned} H &= -2t \sum_{\langle i,j \rangle} \left(c^\dagger_{\vec{i},\uparrow} c^\dagger_{\vec{i},\downarrow} c_{\vec{j},\downarrow} c_{\vec{j},\uparrow} + H.c. \right) + (t \mp 2V) \sum_{\langle i,j \rangle} n_{\vec{i}} n_{\vec{j}} \\ & +4t \sum_{\langle i,j \rangle} \vec{S}_{\vec{i}} \vec{S}_{\vec{j}} - (U + 8V) \sum_{i} n_{\vec{i},\uparrow} n_{\vec{i},\downarrow}. \end{aligned} \tag{160}$$

As apparent, the Hamiltonian conserves the number of singly occupied sites. That is, the operator

$$P = \sum_{\vec{i}} n_{\vec{i}} (2 - n_{\vec{i}}). \tag{161}$$

which counts the singly occupied sites is a conserved quantity:

$$[P, H] = 0. \tag{162}$$

In particular, if one projects onto the subspace with $P = 0$ (P_0) then in this subspace the spin-spin term as well as the Hubbard interaction vanish:

$$\begin{aligned} HP_0 &= -2t \sum_{\langle i,j \rangle} \left(c^\dagger_{\vec{i},\uparrow} c^\dagger_{\vec{i},\downarrow} c_{\vec{j},\downarrow} c_{\vec{j},\uparrow} + H.c. \right) \\ & +(t \mp 2V) \sum_{\langle i,j \rangle} n_{\vec{i}} n_{\vec{j}} - 3(t + V) \sum_{\vec{i}} n_{\vec{i}} \end{aligned} \tag{163}$$

To enforce this constraint (i.e. projection on the P_0 subspace on the imaginary time propagation) within the PQMC one just has to appropriately choose the trial wave function:

$$P_0 |\Psi_T\rangle = |\Psi_T\rangle \tag{164}$$

A possible choice is:

$$|\Psi_T\rangle = |\Psi^\uparrow_T\rangle \otimes |\Psi^\downarrow_T\rangle, \quad |\Psi^\sigma_T\rangle = \prod_{n=1}^{N^\sigma_p} c^\dagger_{\vec{i}_n,\sigma} |0\rangle \tag{165}$$

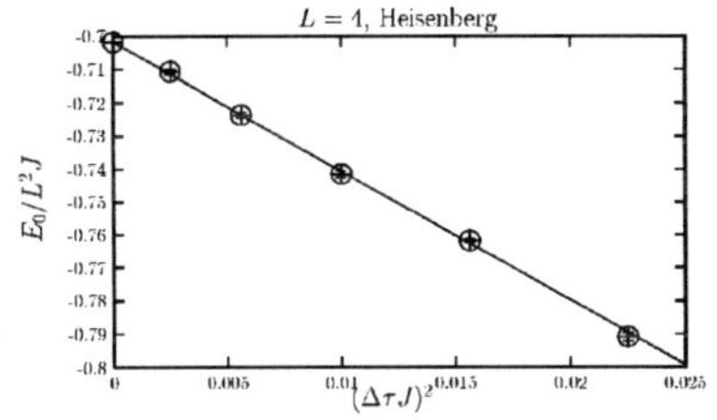

FIGURE 16. Ground state energy of the Heisenberg model on a 4×4 lattice as a function of the imaginary time discretization $\Delta\tau$. The data point at $\Delta\tau = 0$ corresponds to the exact result for this lattice size.

where $N_p^{\uparrow} = N_p^{\downarrow}$ corresponds to the number of electrons in the spin σ sector. Alternatively and in analogy to the Kondo lattice model, the constraint may be imposed with the attractive U term. The inclusion of this term allows the use of the FTQMC.

In the P_0 subspace

$$b^{\dagger}_{\vec{i}} = c^{\dagger}_{\vec{i},\uparrow} c^{\dagger}_{\vec{i},\downarrow} \text{ and } n_{\vec{i}} = 2 b^{\dagger}_{\vec{i}} b_{\vec{i}}. \tag{166}$$

where $b^{\dagger}_{\vec{i}}$ are the desired hard-core bosons. Thus,

$$\begin{aligned} HP_0 &= -2t \sum_{\langle i,j \rangle} \left(b^{\dagger}_{\vec{i}} b_{\vec{j}} + H.c. \right) + 4(t \mp 2V) \sum_{\langle i,j \rangle} b^{\dagger}_{\vec{i}} b_{\vec{i}} b^{\dagger}_{\vec{j}} b_{\vec{j}} \\ &\quad -6(t+V) \sum_{\vec{i}} b^{\dagger}_{\vec{i}} b_{\vec{i}} \end{aligned} \tag{167}$$

which is nothing but a model of hard core bosons with nearest neighbor density-density interaction of arbitrary sign.

We can test the method at the Heisenberg point: $V = 0$. (The explicit mapping, up to a chemical potential term, to Eq. (158) is achieved after the canonical transformation: $b^{\dagger}_{\vec{i}} \rightarrow (-1)^{i_x + i_y} b^{\dagger}_{\vec{i}}$.) Fig. 16 plots the ground state energy as a function of $\Delta\tau$. As apparent, the QMC converges to the exact result. In this formulation, frustrating interactions would amount in inserting terms of the form $t \left(c^{\dagger}_{\vec{i}} c_{\vec{j}} + \text{H.c.} \right)^2$ for $\vec{i}, \vec{j}$ within the same sublattice and with $t > 0$. The reader will easily convince himself that this leads to a sign problem. Finally we note that the use of the auxiliary field QMC for non-frustrated Heisenberg models should be seen as an amusing curiosity since it is not competitive with the loop algorithm approach.

THE HIRSCH-FYE IMPURITY ALGORITHM

A very much related algorithm to the FTQMC and PQMC is the Hirsch-Fye impurity algorithm [60] which is extensively used in the framework of dynamical mean field theories [15, 16]. As it's name suggests, this algorithm is triggered at solving impurity problems and allows simulations directly in the *thermodynamic* limit. However as will be apparent below the CPU time scales as the third power of the inverse temperature β. For simplicity we will consider the single impurity Anderson model for the formulation of the problem and the Kondo model for the example applications.

The Anderson impurity model reads:

$$
\begin{aligned}
H = & \overbrace{\sum_{\vec{k},\sigma} \varepsilon(\vec{k}) c^{\dagger}_{\vec{k},\sigma} c_{\vec{k},\sigma} + V \sum_{\vec{k},\sigma} c^{\dagger}_{\vec{k},\sigma} f_{\sigma} + f^{\dagger}_{\sigma} c_{\vec{k},\sigma} + \varepsilon_f \sum_{\sigma} f^{\dagger}_{\sigma} f_{\sigma}}^{\equiv H_0} \\
& + U \left(f^{\dagger}_{\uparrow} f_{\uparrow} - 1/2 \right) \left(f^{\dagger}_{\downarrow} f_{\downarrow} - 1/2 \right).
\end{aligned}
\tag{168}
$$

Following the procedure introduced for the Hubbard model, the partition function is given by:

$$
Z \equiv \mathrm{Tr}\left[e^{-\beta(H-\mu N)} \right] = \sum_{\vec{s}} \left[\prod_{\sigma} \det \left[1 + B^{\sigma}_{m} B^{\sigma}_{m-1} \cdots B^{\sigma}_{1} \right] \right]. \tag{169}
$$

With the notation

$$
H_0 = \sum_{\sigma, \vec{i}} a^{\dagger}_{\vec{i},\sigma} (h_0)_{\vec{i},\vec{j}} a_{\vec{j},\sigma} \tag{170}
$$

where $a_{\vec{j},\sigma}$ denotes c- of f-operators, $\vec{i}, \vec{j}$ running over all orbitals, and the HS transformation of Eq. (32),

$$
\begin{aligned}
B^{\sigma}_{n} &= e^{-\Delta\tau h_0} e^{V^{\sigma}_{n}} \\
(V^{\sigma}_{n})_{i,j} &= \delta_{i,j} \begin{cases} 0 & \text{if } \vec{i} \neq \text{impurity site} \\ \tilde{\alpha}\sigma s_n & \text{if } \vec{i} = \text{impurity site} \end{cases}
\end{aligned}
\tag{171}
$$

Note that since we are considering a single impurity, we require only a single Hubbard Stratonovich field per time slice, s_n. Finally, $m\Delta\tau = \beta$.

The determinant in a given spin sector may be written as

$$
\det \left[1 + B^{\sigma}_{m} B^{\sigma}_{m-1} \cdots B^{\sigma}_{1} \right] = \det O^{\sigma} \text{ with} \tag{172}
$$

$$
O^{\sigma} = \begin{pmatrix}
1 & 0 & . & . & 0 & B^{\sigma}_{1} \\
-B^{\sigma}_{2} & 1 & 0 & . & . & 0 \\
0 & -B^{\sigma}_{3} & 1 & . & . & 0 \\
. & 0 & -B^{\sigma}_{4} & . & . & . \\
. & . & 0 & . & . & . \\
. & . & . & . & . & . \\
0 & . & . & 0 & -B^{\sigma}_{m} & 1
\end{pmatrix}. \tag{173}
$$

From Eq. (92) we identify:

$$
(O^{\sigma})^{-1} \equiv g^{\sigma} = \begin{pmatrix}
G^{\sigma}(1,1) & G^{\sigma}(1,2) & . & . & G^{\sigma}(1,m) \\
G^{\sigma}(2,1) & G^{\sigma}(2,2) & . & . & G^{\sigma}(2,m) \\
. & & . & . & . \\
G^{\sigma}(m,1) & G^{\sigma}(m,2) & . & . & G^{\sigma}(m,m)
\end{pmatrix} \tag{174}
$$

where $G^\sigma(n_1,n_2)$ are the time displaced Green functions as defined in Eqn. (89) and (90). Given a HS configuration $\vec{s}$ and $\vec{s}'$ and associated matrices V and V' the Green functions g and g' satisfy the following Dyson equation.

$$g' = g + (g-1)(e^{V'-V} - 1)g' \tag{175}$$

To demonstrate the above, we consider

$$\tilde{O}^\sigma = O^\sigma e^{-V^\sigma} = \begin{pmatrix} e^{-V_1^\sigma} & 0 & . & . & 0 & e^{-\Delta\tau h_0} \\ -e^{-\Delta\tau h_0} & e^{-V_2^\sigma} & 0 & . & . & 0 \\ 0 & -e^{-\Delta\tau h_0} & e^{-V_3^\sigma} & . & . & 0 \\ . & 0 & -e^{-\Delta\tau h_0} & . & . & . \\ . & . & 0 & . & . & . \\ . & . & . & . & . & . \\ 0 & . & . & 0 & -e^{-\Delta\tau h_0} & e^{-V_m^\sigma} \end{pmatrix} \tag{176}$$

so that (omitting the spin index σ)

$$\tilde{g}' \equiv \tilde{O}'^{-1} = [\tilde{O} + \underbrace{\tilde{O}' - \tilde{O}}_{\equiv e^{-V'} - e^{-V}}]^{-1} = \tilde{g} - \tilde{g}\left(e^{-V'} - e^{-V}\right)\tilde{g}'. \tag{177}$$

The above equation follows from the identity: $\frac{1}{A+B} = \frac{1}{A} - \frac{1}{A}B\frac{1}{A+B}$. Substitution, $\tilde{g} = e^V g$, leads to the Dyson Eq. (175).

The Green function matrix has dimensions $mN \times mN$ where N is the total number of orbitals and m the number of Trotter slices. Let $x = (\tau_x, i_x)$ with Trotter slice τ_x and orbital i_x, and the site index of the impurity to 0. Since

$$(e^{V'-V} - 1)_{x,y} = (e^{V'-V} - 1)_{x,x}\delta_{x,y}\delta_{i_x,0} \tag{178}$$

we can use the Dyson equation only for the impurity Green function:

$$g'_{f,f'} = g_{f,f'} + \sum_{f''}(g-1)_{f,f''}(e^{V'-V} - 1)_{f'',f''}g'_{f'',f'} \tag{179}$$

with indices $f \equiv (\tau, 0)$ running from $1 \cdots m$.

We now have all the ingredients to carry out a simulation. (Note that the algorithm is free of numerical instabilities.) Let us start from a random HS configuration $\vec{s}'$. For this configuration, we have to compute the impurity Green function. From the $U = 0$ solution, g_0^f (which can be obtained analytically [61]), we compute the impurity Green function for the HS fields $\vec{s}'$ with the use of Eq. (179). This is readily achieved at the cost of an $m \times m$ matrix inversion.

To upgrade a single HS field with the Metropolis method, we have to calculate the ratio of the determinants (see Eq. (169)) as shown in section . This ratio only requires the knowledge of the equal time impurity Green function which is already at hand. If the proposed spin-flip is accepted the impurity Green function may readily be upgraded

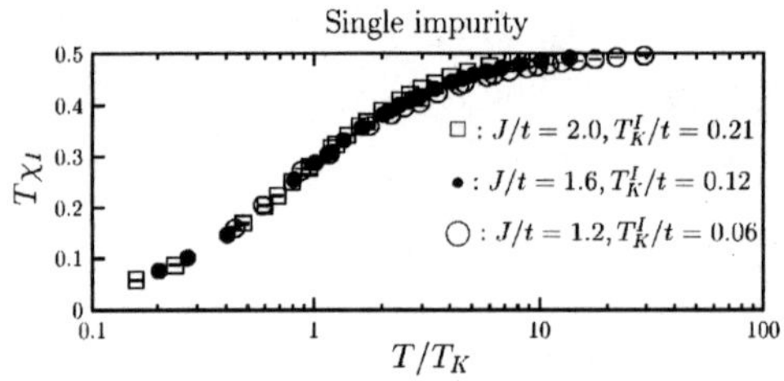

FIGURE 17. Impurity spin susceptibility of the Kondo model as computed with the Hirsch-Fye impurity algorithm. [38]

with the use of Eq. (179). Note that since we are considering a single spin flip, sum over f'' in Eq. (179) has only one non-vanishing contribution.

The attractive feature of the Hirsch-Fye impurity algorithm is that the conduction electrons may be considered directly in the thermodynamic limit. This is not be possible within the previously discussed FTQMC since the dimension of the matrices involved scale as the total number of orbitals. The Hirsch-Fye algorithm is not limited to impurity models. However, when applied to lattice models it is less efficient than the FTQMC and PQMC.

The Hirsch-Fye impurity algorithm may equally be applied to the Kondo model,

$$H = \sum_{\vec{k},\sigma} \varepsilon(\vec{k}) c^{\dagger}_{\vec{k},\sigma} c_{\vec{k},\sigma} + J\vec{S}^{c}_{\vec{I}} \cdot \vec{S}^{f}_{\vec{I}} \tag{180}$$

where the same formulation as for the lattice problem is used. Fig. 17 plots the impurity spin susceptibility

$$\chi_I = \int_0^{\beta} \langle \vec{S}^{f}_{\vec{I}}(\tau) \cdot \vec{S}^{f}_{\vec{I}} \rangle \tag{181}$$

for various values of J/t for a half-filled conduction band. As apparent and at low energies the data collapse to the universal form $\chi_I = \frac{1}{T} f\left(T/T_K^I\right)$ where T_K^I is the Kondo temperature.

CONCLUSION

We have presented in all details the auxiliary field (or determinantal) Quantum Monte Carlo method for lattice problems. The ground state, finite temperature as well as Hirsch-Fye impurity algorithms were considered. The formulation of these algorithms for a range of fermionic as well as bosonic models was discussed. When the sign problem is avoidable the computational effort scales as the volume to the cubed times inverse temperature. For the Hirsch-Fye impurity algorithm – formulated directly in the thermodynamic limit – the computational cost scales as the cubed of the inverse temperature. The algorithms produce thermodynamic, dynamical -in conjunction with the Maximum Entropy method- as well as ground state properties of the model under consideration. They are unique in the sense that the sign problem turns out to be avoidable for particle-hole symmetric models as well as for models with attractive interactions which couple independently to an internal symmetry with an even number of states. It is not clear at present if other symmetries may be put to use so as to avoid the sign problem. Clearly, the sign problem remains the central issue and calls for new numerical approaches to the correlated electron problem.

Acknowledgments Most of the work presented here originates from collaborations with S. Capponi, M. Feldbacher, N. Furukawa, W. Hanke, M. Imada, A. Muramatsu, D.J. Scalapino, S. Sorella, M. Troyer and D. Würtz to which I express my warmest thanks. Financial support of the DFG in terms of a Heisenberg fellowship (grant number AS 120/1-1) is acknowledged.

THE MONTE CARLO METHOD.

In this appendix, we briefly outline the Monte Carlo method. Our aim is to compute:

$$\langle O \rangle_P = \int_\Omega d\vec{x} P(\vec{x}) O(\vec{x}) \tag{182}$$

where Ω denotes the integration space, and $P(\vec{x})$ is a probability distribution,

$$\int_\Omega d\vec{x} P(\vec{x}) = 1 \text{ and } P(\vec{x}) \geq 0 \ \forall \ \vec{x} \varepsilon \Omega. \tag{183}$$

For simplicity, we will assume Ω to be a subspace of $\mathscr{R}^d$ with volume V. Here, d is the dimension. In practice, one does not have access to P, but to the unnormalized probability distribution $g(\vec{x})$ which satisfies

$$P(\vec{x}) = \frac{g(\vec{x})}{Z}, \ Z = \int_\Omega d\vec{x} g(\vec{x}). \tag{184}$$

In the terminology of statistical mechanics, $g(\vec{x})$ corresponds to the Boltzmann weight and Z to the partition function. Hence the evaluation of Eq. (182) boils down to the calculating the quotient of two integrals. One may break up Ω into hypercubes of linear dimension h and approximate the integrals by sums. Depending upon the method used the systematic error will be proportional to h^k. The required number of function evaluations N is of the order V/h^d so that the systematic error scales a $N^{-k/d}$. Clearly, when d is large poor results are obtained. As we will now see, in the large d limit, the Monte Carlo method becomes attractive.

The central limit theorem.

Suppose that we have a set of independent points $\{\vec{x}_i\}$, $i : 1 \cdots N$ distributed according to the probability distribution $P(\vec{x})$ we can approximate $\langle O \rangle_P$ by:

$$\langle O \rangle_P \sim \frac{1}{N} \sum_{\substack{i=1 \\ \vec{x}_i \varepsilon P}}^{N} O(\vec{x}_i) = X. \tag{185}$$

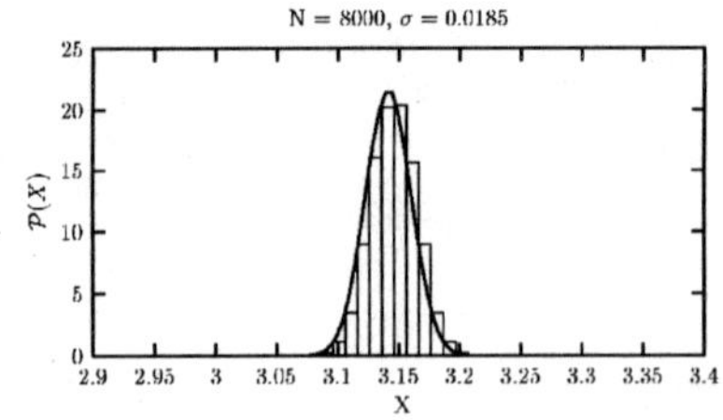

FIGURE 18. Boxes correspond to the distribution results obtained after 10000 simulations. For each simulation we draw $N = 8000$ points. For a single simulation, we obtain $\sigma = 0.0185$. The solid line corresponds to the result of central limit theorem with above value of σ.

Clearly, X will depend upon the chosen series of $\{\vec{x}_i\}$. The central limit theorem, tells us that in the large N limit the probability of obtaining a given value of X, $\mathscr{P}(X)$ reads

$$\mathscr{P}(X) = \frac{1}{\sqrt{2\pi}}\frac{1}{\sigma}\exp\left[-\frac{(X-\langle O\rangle_P)^2}{2\sigma^2}\right] \quad \text{with } \sigma^2 = \frac{1}{N}\left(\langle O^2\rangle_P - \langle O\rangle_P^2\right). \tag{186}$$

Thus independently of the dimension d, the convergence to the exact result scales as $1/\sqrt{N}$. The width of the above normal distribution, σ, corresponds to the statistical error. For practical purposes, one estimates σ by

$$\sigma^2 \approx \frac{1}{N}\left(\frac{1}{N}\sum_{\substack{i=1\\ \vec{x}_i\varepsilon P}} O(\vec{x}_i)^2 - \left(\frac{1}{N}\sum_{\substack{i=1\\ \vec{x}_i\varepsilon P}} O(\vec{x}_i)\right)^2\right) \tag{187}$$

Instead of demonstrating the central limit theorem, we give a simple example: the evaluation of the number π obtained via:

$$\pi = 4\int_0^1 \mathrm{d}x \int_0^1 \mathrm{d}y \Theta(1-x^2+y^2), \tag{188}$$

where Θ is the Heavyside function, $\Theta(x) = 1$ for $x > 0$ and vanishes otherwise. In this example we $P(x,y) \equiv 1$. To generate the a sequence of N points $(x,y)_i$ from this probability distribution, we draw random numbers, x,y, in the interval $[0,1]$. For $N = 8000$ we obtain an error $\sigma = 0.0185$ with the use of Eq. (187). To check the central limit theorem, we repeat the simulation 10000 times with different random numbers. Fig. (18) shows the thus obtained distribution which compares well to the result of the central limit theorem.

The jackknife and bootstrap methods [62] provide alternative ways of estimating the error (187). These methods become particularly useful, if not essential, when one wishes to estimate the error on $f(\langle O_1\rangle,\cdots,\langle O_1\rangle)$ where f is an arbitrary function of n variables. For a given sequence $\vec{x}_i \varepsilon P(\vec{x})$, $i: 1\cdots N$, the jackknife focuses on the samples that leaves out one configuration at a time:

$$f_i^J = f\left(\frac{1}{N-1}\sum_{j\neq i} O_1(\vec{x}_j), \cdots, \frac{1}{N-1}\sum_{j\neq i} O_n(\vec{x}_j)\right). \tag{189}$$

The error estimate on f is then given by:

$$\left(\sigma_f^J\right)^2 \approx N\left(\frac{1}{N}\sum_{i=1}^{N}(f_i^J)^2 - \left(\frac{1}{N}\sum_{i=1}^{N} f_i^J\right)^2\right) \tag{190}$$

One may verify explicitly that for $n = 1$ and $f(x) = x$ Eq. (190) reduces to Eq. (187) up to a factor $(N/(N-1))^2$ which tends to unity in the large N limit.

An alternative method for determining errors of f is the bootstrap algorithm. For a given sample of N configurations $\{\vec{x}_1 \cdots \vec{x}_N\}$ drawn from the probability distribution $P(\vec{x})$, we can construct N^N sets of N configurations, $\left\{\vec{x}_{i_1} \cdots \vec{x}_{i_N}\right\}$ with $i_1 : 1 \cdots N, i_2 : 1 \cdots N, \cdots, i_N : 1 \cdots N$, which correspond to the ideal bootstrap samples. For a given bootstrap sample, defined by the vector $\vec{i} = (i_1, \cdots, i_N)$,

$$f_{\vec{i}}^B = f\left(\frac{1}{N}\sum_{k=1}^{N} O_1(\vec{x}_{i_k}), \cdots, \frac{1}{N}\sum_{k=1}^{N} O_n(\vec{x}_{i_k})\right). \tag{191}$$

The bootstrap estimate of the error is given by:

$$\left(\sigma_f^B\right)^2 \approx \frac{1}{N^N}\sum_{i_1,\cdots 1_N=1}^{N} (f_{\vec{i}}^B)^2 - \left(\frac{1}{N^N}\sum_{i_1,\cdots,i_N=1}^{N} f_{\vec{i}}^B\right)^2. \tag{192}$$

Again, one may check that for the special case, $n = 1$ and $f(x) = x$ Eq. (192) reduces to Eq. (187). Clearly, when N is large, it is numerically out of reach to generate all of the N^N bootstrap samples. Typically, to estimate the right hand side of Eq. (192) 200 or more bootstrap samples are generated stochastically. Since each bootstrap sample is equally probable we can generate them with: $i_k = \text{trunc}\left(N * \xi_k + 1\right)$ where ξ_k is a random number in the interval $[0, 1]$ and the function trunc returns an integer by truncating the numbers after the decimal point.

Generating Markov chains.

Our task is now to generate a set of points $\vec{x}$ distributed according to $P(x)$. We introduce a Monte-Carlo time t, and a time dependent probability distribution $P_t(\vec{x})$ which evolves in time according to a Markov process: the future $(t+1)$ depends only on the present (t). Our aim is to obtain: $P_{t\to\infty}(\vec{x}) = P(\vec{x})$. To define the Markov chain, we introduce a matrix $T_{\vec{y},\vec{x}}$ which corresponds to the probability of going from $\vec{x}$ to $\vec{y}$. The time evolution of $P_t(\vec{x})$ is given by:

$$P_{t+1}(\vec{y}) = \sum_{x} T_{\vec{y},\vec{x}} P_t(\vec{x}) \tag{193}$$

T has to satisfy the following properties.

$$\sum_{x} T_{\vec{y},\vec{x}} = \sum_{y} T_{\vec{y},\vec{x}} = 1, \quad T_{\vec{y},\vec{x}} \geq 0 \tag{194}$$

That is, the probability of reaching a given $\vec{y}$ from any $\vec{x}$ or of landing anywhere in Ω given a initial $\vec{x}$ is of unit. T has to be ergodic:

$$\forall \vec{x}, \vec{y} \varepsilon \Omega \exists s | \, (T^s)_{\vec{y},\vec{x}} > 0. \tag{195}$$

Thus, we are assured to sample the whole phase space provided the above is satisfied. Lastly, the requirement of stationarity:

$$\sum_{\vec{x}} T_{\vec{y},\vec{x}} P(\vec{x}) = P(\vec{y}). \tag{196}$$

Once we have reached the desired distribution, $P(\vec{x})$, we wish to to stay there. Stationarity is automatically satisfied if

$$T_{\vec{y},\vec{x}} P(\vec{x}) = T_{\vec{x},\vec{y}} P(\vec{y}) \tag{197}$$

as may be seen by summing on both sides over $\vec{y}$ or $\vec{x}$. This relation is referred to as detailed balance or microreversibility. However, one has to keep in mind that stationarity and not detailed balance is essential.

That $P_t(\vec{x})$ approaches the equilibrium distribution $P(\vec{x})$ may be seen with:

$$\begin{aligned} ||P_{t+1} - P|| &\equiv \sum_{\vec{y}} |P_{t+1}(\vec{y}) - P(\vec{y})| \\ &= \sum_{\vec{y}} |\sum_{\vec{y}} T_{\vec{y},\vec{x}} P_t(\vec{x}) - \sum_{\vec{x}} T_{\vec{y},\vec{x}} P(\vec{x})| \\ &\leq \sum_{\vec{y}} \sum_{\vec{x}} T_{\vec{y},\vec{x}} |P_t(\vec{x}) - P(\vec{x})| \\ &= \sum_{\vec{x}} |P_t(\vec{x}) - P(\vec{x})| \equiv ||P_t - P||. \end{aligned} \tag{198}$$

Under the assumption of ergodicity, the strict equality holds only when $P_t = P$. Due to Eq. (194), the right eigenvectors of T have eigenvalues $|\lambda| \leq 1$, $\lambda = 1$ corresponding to the stationary distribution P. Starting with an arbitrary distribution $P_0(\vec{x})$ convergence to P will be determined by the eigenvalue of T with second largest absolute value, λ_1. The rate of convergence of $P_t(\vec{x})$ to $P(\vec{x})$ will then scale as $\exp -t/\tau$ where $\tau = -1/\log(|\lambda_1|) > 0$.

Having defined T, we now have to construct it explicitly. Let $T^0_{\vec{y},\vec{x}}$ the probability of proposing a move from $\vec{x}$ to $\vec{y}$ and $a_{\vec{y},\vec{x}}$ the probability of accepting it. $1 - a_{\vec{y},\vec{x}}$ corresponds to the probability of rejecting the move. T_0 is required to satisfy Eq. (194). Since in general we want to propose moves which change the initial configuration, $T^0_{x,x} = 0$. With $a_{\vec{y},\vec{x}}$ and $T^0_{\vec{y},\vec{x}}$ we build $T_{\vec{y},\vec{x}}$ with:

$$T_{\vec{y},\vec{x}} = \begin{cases} T^0_{\vec{y},\vec{x}} a_{\vec{y},\vec{x}} & \text{if } \vec{y} \neq \vec{x} \\ \sum_{\substack{\vec{y} \\ \vec{y} \neq \vec{x}}} T^0_{\vec{y},\vec{x}} \left(1 - a_{\vec{y},\vec{x}}\right) & \text{if } \vec{y} = \vec{x} \end{cases} \tag{199}$$

Clearly $T_{\vec{y},\vec{x}}$ satisfies Eq. (194). To satisfy the stationarity, we impose the detailed balance condition to obtain the equality:

$$T^0_{\vec{y},\vec{x}} a_{\vec{y},\vec{x}} P_{\vec{x}} = T^0_{\vec{x},\vec{y}} a_{\vec{x},\vec{y}} P_{\vec{y}}. \tag{200}$$

Let us set:

$$a_{\vec{y},\vec{x}} = \mathscr{F}(Z) \;\; \text{with} \;\; Z = \frac{T^0_{\vec{x},\vec{y}} P_{\vec{y}}}{T^0_{\vec{y},\vec{x}} P_{\vec{x}}} \tag{201}$$

with $\mathscr{F}(Z) :]0 : \infty[\rightarrow [0,1]$. Since $a_{\vec{x},\vec{y}} = \mathscr{F}(1/Z)$, $\mathscr{F}$ has to satisfy:

$$\frac{\mathscr{F}(Z)}{\mathscr{F}(1/Z)} = Z. \tag{202}$$

There are many possible choices. The Metropolis algorithm is based on the choice:

$$\mathscr{F}(Z) = \min(Z, 1). \tag{203}$$

Thus, one proposes a move from $\vec{x}$ to $\vec{y}$ and accepts it with probability $Z = \frac{T^0_{\vec{x},\vec{y}} P_{\vec{y}}}{T^0_{\vec{y},\vec{x}} P_{\vec{x}}}$. In the practical implementation, one picks a random number r in the interval $[0 : 1]$. If $r < Z$ ($r > Z$) one accepts (rejects) the move. Alternative choices of $\mathscr{F}(Z)$ are for example:

$$\mathscr{F}(Z) = \frac{Z}{1+Z} \tag{204}$$

which is referred to as the Heat bath method.

That the so constructed T matrix is ergodic depends upon the choice of T^0. In many cases, one will wish to combine different types of moves to achieve ergodicity. For a specific move, i we construct $T^{(i)}$ as shown above so that $T^{(i)}$ conditions (194) and (197). The moves may be combined in two ways:

$$T = \sum_i \lambda_i T^{(i)}, \;\; \sum_i \lambda_i = 1 \tag{205}$$

which is referred to as random upgrading since one picks with probability λ_i the move $T^{(i)}$. Clearly, T equally satisfies (194), (197) and if the moves have to be chosen appropriately to satisfy the ergodicity condition. Another choice is sequential upgrading. A deterministic ordering of the moves is chosen to obtain:

$$T = \prod_i T^{(i)}. \tag{206}$$

This choice does not satisfy detailed balance condition, but does satisfy stationarity (196) as well as (194). Again ergodicity has to be *checked* on a case to case basis.

In principle, we could now start with an arbitrary probability distribution $P_0(\vec{x})$ and propagate it along Monte Carlo time t ($P_{t+1} = TP_t$). Convergence to the equilibrium distribution will occur on time scales set by τ. This procedure involves handling many configurations $\vec{x}$ at a given time t. Alternatively one can start with a single configuration $\vec{x}$ and propagate it according to T. That is the probability of having the configuration $\vec{y}$ at the next MC time is given by $T_{\vec{y},\vec{x}}$. This procedure generates a sequence of configuration in MC time. For *large* values of N (see below) $\vec{x}_{t=1}\cdots\vec{x}_{t=N}$ will be distributed according to P. The observable O may now be estimated with:

$$\langle O \rangle_P \approx \frac{1}{N}\sum_{t=1}^{N} O(\vec{x}_t). \tag{207}$$

The required value of N depends autocorrelation time of the observable O:

$$C_O(t) = \frac{\frac{1}{N}\sum_{s=1}^{N} O(\vec{x}_s)O(\vec{x}_{s+t}) - \left(\frac{1}{N}\sum_{s=1}^{N} O(\vec{x}_s)\right)^2}{\frac{1}{N}\sum_{s=1}^{N} O(\vec{x}_s)^2 - \left(\frac{1}{N}\sum_{s=1}^{N} O(\vec{x}_s)\right)^2} \tag{208}$$

One expects $C_O(t) \sim e^{-t/\tau_O}$ where τ_O corresponds to the MC time scale on which memory of the initial configuration is lost. Hence, to obtain meaningful results, $N >> \tau_O$. Note that one should equally take into account a *warm up* time by discarding at least the first τ_O configurations in the MC sequence. Naively, one would expect $\tau_O = \tau$. However, this depends on the overlap of the observable with the slowest mode in the MC dynamics which relaxes as $e^{-t/\tau}$. In particular in a model with spin rotation symmetry the slowest mode may correspond to the rotation of the total spin. In this case, observables which are invariant under a spin rotation will not be effected by the slowest mode of the MC dynamics. Hence in this case $\tau_O < \tau$.

We now consider the estimation of the error. To apply the central limit theorem, we need a set of independent estimates of $\langle O \rangle_P$. This may be done by regrouping the data into *bins* of size $n\tau_O$.

$$\tilde{O}_n(t) = \frac{1}{n\tau_O}\sum_{s=1}^{n\tau_O} O(\vec{x}_{(t-1)n\tau_O+s}) \tag{209}$$

with $t = 1\cdots N/(n\tau_O)$. If n is large enough (i.e. $n \approx 10-20$) then $\tilde{O}_n(t)$ may be considered as an independent estimate, and the error is given by:

$$\sigma_n = \sqrt{\frac{1}{M}\left(\frac{1}{M}\sum_{t=1}^{M} \tilde{O}_n(t)^2 - \left(\frac{1}{M}\sum_{i=1}^{M} \tilde{O}_n(t)\right)^2\right)} \tag{210}$$

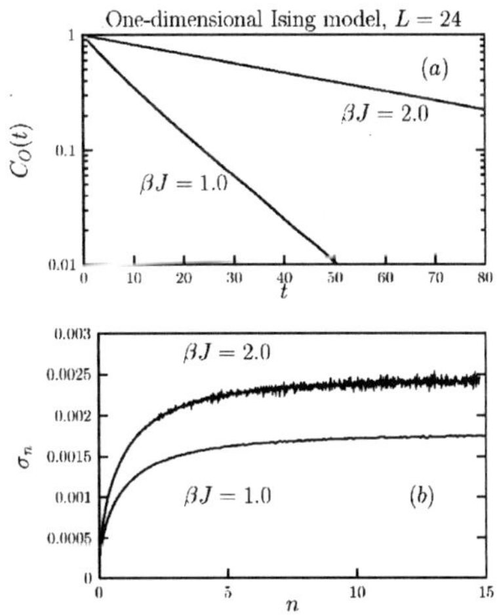

FIGURE 19. One dimensional Ising model on an L=24 site lattice. (a) Autocorrelation time (see Eq. (208)) for $g(r = L/2)$. The time unit is corresponds to a single sweep. (b) Estimate of the error (see Eq. (210)). Here, n corresponds to the size of the bins in units of the autocorrelation time. As apparent $n \sim 10$ is sufficient to obtain a reliable estimate of the error. After 2×10^6 sweeps, our results yield $g(r = L/2) = 0.076 \pm 0.0018$ and 0.909 ± 0.0025 for $\beta t = 1$ and 2 respectively. The exact result reads $g(r = L/2) = 0.0760$ and 0.9106 at $\beta t = 1$ and 2 respectively.

where $M = N/(n\tau_O)$. If n is large enough the error σ_n should be n independent.

We conclude this section with an example of error analysis for the one-dimensional Ising model:

$$H(\{\sigma\}) = -J\sum_{i=1}^{L} \sigma_i \sigma_{i+1} \quad \sigma_{L+1} = \sigma i \tag{211}$$

where $\sigma_i = \pm 1$. This model is easily solved exactly with the transfer matrix method and thus produces a useful testing ground for the MC approach. In particular at zero temperature, $T = 0$, a phase transition to a ferromagnetically ordered phase ($J > 0$) occurs [7]. Spin-spin correlations are given by:

$$g(r) = \frac{1}{L}\sum_{i=1}^{L} \langle \sigma_i \sigma_{i+r} \rangle \text{ with } \langle \sigma_i \sigma_{i+r} \rangle = \frac{\sum_{\{\sigma\}} e^{-\beta H(\{\sigma\})} \sigma_i \sigma_{i+r}}{\sum_{\{\sigma\}} e^{-\beta H(\{\sigma\})}} \tag{212}$$

where β corresponds to the inverse temperature. To simulate the model, we use a simple random site upgrading method: a site (i) is randomly chosen, and the spin is flipped ($\sigma_i \rightarrow -\sigma_i$) with the heat bath algorithm. The MC time unit corresponds to a single sweep meaning that L sites are randomly chosen before a measurement is carried out. Fig. 19 plots the autocorrelation time for $g(r = L/2)$ on an $L = 24$ site lattice at $\beta J = 1$ and $\beta J = 2$. From Fig. 19a one can extract the autocorrelation time: $\tau_O \approx 11, 54$ for for $\beta J = 1, 2$ respectively. Fig. 19b plots the error as function of bin size in units of the τ_O (see Eq. (210)). As apparent, $n \approx 10$ is sufficient to get a reliable estimate of the error.

REFERENCES

1. S. R. White. *Physics Reports*, 301:187, 1998.
2. H. G. Evertz, G. Lana, and M. Marcu. *Phys. Rev. Lett.*, 70:875, 1993.
3. R. Blankenbecler, D. J. Scalapino, and R. L. Sugar. *Phys. Rev. D*, 24:2278, 1981.
4. J. E. Hirsch, D. J. Scalapino, R. L. Sugar, and R. Blankenbecler. *Phys. Rev. B*, 26:5033, 1981.
5. M. Barma and B. S. Shastry. *Phys. Rev. B*, 18:3351, 1978.
6. F. F. Assaad and D. Würtz. *Phys. Rev. B*, 44:2681, 1991.
7. R. J. Baxter. *Exactly solved models in statistical mechanics*. Academic Press Limited, London, 1989.
8. M. Troyer, M. Imada, and K. Ueda. Critical exponents of the quantum phase transition in a planar antiferromagnet. *J. Phys. Soc. Jpn.*, 66:2957, 1997.
9. J. E. Hirsch. *Phys. Rev. B*, 31:4403, 1985.
10. S. R. White, D. J. Scalapino, R. L. Sugar, E. Y. Loh, J. E. Gubernatis, and R. T. Scalettar. *Phys. Rev. B*, 40:506, 1989.
11. E. Loh and J. Gubernatis. Stable numerical simulations of models of interacting electrons in condensed-matter physics. In W. Hanke and Y. V. Kopaev, editors, *Modern Problems of Condensed Matter Physics*, volume 32, page 177. North Holland, Amsterdam, 1992.
12. G. Sugiyama and S.E. Koonin. *Anals of Phys.*, 168:1, 1986.
13. S. Sorella, S. Baroni, R. Car, and M. Parrinello. A novel technique for the simulation of interacting fermion systems. *Europhys. Lett.*, 8:663, 1989.
14. S. Sorella, E. Tosatti, S. Baroni, R. Car, and M. Parrinello. *Int. J. Mod. Phys. B*, 1:993, 1989.
15. M. Jarrell. Hubbard model in infinite dimensions: A quantum monte carlo study. *Phys. Rev. Lett.*, 69:168, 1992.
16. A. Georges, G. Kotliar, W. Krauth, and M. J. Rozenberg. *Rev. of. Mod. Phys.*, 68:13, 1996.
17. R. M. Fye. *Phys. Rev. B*, 33:6271, 1986.
18. B.B. Beard and U.J. Wiese. *Phys. Rev. Lett.*, 77:5130, 1996.
19. H. G. Evertz. The loop algorithm. *cond-mat/9707221*, 1997.
20. M. Brunner, F. F. Assaad, and A. Muramatsu. Single hole dynamics in the *t-j* model on a square lattice. *Phys. Rev. B*, 62:12395, 2000.
21. M. Brunner, S. Capponi, F. F. Assaad, and A. Muramatsu. Single hole dynamics in the $t-j$ model on two- and three-leg ladders. *Phys. Rev. B*, 63:R180511, 2001.
22. M. Troyer, F. F. Assaad, and D. Würtz. Disconnected cluster approach for quantum monte carlo simulations. *Helv. Phys. Acta.*, 64:942, 1991.
23. J. E. Hirsch. *Phys. Rev. B*, 28:4059, 1983.
24. F. F. Assaad. Su(2)-spin invariant auxiliary field quantum monte carlo algorithm for hubbard models. In E. Krause and W. Jäger, editors, *High performance computing in science and engineering*, page 105. Springer, Berlin, 1998. [cond-mat/9806307].
25. F. F. Assaad, M. Imada, and D. J. Scalapino. *Phys. Rev. B*, 56:15001, 1997.
26. F. F. Assaad, M. Imada, and D. J. Scalapino. *Phys. Rev. Lett*, 77:4592, 1996.
27. W. H. Press, S. A. Teukolsky, W. T. Vetterling, and B. P. Flannery. *Numerical Recipes in C*. Cambridge University Press, Cambridge, 1992.
28. F. F. Assaad and M. Imada. *J. Phys. Soc. Jpn.*, 65:189, 1996.
29. F. F. Assaad. *Phys. Rev. Lett.*, 83:796, 1999.
30. M. Jarrell and J.E. Gubernatis. *Physics Reports*, 269:133, 1996.
31. W. von der Linden. *Applied Physics A*, 60:155, 1995.
32. J. E. Hirsch. Stable monte carlo algorithm for fermion lattice systems at low temperatures. *Phys. Rev. B*, 38:12023, 1988.
33. M. Feldbacher and F. F. Assaad. Efficient calculation of imaginary time displaced correlation functions in the projector auxilary field quantum monte-carlo algorithm. *Phys. Rev. B*, 63:73105, 2001.
34. S. Fahy and D. R. Hamann. Positive-projection monte carlo simulation: A new variational approach to strongly interacting fermion systems. *Phys. Rev. Lett.*, 65:3437, 1990.
35. S. Fahy and D. R. Hamann. Diffusive behavior of states in the hubbard-stratonovitch transformation. *Phys. Rev. B*, 43:765, 1991.
36. S. Zhang, J. Carlson, and J. E. Gubernatis. Constrained path quantum monte carlo method for fermion ground states. *Phys. Rev. Lett.*, 74:3652, 1995.

37. S. Zhang, J. Carlson, and J. E. Gubernatis. Constrained path monte carlo method for fermion ground states. *Phys. Rev. B*, 55:7464, 1997.
38. F. F. Assaad. Depleted kondo lattices: mean-field and quantum monte carlo calculations. *Phys. Rev. B*, 65:115104, 2002.
39. E. Fradkin. *Field Theories of condensed matter systems*. Frontiers in Physics. Addison-Wesley Publishing Company, Redwood City, 1991.
40. D. Poilblanc. Twisted boundary conditions in cluster calculations of the optical conductivity in two-dimensional lattice models. *Phys. Rev. B*, 44:9562, 1991.
41. C. Gross. The boundary condition integration technique: Results for the hubbard model in 1d and 2d. *Z. Phys. B*, 86:359, 1992.
42. R. T. Scalettar, E. Y. Loh, J. E. Gubernatis, A. Moreo, S. R. White, D. J. Scalapino, and R. L. Sugar. Phase diagram of the two-dimensional negative-u hubbard model. *Phys. Rev. Lett.*, 62:1407, 1989.
43. F. F. Assaad, W. Hanke, and D. J. Scalapino. Flux quantization and temperature derivative of the superfluid density as criteria for superconductivity in two-dimensional hubbard models. *Phys. Rev. B*, 50:12835, 1994.
44. N. Trivedi and M. Randeria. Deviations from fermi-liquid behavior above t_c in 2d short coherence length superconductors. *Phys. Rev. Lett.*, 75:312, 1995.
45. W. Kohn. *Phys. Rev. A*, 133:171, 1964.
46. M. Imada, A. Fujimori, and Y. Tokura. *Rev. Mod. Phys.*, 70:1039, 1998.
47. S. Sorella and E. Tosatti. Semi-metal-insulator transition of the hubbard model in the honeycomb lattice. *Europhys. Lett.*, 19:699, 1992.
48. G. G. Batrouni and R. T. Scalettar. Anomalous decouplings and the fermion sign problem. *Phys. Rev. B*, 42:2282, 1990.
49. N. Furukawa and M. Imada. Optimization of initial state vector in the ground state algorithm of lattice fermion simulations. *J. Phys. Soc. Jpn.*, 60:3669, 1991.
50. P. A. Lee, T. M. Rice, J. W. Serene, L. J. Sham, and J. W. Wilkins. *Comm. Condens. Matter Phys.*, 12:99, 1986.
51. G. Aeppli and Z. Fisk. *Comm. Condens Matter Phys.*, 16:155, 1992.
52. J. R. Schrieffer and P. A. Wolff. *Phys. Rev.*, 149:491, 1966.
53. H. Tsunetsugu, M. Sigrist, and K. Ueda. *Rev. Mod. Phys.*, 69:809, 1997.
54. M. Vekic, J. W. Cannon, D. J. Scalapino, R. T. Scalettar, and R. L. Sugar. Competition between antiferromagnetic order and spin-liquid behavior in the two-dimensional periodic anderson model at half filling. *Phys. Rev. Lett.*, 74:2367, 1995.
55. R. M. Fye and D. J. Scalapino. *Phys. Rev. Lett.*, 65:3177, 1990.
56. R. M. Fye and D. J. Scalapino. *Phys. Rev. B*, 44:7486, 1991.
57. S. Q. Shen. *Phys. Rev. B*, 53:14252, 1996.
58. H. Tsunetsugu. *Phys. Rev. B*, 55:3042, 1997.
59. S. Capponi and F. F. Assaad. Spin and charge dynamics of the ferromagnetic and antiferromagnetic two-dimensional half-filled kondo lattice model. *Phs. Rev. B*, 63:155113, 2001.
60. J. E. Hirsch and R. M. Fye. *Phys. Rev. Lett.*, 56:2521, 1986.
61. A. C. Hewson. *The Kondo Problem to Heavy Fermions*. Cambridge Studies in Magnetism. Cambridge Universiy Press, Cambridge, 1997.
62. B. Efron. *The jackknife, the bootsrap and other resampling plans*. CBMS-NSF conference series in applied mathematics. J. W. Arrowsmith Ltd., Bristol,England, 1982.

Basic Models in the Quantum Theory of Magnetism

Yu. A. Izyumov

Institute for Metal Physics, Ural Division of the Russian Academy of Sciences, 620219, Ekaterinburg, Russia

Abstract. We regard the Heisenberg model, the Hubbard model, the tJ-model and the sd-model as the basic models of the quantum theory of magnetism in solids. They can describe localized and itinerant magnets and strongly correlated electron systems. This review is devoted to analytical approaches for these models: diagrammatic techniques and the method of generating functional. The diagrammatic techniques are based on a generalization of the Wick theorem for spin and X operators. Peculiarities of such techniques for the basic models appear because the spin and X operators do not commute on a C-value, but their commutator (anticommutator) is an operator itself. The method of generating functional is a generalization of the Kadanoff-Baym approach, developed earlier for usual Fermi systems. The generating functional describes the interaction of a system with fluctuating fields, and different Green's functions can be treated as variational derivatives with respect to these fields. Such approach allows to derive the equation of motion for the Green's functions in each model in terms of functional derivatives. These equations help to find common features in the behavior of the basic models, particularly in finding the multiplicative structure of one-particle Green's functions. Iteration of the equations generates perturbation theory, which is compared with the diagrammatic techniques. Both approaches are applied to the calculation of the quasiparticle spectrum of the models and of collective excitations. A generalized random phase approximation (GRPA) is suggested for calculation of different dynamical susceptibilities. This approximation is developed in both approaches: the diagrammatic technique and the generating functional method.

I. INTRODUCTION

In the quantum theory of magnetism in solids, three groups of models are widely used: localized, itinerant and hybrid. Localized models ignore the electron structure of materials and reduce to considering a set of isolated (atomic) spins occupying lattice sites. The classical representative of this group is just the Heisenberg model [1], and also its strongly anisotropic limit — the Ising model. In itinerant models one starts from the beginning with electrons moving in a lattice of spinless ions and interacting with each other. This interaction leads to the appearance of spontaneous magnetization in the system, to the localization of spin density at lattice sites and sometimes to the localization of the charge carriers themselves.

In the itinerant model one must distinguish two cases: the weak coupling limit (weak Coulomb interaction) and the strong coupling limit. In the latter case we usually refer to it as the Hubbard model [2, 3]. Finally, in the hybrid model we assume the existence, in the solids, of electrons localized on the lattice sites and of conduction electrons, interacting by exchange forces. This is the so called sd-model.

CP678, *Lectures on the Physics of Highly Correlated Electron Systems VII: Seventh Training Course,* edited by A. Avella and F. Mancini

Obviously, the Heisenberg model should describe magnetic insulators, the Hubbard model should be reasonably good for transition metals, where magnetic moments of 3d-shells are partly collectivized in a crystal. The hybrid sd-model describes rather rare earth metals and actinides. Such a model should be better called sf-model, however we shall keep the customary name.

Thus, the three models above encompass the wide region of magnetically ordered materials. When there is a problem for a detailed description of some particular group of materials, the previous models can be extended by including specific interactions, for example magnetic anisotropy. Apart from such complexities, these three models — Heisenberg, Hubbard and sd-model — give the opportunity to describe the most essential phenomena in magnetically ordered materials, and this is why they may be regarded as the basic or fundamental models of magnetism.

The Heisenberg model takes into account only the exchange interaction between localized spins on a lattice, and its Hamiltonian is written in the form

$$\mathscr{H} = -\frac{1}{2}\sum_{i\neq j} J_{ij}\boldsymbol{S}_i\boldsymbol{S}_j. \tag{1}$$

Here indexes i and j label lattice sites ($i \neq j$), $\boldsymbol{S}_i$ is the spin operator at site i, and J_{ij} is the exchange integral between two spins. The model (1) is isotropic, because the scalar product $(\boldsymbol{S}_i\boldsymbol{S}_j) = S_i^x S_j^x + S_i^y S_j^y + S_i^z S_j^z$, is invariant with respect to spin rotation. If $J_{ij} > 0$, spontaneous moment appears in the system, which determines a particular direction z.

An itinerant model can be represented by the Hamiltonian

$$\mathscr{H} = \sum_{ij\sigma} t_{ij} c_{i\sigma}^{+} c_{j\sigma} + U\sum_{i} n_{i\uparrow} n_{j\downarrow}, \tag{2}$$

describing electrons moving on a lattice with on-site Coulomb interaction. In the model (2) electron states are not degenerated, therefore two electrons can occupy the same site if their spins are antiparallel. Here $c_{i\sigma}$ and $c_{i\sigma}^{+}$ are the Fermi annihilation and creation operators of an electron at site i with spin σ, t_{ij} is the matrix element of electrons hopping from site to site. Usually it is supposed that it does not vanish only for nearest neighbors, whose number on the lattice is equal to z. The second term in (2) has a particular form. It has been first suggested by Anderson while he was developing the theory of localized spin states in a metal (the corresponding Hamiltonian is the famous Anderson model [4]). Such a simplified form of the Coulomb interaction turned out to be very convenient when describing magnetism in metals. Fundamental studies made by Hubbard appointed his name to the model (2).

In the Hamiltonian (2) there are two parameters: zt and U (t is the matrix element t_{ij} for nearest neighbors). According to the relations between them one has two cases: the weak coupling $U \ll zt$ and the strong coupling $U \geq zt$. The first case determines a standard itinerant model of magnetism (see for example [5]). The second case corresponds to a theory of strongly correlated systems, which is usually associated with the Hubbard model. In the limit $U \gg zt$ we can transform the Hamiltonian (2) to an effective Hamiltonian, the tJ-model [6]:

$$\mathscr{H} = t\sum_{ij\sigma} \tilde{c}_{i\sigma}^{+} \tilde{c}_{j\sigma} + J\sum_{ij} (\boldsymbol{S}_i^{\mathrm{e}}\boldsymbol{S}_j^{\mathrm{e}}). \tag{3}$$

This Hamiltonian describes the motion of correlated electrons on a lattice (i.e., with the additional constraint that electrons can hop from one site to another only if the latter is not occupied). $\tilde{c}_{i\sigma}$ and $\tilde{c}^+_{i\sigma}$ are just the annihilation and creation operators of the correlated electrons. The second term in (3) describes indirect exchange interaction of electrons on neighbor sites. Its value $J \sim t^2/U$ is just the kinetic Anderson exchange [7]. This interaction determines a tendency of the system towards antiferromagnetic ordering. Spin operators in (3) are expressed through the Fermi operators in the standard way

$$\boldsymbol{S}^{\mathrm{e}}_i = \frac{1}{2} \sum_{\sigma\sigma'} c^+_{i\sigma} \boldsymbol{\tau}_{\sigma\sigma'} c_{i\sigma'}, \tag{4}$$

$\boldsymbol{\tau}$ is a vector with Pauli matrices as components. The Hamiltonian of the tJ-model is a completely electronic one.

The last model is the sd-model, whose Hamiltonian is expressed through Fermi operators for conduction electrons and spin operators at lattice sites:

$$\mathscr{H} = \sum_{ij\sigma} t_{ij} c^+_{i\sigma} c_{j\sigma} - J_{\mathrm{sd}} \sum_{i\sigma\sigma'} (\boldsymbol{S}_i \boldsymbol{\tau}_{\sigma\sigma'}) c^+_{i\sigma} c_{i\sigma'}. \tag{5}$$

Here J_{sd} is the so called sd-exchange integral. It determines the value of the exchange interaction $-J_{\mathrm{sd}} \boldsymbol{S}_i \boldsymbol{S}^{\mathrm{e}}_i$ of a localized spin $\boldsymbol{S}_i$ on site i and electron spin $\boldsymbol{S}^{\mathrm{e}}_i$ at the same site, which is given by relation (4). Thus, Hamiltonian of the sd-model is a quadratic form of the Fermi- operators, and the exchange term is off diagonal with respect to electron spins.

The sd-model contains two parameters, zt and J_{sd}. According to the relation between them one has two cases. When $J_{\mathrm{sd}} \ll zt$ the weak coupling limit takes place, and the sd-term can be treated perturbatively. In the framework of this approximation, important physical results were obtained: magnetization of conduction electrons in a ferromagnetic metal, splitting of the electron band in antiferromagnetic metal, indirect interaction of localized spins via conduction electrons, Kondo effect, coupling of magnetic and transport properties in magnetically ordered metals, suppression of superconductivity by paramagnetic impurities. In the strong coupling limit $J_{\mathrm{sd}} \gg zt$ one can move from Hamiltonian (5) to an effective Hamiltonian of the double exchange model [8, 9], describing ferromagnetism in a large class of compounds like manganites [10], and other unique properties (see the review [11, 12, 13]).

It is natural that at the beginning all these models were investigated by simple methods. In all cases when there is no small parameter ($zt \sim U$ for the Hubbard model, $zt \sim J_{\mathrm{sd}}$ for the sd-model and $T \leq T_C$ for the Heisenberg model) large difficulties appear, especially in the attempts to find analytical solutions of the problem. In these cases the mathematical apparatus was rearranged in order to use some small parameters not connected with parameters of the dimension of energy. Among them we mention electron (hole) concentration when it is small, $1/N$-expansion (N is the number of degenerated states) or $1/z$-expansion (z is the coordination number).

In the last decade the parameter $1/z$ or the similar parameter $1/d$ (d is the dimensionality of the system) obtained an heuristic meaning. Consideration of the limit $d \to \infty$ lies in the framework of the dynamic mean field approximation. With its help all mentioned above electron models were studied under intermediate or strong couplings, and more complicated models as well (see a review [14]). Major improvements were obtained by the method of auxiliary particles — fermions [15, 16, 17, 18, 19], bosons [17, 18, 19, 20, 21], composite operators [22, 23, 24, 25, 26, 27] and others [28, 29, 30, 31, 32, 33, 34, 35, 36, 37, 38, 39, 40, 41, 42, 43].

Our review is devoted to two analytical approaches — diagrammatic technique and the method of generating functional. It is well known that standard diagrammatic technique for Fermi and Bose systems is based on the Wick theorem — a specific algorithm that reduces the calculation of an average of a time-ordered product of a number of Fermi (Bose) operators to products of pair averages. Such an algorithm is possible because an anticommutator (or commutator) of two operators is a C-number. However in the case of spin operators it is not like this: a commutator of two spin operators is some third spin operator. Nevertheless Vaks, Larkin and Pikin (VLP) [44, 45], have shown that an algorithm does exists reducing an average of T-product of some number of spin operators to averages with less number of spin operators. This algorithm may be named the Wick theorem for spin operators. With the help of this algorithm Izyumov and Kassan-Ogly [46, 47, 48] constructed a diagrammatic technique for the Heisenberg model, without using a pseudo-fermionic representation (4) for spin operators, earlier used by VLP [44, 45]. This technique is more convenient in practice then the VLP approach, but both techniques are equivalent in physical sense.

The technique [46, 47, 48] can be generalized to the Hubbard model. The reason is that the Hamiltonian (2) can be expressed in terms of the so called X operators, introduced by Hubbard [49], which are combinations of Fermi operators. Hamiltonian (2) is a quadratic form of X operators, and X operators have commutation relations similar to spin operators, namely, the commutator (or anticommutator) of two X operators is some third X operator. For this reason one can prove a Wick theorem for X operators [50, 51] and construct a diagrammatic technique for the Hubbard model. It was first done by Zaitsev [52], and for the tJ-model by Izyumov and Letfulov [53]. This technique corresponds to perturbation theory near atomic limit, where the one-site solution is taken as the zeroth-order approximation, and the hopping term is taken as the perturbation.

The diagrammatic techniques for spin and X operators are presented in detail in books [54] and [55] respectively (see also monograph [56]). Unfortunately they were published only in Russian, therefore they are not much available for the western readers. The other approach is based on the generating functional, which takes into account the interaction of the system with fluctuating external fields. It was first suggested by Schwinger [57] in quantum electrodynamics, and in statistical mechanics by Kadanoff and Baym [58, 59]. The introduction of the generating functional allows to express different GFs of the system as functional (variational) derivatives over fluctuating fields. This allows to derive exact equations for GFs in terms of variational derivatives.

In the theory of strongly correlated systems the generating functional method was applied by Ruckenstein and Smitt-Rink [60] and later by others [61, 62, 63, 64]. Recently an important development was done for the Heisenberg, Hubbard, tJ and sd models by Izyumov and Chaschin in a series of papers [65, 66, 67, 68, 69, 70] and a monograph [71], published in Russian. In the present review, for the first time in English, we make a systematical presentation of the diagrammatic techniques for spin and X operators and of the method of generating functional as well, with applications to the models: Heisenberg, Hubbard, tJ-model and sd-model.

II. THE HEISENBERG MODEL

A. Diagrammatic technique for spin operators

Wick theorem for spin operators. The Heisenberg model of a magnetic solid with localized spins is given by the Hamiltonian $\mathscr{H} = \mathscr{H}_0 + \mathscr{H}_{\text{int}}$:

$$\mathscr{H}_0 = -g\mu_0 H \sum_i S_i^z, \tag{6}$$

$$\begin{aligned} \mathscr{H}_{\text{int}} &= -\tfrac{1}{2}\sum_{i\neq j} J(i-j)\boldsymbol{S}_i\boldsymbol{S}_j = \\ &= -\tfrac{1}{2}\sum_{i\neq j} J(i-j)\{S_i^z S_j^z + S_i^- S_j^+ + S_i^+ S_j^-\}, \end{aligned} \tag{7}$$

where spin operators for the same site are obey the permutation relations

$$[S^z, S^+] = S^+, \quad [S^z, S^-] = -S^-, \quad [S^+, S^-] = S^z, \tag{8}$$

with

$$S^{\pm} = \frac{1}{\sqrt{2}}(S^x \pm \mathrm{i}S^y).$$

Let us introduce the temperature spin GFs and determine them the same way as in the statistical mechanics of usual Bose systems [72]. We call temperature GF an averaged T-product of spin operators:

$$\langle T\tilde{S}_{i_1}^{\alpha_1}(\tau_1)\tilde{S}_{i_2}^{\alpha_2}(\tau_2)\ldots\tilde{S}_{i_l}^{\alpha_l}(\tau_l)\rangle, \tag{9}$$

where $\tilde{S}_i^{\alpha}(\tau)$ are the operators S_i^{α} in the Heisenberg representation:

$$\tilde{S}_i^{\alpha}(\tau) = \mathrm{e}^{\mathscr{H}\tau} S_i^{\alpha} \mathrm{e}^{-\mathscr{H}\tau}.$$

In (9) $\langle\ldots\rangle$ is the symbol of statistical average with the Hamiltonian $\mathscr{H}$, and T is the symbol of time ordering. It puts operators in order of increasing "time" τ.

In the interaction representation GF (9) is written as

$$\begin{aligned} \langle T\tilde{S}_1^{\alpha_{i_1}}(\tau_1)\ldots\tilde{S}_l^{\alpha_{i_l}}(\tau_l)\rangle &= \frac{\langle TS_1^{\alpha_{i_1}}(\tau_1)\ldots S_l^{\alpha_{i_l}}(\tau_l)\sigma(\beta)\rangle_0}{\langle\sigma(\beta)\rangle_0}, \\ \sigma(\beta) &= T\mathrm{e}^{-\int_0^\beta \mathscr{H}_{\text{int}}\tau \, d\tau} \end{aligned} \tag{10}$$

is the temperature scattering matrix. All operators in (10) are taken in the interaction representation, that is

$$S^{\alpha}(\tau) = \mathrm{e}^{\mathscr{H}_0\tau} S_i^{\alpha} \mathrm{e}^{-\mathscr{H}_0\tau}.$$

Here $\langle\ldots\rangle_0$ means the statistical average with the Hamiltonian $\mathscr{H}_0$:

$$\langle\ldots\rangle_0 = \mathrm{Tr}\{\rho_0\ldots\}, \tag{11}$$

where

$$\rho_0 = \exp(-\beta \mathscr{H}_0)/\mathrm{Tr}\exp(-\beta \mathscr{H}_0) =$$
$$= \exp\left(y_0 \sum_l S_l^z\right) \mathrm{Tr}\exp\left(y_0 \sum_l S_l^z\right),$$

and $y_0 = \beta g \mu_0 H$.

Relations (10) allows to obtain an expansion for the GF in terms of powers of $\mathscr{H}_{\mathrm{int}}$, and the problem of calculation of different order terms reduces to calculations of averages of T-products of operators $S^\alpha(\tau)$.

The Wick theorem for spin operators reads [54, 56]:

$$\begin{gathered}\langle T\{S_1^{\alpha_1}(\tau_1)\dots S_0^-(\tau)\dots S_n^{\alpha_n}(\tau_l)\}\rangle_0 = \\ = K_{01}^0(\tau-\tau_1)\langle T\{[S_1^{\alpha_1}, S_0^-]_{\tau_1} S_2^{\alpha_2}(\tau_2)\dots S_n^{\alpha_n}(\tau_n)\}\rangle_0 + \\ + K_{02}^0(\tau-\tau_2)\langle T\{S_1^{\alpha_1}(\tau_1)[S_2^{\alpha_2}, S_0^-]_{\tau_2}\dots S_n^{\alpha_n}(\tau_n)\}\rangle_0 + \dots + \\ + K_{0n}^0(\tau-\tau_n)\langle T\{S_1^{\alpha_1}(\tau_1)\dots S_{n-1}^{\alpha_{n-1}}(\tau_{n-1})[S_n^{\alpha_n}, S_0^-]_{\tau_n}\}\rangle_0,\end{gathered} \tag{12}$$

where K_{0i} is a GF of zero order approximation.

$$K_{0i}(\tau-\tau_i) \equiv \delta_{0i} K^0(\tau-\tau_i), \tag{13}$$

$$K^0(\tau-\tau_i) = \exp\left(\frac{y_0}{\beta}(\tau-\tau_i)\right)\begin{cases} n_{y_0}, & \tau > \tau_i, \\ 1+n_{y_0}, & \tau < \tau_i, \end{cases}.$$

$$n_{y_0} = \frac{1}{e^{y_0}-1}.$$

By applying relation (12) to expression $\langle TS^-(\tau)S^+(\tau')\rangle_0$, we obtain

$$K^0(\tau-\tau') = \frac{\langle TS^-(\tau)S^+(\tau')\rangle_0}{\langle S^z\rangle_0}.$$

It determines sense of Eq. (13).

Since as a result of the commutation of two spin operators we have only one operator, Eq.(12) reduces the calculation of the averaged T-product of n spin operators to a sum of averages of T-products of $n-1$ operators. Note that, due to the structure of the exchange Hamiltonian (7), averages of type (12) do not vanish if they contain equal numbers of S^- and S^+ operators. This leads to the following procedure.

Single out some operator S^- in a certain average and decompose it according to recipe (12). Each of the obtained averages contains one S^α operator less, but in each of them the numbers of S^- and S^+ operators are equal. After the procedure (12) with each average we exclude from the averages all S^- and S^+ operators. As a result, only averages of S^z operators

$$\langle S_1^z S_2^z \dots S_l^z\rangle_0$$

are left (time dependence in operators S_i^z disappears because S^z commutes with the Hamiltonian $\mathscr{H}_0$). Such averages will be expressed below through Brillouin function and its derivatives with respect to y_0.

Thus, calculation of averaged T-products of spin operators should be done in two stages. First, by use of relation (12) we get out of S^- and S^+ operators in the given average. At the end of the first stage our average is decomposed into a sum of terms with products of K^0-functions, multiplied by some average of S^z operators only. On the second stage these averages should be calculated.

The result of the first stage can be described as taking all possible pairing of S^- operators with S^+ and S^z. To such pairings the following change of a pair of operators in our average for one operator should be done:

$$S_1^-(\tau_1)S_2^z(\tau_2) = -K_{12}^0(\tau_1 - \tau_2)S_2^-(\tau_2), \tag{14}$$

$$S_1^-(\tau_1)S_2^+(\tau_2) = K_{12}^0(\tau_1 - \tau_2)S_2^z(\tau_2), \ . \tag{15}$$

These relations follow immediately from Eq.(12) and commutators (8). Of course Eqs.(14) – (15) make sense only under the operation of averaging $\langle \ldots \rangle_0$.

We see that in further pairings, index of site 1 attached to operator S_1^- does not enter. Now, instead of the operator S_1^-, stands the operator S_2^- or S_2^z. Operator S_2^- in (14) has to participate in other pairings, however index 2 will enter only one more pairing. As a result, the following system of pairings appears:

$$S_1^-(\tau_1)S_2^z(\tau_2)S_3^+ = -K_{12}^0(\tau_1 - \tau_2)K_{23}^0(\tau_2 - \tau_3)S_3^z(\tau_3). \tag{16}$$

Like the operator S_2^z in (15), it can be left either unpaired or can be paired with some operator S_3. As a result of the second possibility an operator S_2^- appears, which will participate in further pairings, that is index 2 will enter in three (and only three) pairings. So, the following system of pairings appears:

$$\begin{gathered} S_1^-(\tau_1)S_2^+(\tau_2)S_3^-(\tau_3)S_4^+(\tau_4) = \\ = -K_{12}^0(\tau_1 - \tau_2)K_{32}^0(\tau_3 - \tau_2)K_{24}^0(\tau_2 - \tau_4)S_4^z(\tau_4). \end{gathered} \tag{17}$$

Subsequent application of relation (12) leads to the total disappearance of S^- and S^+ operators. It is possible to describe the algorithm this way: an average of a T-product is decomposed into a sum of terms with all possible pairings of the spin operators. This result may be regarded as the Wick theorem for spin operators. With its help, corrections to a GF can be written in any order of perturbation theory. For the presentation of these correction terms it is convenient to introduce a diagrammatic technique.

We denote the GF $K_{12}^0(\tau - \tau')$ by a solid line $\underset{\mathbf{1}}{}\!\longrightarrow\!\underset{\mathbf{2}}{}$. Systems of pairing (15) – (17) correspond to graphic elements with all sites being coincident because of δ-symbols prescribed to K_{12}^0. Any system of pairing is presented as a combination of elements in Fig. 1.

It is easy to do also an inverse operation: by a given graphic element we can restore the system of pairings. One has only to remember that a vertex with outgoing Green's line

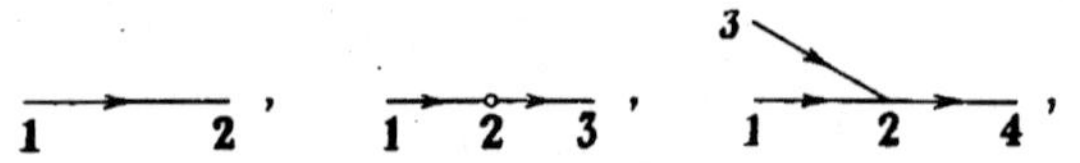

FIGURE 1. Graphic depiction of the pairing systems (15) — (17)

corresponds to operator S^-; a vertex with one ingoing or three Green's lines corresponds to operator S^+; and, finally, a vertex with one ingoing and one outgoing lines corresponds to operator S^z. Thus we come to the conclusion that, in diagrams, only five types of different vertexes can exist (Fig. 2).

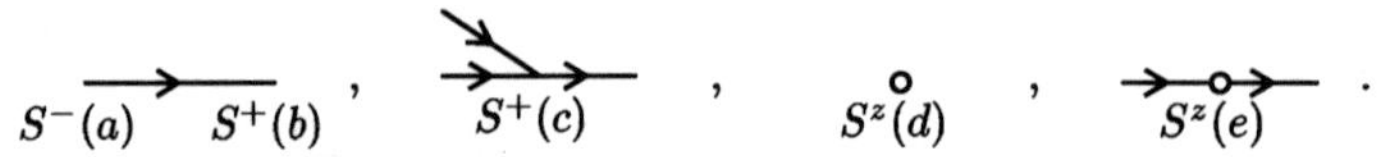

FIGURE 2. Type of vertexes for the exchange Hamiltonian. (Vertexes d and e are denoted by light points)

Vertex of d type corresponds to operator S^z, which do not participate in pairings. We call vertexes c and e anomalous ones. Each of these vertexes introduces in the analytical expression sign minus, as one can see from relation (16) and (17). After taking all pairings in a T-product only S^z operators are left under symbol of averaging.

Calculation of averages for S^z-operators product. Because of additive operator $\mathscr{H}_0$, an average $\langle S_1^z S_2^z \ldots S_l^z \rangle_0$ is decomposed into a product of averages referred to the same site, and

$$\langle S_i^z \rangle_0 = \mathrm{Tr}\left(S_i^z \mathrm{e}^{y_0 S_i^z}\right) / \mathrm{Tr}\,\mathrm{e}^{y_0 S_i^z} = S B_S(S y_0) \equiv b(y_0), \tag{18}$$

where $B_S(x)$ is the Brillouin function for spin S:

$$B_S(x) = \left(1 + \frac{1}{2S}\right) \mathrm{cth}\left(1 + \frac{1}{2S}\right) x - \frac{1}{2S} \mathrm{cth} \frac{x}{2S}.$$

An average of an arbitrary power of S^z is expressed through the function $b(y_0)$ and its derivatives $b^{[n]}$ by means of the recursive relation [44, 45]

$$\langle (S^z)^N \rangle_0 = \sum_{n=0}^{N-1} \frac{(N-1)!}{n!(N-1-n)!} b^{[n]} \langle (S^z)^{N-n-1} \rangle_0. \tag{19}$$

From (19), particular expressions follow:

$$\begin{aligned}
\langle S^z \rangle_0 &= b, \\
\langle (S^z)^2 \rangle_0 &= b^2 + b', \\
\langle (S^z)^3 \rangle_0 &= b^3 + 3b'b + b'', \\
\langle (S^z)^4 \rangle_0 &= b^4 + 6b'b^2 + 3b'^2 + 4b''b + b''', \\
&\ldots,
\end{aligned}$$

where the argument of the function b and its derivatives is the quantity y_0. [46, 47, 48]:

$$
\begin{aligned}
\langle S^z \rangle_0 &= b, \\
\langle S_1^z S_2^z \rangle_0 &= b^2 + b' \delta_{12}, \\
\langle S_1^z S_2^z S_3^z \rangle_0 &= b^3 + b'b(\delta_{12} + \delta_{13} + \delta_{23}) + b'' \delta_{12}\delta_{23}, \\
\langle S_1^z S_2^z S_3^z S_4^z \rangle_0 &= b^4 + b'b^2(\delta_{12} + \delta_{13} + \delta_{23} + \delta_{24} + \delta_{34} + \delta_{14}) + \\
&+ b'^2(\delta_{12}\delta_{34} + \delta_{13}\delta_{24} + \delta_{14}\delta_{23}) + \\
&+ b''b(\delta_{12}\delta_{23} + \delta_{12}\delta_{24} + \delta_{13}\delta_{34} + \delta_{23}\delta_{34}) + b''' \delta_{12}\delta_{23}\delta_{34}, \\
& \ldots,
\end{aligned} \tag{20}
$$

where δ_{ik} is the Kronecker symbol of site indexes.

Now we can state a rule for the calculation of averages (20): first, write down all δ-symbols of the site indexes $1, 2, \ldots, p$, which correspond to pair indexes. Second, write all the independent products of two, three and so forth δ-functions up to the connection all p site indexes. Each connection of two indexes is given a factor b', each connection of three indexes a factor b'', etc. Any index of site not connected with other indexes corresponds to b. As a result, the average is equal to the sum of all the terms mentioned above. It is easy to check that every formula of the type (20) is constructed according to this rule.

Average decomposition of the type (20) allow a graphic interpretation. We denote indexes $1, 2, \ldots, p$ by points (vertexes) on a plane. The recipe of the graphic representation of formulas (20) is to include these points in ovals containing two points, three points and so on up to p points. We call them blocks. Points included in an oval have the same site indexes according to the definition. If we agree that an oval with n points inside of it corresponds to $b^{[n-1]}$, we obtain graphic representations of formulas (20). Thus, for the first three terms we have the following representation (Fig. 3):

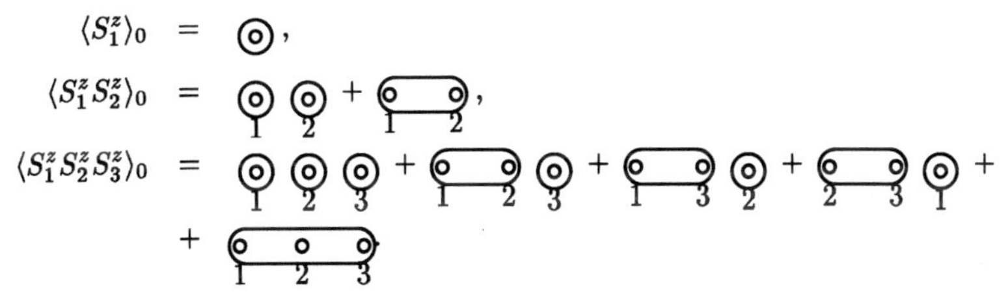

FIGURE 3. Graphic depiction of averages of S^z operators product

Construction of diagrammatic technique. As a result of the application of the Wick theorem and of the calculation of the averages of S^z-operators, the initial average of T-product of spin operators is decomposed into sum of terms which are products of GFs K^0 and different derivatives of Brillouin functions. Let us take an example: a particular system of pairings:

$$\begin{aligned}&\langle TS_1^-(\tau_1)S_2^z(\tau_2)S_3^+(\tau_3)S_4^z(\tau_4)S_5^-(\tau_5)S_6^+(\tau_6)S_7^-(\tau_7)S_8^+(\tau_8)\rangle_0 = \\ &= (-1)^2 K_{12}^0(\tau_1-\tau_2)K_{23}^0(\tau_2-\tau_3)K_{56}^0(\tau_5-\tau_6)K_{76}^0(\tau_7-\tau_6)\times \\ &\times K_{68}^0(\tau_6-\tau_8)\langle S_3^z S_4^z S_8^z\rangle_0.\end{aligned} \tag{21}$$

The result of the first stage of calculation is showed in Fig. 4.

FIGURE 4. Graphic representation of expression (21) before averaging of S^z operators

In the second stage we have to calculate the average $\langle S_3^z S_4^z S_8^z\rangle$, according to formulas (20). Our average is made up by five terms (Fig. 5).

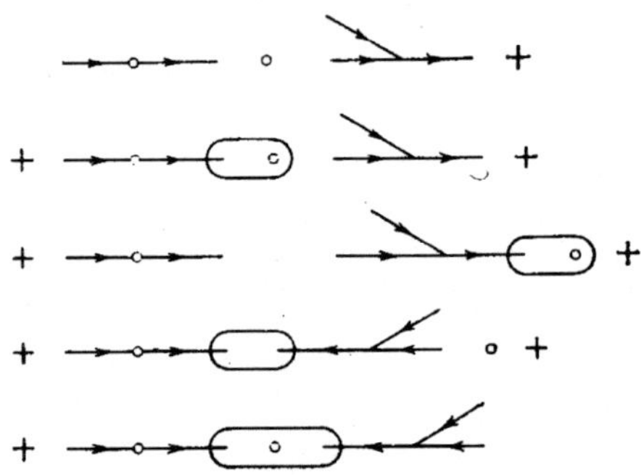

FIGURE 5. Graphic depiction of contribution in the T-product (21) from the indicated system of pairing

This example shows how an average of aT-product is presented by graphs. We come to the following rules for the graphic representation of perturbation theory corrections for a GF (10) [46, 47, 48, 54]. In order to write down a correction of n-th order for the GF with l external vertexes one must:

1. Draw all topologically inequivalent skeleton graphs containing $2n+1$ vertexes (with l prescribed external vertexes) and $2n$ internal vertexes joined in pairs by interaction lines; and one must take into account the fact that the vertexes of type a are joined only to vertexes of type b and c, while vertexes of type d and e are joined only with vertexes of the same type (Fig.6);
2. Multiply each skeleton graph by the collection of diagrams with thickened blocks;
3. Assign to each diagram the analytical expression:

FIGURE 6. Bare vertex parts in the isotropic Heisenberg model

$$(-1)^L C_n \prod_\alpha b^{[\beta_\alpha - 1]} \int_0^\beta \ldots \int_0^\beta d\tau_1 \ldots d\tau_n \sum_{\mathbf{1} \neq \mathbf{1}'} \ldots \sum_{\mathbf{n} \neq \mathbf{n}'} \times \\ J(\mathbf{1}-\mathbf{1}') \ldots J(\mathbf{n}-\mathbf{n}') K^0 \ldots K^0 \underbrace{\delta \ldots \delta}_{\beta_\alpha - 1}. \tag{22}$$

The product $\prod_\alpha$ is taken over all blocks of the diagram; β_α is the number of distinct parts in the α-th block; $\delta \ldots \delta$ is the product of δ-symbols that takes into account the equality of all site indexes in a block; L is the number of anomalous vertexes of type c and e in the diagram; the numerical coefficient C_n is defined by the formula

$$C_n = \frac{1}{n_a! n_b! \ldots}, \tag{23}$$

where n_a, n_b are the numbers of indistinguishable parts in the diagram of type $a, b, \ldots$.

Let us give the rule for writing diagrams in the momentum representation. The zeroth approximation spin GF $K^0(\tau - \tau')$ is defined on a finite interval $-\beta < \tau < \beta$, and thus it can be expanded in a Fourier series with discrete frequencies:

$$K^0(\tau) = \frac{1}{\beta} \sum_n K^0(\omega_n) e^{-i\omega_n \tau},$$

where

$$K^0(\omega_n) = \frac{1}{2} \int_{-\beta}^{\beta} e^{i\omega_n \tau} K^0(\tau) d\tau.$$

and $\omega_n = 2\pi n/\beta$. After integrating on τ we obtain

$$K^0(\omega_n) = \frac{1}{i\omega_n + y_0/\beta}. \tag{24}$$

In graphs for GFs singletail diagrams can appear (Fig. 7). The singletail diagram corresponding to the expression $x_1 = \beta \sum_{\mathbf{1}'} J(\mathbf{1}-\mathbf{1}') \langle S^z_{\mathbf{1}'} \rangle$ can be attached to a vertex of type d or e.

Consider the attachment of singletail diagrams to the vertexes of type e on the Green's lines forming the infinite series (Fig. 7). To the sum of the series represented by the thickened line corresponds an analytic expression $K(\omega_n)$ which is to be found. According to

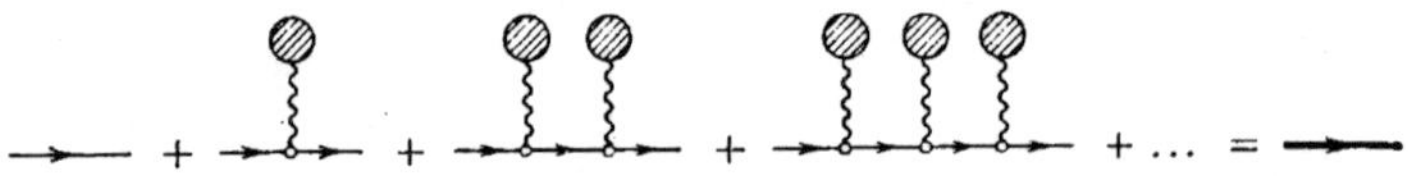

FIGURE 7. Summing of the singletail diagrams in series for spin Green's functions

the rules introduced above, to the graphic series an analytic series corresponds

$$K(\omega_n) = K^0(\omega_n) - K^{0^2}(\omega_n)x_1 + K^{0^3}(\omega_n)x_1^2 - K^{0^4}(\omega_n)x_1^3 + \ldots = \\ = \frac{K^0(\omega_n)}{1 + x_1 K^0(\omega_n)} = \frac{1}{\mathrm{i}\omega_n + y/\beta}. \tag{25}$$

Thus, taking into account the singletail diagrams on the elementary Green's lines is equivalent to the replacement of the quantity y_0 by y in $K^0(\omega_n)$. We shall assume that the contribution of the singletail diagrams is counted and that the expression (25) corresponds to the GF. It is obvious that this is equivalent to the following regrouping of our Hamiltonian:

$$\mathscr{H}_0 = -\left[g\mu_0 H + \sum_{\mathbf{1}'} J(\mathbf{1}-\mathbf{1}')\langle S^z_{\mathbf{1}'}\rangle\right] \sum_{\mathbf{1}} S^z_{\mathbf{1}},$$
$$\mathscr{H}_{\mathrm{int}} = -\tfrac{1}{2} \sum_{\mathbf{1}\neq\mathbf{1}'} J(\mathbf{1}-\mathbf{1}') \Big[(S^z_{\mathbf{1}} - \langle S^z_{\mathbf{1}}\rangle)(S^z_{\mathbf{1}'} - \langle S^z_{\mathbf{1}'}\rangle) + \\ + S^-_{\mathbf{1}} S^+_{\mathbf{1}'} + S^+_{\mathbf{1}} S^-_{\mathbf{1}'}\Big]. \tag{26}$$

This regrouping includes a self-consistent field in the zeroth approximation.

It is possible to show that accounting for the infinite series of singletail diagrams inserted in a block, containing $n+1$ separated parts (analytic expression $b^{[n]}(y_0)$ corresponds to it), leads to the replacement $y_0 \to y$ [44, 45] in the argument of the Brillouin function. Thus if we include a self-consistent field in the zeroth approximation Hamiltonian, singletail diagrams do not appear neither on Green's lines nor in blocks. Below we shall use definition of $\mathscr{H}_0$ and $\mathscr{H}_{\mathrm{int}}$ as written in (26).

Graphic depiction of pair Green's functions. The spectrum of elementary excitations of a Heisenberg ferromagnet and the kinetic coefficients are determined by pair GFs

$$K^{-+}(\mathbf{1}\tau, \mathbf{1}'\tau') = \langle T S^-_{\mathbf{1}}(\tau) S^+_{\mathbf{1}'}(\tau')\rangle, \tag{27}$$

$$K^{zz}(\mathbf{1}\tau, \mathbf{1}'\tau') = \langle T(S^z_{\mathbf{1}}(\tau) - \langle S^z_{\mathbf{1}}\rangle)(S^z_{\mathbf{1}'}(\tau') - \langle S^z_{\mathbf{1}'}\rangle)\rangle, \tag{28}$$

We call them transversal and longitudinal GFs, respectively. According to the rules formulated earlier, their graphic series for them are shown in Fig. 8 and Fig. 9.

Consider first an infinite series of graphs which are the continuation of the first row in Fig. 8. In momentum representation the above expression corresponds to

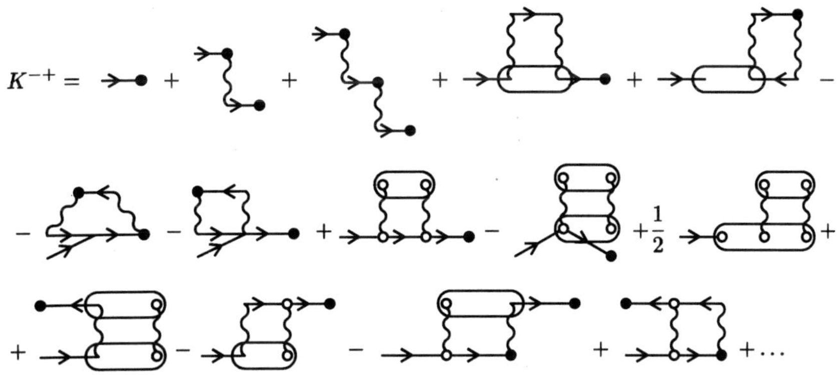

FIGURE 8. Graphic series for the transversal Green's function

$$K^{-+}(\boldsymbol{q},\omega_n) = bK(\omega_n) + (bK(\omega_n))^2 J(\boldsymbol{q}) +$$
$$+ (bK(\omega_n))^3 (J(\boldsymbol{q}))^2 + \ldots = \frac{bK(\omega_n)}{1 - bK(\omega_n) J(\boldsymbol{q})}.$$

With the use of formula (25) we can rewrite the result in the form

$$K^{-+}(\boldsymbol{q},\omega_n) = b(y)K(\boldsymbol{q},\omega_n), \quad K(\boldsymbol{q},\omega_n) = \frac{1}{\mathrm{i}\omega_n + \varepsilon_{\boldsymbol{q}}}, \tag{29}$$

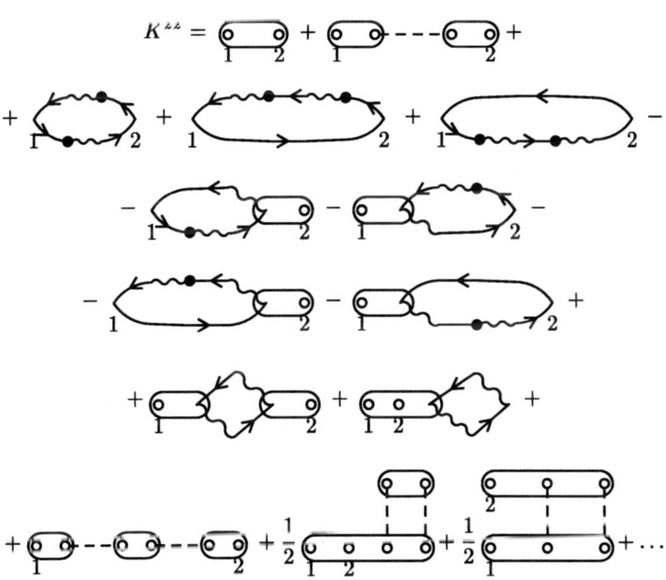

FIGURE 9. Graphic series for the longitudinal Green's function

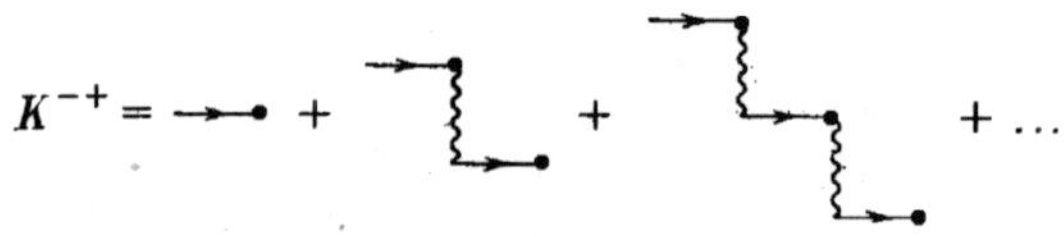

FIGURE 10. Simple series for the transversal Green's function. The Green's lines here and after correspond to expression (25)

where

$$\varepsilon_q = g\mu_0 H + b(y)\{J(0) - J(\boldsymbol{q})\}. \tag{30}$$

At low temperature ($T \ll T_C$) $b(y) \to S$, and the expression (30) is the energy of a spin wave in a ferromagnet. At finite temperature (30) practically coincides with spin wave energy

$$\varepsilon_q = g\mu_0 H + \langle S^z \rangle \{J(0) - J(\boldsymbol{q})\}, \tag{31}$$

has been obtained by the decoupling method of double-time GF's [73]. Thus the infinite series (Fig. 10) corresponds to introducing collective modes of the system, that is spin waves.

Similarly, consider an infinite series of graphs for K^{zz} (Fig. 11), which is the continuation of the first row in Fig. 9.

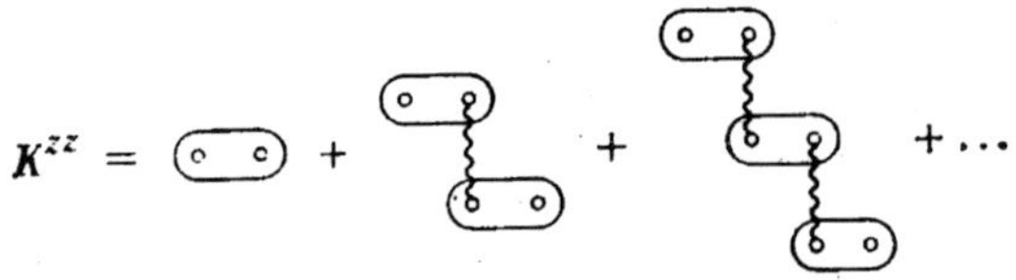

FIGURE 11. Simple series for the longitudinal Green's function. Two-points blocks correspond to expression $b'(y)$

In momentum representation we have from here [44, 45]:

$$\begin{aligned} K^{zz}(\boldsymbol{q}, \omega_n) &= \beta b' \delta_{n0} + \beta b'^2 J(\boldsymbol{q}) \delta_{n0} + \beta b'^3 J^2(\boldsymbol{q}) \delta_{n0} + \ldots = \\ &= \frac{\beta b'(y))}{1 - \beta b' J(\boldsymbol{q})} \delta_{n0}, . \end{aligned} \tag{32}$$

This result gives the Ornstein-Zernike correlation function for the Ising model.

Return to the general expansions for GFs depicted in Fig. 8 and Fig. 9. It is possible to write general equations for them using a definition of *irreducible* diagram *not cut over an interaction line* [44, 45]. Any graph for GF K^{-+} is constructed of several irreducible parts, connected by interaction lines. Denote the total set of irreducible parts by Σ^{-+}. It is obvious that the infinite series for the GF can be regrouped in such a way:

$$K^{-+} = \Sigma^{-+} + \Sigma^{-+} J \Sigma^{-+} + \Sigma^{-+} J \Sigma^{-+} J \Sigma^{-+} + \ldots$$

(J is nothing but an interaction line). Integral equation for K^{-+} is equivalent to this series

$$K^{-+} = \Sigma^{-+} + \Sigma^{-+} J K^{-+}. \tag{33}$$

A similar equation holds for the longitudinal GF:

$$K^{zz} = \Sigma^{zz} + \Sigma^{zz} J K^{zz}. \tag{34}$$

Equations of type (33) and (34) for pair GFs in the theory of Fermi systems were introduced by Larkin. In the problem of Heisenberg ferromagnet such equations were obtained in [45]. We call them Larkin's equations.

In momentum representation these equations can be solved, and give expressions for pair GFs through the irreducible part

$$K^{-+}(\boldsymbol{q}, \omega_n) = \frac{\Sigma^{-+}(\boldsymbol{q}, \omega_n)}{1 - J(\boldsymbol{q})\Sigma^{-+}(\boldsymbol{q}, \omega_n)},$$
$$K^{zz}(\boldsymbol{q}, \omega_n) = \frac{\Sigma^{zz}(\boldsymbol{q}, \omega_n)}{1 - J(\boldsymbol{q})\Sigma^{zz}(\boldsymbol{q}, \omega_n)}.$$

If we take the zeroth order approximation for Σ^{-+}:

$$\Sigma_0^{-+}(\boldsymbol{q}, \omega_n) =$$  $$= b(y) K^0(\omega_n),$$

then the expression for K^{-+} coincides with the one obtained by direct summation of diagrams. Similarly, by taking the zeroth approximation for Σ^{zz}:

$$\Sigma_0^{zz}(\boldsymbol{q}, \omega_n) =$$ 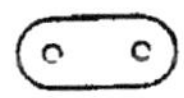 $$= \beta b'(y) \delta_{n0},$$

we come to equation (32).

Larkin equations (33) and (34) and, following from them, general expressions for the transversal and longitudinal GFs, reduce the problem of the calculation of pair GFs to the calculation of their irreducible parts.

Figs. 8 and 9 show that in fact the rules of connection of vertexes shown earlier in Fig. 6 are fulfilled. Notice that each graph for the transversal GF $K^{-+}(12)$ has two external vertexes, one of them corresponding to operator S_1^-, and another to S_2^+. Since S^+ corresponds to vertexes of two types (usual and anomalous), graphs in Fig. 8 take into account all possible combinations of external vertexes. Notice also that if the end of a Green's line does not stand in a block, always a thick dot is attached to the end, which corresponds to b. It is easy to check that the set of graphs in Fig. 8 is complete in the sense that it takes into account all topologically inequivalent graphs in the limit of second order perturbation theory.

Similar comments can be done for Fig. 9. Each graph for the longitudinal GF has two external vertexes of type d and e. The sign of a graph is determined by the parity of the total number of anomalous vertexes, as in the case of the transversal GF. Coefficient $\frac{1}{2}$ for the two last diagrams in Fig. 9 is determined by the number of indistinguishable parts in the diagram.

An analysis of transversal GF in a wide interval of temperatures up to vicinity of Curie point T_C was done by VLP [44, 45]. They are discussed in details in books [54, 56]. Now we concentrate on the calculation of longitudinal GF, in order to study fluctuations of longitudinal spin components in a ferromagnet.

Fluctuations of longitudinal spin components. Consider carefully the graphic representation for GF $K^{zz}(12)$, Fig. 9. In this graphs it is possible to thicken Green's lines shown in Fig. 8. Three graphs from the second row can be regarded as the thickening of a loop  by changing thin lines to thick ones (spin wave propagators). As a result of such thickening we come to the graphic series for longitudinal GF (Fig. 12), where all Green's lines are propagators of spin waves.

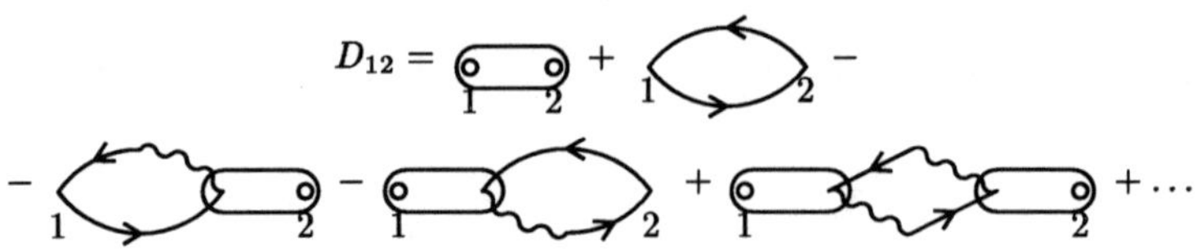

FIGURE 12. Graphic series for longitudinal GF with spin waves Green's lines

We see that the zeroth approximation for K^{zz} is given by two terms: a cumulant and a loop. The first one describes fluctuations of spontaneous moment, the second one describes the contribution of virtual creation and annihilation of spin waves in the system. The next three graphs formally correspond to a first order approximation for K^{zz}. They have a complex character in the sense that they include vertexes of two types: usual and anomalous. Thus even from Fig. 12 one can see that longitudinal GF depends on cumulants and four types of loops (Fig. 13). The cumulant 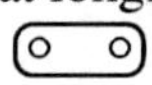 $= \pi(12)$ determines

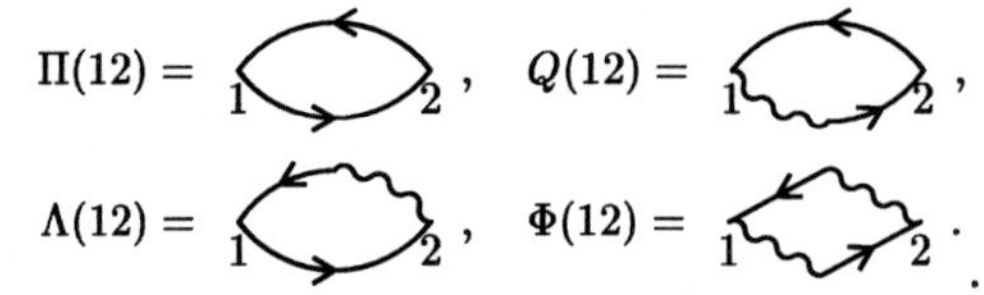

FIGURE 13. Four different loops in graphic series for longitudinal GF

the expression we came across

$$\pi(k) = \beta b' \delta_{n0}. \tag{35}$$

Zeroth approximation for K^{zz} (in form of a loop $\Pi(\boldsymbol{k})$) has been already obtained by VLP, who showed that $\pi(k) = \pi(\boldsymbol{k}, \mathrm{i}\omega_n)$ has a logarithmic singularity on a spin wave frequency $\varepsilon_{\boldsymbol{k}}$. It indicates the necessity to continue the series (Fig. 12), taking into account loops contributions in all orders. Such a summing was done in papers [65, 66, 67], where it was shown that higher order contribution strongly renormalize K^{zz} and lead to the rather peculiar behavior of the longitudinal fluctuations in a ferromagnet.

By means of the diagrammatic technique rules we can continue the series in Fig. 12 and see that in higher orders different combinations of loops (Fig. 13), should appear [66]. It is possible to sum all such diagrams by using the method applied earlier in the diagrammatic technique for the tJ-model [53].

Central place of such a summing is the calculation of a 4-point vertex part (Fig. 14), describing the interaction of spin waves. We perform it by the Bethe-Salpether equation. It is easy to verify by the iteration of this equation that we take into account all loop type diagrams and only them. The graphic equation (Fig. 14) corresponds to the analytic one:

$\Gamma(k_1 - q, k_1; k_2 + q, k_2) =$ 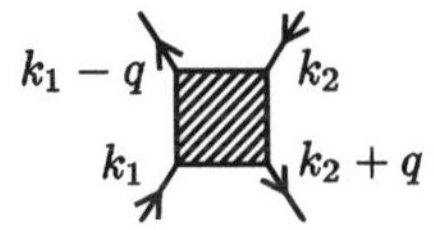

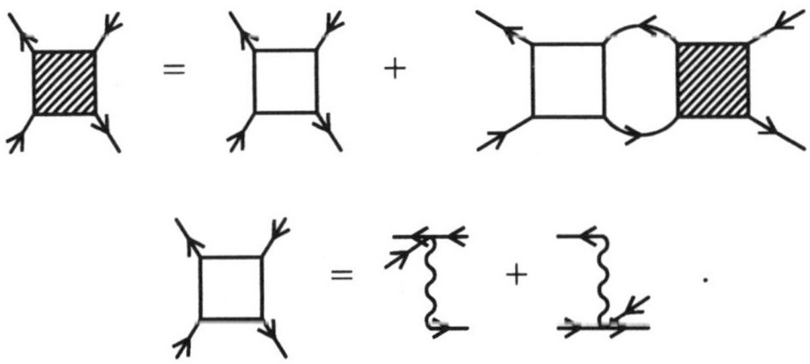

FIGURE 14. 4-point vertex part and equation for it

$$\begin{aligned}\Gamma(k_1 - k, k_1; k_2 + k, k_2) = J(\boldsymbol{k}_2 + \boldsymbol{k}) + J(\boldsymbol{k}_1 - \boldsymbol{k}) + \\ + \tfrac{1}{N}\sum_{\boldsymbol{k}_3} J(\boldsymbol{k}_3) k(k_3 - k) k(k_3) \Gamma(k_3 - k, k_3; k_2 + k, k_2) + \\ + J(\boldsymbol{k}_1 - \boldsymbol{k}) \tfrac{1}{N}\sum_{\boldsymbol{k}_3} k(k_3 - k) k(k_3) \Gamma(k_3 - k, k_3; k_2 + k, k_2).\end{aligned} \tag{36}$$

Here

$$k(q) = \frac{1}{\mathrm{i}\omega_n + \varepsilon_{\boldsymbol{q}}}, \quad \varepsilon(\boldsymbol{q}) = \langle S^z \rangle (J(\boldsymbol{0}) - J(\boldsymbol{q})), \tag{37}$$

is the propagator of a spin wave with momentum $\boldsymbol{q}$ and frequency $\omega_n = 2n\pi T$.

Since the integral equation (36) has an degenerated kernel, it reduces to an algebraic one. To this purpose we multiply (36) by $k(k_1 - k)k(k_1)$ and sum both sides of the equation over k_1. Next we multiply (36) by $J(\boldsymbol{k}_1)k(k_1 - k)k(k_1)$ and sum over k_1 also. As a result we get a pair of linear algebraic equations. The solution gives:

$$\begin{aligned}\Gamma(k_1 - k, k_1; k_2 + k, k_2) = \\ = \frac{1}{d(k)}\{\Phi(k) + J(\boldsymbol{k}_1 - \boldsymbol{k})(1 - Q(k)) + J(\boldsymbol{k}_2 + \boldsymbol{k})(1 - \Lambda(k)) + \\ + J(\boldsymbol{k}_1 - \boldsymbol{k}) J(\boldsymbol{k}_2 + \boldsymbol{k}) \Pi(k)\},\end{aligned} \tag{38}$$

where

$$d(k) = (1 - \Lambda(k))(1 - Q(k)) - \Pi(k)\Phi(k). \tag{39}$$

Here we introduced four quantities

$$\begin{pmatrix} \Pi(k) \\ Q(k) \\ \Lambda(k) \\ \Phi(k) \end{pmatrix} = \frac{1}{N}\sum_{k_1} \begin{pmatrix} 1 \\ J(\boldsymbol{k}_1) \\ J(\boldsymbol{k}_1 - \boldsymbol{k}) \\ J(\boldsymbol{k}_1) J(\boldsymbol{k}_1 - \boldsymbol{k}) \end{pmatrix} k(k_1 - k) k(k_1). \tag{40}$$

All results will be expressed through them. These quantities correspond to four loops (Fig. 13).

Then we have to calculate 3-point vertex parts, describing the interaction of spin waves with longitudinal fluctuations. In the loop approximation they are expressed through 4-point vertex parts, Fig. 15 . We find

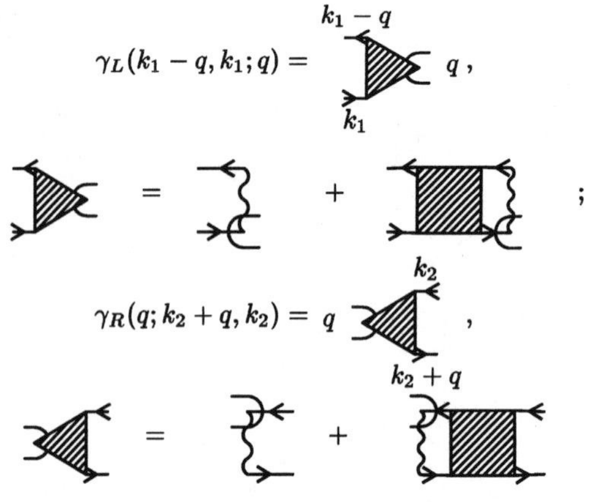

FIGURE 15. Expression 3-point vertex part through 4-point part

$$\gamma_L(k_1-k,k_1;k) = \frac{1}{d(k)}\{\Phi(k)+J(\boldsymbol{k}_1-\boldsymbol{k})(1-Q(k))\}, \tag{41}$$

$$\gamma_R(k;k_2+k,k) = \frac{1}{d(k)}\{\Phi(k)+J(\boldsymbol{k}_2+\boldsymbol{k})(1-\Lambda(k))\}. \tag{42}$$

The expressions found for Γ, γ_L and γ_R, allow to calculate the longitudinal GF $K^{zz}(12)$. To this purpose we must use the graphic representation for its irreducible part Σ^z_{12} (Fig. 16). According to definition it is a set of diagrams for D_{12} can not be cut over

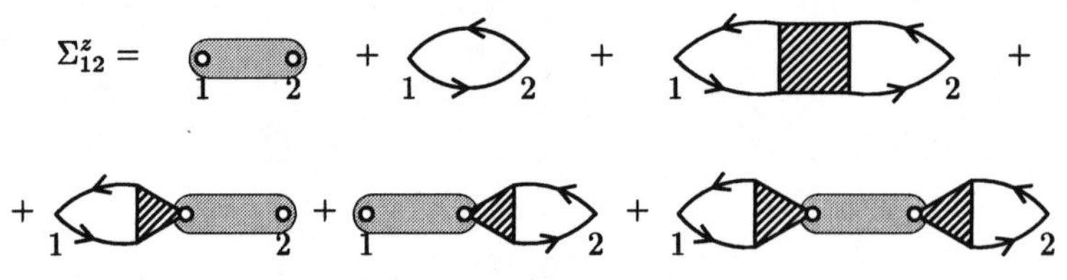

FIGURE 16. Graphic representation of irreducible part of longitudinal Green's function

an interaction line J^z. Here symbol 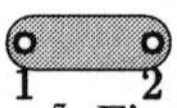 means a thickened object obeyed Dyson equation with irreducible self-energy m^z, Fig. 17. We call it propagator k^z.

In Fig. 16 we take into account all loops, giving contributions in K^{zz}. Two external vertexes 1 and 2 in right hand side are presented by two types of S^z_i vertexes. One is denoted by a light point, another by a point with an ingoing and an outgoing line.

The cumulant  is determined by expression (35). From Fig. 17 follows the analytic expression for the propagator

$$k^z(k) = \frac{\beta b'\delta_{\omega_n 0}}{1-\beta b' m^z(k)}, \tag{43}$$

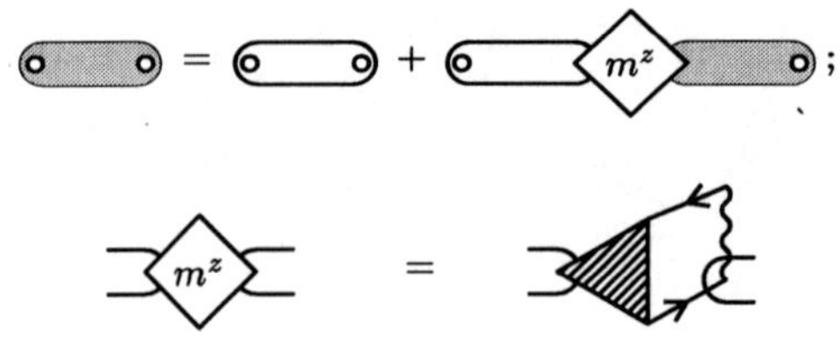

FIGURE 17. Dyson equation for propagator k^z of longitudinal Green's function

where

$$m^z(k) = \frac{\Phi(k)}{d(k)}.$$

The contribution of the third and fourth graphs in Fig. 16 is given by an expression

$$\Pi(k) + \frac{1}{N^2}\sum_{k_1k_2} k(k_1-k)k(k_1)\Gamma(k_1-k,k_1;k_2+k,k_2)k(k_2)k(k_2+k) = \frac{\Pi(k)}{d(k)}.$$

Similarly we find the contribution of the remaining graphs in Fig. 16. Summing all contributions we come to the expression

$$\Sigma^z(k) = \frac{\pi(k)+\Pi(k)}{(1-\Lambda(k))(1-Q(k))-[\pi(k)+\Pi(k)]\Phi(k)}.$$

Remember now that in graphs expressions in Fig. 16 we did not take into account contributions including interactions lines J^z. In that sense, all graphs for Σ^z in Fig. 16 can be regarded as not cut over line J^z, connecting vertexes of S^z type (we denote such lines by dashed lines). The contribution of all graphs for $K^{zz}(k)$ is found with the help of Larkin equation:

$$K^{zz}(k) = \Sigma^z(k) + \Sigma^z(k)J^z(\boldsymbol{k})K^{zz}(k),$$

If we substitute here the expression for $\Sigma^z(k)$, we come to the final expression for longitudinal GF [67, 70]:

$$K^{zz}(k) = \frac{\pi(k)+\Pi(k)}{(1-\Lambda(k))(1-Q(k))-[\pi(k)+\Pi(k)](\Phi(k)+J^z(\boldsymbol{k}))}. \tag{44}$$

Thus, taking all graphs with the lines J^z is reduced to change in Σ^z $\Phi(k) \to \Phi(k)+J(\boldsymbol{k})$.

Notice that expression (44) contains a singularity in the frequency contribution due to the term $\pi(k) = \beta b' \delta_{\omega_n,0}$. The appearance of this contribution in a temperature GF reflects the difference between isolated and isothermic susceptibility of a system. This issue was discussed from a general standpoint in papers [74, 75]. First of all we separate a singular term in the expression (44) writing it in the form

$$K^{zz}(\boldsymbol{q},\mathrm{i}\omega_n) = K_I^{zz}(\boldsymbol{q},\mathrm{i}\omega_n) + \delta_{\omega_n,0}[K_T^{zz}(\boldsymbol{q},0) - K_I^{zz}(\boldsymbol{q},0)], \tag{45}$$

where

$$K_I^{zz}(\boldsymbol{q},\mathrm{i}\omega_n) \equiv K_I^{zz}(q) = \frac{\Pi(q)}{[1-\Lambda(q)][1-Q(q)]-\Pi(q)[\Phi(q)+J^z(\boldsymbol{q})]},$$
$$K_T^{zz}(\boldsymbol{q},0) = \frac{\beta b' + \Pi(\boldsymbol{q},0)}{[1-\Lambda(\boldsymbol{q},0)][1-Q(\boldsymbol{q},0)]-[\beta b' + \Pi(\boldsymbol{q},0)][\Phi(\boldsymbol{q})+J^z(\boldsymbol{q})]}. \tag{46}$$

The intensity of the singular contribution coincides with a difference between isolated K_T^{zz} and isotherm K_I^{zz} susceptibility at zero frequency. It indicates the nonergodicity of a system [74, 75].

We are interested in the isolated susceptibility following from $K_I^{zz}(\boldsymbol{q},\mathrm{i}\omega_n)$ by means of the analytic continuation $\mathrm{i}\omega_n \to \omega + \mathrm{i}\delta$. To calculate momentum and frequency dependence of spin susceptibility $\chi^{zz}(\boldsymbol{q},\omega)$ with the help of expression (46) we must calculate four quantities: $\Lambda(q)$, $Q(q)$, $\Pi(q)$ and $\Phi(q)$, determined by general formulas (40).

An example of numerical calculation of the frequency dependence of the dynamical structure factor $S^{zz}(\boldsymbol{q},\omega) = \frac{J}{\omega}\mathrm{Im}\chi^{zz}(\boldsymbol{q},\omega)$ is given in Fig. 18 for different wave vectors and fixed temperature. We can see a three-peak structure with two wide maxima at frequencies $\pm\omega_{\boldsymbol{q}}$ and a rather narrow central peak of lower intensity. When increasing $\boldsymbol{q}$, the height of the central peak is approaching to the height of $\omega_{\boldsymbol{q}}$-picks, so they have a tendency to form a wide distribution in the spectrum. Positions of the resonance peaks $\omega_{\boldsymbol{q}}$ and its width are linear to q.

The intensity of the $\omega_{\boldsymbol{q}}$-peak depends on temperature and wave vector. It sharply increases when approaching to T_C the more the less q. The tendency to singular behavior of $S^{zz}(\boldsymbol{q},\omega)$ when $q \to 0$ and $T \to T_C$ indicates proximity of hydrodynamic regime. Our GRPA results are valid beyond the hydrodynamic regime, determined the condition

$$q \ll \kappa; \quad \kappa = 1/\xi = \sqrt{1-\tau} \quad (\tau \to 1),$$

meaning that the wave vector is much less than the inverse of the correlation length ξ. The physics of fluctuation in the hydrodynamic regime is determined by conservation laws (spontaneous magnetization) [76]. Therefore behavior of fluctuations with large q, which we describe in framework of the loop approximation, should not be valid at small q.

Notice that in spectrum distribution for a small $q = 0,1$ we have negative region of small frequencies. However for this case $\kappa \approx 0,45$. Thus, it turns out that we are inside of hydrodynamic regime and our GRPA approximation does not work here.

We come to the conclusion that in the region of temperatures under study, $0,6 < \tau < 0,95$, in the dynamical structure factor for longitudinal fluctuations the three-peak structure appears with two close wide picks, which are bringing together and form compact distribution when T tends to T_C. It can be seen in neutron scattering experiments as a single wide peak centered at zero frequency. Perhaps contradictory results in neutron experiments on the study longitudinal fluctuations in ferromagnet (when ones researchers see two peaks and others – one broad maximum) [77] are explained by get different regimes, determined by temperature and parameters of the systems. Obtained results [67] are compatible with numerical calculations of spectral density of longitudinal fluctuations in classic Heisenberg model by Monte-Carlo method [78].

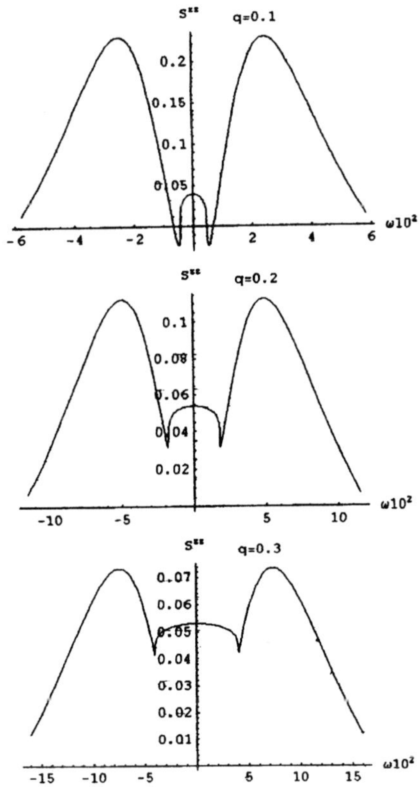

FIGURE 18. Dynamical structure factor for longitudinal fluctuations at $\tau = T/T_C = 0{,}8$ and $q = 0{,}1; 0{,}2; 0{,}3$

B. Method of generating functional

Introduction of the generating functional. Now it is more convenient not to study the isotropic model with Hamiltonian (6), (7), but to include an anisotropy. Then Hamiltonian of model with inhomogeneous field applied along z-axis and exchange anisotropy is written as

$$\mathscr{H} = -\sum_i h_i S_i^z - \sum_{ij} S_i^+ J_{ij} S_j^- - \frac{1}{2}\sum_{ij} S_i^z J_{ij}^z S_j^z. \qquad (47)$$

When $h_i = 0$ and $J_{ij}^z = J_{ij}$ we come back to the initial Hamiltonian.

The most important quantity in statistical mechanics is the partition function

$$Z = \mathrm{Tr}(\mathrm{e}^{-\beta\mathscr{H}}).$$

For the calculation of spin GF in our model it is useful to introduce a generating functional $Z[V]$, which is a generalization of the partition function Z:

$$Z[V] = \mathrm{Tr}(\mathrm{e}^{-\beta\mathscr{H}} T \mathrm{e}^{V}), \tag{48}$$

where V takes into account some fluctuating fields acting on the system. For our can be chosen in the form [66]

$$V = v_{1'} S^z_{1'} + S^+_{1'} u_{1'2'} S^-_{2'}. \tag{49}$$

Here figure indexes, as earlier, denote site number and time, so the introduced fields v_1 and u_{12} depend not only on sites but on time as well. This implies a sum over repeated indexes, more correctly a sum over sites and integration over time. Moreover, the repeated indexes will be marked, as a rule, by primes.

The usefulness of the generating functional (48) is that variational derivatives of it over fields v_1 and u_{12} give spin GF:

$$\begin{aligned} \frac{\delta Z}{\delta v_1} &= \mathrm{Tr}(\mathrm{e}^{-\beta\mathscr{H}} T S^z_1 \mathrm{e}^{V}) \equiv Z[V]\langle T S^z_1 \rangle_V, \\ \frac{\delta Z}{\delta u_{12}} &= \mathrm{Tr}(\mathrm{e}^{-\beta\mathscr{H}} T S^-_1 S^+_2 \mathrm{e}^{V}) \equiv Z[V]\langle T S^-_1 S^+_2 \rangle_V, \\ \frac{\delta^2 Z}{\delta v_1 \delta v_2} &= \mathrm{Tr}(\mathrm{e}^{-\beta\mathscr{H}} T S^z_1 S^z_2 \mathrm{e}^{V}) \equiv Z[V]\langle T S^z_1 S^z_2 \rangle_V, \end{aligned} \tag{50}$$

where

$$\langle \ldots \rangle_V = \frac{\mathrm{Tr}(\mathrm{e}^{-\beta\mathscr{H}} \ldots \mathrm{e}^{V})}{\mathrm{Tr}(\mathrm{e}^{-\beta\mathscr{H}} \mathrm{e}^{V})}.$$

Transversal and longitudinal GF can be obtained putting in (50) the following expressions for the fields v_1 and u_{12}:

$$v_1 = h_{i_1}, \quad u_{12} = J_{i_1 i_2} \delta(\tau_1 - \tau_2).$$

It is convenient to make a substitution

$$Z = \mathrm{e}^{\Phi}, \tag{51}$$

using the functional Φ, which is an analog of the free energy. Then relations (50) are rewritten in the form:

$$\begin{aligned} \frac{\delta \Phi}{\delta v_1} &= \langle S^z_1 \rangle, \\ \frac{\delta \Phi}{\delta u_{12}} &= \langle T S^-_1 S^+_2 \rangle, \\ \frac{\delta^2 \Phi}{\delta v_1 \delta v_2} &= \langle T (S^z_1 - \langle S^z_1 \rangle)(S^z_2 - \langle S^z_2 \rangle) \rangle \equiv \langle T S^z_1 S^z_2 \rangle_c. \end{aligned} \tag{52}$$

In the last expression, the symbol $\langle \ldots \rangle_c$ means the connected GF. Obviously, any GF containing any numbers of operators under the sign of T-product can be obtained by differentiating Φ over fields v and u.

Equation for transversal Green's function. The equation of motion for the operator S_1^- allows to obtain equation for the transversal GF $\langle TS_1^- S_2^+\rangle_V$ in the fluctuating fields. According to recipe in Appendix, one can write

$$\frac{\partial}{\partial\tau_1}((TS_1^- S_2^+ e^V)) = ((T[S_1^-, S_2^+]e^V)) + ((T\{S_1^-, V\}S_2^+ e^V)) + ((T\dot{S}_1^- S_2^+ e^V))$$

Here the double brackets symbol means taking the trace, that is,

$$((TS_1^- S_2^+ e^V)) = \mathrm{Tr}\{e^{-\beta\mathscr{H}} TS_1^- S_2^+ e^V\}.$$

Let us use the equation of motion

$$\dot{S}_i^- = -[S_i^-, \mathscr{H}] = h_i S_i^- - S_i^z \sum_j J_{ij} S_j^- + S_i^- \sum_j J_{ij} S_j^z$$

for operator S_1^- and rewrite it in the form with figure arguments

$$\dot{S}_1^- = h_i S_1^- - \alpha_0(11';3') S_{3'}^z S_{1'}^-, \tag{53}$$

where

$$\alpha_0(12;3) = \delta_{13} J_{12} - \delta_{12} J_{13}^z, \tag{54}$$

and $J_{12} = \delta(\tau_1 - \tau_2) J_{i_1 i_2}$. Use also expression

$$\{S_1^-, V\} = v_1 S_1^- - S_1^z u_{11'} S_{1'}^-,$$

then we can write equation for the transversal GF in the form:

$$-\left(\frac{\partial}{\partial\tau_1} - \tilde{v}_1\right)((TS_1^- S_2^+ e^V)) = \delta_{12}((TS_1^z e^V)) + \tilde{\alpha}_0(11';3')((TS_{1'}^- S_2^+ S_{3'}^z e^V)). \tag{55}$$

Here $\tilde{v}_1 = v_1 + h_1$, and $\tilde{\alpha}_0$ is given by expression (54), in which J_{12} must be changed to $\mathscr{J}_{12} = J_{12} + u_{12}$. All averages of T-products we can change on the derivatives over field v_1 and u_{12}. Because of additivity in expressions for $\tilde{v}_1$ and $\mathscr{J}_{12}$, we can use the derivatives with respect quantity h_1 and $\mathscr{J}_{12}$, and then put $v_1 = u_{12} = 0$:

$$\begin{aligned} ((TS_1^z e^V)) &= \frac{\delta Z}{\delta h_1}, \\ ((TS_1^- S_2^+ e^V)) &= \frac{\delta Z}{\delta \mathscr{J}_{12}}, \\ ((TS_1^- S_2^+ S_3^z e^V)) &= \frac{\delta^2 Z}{\delta \mathscr{J}_{12} \delta h_3}. \end{aligned}$$

Then equation (55) is written in terms of functional derivatives

$$-\left(\frac{\partial}{\partial\tau_1} - h_1\right)\frac{\delta Z}{\delta \mathscr{J}_{12}} = \delta_{12}\frac{\delta Z}{\delta h_1} + \alpha_0(11',3')\frac{\delta^2 Z}{\delta \mathscr{J}_{1'2}\delta h_3'} \tag{56}$$

After substitution (51) we come to the final form of equation for transversal GF

$$\left[k^{(0)-1}_{11'}-\left(\alpha_0\frac{\delta\Phi}{\delta h}\right)_{11'}-\left(\alpha_0\frac{\delta}{\delta h}\right)_{11'}\right]\frac{\delta\Phi}{\delta\mathscr{J}_{1'2}}=\delta_{12}\frac{\delta\Phi}{\delta h_2}, \tag{57}$$

where

$$k^{(0)-1}_{11'}=-\left(\frac{\partial}{\partial\tau_1}-h_1\right)\delta_{11'} \tag{58}$$

(we use in the following a convenient notation: for example, for the quantity X depending only on a single index, $(\alpha_0 X)_{12}\equiv\alpha_0(12;3')X_{3'}$). In (57) h_1 and $\mathscr{J}_{12}$ are fields of general type. Real GF is found from the solution of equation (57). After differentiation over fields, they should be taken equal to particular values, corresponding to the Hamiltonian, that is $\mathscr{J}_{12}=\delta(\tau_1-\tau_2)J_{i_1i_2}$ and $h_1=h_i$.

Thus, equation (57) may be regarded as an equation for the generating functional $\Phi=\Phi[h,J,J^z]$ in arbitrary fluctuating fields. Derivatives over h_1 and $\mathscr{J}_{12}$ determine different spin GFs of the system in the fields:

$$\frac{\delta\Phi}{\delta h_1} = \langle S^z_1\rangle_V, \tag{59}$$

$$\frac{\delta\Phi}{\delta\mathscr{J}_{12}} = \langle TS^-_1S^+_2\rangle_V, \tag{60}$$

$$\frac{\delta^2\Phi}{\delta h_1\delta h_2} = \langle TS^z_1S^z_2\rangle_{cV}. \tag{61}$$

The basic equation (57) can be differentiated over h, and then a chain of subsequent equations is generated. Dealing with this system of equation can be one of the possible approaches to the analysis of the model. The other approach is based directly on the basic equation.

Let us look for a solution of equation (57) in the form [65, 70]

$$\frac{\delta\Phi}{\delta\mathscr{J}_{12}}=k_{11'}P_{1'2}. \tag{62}$$

By substitution of (62) in equation (57), we obtain a couple of equations for the determination of the quantities k_{12} and P_{12}

$$k^{-1}_{12}=k^{(0)-1}_{12}-\left(\alpha_0\frac{\delta\Phi}{\delta h}\right)_{12}+k_{4'3'}\left(\alpha_0\frac{\delta}{\delta h}\right)_{14'}k^{-1}_{3'2}, \tag{63}$$

$$P_{12}=\delta_{12}\frac{\delta\Phi}{\delta h_2}+k_{4'3'}\left(\alpha_0\frac{\delta}{\delta h}\right)_{14'}P_{3'2}. \tag{64}$$

By iterating this equations we obtain a series for P_{12} (the terminal part) and one for k_{12} (the Dyson propagator). The reason for the latter definition is the expression for $k^{(0)}_{12}$, which is a propagator for the transversal GF without interaction.

For the propagator k_{12} we introduce the self energy part M_{12} with the help of standard definition.

$$k_{12}^{-1} = k_{12}^{(0)-1} - M_{12}. \tag{65}$$

Then from (63) equation for M follows:

$$M_{12} = \left(\alpha_0 \frac{\delta\Phi}{\delta h}\right)_{12} - \alpha_0(11';2)k_{1'2} + k_{4'3'}\left(\alpha_0 \frac{\delta}{\delta h}\right)_{14'} M_{3'2}. \tag{66}$$

By combining equations (63) and (64) one can show that M_{12} may be written in the form

$$M_{12} = m_{12} + P_{11'}J_{1'2}. \tag{67}$$

In terms of a diagrammatic technique we call the value m_{12} not-cut over the interaction line J_{12}. Contribution of cut-able diagrams is given by the second term in (67).

It is easy to show that equation (62) may be written in the form

$$K_{12} = \Sigma_{12} + \Sigma_{11'}J_{1'2'}K_{2'2}, \tag{68}$$

where we denote $K_{12} \equiv K_{12}^{-+}$ the transversal GF and introduce the quantity Σ_{12} determined by relations

$$\Sigma_{12} = (k^{ir}P)_{12}, \quad (k^{ir})_{12}^{-1} = k_{12}^{(0)-1} - m_{12}.$$

Obviously Σ_{12} is not cut-able over the interaction line, therefore equation (68) is the Larkin equation for transversal GF.

Expansion of the generating functional and connection with diagrammatic technique for spin operators. To construct a perturbation theory for spin GFs we must to develop a procedure, consisting of three stages [66].

1. It is necessary at first to obtain a formal expansions for spin GFs in powers of the exchange interactions $\mathscr{J}$ and $\mathscr{J}^z$. For the transversal GF it is done by iteration of equations (63) and (64), however it does not solve the problem, because the iteration terms contains variational derivatives of Φ over h.
2. For calculating them we suggest the expansion of the generating functional into a Taylor series in powers of $\mathscr{J}$ and $\mathscr{J}^z$. In the proximity of $\mathscr{J} = \mathscr{J}^z = 0$ coefficients in this series are variational derivatives Φ with respect to fields $\mathscr{J}$ and $\mathscr{J}^z$, evaluated then in a point $\mathscr{J} = \mathscr{J}^z = 0$. We can calculate them by taking derivatives of equation (62) over fields $\mathscr{J}$ and $\mathscr{J}^z$. Then we must substitute obtained expressions for the coefficients in the Taylor series.
3. By subsequent differentiation over the field h with the help of definition (59) and (61) we find an expansion for average spin and longitudinal GF. If now we substitute expansions of that quantities in the iteration terms of equations (63) and (64), we finally obtain expansions for transversal GF as well.

Such a sophisticated way is an absolute necessity because all involved quantities are tightly connected with each other through their equation of motion. In this chapter we achieve this procedure for the calculation of quantities (59) – (61) up to the second order with respect to $\mathscr{J}$. We shall see that the obtained series are exactly equivalent to the diagrammatic technique for spin operators, constructed by standard way with use of the Wick theorem [46, 54]. By the method of the generating functional the rule for the diagrammatic technique is derived in a much easier way. At the same time the approach with variational derivatives operates with "exact" (dressed) GF, but not with bare ones as in the standard diagrammatic technique.

The generating functional $\Phi[V]$ depends on two fluctuating fields h and $\mathscr{J}$, related to a real magnetic field h and the exchange interaction J_{ij}; both are parameters of Hamiltonian. Moreover, the Hamiltonian of the anisotropic model depends also on the exchange parameter J^z_{ij}, and for this reason we introduce a third fluctuating field $\mathscr{J}^z$ by including in V a term $S^z_{1'}\mathscr{J}^z_{1'2'}S^z_{2'}$. Now $\Phi[V]$ depends on three fields h, $\mathscr{J}$ and $\mathscr{J}^z$ and we expand $\Phi[V]$ in Taylor series with respect to $\mathscr{J}$ and $\mathscr{J}^z$

$$\begin{aligned}\Phi[h,\mathscr{J},\mathscr{J}^z] = \Phi_0[h] + \mathscr{J}_{2'1'}\left(\frac{\delta\Phi}{\delta\mathscr{J}_{1'2'}}\right)^0 + \mathscr{J}^z_{2'1'}\left(\frac{\delta\Phi}{\delta\mathscr{J}^z_{1'2'}}\right) + \\ +\frac{1}{2!}\left[\mathscr{J}_{2'1'}\mathscr{J}_{4'3'}\left(\frac{\delta^2\Phi}{\delta\mathscr{J}_{1'2'}\delta\mathscr{J}_{3'4'}}\right)^0 + \mathscr{J}^z_{2'1'}\mathscr{J}^z_{4'3'}\left(\frac{\delta^2\Phi}{\delta\mathscr{J}^z_{1'2'}\delta\mathscr{J}^z_{3'4'}}\right)^0 + \right. \\ \left. +2\mathscr{J}_{2'1'}\mathscr{J}^z_{4'3'}\left(\frac{\delta^2\Phi}{\delta\mathscr{J}_{1'2'}\delta\mathscr{J}^z_{3'4'}}\right)^0\right] + \frac{1}{3!}\cdots\end{aligned} \tag{69}$$

The symbol $(\ldots)^0$ means that after differentiation we put fluctuating fields $\mathscr{J}$ and $\mathscr{J}^z$ equal to the exchange interaction in the system.

It is convenient to express the derivative over the field $\mathscr{J}^z$ through the field h, with the help of the equation of connection

$$\frac{\delta\Phi}{\delta\mathscr{J}^z_{12}} = \frac{1}{2}\left(\frac{\delta^2\Phi}{\delta h_1\delta h_2} + \frac{\delta\Phi}{\delta h_1}\frac{\delta\Phi}{\delta h_2}\right),$$

which can be verified by direct differentiation of (48). By differentiating this equation once more we obtain the expression for the second derivative over $\mathscr{J}^z$:

$$\frac{\delta^2\Phi}{\delta\mathscr{J}^z_{12}\delta\mathscr{J}^z_{34}} = \frac{1}{4}\left(\frac{\delta^2}{\delta h_1\delta h_2} + \frac{\delta\Phi}{\delta h_1}\frac{\delta}{\delta h_2} + \frac{\delta\Phi}{\delta h_2}\frac{\delta}{\delta h_1}\right)\left(\frac{\delta^2\Phi}{\delta h_3\delta h_4} + \frac{\delta\Phi}{\delta h_3}\frac{\delta\Phi}{\delta h_4}\right),$$

and so on. Thus in the series (69) only derivatives with respect $\mathscr{J}$ and h are left.

Derivatives of Φ over $\mathscr{J}$ are calculated by differentiation of equation (62), derivatives of Φ over h are found by differentiation of the Taylor series (69). In doing this calculation one has to use equations (63) and (64), and the identities

$$\frac{\delta k_{12}}{\delta h_3} = -k_{11'}\frac{\delta k^{-1}_{1'2'}}{\delta h_3}k_{2'2}, \quad \frac{\delta k^{(0)}_{12}}{\delta h_3} = -k^{(0)}_{13}k^{(0)}_{32}; \tag{70}$$

$$\frac{\delta}{\delta\mathscr{J}_{34}}\alpha_0(12;5) = \delta_{15}\delta_{14}\delta_{32}.$$

We shall show how this algorithm works on an example of calculation of the average spin $\langle S_1^z \rangle_V$ with accuracy up to second order with respect to $\mathcal{J}$ and $\mathcal{J}^z$. Use relation (59) and substitute in it the expansion (69). We have

$$\langle S_1^z \rangle = \frac{\delta \Phi}{\delta h_1} = \frac{\delta \Phi_0}{\delta h_1} + \mathcal{J}_{2'1'} \frac{\delta}{\delta h_1} \left(\frac{\delta \Phi}{\delta \mathcal{J}_{1'2'}} \right)^0 + \mathcal{J}^z_{2'1'} \frac{\delta}{\delta h_1} \left(\frac{\delta \Phi}{\delta \mathcal{J}^z_{1'2'}} \right)^0 +$$
$$+ \frac{1}{2!} \left[\mathcal{J}_{2'1'} \mathcal{J}_{4'3'} \frac{\delta}{\delta h_1} \left(\frac{\delta^2 \Phi}{\delta \mathcal{J}_{1'2'} \delta \mathcal{J}_{3'4'}} \right)^0 + 2 \mathcal{J}_{2'1'} \mathcal{J}^z_{4'3'} \frac{\delta}{\delta h_1} \left(\frac{\delta^2 \Phi}{\delta \mathcal{J}_{1'2'} \delta \mathcal{J}^z_{3'4'}} \right)^0 + \right.$$
$$\left. + \mathcal{J}^z_{2'1'} \mathcal{J}^z_{4'3'} \frac{\delta}{\delta h_1} \left(\frac{\delta^2 \Phi}{\delta \mathcal{J}^z_{1'2'} \delta \mathcal{J}^z_{3'4'}} \right)^0 \right] + \dots \tag{71}$$

Notice that the GF of zeroth order is local in the site space ($k_{12}^{(0)} \sim \delta_{12}$), therefore the second, third and fifth terms in the right hand side are equal to zero, because exchange integrals J_{ij} and J_{ij}^z are not local. The second derivative over $\mathcal{J}$ in the zeroth order approximation is calculated by differentiation of equation (62) over $\mathcal{J}$

$$\frac{\delta^2 \Phi}{\delta \mathcal{J}_{12} \delta \mathcal{J}_{34}} = -k_{11'} \left(\frac{\delta k^{-1}_{1'2'}}{\delta \mathcal{J}_{34}} \right) (kP)_{2'2} + k_{11'} \frac{\delta P_{1'2}}{\delta \mathcal{J}_{34}}. \tag{72}$$

Derivatives of k^{-1} and P over $\mathcal{J}$ can be find from equations, obtained by differentiation of equation (63)) and (64)):

$$\begin{aligned} \frac{\delta k^{-1}_{12}}{\delta \mathcal{J}_{34}} &= -\delta_{14}\delta_{32} \frac{\delta \Phi}{\delta h_1} + \delta_{14} k_{33'} \frac{\delta k^{-1}_{3'2}}{\delta h_4} - k_{4'1'} \frac{\delta k^{-1}_{1'2'}}{\delta \mathcal{J}_{34}} k_{2'3'} \left(\alpha_0 \frac{\delta}{\delta h} \right)_{14'} k^{-1}_{3'2} + \\ &+ k_{4'3'} \left(\alpha_0 \frac{\delta}{\delta h} \right)_{14'} \frac{k^{-1}_{3'2}}{\delta \mathcal{J}_{34}} - \alpha_0(12;3') \frac{\delta^2 \Phi}{\delta h_{3'} \delta \mathcal{J}_{34}}, \end{aligned}$$

$$\begin{aligned} \frac{\delta P_{12}}{\delta \mathcal{J}_{34}} &= -\delta_{12} \left(k_{33'} \frac{\delta k^{-1}_{3'4'}}{\delta h_2} \frac{\delta \Phi}{\delta \mathcal{J}_{4'4}} - k_{33'} \frac{\delta P_{3'4}}{\delta h_2} \right) + \delta_{14}\delta_{33'} \frac{\delta P_{3'2}}{\delta h_4} - \\ &- k_{4'1'} \frac{\delta k^{-1}_{1'2'}}{\delta \mathcal{J}_{34}} k_{2'3'} \left(\alpha_0 \frac{\delta}{\delta h} \right)_{14'} P_{3'2} + k_{4'3'} \left(\alpha_0 \frac{\delta}{\delta h} \right)_{14'} \frac{P_{3'2}}{\delta \mathcal{J}_{34}}. \end{aligned}$$

Here we used an identity (70). Now from (72) it follows

$$\begin{aligned} \left(\frac{\delta^2 \Phi}{\delta \mathcal{J}_{12} \delta \mathcal{J}_{34}} \right)^0 &= k^{(0)}_{14} \left(\frac{\delta \Phi}{\delta h_4} \right)^0 k^{(0)}_{32} \left(\frac{\delta \Phi}{\delta h_2} \right)^0 - k^{(0)}_{14} k^{(0)}_{34} k^{(0)}_{42} \left(\frac{\delta \Phi}{\delta h_2} \right)^0 - \\ &- k^{(0)}_{12} k^{(0)}_{32} k^{(0)}_{24} \left(\frac{\delta \Phi}{\delta h_4} \right)^0 + (k^{(0)}_{12} k^{(0)}_{34} + k^{(0)}_{14} k^{(0)}_{32}) \left(\frac{\delta^2 \Phi}{\delta h_2 \delta h_4} \right)^0. \end{aligned} \tag{73}$$

The second derivative of Φ over $\mathcal{J}^z$ in (71) is expressed through the derivatives of Φ over field h, which at zeroth approximation are local as well. As a result, expression (71) is given by

$$\begin{aligned}\langle S^z_1\rangle = \frac{\delta\Phi_0}{\delta h_1} + \mathscr{J}^z_{2'1'}\left(\frac{\delta^2\Phi}{\delta h_1\delta h_{1'}}\right)^0\frac{\delta\Phi_0}{\delta h_{2'}} - k^{(0)}_{11'}\frac{\delta\Phi_0}{\delta h_{1'}}\mathscr{J}_{1'2'}k^{(0)}_{2'3}\frac{\delta\Phi_0}{\delta h_{3'}}\mathscr{J}_{3'4'}k^{(0)}_{4'1}+ \\ +\left(\frac{\delta^2\Phi}{\delta h_1\delta h_{1'}}\right)^0(\mathscr{J}k^{(0)})_{1'2'}\frac{\delta\Phi_0}{\delta h_{2'}}(\mathscr{J}k^{(0)})_{2'1'}+ \\ +\mathscr{J}^z_{2'1'}\mathscr{J}^z_{4'3'}\left[\frac{1}{2}\left(\frac{\delta^3\Phi}{\delta h_1\delta h_{1'}\delta h_{3'}}\right)^0\left(\frac{\delta^2\Phi}{\delta h_{2'}\delta h_{4'}}\right)^0+\right. \\ \left.+\left(\frac{\delta^2\Phi}{\delta h_1\delta h_{1'}}\right)^0\frac{\delta\Phi_0}{\delta h_{3'}}\left(\frac{\delta^2\Phi}{\delta h_{2'}\delta h_{4'}}\right)^0+\frac{1}{2}\frac{\delta\Phi_0}{\delta h_{1'}}\frac{\delta\Phi_0}{\delta h_{3'}}\left(\frac{\delta^3\Phi}{\delta h_1\delta h_{2'}\delta h_{4'}}\right)^0\right]+\ldots\end{aligned} \tag{74}$$

Here and further the first derivative of Φ over h in the zeroth approximation is denoted by $\frac{\delta\Phi_0}{\delta h_1}$, where $\Phi_0(h)$ is the first term in the Taylor series (69).

We introduce a graphic representation for the obtained series, Fig. 19. Then series

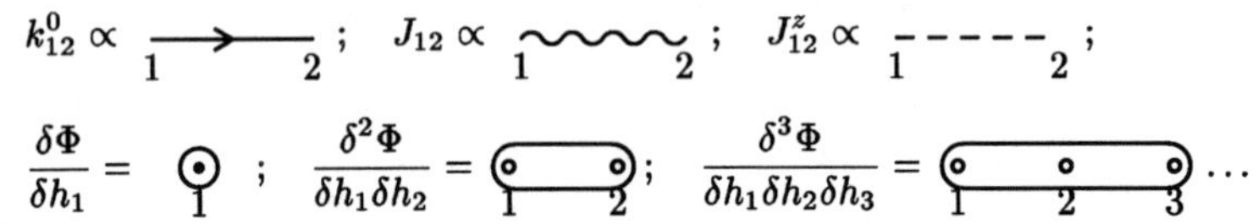

FIGURE 19. Graphic elements for Green's functions

(74) is given in Fig. 20. In some graphs here singletail diagrams exists. From chapter

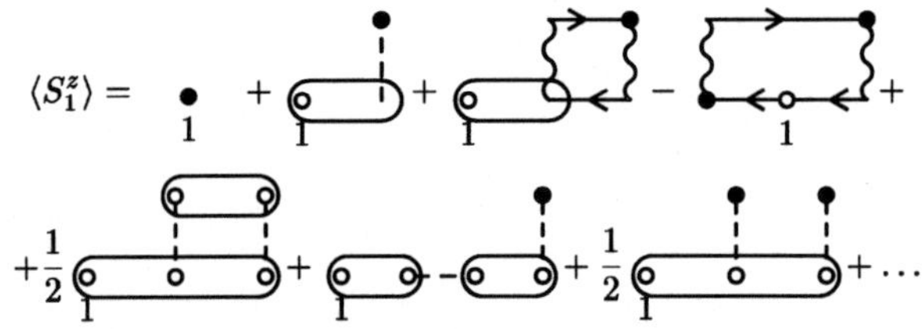

FIGURE 20. Series (74) for average spin

IIA we know that all such graphs can be summed and this leads to the renormalization of the argument of the quantity $\Phi_0(h)$, namely, the external field h should be changed to an effective field $\tilde{h}$. Thus graphs in the series for S^z_1 with singletail diagrams can be dropped, but everywhere we must replace $h \to \tilde{h}$.

The relation determines the longitudinal GF

$$\langle TS^z_1S^z_2\rangle_c = \langle T(S^z_1 - \langle S^z_1\rangle)(S^z_2 - \langle S^z_2\rangle)\rangle = \frac{\delta^2\Phi}{\delta h_1\delta h_2}. \tag{75}$$

The expansion for it can be obtained by differentiation of the series (74) over field h. The differentiation of functions $k^{(0)}_{12}$ over h is given with the help of one of the relations (70), but differentiation of Φ simply increase the order of derivatives. As a result we come to an expression where all singletail diagrams are absent:

$$\langle TS^z_1S^z_2\rangle_c = \frac{\delta^2\Phi_0}{\delta h_1\delta h_2} + \mathscr{J}^z_{2'1'}\left(\frac{\delta^2\Phi}{\delta h_1\delta h_{1'}}\right)^0\left(\frac{\delta^2\Phi}{\delta h_2\delta h_{2'}}\right)^0 + \tag{76}$$

$$
+\left(k^{(0)}\frac{\delta\Phi_0}{\delta h}\mathscr{I}k^{(0)}\right)_{12}\left(k^{(0)}\frac{\delta\Phi_0}{\delta h}\mathscr{I}k^{(0)}\right)_{21}+
$$

$$
+k^{(0)}_{12}\left(k^{(0)}\frac{\delta\Phi_0}{\delta h}\mathscr{I}k^{(0)}\frac{\delta\Phi_0}{\delta h}\mathscr{I}k^{(0)}\right)_{21}+\left(k^{(0)}\frac{\delta\Phi_0}{\delta h}\mathscr{I}k^{(0)}\frac{\delta\Phi_0}{\delta h}\mathscr{I}k^{(0)}\right)_{12}k^{(0)}_{21}-
$$

$$
-\left(k^{(0)}\frac{\delta\Phi_0}{\delta h}\mathscr{I}k^{(0)}\right)_{11'}\left(\frac{\delta^2\Phi}{\delta h_2\delta h_{1'}}\right)^0(\mathscr{I}k^{(0)})_{1'1}-
$$

$$
-\left(\frac{\delta^2\Phi}{\delta h_1\delta h_{1'}}\right)^0(\mathscr{I}k^{(0)})_{1'2}\left(k^{(0)}\frac{\delta\Phi_0}{\delta h}\mathscr{I}k^{(0)}\right)_{21'}-
$$

$$
-k^{(0)}_{11'}\left(\frac{\delta^2\Phi}{\delta h_2\delta h_{1'}}\right)^0\left(\mathscr{I}k^{(0)}\frac{\delta\Phi_0}{\delta h}\mathscr{I}k^{(0)}\right)_{1'1}-
$$

$$
-\left(\frac{\delta^2\Phi}{\delta h_1\delta h_{1'}}\right)^0\left(\mathscr{I}k^{(0)}\frac{\delta\Phi_0}{\delta h}\mathscr{I}k^{(0)}\right)_{1'2}k^{(0)}_{21'}+
$$

$$
+\left(\frac{\delta^2\Phi}{\delta h_1\delta h_{1'}}\right)^0(\mathscr{I}k^{(0)})_{1'2'}\left(\frac{\delta^2\Phi}{\delta h_2\delta h_{2'}}\right)^0(\mathscr{I}k^{(0)})_{2'1'}+
$$

$$
+\left(\frac{\delta^3\Phi}{\delta h_1\delta h_2\delta h_{1'}}\right)^0(\mathscr{I}k^{(0)})_{1'2'}\left(\frac{\delta\Phi_0}{\delta h_{2'}}\right)(\mathscr{I}k^{(0)})_{2'1'}+
$$

$$
+\mathscr{I}^z_{2'1'}\mathscr{I}^z_{4'3'}\left[\left(\frac{\delta^2\Phi}{\delta h_1\delta h_{1'}}\right)^0\left(\frac{\delta^2\Phi}{\delta h_2\delta h_{3'}}\right)^0\left(\frac{\delta^2\Phi}{\delta h_{2'}\delta h_{4'}}\right)^0+\right.
$$

$$
+\frac{1}{2}\left(\frac{\delta^4\Phi}{\delta h_1\delta h_2\delta h_{4'}\delta h_{3'}}\right)^0\left(\frac{\delta^2\Phi}{\delta h_{2'}\delta h_{4'}}\right)^0+
$$

$$
\left.+\frac{1}{2}\left(\frac{\delta^3\Phi}{\delta h_1\delta h_{1'}\delta h_{3'}}\right)^0\left(\frac{\delta^3\Phi}{\delta h_2\delta h_{2'}\delta h_{4'}}\right)^0\right]+\ldots
$$

One can verify that the graphic representation of this expression coincides with that shown in Fig. 9.

Expand now the transversal GF. From definition (60) it follows that we need to differentiate the initial Taylor series (69) over $\mathscr{I}$. Up to the second order we have an expression

$$\frac{\delta\Phi}{\delta\mathscr{I}_{12}}=\left(\frac{\delta\Phi}{\delta\mathscr{I}_{12}}\right)^0+\mathscr{I}_{2'1'}\left(\frac{\delta^2\Phi}{\delta\mathscr{I}_{12}\delta\mathscr{I}_{1'2'}}\right)^0+\mathscr{I}^z_{4'3'}\left(\frac{\delta^2\Phi}{\delta\mathscr{I}_{12}\delta\mathscr{I}^z_{3'4'}}\right)^0+$$
$$+\frac{1}{2}\left[\mathscr{I}_{4'3'}\mathscr{I}_{6'5'}\left(\frac{\delta^3\Phi}{\delta\mathscr{I}_{12}\delta\mathscr{I}_{3'4'}\delta\mathscr{I}_{5'6'}}\right)^0+2\mathscr{I}_{4'3'}\mathscr{I}^z_{6'5'}\left(\frac{\delta^3\Phi}{\delta\mathscr{I}_{12}\delta\mathscr{I}_{3'4'}\delta\mathscr{I}^z_{5'6'}}\right)^0+\right.$$
$$\left.+\mathscr{I}^z_{4'3'}\mathscr{I}^z_{6'5'}\left(\frac{\delta^3\Phi}{\delta\mathscr{I}_{12}\delta\mathscr{I}^z_{3'4'}\delta\mathscr{I}^z_{5'6'}}\right)^0\right]+\ldots$$

Derivatives of Φ over $\mathscr{I}$ are calculated by the same method we have used earlier, namely, we use the multiplicative representation (62) for GF $\frac{\delta\Phi}{\delta\mathscr{I}_{12}}$ and take the derivative of the product of the two functions $k_{11'}$ and $P_{1'2}$. Since we must calculate the third derivative over $\mathscr{I}$, the calculation of the transversal GF is longer. As a result of now standard calculations we come to the following expression

$$\frac{\delta\Phi}{\delta\mathscr{I}_{12}}=$$
$$=k^{(0)}_{12}\frac{\delta\Phi_0}{\delta h_2}+\left(k^{(0)}\frac{\delta\Phi_0}{\delta h}\mathscr{I}k^{(0)}\frac{\delta\Phi_0}{\delta h}\right)_{12}+\left(k^{(0)}\frac{\delta\Phi_0}{\delta h}\mathscr{I}k^{(0)}\frac{\delta\Phi_0}{\delta h}\mathscr{I}k^{(0)}\frac{\delta\Phi_0}{\delta h}\right)_{12}-$$
$$-k^{(0)}_{11'}\left(\mathscr{I}k^{(0)}\frac{\delta\Phi_0}{\delta h}\mathscr{I}k^{(0)}\right)_{1'1'}k^{(0)}_{1'2}\frac{\delta\Phi_0}{\delta h_2}-k^{(0)}_{12}\left(k^{(0)}\frac{\delta\Phi_0}{\delta h}\mathscr{I}k^{(0)}\frac{\delta\Phi_0}{\delta h}\mathscr{I}k^{(0)}\right)_{22}+$$
$$+k^{(0)}_{12}\frac{\delta^2\Phi_0}{\delta h_2\delta h_{2'}}\left(\mathscr{I}k^{(0)}\frac{\delta\Phi_0}{\delta h}\mathscr{I}k^{(0)}\right)_{2'2'}+k^{(0)}_{11'}\frac{\delta^2\Phi_0}{\delta h_{1'}\delta h_2}\left(\mathscr{I}k^{(0)}\frac{\delta\Phi_0}{\delta h}\mathscr{I}k^{(0)}\right)_{1'2}+$$
$$+\mathscr{I}^z_{6'5'}k^{(0)}_{15'}\left(k^{(0)}\frac{\delta\Phi_0}{\delta h}\mathscr{I}k^{(0)}\right)_{5'6'}k^{(0)}_{6'2}\frac{\delta\Phi_0}{\delta h_2}-$$
$$-\mathscr{I}^z_{6'5'}k^{(0)}_{15'}\left(k^{(0)}\frac{\delta\Phi_0}{\delta h}\mathscr{I}k^{(0)}\right)_{5'2}\frac{\delta^2\Phi_0}{\delta h_2\delta h_{6'}}-$$
$$-k^{(0)}_{14'}\frac{\delta^2\Phi_0}{\delta h_{4'}\delta h_{5'}}\mathscr{I}_{4'3'}\mathscr{I}^z_{6'5'}k^{(0)}_{3'5'}k^{(0)}_{5'2}\frac{\delta\Phi_0}{\delta h_2}+$$
$$+k^{(0)}_{14'}\frac{\delta^2\Phi_0}{\delta h_{4'}\delta h_{5'}}\mathscr{I}^z_{4'3'}\mathscr{I}^z_{6'5'}k^{(0)}_{3'5'}k^{(0)}_{5'2}\frac{\delta^2\Phi_0}{\delta h_2\delta h_{6'}}+$$
$$+k^{(0)}_{11'}k^{(0)}_{1'3'}k^{(0)}_{3'2}\frac{\delta\Phi_0}{\delta h_2}\mathscr{I}^z_{1'2'}\frac{\delta^2\Phi_0}{\delta h_{2'}\delta h_{4'}}\mathscr{I}^z_{4'3'}-\mathscr{I}^z_{2'1}\mathscr{I}^z_{4'3'}k^{(0)}_{13'}k^{(0)}_{3'2}\frac{\delta^2\Phi_0}{\delta h_{2'}\delta h_{4'}}\frac{\delta^2\Phi_0}{\delta h_{1'}\delta h_2}+$$
$$+\frac{1}{2}k^{(0)}_{12}\frac{\delta^3\Phi_0}{\delta h_2\delta h_{1'}\delta h_{3'}}\mathscr{I}^z_{3'4'}\mathscr{I}^z_{1'2'}\frac{\delta^2\Phi_0}{\delta h_{2'}\delta h_{4'}}+\ldots \qquad (77)$$

Graphic representation of the obtained series coincides with that shown in Fig. 8.

We have shown that the graphic representation of the analytical contribution in quantities $\langle S_1^z \rangle$, $\langle T S_1^z S_2^z \rangle_c$ and $\langle T S_1^- S_2^+ \rangle$ practically coincides with that, obtained with use of the Wick theorem for the Heisenberg model [54]. In the standard approach based on the Wick theorem the process of writing perturbation terms in GF reduces to completely taking all the systems of the operator pairings. In the method of the generating functional the process is reduced to the differentiation of products of GFs, which is rather automatic and is less risky to lose something. Moreover, in this approach topologically equivalent diagrams do not appear: their total contribution is taken into account automatically.

Self-consistent perturbation theory for transversal Green's function. Let us look now which way the GF $\frac{\delta \Phi}{\delta \mathscr{J}_{12}}$ can be calculated directly from exact equations (64) and (66). Upon iteration this equations produce formal series in powers of interactions. In the limit of second order we obtain expansions

$$
\begin{aligned}
M_{12} &= \alpha_0(12;4')\frac{\delta \Phi}{\delta h_{4'}} - \alpha_0(12';2)k_{2'2} + \alpha_0(12';4')k_{2'3}\alpha_0(3'2;6')\frac{\delta^2 \Phi}{\delta h_{4'}\delta h_{6'}} + \\
&+ \alpha_0(12';4')k_{2'3'}\alpha_0(3'1';2)k_{1'4'}k_{4'2} + \ldots,
\end{aligned} \tag{78}
$$

$$
\begin{aligned}
P_{12} &= \delta_{12}\frac{\delta \Phi}{\delta h_2} + \alpha_0(11';2')\frac{\delta^2 \Phi}{\delta h_{2'}\delta h_2}k_{1'2} - \\
&- k_{4'1'}\alpha_0(13';2')\alpha_0(14';6')\frac{\delta^2 \Phi}{\delta h_{2'}\delta h_2}k_{3'6'}k_{6'2} + \\
&+ k_{4'1'}\alpha_0(14';6')\alpha_0(1'3';2')\frac{\delta^3 \Phi}{\delta h_{2'}\delta h_{6'}\delta h_2}k_{3'2} + \ldots
\end{aligned} \tag{79}
$$

Terms of this expansions include exact propagators k and many particles longitudinal GFs. They should be calculated with the help of some approximate procedure.

After unpacking the expression (54) for α_0, expansions (78) and (79) have the form

$$
\begin{aligned}
M_{12} = &\frac{\delta \Phi}{\delta h_1}J_{12} - \delta_{12}J_{12'}\frac{\delta \Phi}{\delta h_{2'}} - \delta_{12}(Jk)_{11} + J^z_{12}k_{12} + \\
&+ J_{12'}\frac{\delta^2 \Phi}{\delta h_1 \delta h_{3'}}k_{2'3'}J_{3'2} - J_{12'}k_{2'2}J^z_{23'}\frac{\delta^2 \Phi}{\delta h_1 \delta h_{3'}} - \\
&- J^z_{12'}\frac{\delta^2 \Phi}{\delta h_{2'}\delta h_{3'}}k_{13'}J_{3'2} + J^z_{12'}k_{12}\frac{\delta^2 \Phi}{\delta h_{2'}\delta h_{3'}}J^z_{23'} + \\
&+ J_{12'}k_{2'2}J_{23'}k_{3'1}k_{12} - J_{12'}k_{2'1'}J^z_{1'2}k_{1'1}k_{12} - \\
&- J^z_{12'}k_{12}J_{23'}k_{3'2'}k_{2'2} + J^z_{12'}k_{2'2}k_{13'}J^z_{3'2}k_{3'2'} + \ldots
\end{aligned} \tag{80}
$$

We write down the final part only up to the first order

$$
P_{12} = \delta_{12}\frac{\delta \Phi}{\delta h_2} + (Jk)_{12}\frac{\delta^2 \Phi}{\delta h_1 \delta h_2} - k_{12}J^z_{12'}\frac{\delta^2 \Phi}{\delta h_{2'}\delta h_2} + \ldots \tag{81}
$$

It easy to verify that the graphs for M and P, Fig.21, satisfy all rules of the standard diagrammatic technique derived with the help of the Wick theorem.

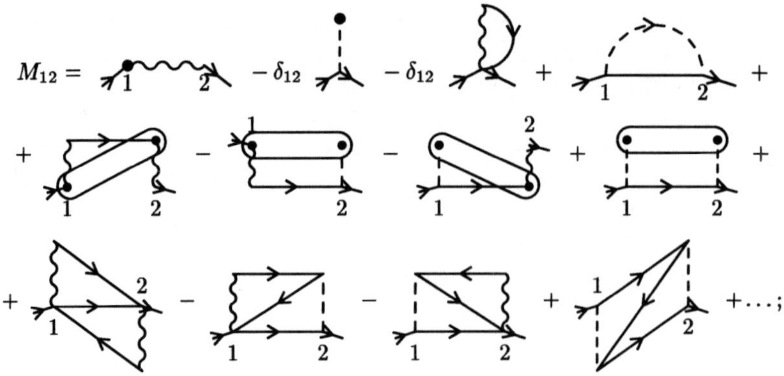

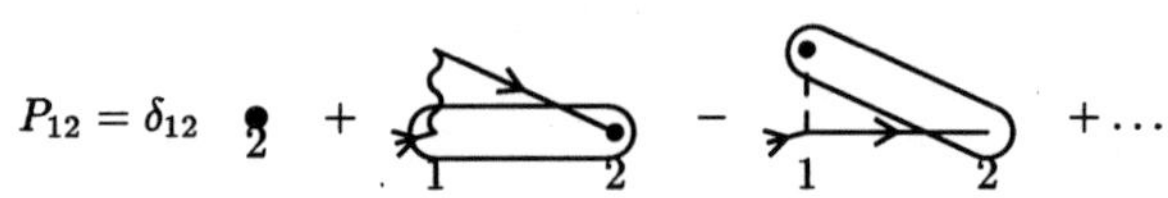

FIGURE 21. Graphic representation of self energy and terminal part for the transversal Green's function.
(Ingoing and outgoing Green's line do not belong M_{12}, as well as ingoing line do not belong P_{12})

In momentum representation the second order correction is given by

$$M_k^{(2)} = \sum_{k_1}(J(\boldsymbol{k}+\boldsymbol{k}_1) - J^z(\boldsymbol{k}_1))(J(\boldsymbol{k}) - J^z(\boldsymbol{k}_1))k(k+k_1)K^{zz}(k_1) - \\ - \sum_{k_1 k_2}(J(\boldsymbol{k}_1) - J^z(\boldsymbol{k}_1 - \boldsymbol{k}))(J^z(\boldsymbol{k}_2 - \boldsymbol{k}) - J(\boldsymbol{k}_1 + \boldsymbol{k}_2 - \boldsymbol{k}))\times \\ \times k(k_1)k(k_2)k(k_1+k_2-k). \tag{82}$$

Here $k(k)$ and $K^{zz}(k)$ are the Fourier components of the Dyson propagator of transversal GF k_{12} and the longitudinal GF $\frac{\delta^2\Phi}{\delta h_1 \delta h_2}$, respectively. For the final calculation of (82) it is necessary to use an approximation for this GFs. We take $k(k)$ in Hartree-Fock approximation. The first order correction of $M_{\boldsymbol{k}}^{(1)}$ just determines the transversal GF in this approximation:

$$k(k) \equiv k_{\boldsymbol{k}}(\mathrm{i}\omega_n) = \frac{1}{\mathrm{i}\omega_n + \varepsilon_{\boldsymbol{k}}}, \qquad \omega_n = 2\pi n T. \tag{83}$$

It describes the propagation of a spin wave with momentum $\boldsymbol{k}$ and energy $\varepsilon_{\boldsymbol{k}}$ (see expression (37)). We find the longitudinal GF $K^{zz}(k)$ at the same approximation from Larkin equation. According to the graphic series, Fig. 12, the zeroth approximation for Σ^z with dressed transversal GF, corresponding to expression (83), is determined by two terms:

$$\Sigma^z(k) \equiv \Sigma^z_{\boldsymbol{k}}(i\omega_n) = \beta b' \delta_{\omega_n 0} + \Pi(k), \tag{84}$$

where

$$\Pi(k) \equiv \sum_q k(q)k(q-k)$$

is the expression for a loop graph, giving contribution from spin waves in the longitudinal spin fluctuations, and b' corresponds to the first graph in Fig. 12.

The first term in (84) determines the static fluctuations and second term the dynamic ones. If the dynamic term $K^{zz}(q) \approx \Sigma^z_q = \Pi_q$, we can substitute it in the expression (82) and see that the first term has the same structure of the second. By combining both terms after summation over frequencies we come to the following expression for the spin wave self energy correction [45, 79]:

$$\begin{aligned} M^{(2)}_{\boldsymbol{k}}(i\omega_n) = \frac{1}{2N^2} \sum_{\boldsymbol{k}_1 \boldsymbol{k}_2} (J^z(\boldsymbol{k}_1 - \boldsymbol{k}) + J^z(\boldsymbol{k}_2 - \boldsymbol{k}) - J(\boldsymbol{k}_1) - J(\boldsymbol{k}_2)) \times \\ \times (J^z(\boldsymbol{k}_1 - \boldsymbol{k}) + J^z(\boldsymbol{k}_2 - \boldsymbol{k}) - J(\boldsymbol{k}_1 + \boldsymbol{k}_2 - \boldsymbol{k}) - J(\boldsymbol{k})) \times \\ \times \frac{n(\varepsilon_{\boldsymbol{k}_1 + \boldsymbol{k}_2 - \boldsymbol{k}})[1 + n(\varepsilon_{\boldsymbol{k}_1}) + n(\varepsilon_{\boldsymbol{k}_2})] - n(\varepsilon_{\boldsymbol{k}_1}) n(\varepsilon_{\boldsymbol{k}_2})}{i\omega_n - \varepsilon_{\boldsymbol{k}_1} - \varepsilon_{\boldsymbol{k}_2} + \varepsilon_{\boldsymbol{k}_1 + \boldsymbol{k}_2 - \boldsymbol{k}}}. \end{aligned} \tag{85}$$

After the analytic continuation $i\omega_n \to \omega + i\delta$, we find from (85) the value of attenuation of a spin wave $\Gamma_{\boldsymbol{k}}(\omega) = -\mathrm{Im} M^{(2)}_{\boldsymbol{k}}(\omega)$ near the mass surface $\omega \approx \varepsilon_{\boldsymbol{k}}$:

$$\begin{aligned} \Gamma_{\boldsymbol{k}}(\omega) == \tfrac{\pi}{2}(e^{\beta\omega} - 1)\tfrac{1}{N^2} \sum_{\boldsymbol{k}_1 \boldsymbol{k}_2} (J^z(\boldsymbol{k}_1 - \boldsymbol{k}) + J^z(\boldsymbol{k}_2 - \boldsymbol{k}) - J(\boldsymbol{k}_1) - J(\boldsymbol{k}_2))^2 \times \\ \times n(\varepsilon_{\boldsymbol{k}_1}) n(\varepsilon_{\boldsymbol{k}_2})[1 + n(\varepsilon_{\boldsymbol{k}_1 + \boldsymbol{k}_2 - \boldsymbol{k}})] \delta(\omega - \varepsilon_{\boldsymbol{k}_1} - \varepsilon_{\boldsymbol{k}_2} + \varepsilon_{\boldsymbol{k}_1 + \boldsymbol{k}_2 - \boldsymbol{k}}). \end{aligned} \tag{86}$$

The linear combination of four values $J(\boldsymbol{q})$ is the amplitude of two spin waves interactions.

If we take into account the static contribution in (84) we obtain an additional contribution for spin wave damping due to the scattering on fluctuations of magnetization [45]:

$$\Gamma^z_{\boldsymbol{k}}(\omega) = \pi b' \sum_{\boldsymbol{q}} \frac{(J(\boldsymbol{q} - \boldsymbol{k}) - J(\boldsymbol{q}))^2}{1 - \beta J(\boldsymbol{k} - \boldsymbol{q}) b'} \delta(\varepsilon_{\boldsymbol{q}} - \omega). \tag{87}$$

VLP [45] studied in details both contributions in spin wave damping in a wide temperature interval up to the proximity of the critical region $T < T_K$. They show that both mechanisms lead to small damping of spin wave for small enough wave vectors $\boldsymbol{k}$.

III. THE HUBBARD MODEL

A. Diagrammatic technique

Perturbation theory with Hubbard operators. Consider a standard Hubbard model for non degenerate states. The Hamiltonian of the model in terms of Fermi operators

$$\mathscr{H} = \sum_{ij\sigma} t_{ij} c_{i\sigma}^{+} c_{j\sigma} + U \sum_{i} n_{i\uparrow} n_{i\downarrow} \tag{88}$$

consists of two terms, describing, respectively, the hopping of electrons on a lattice and the Coulomb repulsion on a site. In the limit $U \gg zt$ it is convenient to set the Coulomb term as the zeroth order approximation Hamiltonian, so that

$$\begin{gathered} \mathscr{H} = \mathscr{H}_0 + \mathscr{H}_{\text{int}}, \\ \mathscr{H}_0 = \sum_i \left\{ -\tfrac{h}{2}(n_{i\uparrow} - n_{i\downarrow}) - \mu(n_{i\uparrow} + n_{i\downarrow}) + U n_{i\uparrow} n_{i\downarrow} \right\}, \\ \mathscr{H}_{\text{int}} = \sum_{ij\sigma} t_{ij} c_{i\sigma}^{+} c_{j\sigma}. \end{gathered}$$

In $\mathscr{H}_0$ we included the magnetic field H, $(h = g\mu_B H)$ and the chemical potential μ.

A site contribution in $\mathscr{H}_0$ is a diagonal matrix

$$\mathscr{H}_0 = \text{diag}\{0, -h/2 - \mu, h/2 - \mu, U - 2\mu\}$$

in the space of states

$$p = |0\rangle, |+\rangle, |-\rangle, |2\rangle,$$

where symbols $|+\rangle$ and $|-\rangle$ denote one-electron states on a site with spin $\uparrow$ or $\downarrow$, while $|0\rangle$ and $|2\rangle$ describe the empty and the double occupied states.

The Hamiltonian (88), under the condition $U \gg zt$, can be conveniently expressed in terms of Hubbard operators X^{pq} [49]. Any operator X_i^{pq}, at site i, describes transfers between the four possible states p. X operators can be expressed through the usual electron operators by means of the relations

$$\begin{gathered} X_i^{\sigma 0} = c_{i\sigma}^{+}(1 - n_{i\bar{\sigma}}), \quad X_i^{2\sigma} = \sigma c_{i\bar{\sigma}}^{+} n_{i\sigma}, \\ X_i^{\sigma\bar{\sigma}} = c_{i\sigma}^{+} c_{i\bar{\sigma}}, \quad X_i^{20} = \sigma c_{i\bar{\sigma}}^{+} c_{i\sigma}^{+}, \\ X_i^{00} = (1 - n_{i\uparrow})(1 - n_{i\downarrow}), \\ X_i^{\sigma\sigma} = n_{i\sigma}(1 - n_{i\bar{\sigma}}), \quad X_i^{22} = n_{i\uparrow} n_{i\downarrow} \end{gathered} \tag{89}$$

and their conjugates $(X_i^{pq})^{+} = X_i^{qp}$, so there are 16 different operators describing all transfers between four on-site states. In a site basis to each operator X_i^{pq} corresponds a 4×4 matrix, whose entries are equal to zero except one which equals 1, placed at the cross of the p-th row and q-th column. Such operators obey particular relations, forming their algebra:

$$X_i^{rs} X_i^{pq} = \delta_{sp} X_i^{rq}.$$

It can be seen from (89) that $X_i^{\sigma 0}$ describes the creation of an electron on site i with spin σ, if the site was empty. $X_i^{2\sigma}$ describes the creation of an electron if the site i was occupied by an electron with spin $\bar{\sigma}$. $X_i^{\sigma\bar{\sigma}}$ switches the electron spin and X_i^{20} provides the creation of a pair of electrons. Thus the operators $X_i^{\sigma 0}$ and $X_i^{2\sigma}$ change the number of electrons on a site by 1, that is why they can be called Fermi-like operators (f-operators). Operators $X_i^{\sigma\bar{\sigma}}$ and X_i^{20} change the number of electrons by 0 or 2, therefore they are called Bose-like operators (b-operators). Notice that the usual electron operators are linear combinations of two Fermi-like X operators

$$c_{i\sigma}^{+} = X_i^{\sigma 0} + \sigma X_i^{2\bar{\sigma}}, \quad c_{i\sigma} = X_i^{0\sigma} + \sigma X_i^{\sigma 2}. \tag{90}$$

In terms of X operators $\mathcal{H}_0$ is a linear form, but $\mathcal{H}_{\text{int}}$ becomes a quadratic form

$$\begin{aligned}\mathcal{H}_0 &= \sum_{ip} \varepsilon_p X_i^{pp}, \\ \mathcal{H}_{\text{int}} &= \sum_{ij} t_{ij} \Big[(X_i^{+0} + X_i^{2-})(X_j^{0+} + X_j^{-2}) + \\ &\quad + (X_i^{-0} + X_i^{2+})(X_j^{0-} + X_j^{+2}) \Big],\end{aligned} \tag{91}$$

where ε_p is an eigenvalue of the diagonal matrix $\mathcal{H}_0$.

X operators satisfy the commutation relations

$$[X_i^{pq}, X_j^{rs}]_{\pm} = (\delta_{qr} X_i^{ps} \pm \delta_{ps} X_i^{rq}) \delta_{ij}, \tag{92}$$

the upper sign should be taken when both operators are f-like, the lower sign is for the case when both or only one operator are of b-type.

The statistical mechanics of the model is described by GFs, which are averages of T-products of X operators

$$\langle T \left(\tilde{X}_{i_1}^{p_1 q_1}(\tau_1) \ldots \tilde{X}_{i_n}^{p_n q_n}(\tau_n) \right) \rangle,$$

where $\tilde{X}_i^{pq}(\tau)$ are the Heisenberg representation of X_i^{pq} with exact Hamiltonian $\mathcal{H}$. This average is calculated in the interaction representation

$$\langle T \left(\tilde{X}_{i_1}^{p_1 q_1}(\tau_1) \ldots \tilde{X}_{i_n}^{p_n q_n}(\tau_n) \right) \rangle = \frac{\langle T \left(X_{i_1}^{p_1 q_1}(\tau_1) \ldots X_{i_n}^{p_n q_n}(\tau_n) \sigma(\beta) \right) \rangle_0}{\langle \sigma(\beta) \rangle_0}, \tag{93}$$

where $\sigma(\beta)$ is the temperature scattering matrix, and $X_i^{pq}(\tau)$ is the operator X_i^{pq} in the interaction representation

$$X_i^{pq}(\tau) = \mathrm{e}^{\tau \mathcal{H}_0} X_i^{pq} \mathrm{e}^{-\tau \mathcal{H}_0}.$$

Averaging in (93) is taken over Gibbs ensemble with Hamiltonian $\mathcal{H}_0$.

It is easy to show that the relation

$$[\mathcal{H}_0, X_i^{pq}] = (\varepsilon_p - \varepsilon_q) X_i^{pq},$$

holds, from which follows the expression for any operator X_i^{pq} in the interaction representation

$$X_i^{pq}(\tau) = \mathrm{e}^{(\varepsilon_p - \varepsilon_q)\tau} X_i^{pq}. \tag{94}$$

Due to the commutation relations (92), the Wick theorem for X operators is fulfilled [52, 51, 50]. This allows to reduce an average of n X operators to a sum of averages with $n-1$ operators. This is similar to relation (12) for spin operators:

$$\begin{aligned}&\langle T\left(X_{i_1}^{p_1q_1}(\tau_1)\ldots X_i^{pq}(\tau)\ldots X_{i_n}^{p_nq_n}(\tau_n)\right)\rangle_0=\\&=\sum_{k=1}^{n}(-1)^{Q_k}g_{ii_k}^{0pq}(\tau-\tau_k)\langle T\left(X_{i_1}^{p_1q_1}(\tau_1)\ldots X_{i_{k-1}}^{p_{k-1}q_{k-1}}(\tau_{k-1})\times\right.\\&\left.\times[X_{i_k}^{p_kq_k},X_i^{pq}]_\pm(\tau_k)X_{i_{k+1}}^{p_{k+1}q_{k+1}}(\tau_{k+1})\ldots X_{i_n}^{p_nq_n}(\tau_n)\right)\rangle_0.\end{aligned}\tag{95}$$

In the right hand side, under the symbol of average, instead of pair operators there is an anticommutator if the operator $X_{i_k}^{p_kq_k}$ is of f-type, or a commutator, if the operator $X_{i_k}^{p_kq_k}$ is of b-type. Here Q_k is the number of permutations of the operator $X_i^{pq}(\tau)$ through all f-operators, if X_i^{pq} is an f-operator. In the case of b-operator we must take $Q_k=0$. Finally, a factor $g_{ii_k}^{0pq}(\tau-\tau_k)$ is a propagator of Fermi or Bose type. Both are diagonal with respect to site indexes:

$$g_{ii'}^{0pq}(\tau)=\delta_{ii'}g^{0pq}(\tau).\tag{96}$$

For f-operators we denote the function $g^{0pq}(\tau)$ by $G_{pq}^0(\tau)$, and for b-operators by $D_{pq}^0(\tau)$. They are, respectively

$$G_{pq}^0(\tau)=-\mathrm{e}^{(\varepsilon_p-\varepsilon_q)\tau}\begin{cases}1-f(\varepsilon_p-\varepsilon_q), & \tau>0,\\ -f(\varepsilon_p-\varepsilon_q), & \tau<0,\end{cases}\tag{97}$$

$$D_{pq}^0(\tau)=-\mathrm{e}^{(\varepsilon_p-\varepsilon_q)\tau}\begin{cases}1-N(\varepsilon_p-\varepsilon_q), & \tau>0,\\ N(\varepsilon_p-\varepsilon_q), & \tau<0.\end{cases}\tag{98}$$

Let's return to relation (95). It is useful to introduce the pairing of two operators under the symbol of T-product, given by relations

$$\overrightarrow{X^{pq}(\tau)X^{p'q'}}(\tau')=\mp G_{pq}(\tau-\tau')[X^{p'q'},X^{pq}]_\pm(\tau'),\tag{99}$$

$$\overrightarrow{X^{pq}(\tau)X^{p'q'}}(\tau')=D_{pq}^0(\tau-\tau')[X^{p'q'},X^{pq}]_-(\tau').\tag{100}$$

The first relation is valid when X_i^{pq} is an f-operator, and second when X_i^{pq} is a b-operator. In (99) the upper sign is taken when the operator $X_i^{p'q'}$ is also of f-type, while the lower sign when $X_i^{p'q'}$ is b-type. With the operation of pairing, the result coming from the application of the Wick theorem is expressed as a first step of calculation of the initial average. This average is decomposed into a sum of terms with all possible pairing.

Relation (95) is interpreted as a set of pairings of the initial operator X_i^{pq} with all the others. In the next step one must apply again the Wick theorem, and then again until in all terms only diagonal operators are left.

Let us denote the diagonal operator by Z_i^{pp}. Averages of Z_i^{pp} can be expressed as derivatives of the generating functional

$$W = \mathrm{Tr}\left\{\exp\left[-\beta\mathscr{H}_0 + \sum_i\sum_p \lambda_i^p Z_i^{pp}\right]\right\}.$$

By using the identity

$$\langle Z_{i_1}^{p_1p_1}\dots Z_{i_n}^{p_np_n}\rangle_0 = \frac{1}{W}\frac{\partial^n W}{\partial\lambda_{i_1}^{p_1}\dots\partial\lambda_{i_n}^{p_n}}\bigg|_{(\lambda)=0},$$

one can obtain the following relations

$$\begin{gathered}\langle Z_i^{pp}\rangle_0 = Z_0^p,\\ \langle Z_{i_1}^{p_1p_1}Z_{i_2}^{p_2p_2}\rangle_0 = Z_0^{p_1}Z_0^{p_2} + \frac{\partial Z_0^{p_1}}{\partial\lambda^{p_2}}\delta_{i_1i_2},\\ \dots,\end{gathered} \tag{101}$$

where

$$Z_0^p = \frac{1}{W}\frac{\partial W}{\partial\lambda^p}\bigg|_{(\lambda)=0}.$$

Derivatives of $(k-1)$-order of W are called cumulants of k-th order. Obviously,

$$\frac{\partial^{k-1}Z_0^{p_1}}{\partial\lambda^{p_2}\dots\partial\lambda^{p_k}} = \langle\left(Z^{p_1p_1} - Z_0^{p_1}\right)\left(Z^{p_2p_2} - Z_0^{p_2}\right)\dots\left(Z^{p_kp_k} - Z_0^{p_k}\right)\rangle_0. \tag{102}$$

In the general case an average of n operators Z_i^{pp} is decomposed into a sum of terms, each of which is a product of different cumulants, where the total number of the orders of cumulants is equal to n. A cumulant of the k-th order is accompanied by a product of $k-1$ δ-symbols of site indexes, providing the equality of all these indexes, in full similarity with the expressions for averages $\langle S_1^z S_2^z \dots S_p^z\rangle$ in the Heisenberg model.

First of all, we calculate the fermion GF of zeroth approximation. By averaging both sides of equation (99) we obtain

$$-\langle T(\overrightarrow{X_i^{0+}(\tau)X_{i'}^{+0}}(\tau'))\rangle_0 = \delta_{ii'}G_\uparrow^0(\tau-\tau')\langle F_{i'}^{+0}\rangle_0,$$

$$-\langle T(\overrightarrow{X_i^{0-}(\tau)X_{i'}^{-0}}(\tau'))\rangle_0 = \delta_{ii'}G_\downarrow^0(\tau-\tau')\langle F_{i'}^{-0}\rangle_0.$$

One can see from here that functions $G_\uparrow^0$ and $G_\downarrow^0$ in (95) coincide, apart from the factors $\langle F^{+0}\rangle$ and $\langle F^{-0}\rangle$, with one-particle GFs, constructed with X operators. Similarly we have an expression

$$-\langle T(\overrightarrow{X_i^{+-}(\tau)X_{i'}^{-+}}(\tau'))\rangle_0 = \delta_{ii'}D^0(\tau-\tau')\langle B_{i'}^{+-}\rangle_0,$$

showing a sense of bosonic function D^0.

With the help of relation (95) an average in each order of perturbation theory is decomposed into a sum of terms which are products of the zeroth order GFs $G^0_{\uparrow}, G^0_{\downarrow}, D^0$ and cumulants. We denote GFs by lines with an arrow, and $t_{ii'}$ by a wavy line (Fig. 22). Cumulants of different order are denoted by points included in ovals. Thick point

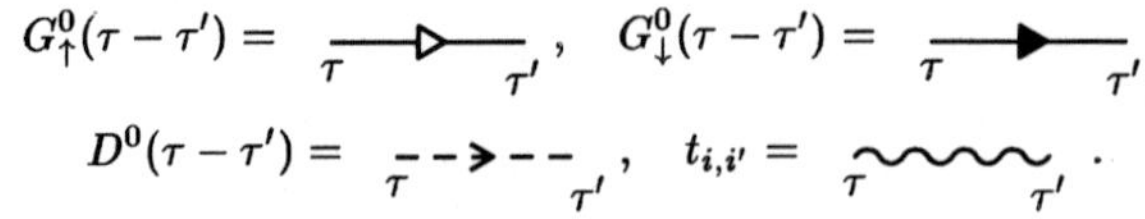

FIGURE 22. Graphic representation of propagators and matrix elements of hopping

corresponds to an average of a diagonal operator Z_i^{pp}, a cumulant of zeroth order. We use the following correspondence between analytic and graphic notation for cumulants

$$\langle Z_i^{pp}\rangle_0 = \bullet, \quad \langle (Z_i^{pp} - \langle Z_i^{pp}\rangle_0)(Z_{i'}^{p'p'} - \langle Z_{i'}^{p'p'}\rangle_0)\rangle_0 = (\bullet \;\; \bullet), \ldots \tag{103}$$

Like in any diagrammatic technique, graphs for the quantity (93) can be connected and disconnected. The theorem of connection holds, allowing to go from expression (93) to the following expression

$$\langle \Big(\tilde{X}_{i_1}^{p_1 q_1}(\tau) \ldots \tilde{X}_{i_n}^{p_n q_n}(\tau_n)\Big)\rangle = \langle \Big(X_{i_1}^{p_1 q_1}(\tau_1) \ldots X_{i_n}^{p_n q_n}(\tau_n)\sigma(\beta)\Big)\rangle_0^{\mathrm{connect}}, \tag{104}$$

stating that, in the series of graphs representing our quantity, only connected graphs should be taken.

Diagrammatic representation of Green's functions in the limit $U \to \infty$. We start from the limit $U \to \infty$, when a state $|2\rangle$ can not exist, so the total Hamiltonian (91) reduces to

$$\begin{aligned} \mathscr{H}_0 &= \varepsilon_+ \sum_i X_i^{++} + \varepsilon_- \sum_i X_i^{--}, \\ \mathscr{H}_{\mathrm{int}} &= \sum_{ij} t_{ij}(X_i^{+0}X_j^{0+} + X_i^{-0}X_j^{0-}), \end{aligned} \tag{105}$$

excluding double occupied states. Here $\varepsilon_\pm = \mp\frac{1}{2}h - \mu$. Thus, from 16 X operators only 9 are now involved:

$$X^{+0}, X^{-0}, X^{0+}, X^{0-}, X^{+-}, X^{-+}, X^{00}, X^{--}, X^{++}.$$

The first four operators are f-type, X^{-+} and X^{+-} are b-type. Moreover, there are three diagonal operators X^{00}, X^{--}, X^{++}.

It is convenient to use, instead of these, three linear combinations

$$F_i^{+0} = X_i^{00} + X_i^{++}, \quad F_i^{-0} = X_i^{00} + X_i^{--}, \quad B_i^{+-} = X_i^{++} - X_i^{--}.$$

All of them commute with each others. The complete list of permutation relations is given in Table 1.

TABLE 1. Permutation relations for operators X, F and B (The first 4×4 block with f-operators implies anticommutators, in all other cases, commutators. In the anticommutator block a table is symmetrical, and in the rest part it is antisymmetrical)

	X^{+0}	X^{0+}	X^{-0}	X^{0-}	X^{+-}	X^{-+}	F^{+0}	F^{-0}	B^{+-}
X^{+0}	0	F^{+0}	0	X^{+-}	0	$-X^{-0}$	0	X^{+0}	$-X^{+0}$
X^{0+}	F^{+0}	0	X^{-+}	0	X^{0-}	0	0	$-X^{0+}$	X^{0+}
X^{-0}	0	X^{-+}	0	F^{-0}	$-X^{+0}$	0	X^{-0}	0	X^{-0}
X^{0-}	X^{+-}	0	F^{-0}	0	0	X^{0+}	$-X^{0-}$	0	$-X^{0-}$
X^{+-}	0	$-X^{0-}$	X^{+0}	0	0	B^{+-}	$-X^{+-}$	X^{+-}	$-2X^{+-}$
X^{-+}	X^{-0}	0	0	$-X^{0+}$	$-B^{+-}$	0	X^{-+}	$-X^{-+}$	$2X^{-+}$
F^{+0}	0	0	$-X^{-0}$	X^{0-}	X^{+-}	$-X^{-+}$	0	0	0
F^{-0}	$-X^{+0}$	X^{0+}	0	0	$-X^{+-}$	X^{-+}	0	0	0
B^{+-}	X^{+0}	$-X^{0+}$	$-X^{-0}$	X^{0-}	$2X^{+-}$	$-2X^{-+}$	0	0	0

Like in any diagrammatic technique, the possible types of vertexes and the rules of their connections by wavy lines are determined by the structure of Hamiltonian $\mathscr{H}_{\text{int}}$, or more precisely, by the quantity

$$\int_0^\beta d\tau \mathscr{H}_{\text{int}}(\tau) = \int d1'd2't(1'-2')(X^{+0}(1')X^{0+}(2') + X^{-0}(1')X^{0-}(2')).$$

Here four-component variables are introduced: $1 = (i_1\tau_1)$ and $t(1-2) = t_{i_1i_2}\delta(\tau_1 - \tau_2)$. The types of vertex parts should not depend on calculated GFs. Here we keep in mind internal vertexes, connected by interaction lines. As to external vertexes they naturally are determined by the operators, involving in the definition of GF. Thus the main structural elements of the graphic technique are internal vertexes. Since the Hamiltonian includes two pairs of operators $X^{+0}(1')X^{0+}(2')$ and $X^{-0}(1')X^{0-}(2')$, we have the following types of vertexes, Fig. 23. They appear for example when calculating an average

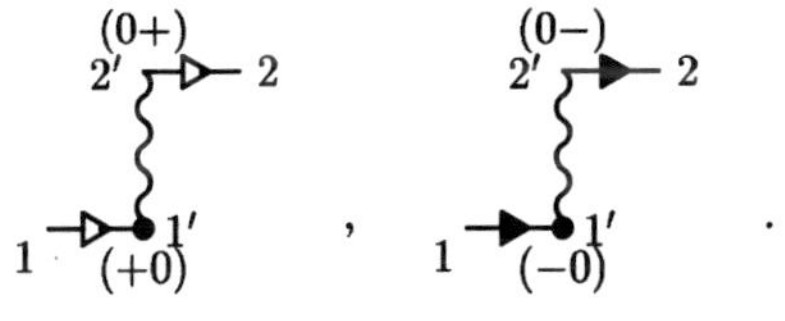

FIGURE 23. Simple vertexes with one electron line

$$t(1'-2')\langle T(X^{0+}(1)X^{+0}(2)X^{+0}(1')X^{0+}(2'))\rangle_0 =$$

$$= -t(1'-2')G^0_\uparrow(1-1')G^0_\uparrow(2'-2)\langle F^{+0}(1')F^{+0}(2)\rangle_0,$$

and a similar average of operators with opposite spins. However when calculating more complicated averages other types of vertexes with a few ingoing Green's lines can appear, but with the same "vertex charge" (+0) or (–0).

Consider for example two expressions:

$$t(1'-2')\langle T(X^{0-}(1)X^{0+}(2)X^{-+}(3)X^{+0}(1')X^{0+}(2'))\rangle_0 =$$

$$= -t(1'-2')G^0_{\uparrow}(2'-2)G^0_{\downarrow}(1-1')D^0(1'-3)\langle F^{+0}(2)B^{+-}(3)\rangle_0;$$

$$t(1'-2')\langle T(X^{0-}(1)X^{0+}(2)X^{+0}(4)X^{-0}(3)X^{-0}(1')X^{0-}(2'))\rangle_0 =$$

$$= -t(1'-2')G^0_{\downarrow}(2'-3)G^0_{\uparrow}(2-1')G^0_{\downarrow}(1-1')G^0_{\uparrow}(1'-4)\langle F^{+0}(4)F^{-0}(3)\rangle_0.$$

Pairings indicated by arrows should be taken in the sequence from up to down. We see that in the first example the operator prescribed at point $1'$ takes part in two pairings, whilst in the second example in three pairings. This means that two Green's lines should be attached to point $1'$ in the first case and three lines in the second one. This leads to the appearance of double lines vertexes (+0) and three lines vertexes (–0), Fig. 24.

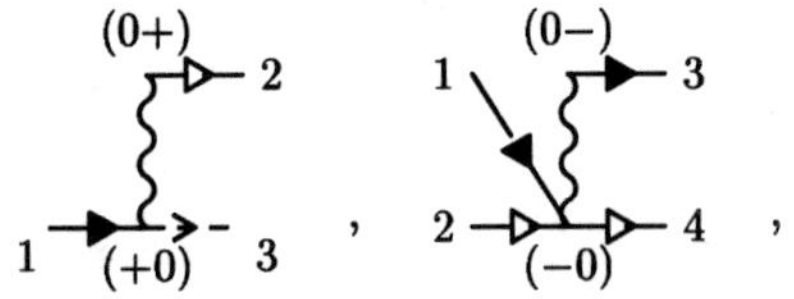

FIGURE 24. Vertexes with two and three Green's lines

It is easy to find the "charge" of a complicated vertex. To this purpose we write down the operators corresponding to all ingoing and outgoing Green's lines and multiply them in accordance with the rules of multiplication for X operators. By this way we get for double and triple lines vertexes in the graphs of Fig. 24:

$$X^{+-}X^{-0} = X^{+0}, \quad X^{-0}X^{0+}X^{+0} = X^{-0}.$$

This is confirmed also by the charges of vertexes on opposite ends of interaction lines.

A comment of general character. When applying Wick theorem to some average of T-product of X-operators we meet some ambiguity, resulting from the existence of not only one annihilation operator, like in the Heisenberg model, but of a number of them (X^{0+}, X^{0-}, X^{+-}). One may start from any of them, and this leads to different form of graphic representation for a particular GF. To make the diagrammatic technique unique, we can introduce a system of priority for the annihilation operators X^{pq}, which should fix the sequence, that is an agreement from which kind of operator we will start the pairing. A system of priority is arbitrary, but it should be fixed along all the calculations. This topic was discussed in [80] and in the book [55], where calculations of a particular GF with different systems of priority are compared. This ambiguity in the diagrammatic technique for X operators is a disadvantage of this technique. However, in the method of generating functional there is not such a difficulty.

Let's introduce one-particle GFs, describing the propagation of an electron with a given spin:

$$\begin{aligned}\mathscr{G}_{\uparrow}(1-1') &= -\langle T(\tilde{X}^{0+}(1)\tilde{X}^{+0}(1'))\rangle,\\ \mathscr{G}_{\downarrow}(1-1') &= -\langle T(\tilde{X}^{0\ }\ (1)\tilde{X}^{-0}(1'))\rangle.\end{aligned} \tag{106}$$

In the zeroth approximation we have

$$\mathscr{G}^{0}_{\uparrow}(1-1') = G^{0}_{\uparrow}(1-1')\langle F^{+0}\rangle_0 = $$

$$\mathscr{G}^{0}_{\downarrow}(1-1') = G^{0}_{\downarrow}(1-1')\langle F^{-0}\rangle_0 = \tag{107}$$

In momentum representation electron propagators are equal to

$$G^{0}_{\sigma}(\mathrm{i}\omega_n) = \frac{1}{\mathrm{i}\omega_n - \varepsilon_\sigma}, \qquad \omega_n = (2n+1)\pi T. \tag{108}$$

The corresponding cumulants are expressed through the average population values with a given spin on a site::

$$\langle F^{+0}\rangle = 1 - n_{\downarrow}, \quad \langle F^{-0}\rangle = 1 - n_{\uparrow}.$$

On-site GFs (107) describe localized excitation on a site. The transport of the excitation from one site to another is described by an infinite series of graphs. In the simplest case we have a series, Fig. 25, (similar for opposite spin), which gives a propagator

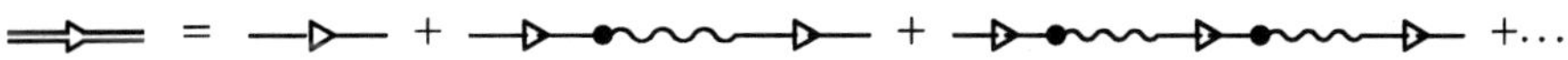

FIGURE 25. Hubbard-I approximation for electron Green's function

$$G_\sigma(\boldsymbol{k};\mathrm{i}\omega_n) = \frac{1}{\mathrm{i}\omega_n - E_\sigma(\boldsymbol{k})}, \tag{109}$$

where

$$E_\sigma(\boldsymbol{k}) = \langle F^{\sigma 0}\rangle_0 \varepsilon(\boldsymbol{k}) + \varepsilon_\sigma. \tag{110}$$

The series in Fig. 25 corresponds to the Hubbard-I approximation [2], and $E_\sigma(\boldsymbol{k})$ is the energy of an electron of the lower Hubbard band in the limit $U \to \infty$.

Consider now a spin GF

$$D_{\perp}(1-2) = -\langle T(\tilde{X}^{+-}(1)\tilde{X}^{-+}(2))\rangle, \tag{111}$$

which describes the correlation of transversal spin components, because operators X^{+-} and X^{-+} are equivalent to operators S^{+} and S^{-}. This GF will give in particular the spin wave spectrum in a magnetically ordered system. In the zeroth approximation

$$D^{0}_{\perp}(1-2) = D^{0}(1-2)\langle B^{+-}\rangle_0 \equiv \tag{112}$$
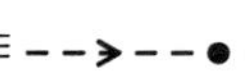

Write down the series of diagrams for $D_{\perp}$ up to the first order of perturbation theory, Fig. 26 [53].

All graphs of this series may be collected in groups, characterized by the type of left and right hand side external vertexes. This leads to the structure for all series, displayed

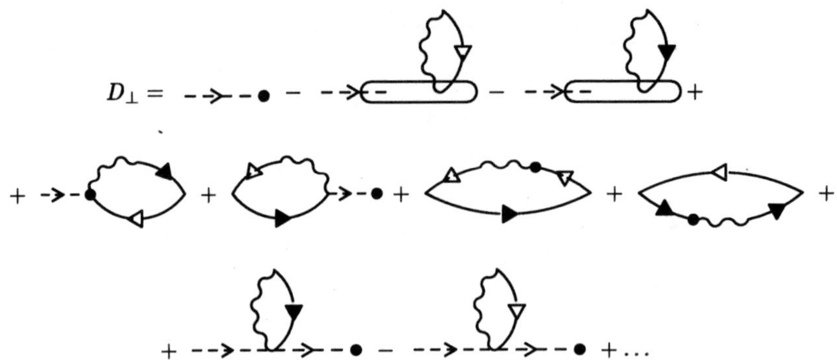

FIGURE 26. Expansion for transversal spin GF in the system of priority $X^{0+} > X^{0-} > X^{+-}$

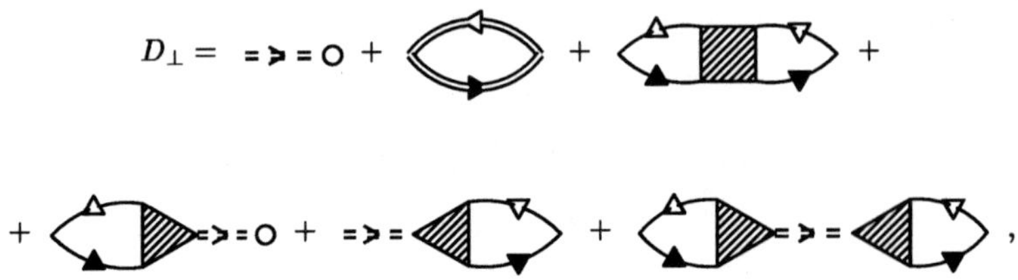

FIGURE 27. Graph representation for the transversal Green's function

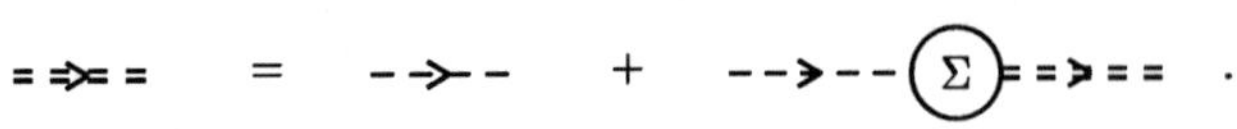

FIGURE 28. Dyson equation for propagator of the transversal Green's function

in Fig. 27. Here the thickened propagator obeys Dyson equation, Fig. 28. In Fig. 27 all electron lines represent electron propagators.

Thus, we see that the GF $D_\perp$ is expressed through vertex parts of three types: electron 4-point and two 3-point ones. These vertex parts cannot be cut over a spin line. At the end of the spin line in Fig. 27 a light small circle stands, corresponding to the not completely thicken quantity $\langle B^{+-}\rangle$. The problem of calculation of a spin GF reduces now to the calculation of 3-point and 4-point vertex parts and of the self energy of the spin propagator.

Generating Random Phase Approximation. In the theory of magnetism of usual Fermi systems, the random phase approximation (RPA) is used very often. It takes into account the sum of loop-type diagrams. Each loop is formed by two electron Green's lines with opposite spins if we consider transversal GF, and with parallel spins for longitudinal GF. For strongly correlated systems a similar approximation can be suggested when spin susceptibilities are calculated. However, the noncommutativity of electron operators on a C-value leads to a severe complexity. In particular, four different loops appear, which enter in the expansion series in different combinations. Summing up all such diagrams in all orders of perturbation theory corresponds to the generalized random phase approximation (GRPA) [53, 81].

GRPA for the Hubbard model has been first applied for the case $U \to \infty$. One can see in Fig. 24 that there are two bare graphs for electron-electron interaction (the second graph is obtained from the first by changing the colors of electron lines). The expansion of the two-particles vertex part with antiparallel lines up to the second order is shown in Fig. 29. The total series for the 4-points vertex part is equivalent to the equation in Fig.

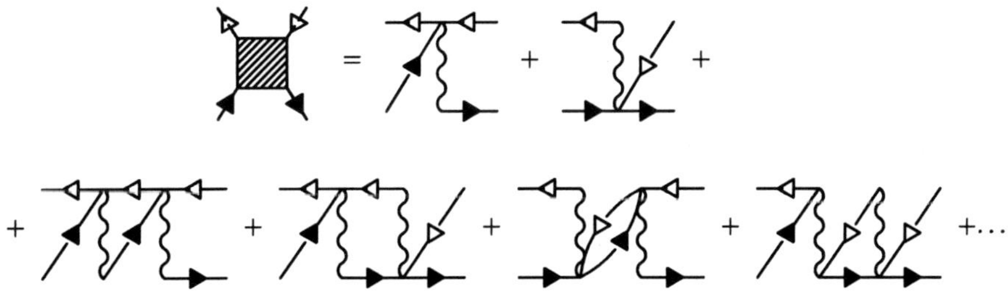

FIGURE 29. Series of diagrams for electron-electron vertex part with antiparallel lines

30. We can rewrite this equation in the form of Bethe-Salpether equation with a bare 4-point vertex part, Fig. 31.

After iterating the equation of Fig. 31, we can see that it corresponds to a series of graphs, consisting of the loops shown in Fig. 29. Introduce an analytical notation for 4-point vertex part of Fig. 32. The graphic equation (Fig. 31) is equivalent to an analytical one:

$$\begin{gathered}\Gamma(k_1 - kk_1; k_2 + kk_2) = \varepsilon(\boldsymbol{k}_2 + \boldsymbol{k}) + \varepsilon(\boldsymbol{k}_1 - \boldsymbol{k}) + \\ + \frac{1}{N}\sum_{k_3} \varepsilon(\boldsymbol{k}_3) G_\uparrow(k_3 - k) G_\downarrow(k_3) \Gamma(k_3 - kk_3; k_2 + kk_2) + \\ + \varepsilon(\boldsymbol{k}_1 - \boldsymbol{k}) \frac{1}{N}\sum_{k_3} G_\uparrow(k_3 - k) G_\downarrow(k_3) \Gamma(k_3 - kk_3; k_2 + kk_2).\end{gathered} \tag{113}$$

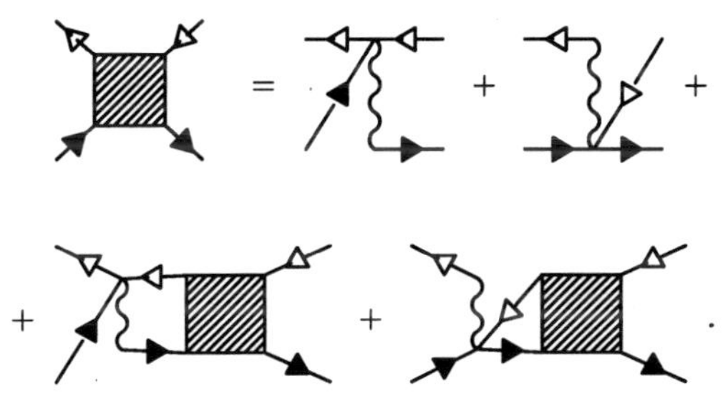

FIGURE 30. Equation for the vertex part equivalent to the series in Fig. 29

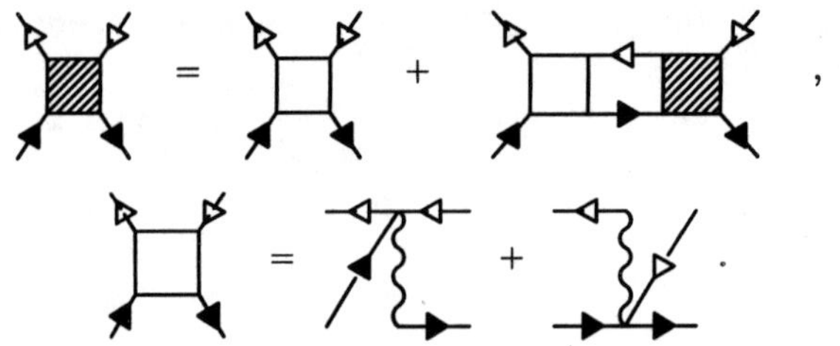

FIGURE 31. Bethe-Salpether equation for the effective electron-electron interaction

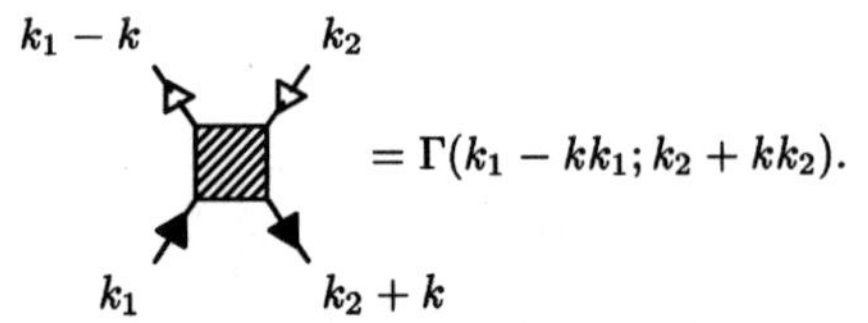

FIGURE 32. Notation for 4-point vertex part

Since this integral equation has a degenerate kernel, it reduces to a system of two algebraic equations having a solution

$$\begin{aligned} &\Gamma(k_1 - kk_1; k_2 + kk_2) = \\ &= \frac{1}{d_\perp(k)}\{\Phi_\perp(k) + \varepsilon(\boldsymbol{k}_1 - \boldsymbol{k})(1 - Q_\perp(k)) + \\ &+ \varepsilon(\boldsymbol{k}_2 + \boldsymbol{k})(1 - \Lambda_\perp(k)) + \varepsilon(\boldsymbol{k}_1 - \boldsymbol{k})\varepsilon(\boldsymbol{k}_2 + \boldsymbol{k})\Pi_\perp(k)\}, \end{aligned} \tag{114}$$

where

$$d_\perp(k) = (1 - \Lambda_\perp(k))(1 - Q_\perp(k)) - \Pi_\perp \Phi_\perp(k). \tag{115}$$

Here we introduce four quantities

$$\begin{pmatrix} \Pi_\perp(k) \\ Q_\perp(k) \\ \Lambda_\perp(k) \\ \Phi_\perp(k) \end{pmatrix} = \frac{1}{N}\sum_{k_1} \begin{pmatrix} 1 \\ \varepsilon(\boldsymbol{k}_1) \\ \varepsilon(\boldsymbol{k}_1 - \boldsymbol{k}) \\ \varepsilon(\boldsymbol{k}_1)\varepsilon(\boldsymbol{k}_1 - \boldsymbol{k}) \end{pmatrix} G_\uparrow(k_1 - k)G_\downarrow(k_1), \tag{116}$$

corresponding to the four loops appearing in the expansion of Fig. 29. They are shown in Fig. 33.

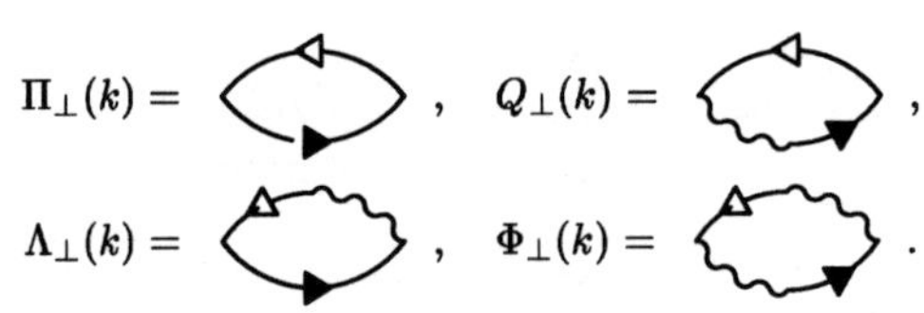

FIGURE 33. Four types of the electron loops

In GRPA 3-point vertex parts are expressed via 4-point parts and bare 3-point parts

 $= \gamma_L(k_1 - kk_1; k),$ 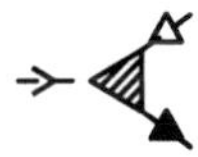 $= \gamma_R(k; k_2 + kk_2).$ (117)

(see Fig. 34).

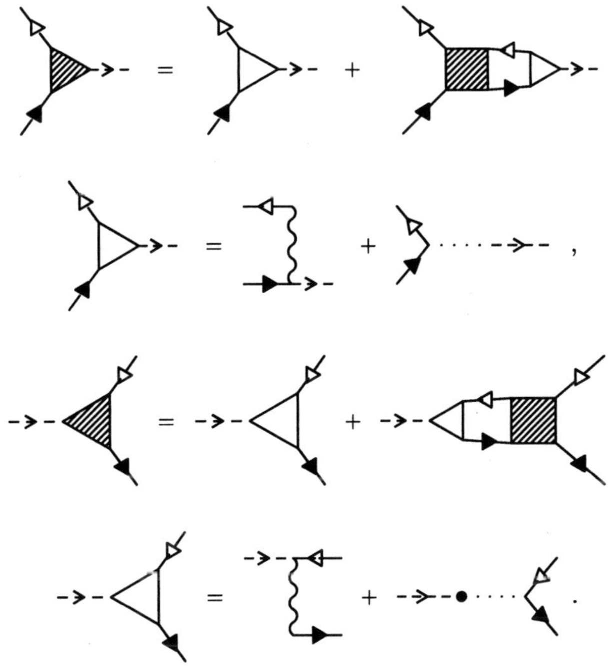

FIGURE 34. Expressions 3-point vertex parts through 4-point parts

The analytical form of γ_L is

$$\gamma_L(k_1 - kk_1; k) =$$
$$= \varepsilon(\boldsymbol{k}_1 - \boldsymbol{k}) + \frac{1}{N}\sum_{k_2} \varepsilon(\boldsymbol{k}_2) G_\uparrow(k_2) G_\downarrow(k_2 + k)\Gamma(k_1 - kk_1; k_2 + kk_2),$$

and a similar relation for γ_R. After the substitution of (114) for Γ, we find

$$\gamma_L(k_1 - kk_1; k) = \frac{1}{d_\perp(k)}\{\Phi_\perp(k) + \varepsilon(\boldsymbol{k}_1 - \boldsymbol{k})(1 - Q_\perp(k))\}, \quad (118)$$

$$\gamma_R(k; k_2 + kk_2) = \frac{\langle B^{+-}\rangle_0}{d_\perp(k)}\{\Phi_\perp(k) + \varepsilon(\boldsymbol{k}_2 + \boldsymbol{k})(1 - \Lambda_\perp(k))\}. \quad (119)$$

Now we can calculate GF $D_\perp$, determined by the graphic expression in Fig. 27. First, we find the expressions

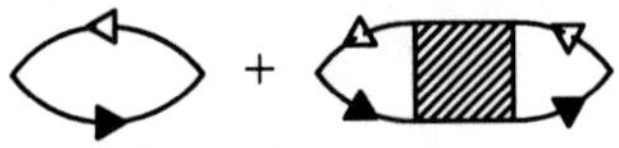

$$\equiv \Pi(k) + \frac{1}{N^2}\sum_{k_1k_2} G_\uparrow(k_1-k)G_\downarrow\Gamma(k_1-kk_1;k_2+kk_2)G_\uparrow(k_2)G_\downarrow(k_2+k) = \frac{1}{d_\perp(k)}\Pi(k). \tag{120}$$

$$\equiv \mathscr{R}(k) = +\frac{\langle B^{+-}\rangle_0}{d_\perp(k)}\{Q_\perp(k)(1-\Lambda(k))+\Pi_\perp(k)\Phi_\perp(k)\}, \tag{121}$$

$$\equiv \mathscr{L}(k) = +\frac{1}{d_\perp(k)}\{\Lambda(k)(1-Q_\perp(k))+\Pi_\perp(k)\Phi_\perp(k)\}, \tag{122}$$

Finally, the magnon GF obeys to Dyson equation

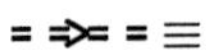

$$\equiv D(k) = \frac{1}{\mathrm{i}\omega_n - h - M(k)}, \tag{123}$$

where the self-energy part, Fig. 35, is given by the expression

$M =$ 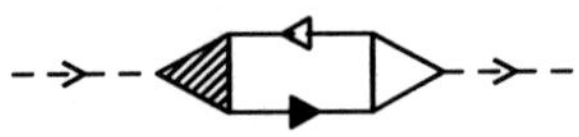

FIGURE 35. Self-energy part for magnon Green's function

$$M(k) = \frac{\langle B^{+-}\rangle_0}{d_\perp(k)}\Phi(k). \tag{124}$$

Substituting expressions for quantities $\Gamma, \gamma_L, \gamma_R, \Sigma$ in the expression corresponding to Fig. 27, we find the GF for transversal spin components

$$D_\perp(\boldsymbol{k};\mathrm{i}\omega_n) = -\frac{\pi_\perp(k)+\Pi_\perp}{(1-\Lambda_\perp(k))(1-Q_\perp(k))-(\pi_\perp(k)+\Pi_\perp(k))\Phi_\perp(k)}. \tag{125}$$

Here

$$\pi_\perp(k) = \langle B^{+-}\rangle_0 D^0(\mathrm{i}\omega_n), \tag{126}$$

and $D^0(\mathrm{i}\omega_n)$ is the zeroth approximation of the magnon GF

$$D^0(\mathrm{i}\omega_n) = \frac{1}{\mathrm{i}\omega_n - h}.$$

Expression (125) for the transversal GF in the Hubbard model is isomorphic to expression (44) for longitudinal GF in the Heisenberg model. In both cases quantities corresponding to loops consisted by two Green's lines. The essential difference is that in the Heisenberg model loops are formed by spin wave Green's lines while in the Hubbard model they are formed by electron Green's lines (see (116)). The reason for such an isomorphism is in the similarity of the permutation relations of spin operators S^+ and S^- and b-like operators X^{+-} and X^{-+}.

In the case of the Heisenberg model, the quantity $\pi(k)$ has a singularity $\pi(k) \sim \delta_{\omega_n,0}$. We show that expression (125) for the Hubbard model at zero field has a singularity as well $\pi_\perp(\mathrm{i}\omega_n) \sim \delta_{\omega_n,0}$. To this purpose we calculate the limit of the quantity $\langle B^{+-}\rangle_0 D^0(\mathrm{i}\omega_n)$ at $h \to 0$. We start from the expression

$$\langle B^{+-}\rangle_0 = \langle F^{+0}\rangle_0 - \langle F^{-0}\rangle_0 = \frac{\mathrm{e}^{-\beta\varepsilon_+} - \mathrm{e}^{-\beta\varepsilon_-}}{1 + \mathrm{e}^{-\beta\varepsilon_+} + \mathrm{e}^{-\beta\varepsilon_-}},$$

After differentiation we find

$$\left.\frac{\partial\langle B^{+-}\rangle_0}{\partial h}\right|_{h\to 0} = \frac{1}{2}\beta n_0 \equiv \beta B'_\perp,$$

$$n_0 = \frac{2\mathrm{e}^{\beta\mu}}{1 + 2\mathrm{e}^{\beta\mu}}.$$

Calculating $\langle B^{+-}\rangle_0 D^0(\mathrm{i}\omega_n)$, we find the limit

$$[\langle B^{+-}\rangle_0 D^0(\mathrm{i}\omega_n)]_{h\to 0} = -\beta B'_\perp \delta_{\omega_n,0}. \tag{127}$$

Thus, at zero field, expression (125) for the transversal GF reduces to the following [53]:

$$D_\perp(\boldsymbol{k}, \mathrm{i}\omega_n) = \frac{\beta B'_\perp \delta_{\omega_n,0} - \Pi_\perp(k)}{(1 - \Lambda_\perp(k))(1 - Q_\perp(k)) + [\beta B'_\perp \delta_{\omega_n,0} - \Pi_\perp(k)]\Phi_\perp(k)}. \tag{128}$$

Now we can separate a singular part in (128)

$$D_\perp(k) = D^I_\perp(k) + [D^T_\perp(\boldsymbol{k},0) - D^I_\perp(\boldsymbol{k},0)]\delta_{\omega_n,0}, \tag{129}$$

where

$$D^I_\perp(k) = \frac{-\Pi_\perp(k)}{(1 - \Lambda_\perp(k))(1 - Q_\perp(k)) - \Pi_\perp(k)\Phi_\perp(k)}, \tag{130}$$

$$D^T_\perp(\boldsymbol{k}) = \frac{\beta B'_\perp - \Pi_\perp(\boldsymbol{k},0)}{(1 - \Lambda_\perp(\boldsymbol{k},0))(1 - Q_\perp(\boldsymbol{k},0)) + (\beta B'_\perp - \Pi_\perp(\boldsymbol{k},0))\Phi_\perp(\boldsymbol{k},0)}. \tag{131}$$

After the analytical continuation $\mathrm{i}\omega_n \to \omega + \mathrm{i}\delta$, the quantity $D^I_\perp(k)$ reduces to isolated susceptibility, and $D^T_\perp(\boldsymbol{k},0)$ to isothermal susceptibility. In a similar way we can calculate in GRPA longitudinal GF and also density-density GF in the Hubbard model [53].

The tJ-model. Up to now we studied the Hubbard model in the limit $U \to \infty$. At finite but large $U \gg zt$ one can change the total Hamiltonian (91) into an effective Hamiltonian

[6], corresponding to the tJ-model. It describes the motion of correlated electrons on a lattice with indirect exchange interaction between nearest neighbors. It reads

$$\mathscr{H} = t\sum_{ij\sigma} X_i^{\sigma 0} X_j^{0\sigma} + J\sum_{ij}\left(\boldsymbol{S}_i\boldsymbol{S}_j - \frac{1}{4}n_i n_j\right). \tag{132}$$

It is possible to give another expression (132)

$$\mathscr{H} = t\sum_{ij\sigma} X_i^{\sigma 0} X_j^{0\sigma} + J\sum_{ij}\left(X_i^{-+}X_j^{+-} - X_i^{++}X_i^{--}\right). \tag{133}$$

Here $J \sim t^2/U$. It is convenient to generalize it by introducing hopping matrix elements between any two sites: $t \to t_{ij}, \quad J \to J_{ij}$.

In order to construct the diagrammatic technique we must now write an additional term in the expression determining the scattering matrix for the tJ-model:

$$\begin{aligned}\int_0^\beta d\tau \mathscr{H}_{\text{int}}(\tau) = \int d1'd2' t(1'-2')\left(X^{+0}(1')X^{0+}(2') + X^{-0}(1')X^{0-}(2')\right) + \\ + \int d1'd2' J(1'-2')\left(X^{-+}(1')X^{+-}(2') - X^{++}(1')X^{--}(2')\right),\end{aligned} \tag{134}$$

where

$$t(1-2) = t_{i_1 i_2}\delta(\tau_1 - \tau_2), \quad J(1-2) = J_{i_1 i_2}\delta(\tau_1 - \tau_2).$$

Thus, along with wavy lines denoting hopping matrix elements we should include point lines, denoting the exchange interaction J_{12}. A point line can connect vertexes of type (–+) and (+–) or vertexes (++) and (– –), corresponding to the second term in (134).

Describe first the effects of the self consistent field in the exchange interaction. Consider a series of graphs for the electron GF, Fig. 36. If we admit that Green's line

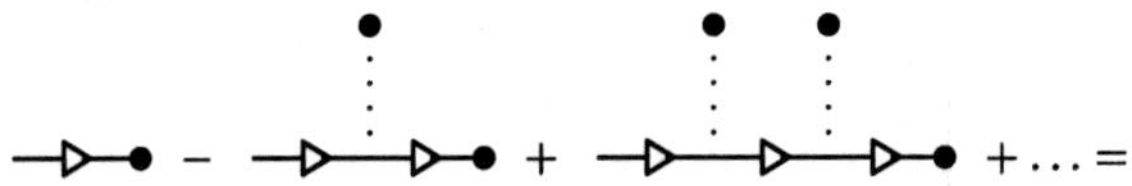

FIGURE 36. Series for electron GF for the account of the exchange interaction

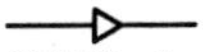 is taken in the Hubbard-I approximation to account for the hopping term in (134), then the series in Fig. 36 leads to the propagator

$$G_\sigma(k) = \frac{1}{\mathrm{i}\omega_n - E_\sigma(\boldsymbol{k}) + \mu},$$

where

$$E_\sigma(\boldsymbol{k}) = (1 - n_{\bar\sigma})\varepsilon(\boldsymbol{k}) - J(0)\langle X^{\bar\sigma\bar\sigma}\rangle_0 - \frac{1}{2}\sigma h.$$

In the same way it is possible to calculate the magnon GF, Fig. 37. A thick point at the end of point lines corresponds to $\langle B^{+-}\rangle_0$. Thus the series in Fig. 37, determines the spin GF

$$\tilde{D}^0(\mathrm{i}\omega_n) = \frac{1}{\mathrm{i}\omega_n - h + J(0)\langle B^{+-}\rangle_0}.$$

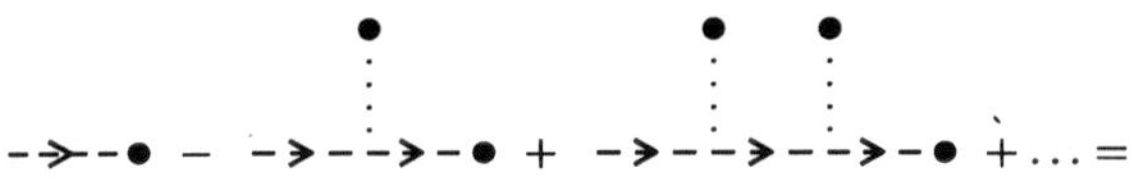

FIGURE 37. Series for spin GF for account of exchange interaction

Thus diagrams corresponding to mean field corrections are summed out. Now we consider a series, transforming a localized spin GF into a propagator. According to Fig. 38 we obtain the propagator for the magnon GF

FIGURE 38. Series for the magnon GF

$$\tilde{D}(\boldsymbol{k};\mathrm{i}\omega_n) = \frac{1}{\mathrm{i}\omega_n - h - \langle B^{+-}\rangle (J(0) - J(\boldsymbol{k}))},$$

which appears in the Heisenberg model. We took only the exchange term. The hopping term must be taken in GRPA. In this case we must add to the bare electron-electron vertex part (Fig. 31) two exchange terms, Fig. 39. To begin with, let us take only the first

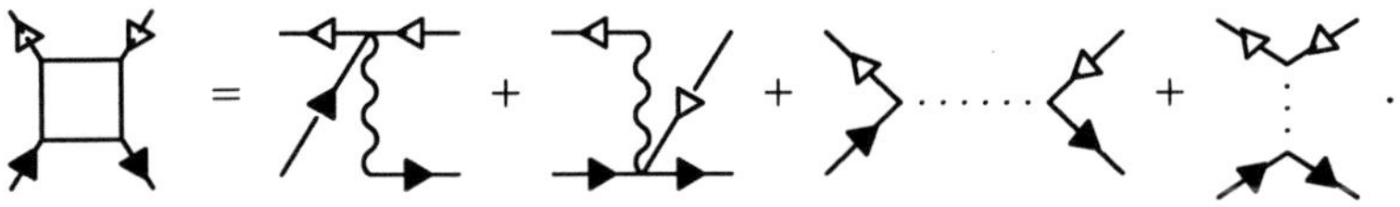

FIGURE 39. Bare electron-electron vertex part for the tJ-model

of the two exchange terms. It is possible to show that the Bethe-Salpether equation in this case has a solution of (114)-type, where one must change $\Phi_\perp(k) \to \Phi_\perp(k) + J(k)$. It leads to an expression for the transversal GF of the form (125) or (128) with the same change. So, instead of expression (130), obtained for the Hubbard model with $U \to \infty$, we have for the tJ-model [53]

$$D_\perp^T(\boldsymbol{k}) = \frac{-\Pi_\perp(k)}{(1-\Lambda_\perp(k))(1-Q_\perp(k)) - \Pi_\perp(k)(\Phi_\perp(k) + J(\boldsymbol{k}))}. \tag{135}$$

The second exchange term in Fig. 39 leads to thickened loops and does not correspond to GRPA. Nevertheless Bethe-Salpether equation can be solved exactly even in that case. Effects of the second exchange term have been studied in [82].

B. Method of generating functional

Electron Green's function in the limit $U = \infty$. Now we develop an approach to the Hubbard model similar to the one developed in Section II for the Heisenberg model. Since the Hamiltonian of the Hubbard model is much more complicated than that of the Heisenberg model we start from the limit case $U = \infty$

$$\mathscr{H} = \sum_{i\sigma} \varepsilon_\sigma X_i^{\sigma\sigma} + \sum_{ij} t_{ij} X_i^{\sigma 0} X_j^{0\sigma}. \tag{136}$$

Let us introduce the generating functional

$$Z[V] = \mathrm{Tr}\,(\mathrm{e}^{-\beta \mathscr{H}} T\mathrm{e}^{V}) \equiv ((T\mathrm{e}^{V}))$$

and define V as follows [68]

$$V = v_{1'}^{0} X_{1'}^{00} + v_{1'}^{\sigma_2' \sigma_1'} X_{1'}^{\sigma_1' \sigma_2'} + X_{2'}^{\sigma_2' 0} \ell_{2'1'}^{\sigma_2' \sigma_1'} X_{1'}^{0\sigma_1'}. \tag{137}$$

As earlier (see expression (49) in the case of the Heisenberg model), we introduce one-point and two-point fluctuating fields depending now on electron spin indexes. We have three fields v_1^0, $v_1^{\sigma\sigma}$ and $v_1^{\sigma\bar{\sigma}}$, standing as coefficients for diagonal and Bose-like operators X_1^{00}, $X_1^{\sigma\sigma}$ and $X_1^{\sigma\bar{\sigma}}$, and a field matrix $\ell_{12}^{\sigma_1\sigma_2}$, standing in quadratic form of fermi-like operators $X_1^{\sigma 0}$ and $X_2^{0\sigma}$. Thus expression (137) for V includes all operators forming algebra of X-operators for the tJ-model. Notice that V is valid for the total Hamiltonian of the tJ-model, but not only for the limit case (136). Remind that summing over all repeated indexes in (137) is implied.

Variational derivatives of $Z[V]$ over the fluctuating fields determine averages of T-products of X-operators

$$\langle T X_1^{00} \rangle_V = -\frac{1}{Z[V]} \frac{\delta Z[V]}{\delta v_1^0}, \quad \langle T X_1^{\sigma_1\sigma_2} \rangle_V = -\frac{1}{Z[V]} \frac{\delta Z[V]}{\delta v_1^{\sigma_1\sigma_2}};$$

$$\langle T X_1^{00} X_2^{00} \rangle_V = \frac{1}{Z[V]} \frac{\delta^2 Z[V]}{\delta v_1^0 \delta v_2^0}, \quad \langle T X_1^{\sigma\bar{\sigma}} X_2^{\bar{\sigma}\sigma} \rangle_V = \frac{1}{Z[V]} \frac{\delta^2 Z[V]}{\delta v_1^{\sigma\bar{\sigma}} \delta v_2^{\bar{\sigma}\sigma}};$$

$$\langle T X_1^{0\sigma_1} X_2^{\sigma_2 0} \rangle_V = \frac{1}{Z[V]} \frac{\delta Z[V]}{\delta \ell_{12}^{\sigma_1\sigma_2}}.$$

and so on.

When the fluctuating fields are absent (V=0) this expressions determine bosonic and fermionic GFs. If we make a substitution $Z = \mathrm{e}^{\Phi}$, we have the following expressions for these GFs:

$$D_{12}^{\sigma\bar{\sigma}} = -\langle T X_1^{\sigma\bar{\sigma}} X_2^{\bar{\sigma}\sigma} \rangle = -\left.\frac{\delta^2 \Phi}{\delta v_1^{\sigma\bar{\sigma}} \delta v_2^{\bar{\sigma}\sigma}}\right|_{V=0}, \tag{138}$$

$$\mathscr{G}_{12}^{\sigma_1\sigma_2} = -\langle T X_1^{0\sigma_1} X_2^{\sigma_2 0} \rangle_V = -\left.\frac{\delta \Phi}{\delta \ell_{12}^{\sigma_1\sigma_2}}\right|_{V=0}, \tag{139}$$

and for averages of X operators

$$\langle X_1^{00} \rangle = -\left.\frac{\delta \Phi}{\delta v_1^0}\right|_{V=0}, \quad \langle X_1^{\sigma_1\sigma_2} \rangle = -\left.\frac{\delta \Phi}{\delta v_1^{\sigma_1\sigma_2}}\right|_{V=0}. \tag{140}$$

Since in $\mathscr{H}$ and V there is a quadratic form of f-operators, we can change the differentiation over $\ell_{12}^{\sigma_1\sigma_2}$ by a differentiation over $t_{12}^{\sigma_1\sigma_2}$ and instead of (139) we can write

$$\mathscr{G}_{12}^{\sigma_1\sigma_2} = -\frac{\delta \Phi}{\delta t_{12}^{\sigma_1\sigma_2}}, \tag{141}$$

with the provision that after the differentiation we have to put

$$t_{12}^{\sigma_1\sigma_2} = \delta(\tau_1 - \tau_2)\delta_{\sigma_1\sigma_2} t_{i_1 i_2}.$$

Of course the GF $\mathscr{G}_{12}^{\sigma_1\sigma_2}$ (139) is diagonal with respect to spin indexes. However in the fluctuating fields it is not like this, that is why we keep indexes σ_1 and σ_2 for electron GF. We write an equation for it using Appendix. We have

$$\frac{\partial}{\partial \tau_1}((TX_1^{0\sigma_1}X_2^{\sigma_2 0}\mathrm{e}^{-V})) = ((T\{X_1^{0\sigma_1},X_2^{\sigma_2 0}\}_+\mathrm{e}^{-V}))+ \\ +((T\dot{X}_1^{0\sigma_1}X_2^{\sigma_2 0}\mathrm{e}^{-V})) - ((T\{X_1^{0\sigma_1},V\}X_2^{\sigma_2 0}\mathrm{e}^{-V})).$$

Write down the permutation relations

$$\{X_1^{0\sigma_1},X_2^{\sigma_2 0}\}_+ = \delta_{12}(X_1^{00}\delta_{\sigma_1\sigma_2} + X_1^{\sigma_2\sigma_1}),$$

$$\{X_1^{0\sigma_1},V\} = -v_1^0 X_1^{0\sigma_1} + X_1^{0\sigma_1'} v_1^{\sigma_1'\sigma_1} + (X_1^{00}\delta_{\sigma_1\sigma_2'} + X_1^{\sigma_2'\sigma_1})\ell_{11'}^{\sigma_2'\sigma_1'} X_{1'}^{0\sigma_1'},$$

$$\dot{X}_1^{0\sigma_1} = -\varepsilon^{\sigma_1} X_1^{0\sigma_1} - (X_1^{00}\delta_{\sigma_1\sigma_1'} + X_1^{\sigma_2'\sigma_1}) t_{11'}^{\sigma_2'\sigma_1'} X_{1'}^{0\sigma_1'}.$$

As a result, our equation has the form

$$\left[\left(\frac{\partial}{\partial \tau_1} + \varepsilon_1^{\sigma_1} - v_1^0\right)\delta_{11'}\delta_{\sigma_1\sigma_1'} + v_1^{\sigma_1'\sigma_1}\delta_{11'}\right]((TX_{1'}^{0\sigma_1'}X_2^{\sigma_2 0}))_V = \\ = \delta_{12}[\delta_{\sigma_1\sigma_2}((TX_2^{00}))_V + ((TX_2^{\sigma_2\sigma_1}))_V] - \\ -(t_{11'}^{\sigma_1\sigma_1'} + \ell_{11'}^{\sigma_1\sigma_1'})((TX_{1'}^{0\sigma_1'}X_2^{\sigma_2 0}X_1^{00}))_V - (t_{11'}^{\sigma_2'\sigma_1'} + \ell_{11'}^{\sigma_2'\sigma_1'})((TX_{1'}^{0\sigma_1'}X_2^{\sigma_2 0}X_1^{\sigma_2'\sigma_1}))_V.$$

We see that ℓ and t enter additively, therefore we can put $\ell_{12}^{\sigma_1\sigma_2} = 0$, implying that $t_{12}^{\sigma_1\sigma_2}$ is a general fluctuating field. After differentiation, we have to connect it with the parameter t_{ij} of the Hamiltonian.

Express all quantities of type $((\ldots))_V$ through variational derivatives; then the last equation is written in the form:

$$(G_{0V}^{-1})_{11'}^{\sigma_1\sigma_1'}\frac{\delta Z}{\delta t_{1'2}^{\sigma_1'\sigma_2}} = \delta_{12}\left(\delta_{\sigma_1\sigma_2}\frac{\delta Z}{\delta v_1^0} + \frac{\delta Z}{\delta v_1^{\sigma_2\sigma_1}}\right) - \\ -t_{11'}^{\sigma_1\sigma_1'}\frac{\delta}{\delta v_1^0}\frac{\delta Z}{\delta t_{1'2}^{\sigma_1'\sigma_2}} - t_{11'}^{\sigma_2'\sigma_1'}\frac{\delta}{\delta v_1^{\sigma_2'\sigma_1}}\frac{\delta Z}{\delta t_{1'2}^{\sigma_1'\sigma_2}}. \tag{142}$$

Here we introduce an expression

$$(G_{0V}^{-1})_{12}^{\sigma_1\sigma_2} = -\left(\frac{\partial}{\partial \tau_1} + \varepsilon_1^{\sigma_1} - v_1^0\right)\delta_{12}\delta_{\sigma_1\sigma_2} - v_1^{\sigma_2\sigma_1}\delta_{12}, \tag{143}$$

which defines the inverse propagator for an electron in a fluctuating field, but without hopping between sites. After the substitution $Z = \mathrm{e}^{\Phi}$, equation (142) is written in the final form

$$\left[(G_{0V}^{-1})_{11'}^{\sigma_1\sigma_1'}+t_{11'}^{\sigma_1\sigma_1'}\frac{\delta\Phi}{\delta v_1^0}+t_{11'}^{s'\sigma_1'}\frac{\delta\Phi}{\delta v_1^{s'\sigma_1}}+\right.$$

$$\left.+t_{11'}^{\sigma_1\sigma_1'}\frac{\delta}{\delta v_1^0}+t_{11'}^{s'\sigma_1'}\frac{\delta}{\delta v_1^{s'\sigma_1}}\right]\frac{\delta\Phi}{\delta t_{1'2}^{\sigma_1'\sigma_2}}=\delta_{12}\left(\delta_{\sigma_1\sigma_2}\frac{\delta\Phi}{\delta v_1^0}+\frac{\delta\Phi}{\delta v_1^{\sigma_2\sigma_1}}\right). \tag{144}$$

Thus we obtain the equation of motion for the electron GF in fluctuating fields, that is

$$\mathcal{G}_{12}^{\sigma_1\sigma_2}[V]\equiv-\langle TX_1^{0\sigma_1}X_2^{\sigma_2 0}\rangle_V=-\frac{\delta\Phi[V]}{\delta t_{12}^{\sigma_1\sigma_2}}. \tag{145}$$

The structure of this equation is similar to equation (57) for transversal GF in the Heisenberg model, and equation (143) is similar to equation (58) for propagator GF of zeroth approximation. It follows from here that the methods of solution of our equation (144) should be similar to those for the Heisenberg model.

Similarly to multiplicative representation (62) for spin GF we introduce a multiplicative representation for electron GF

$$-\frac{\delta\Phi[V]}{\delta t_{12}^{\sigma_1\sigma_2}}=(G\Lambda)_{12}^{\sigma_1\sigma_2}, \tag{146}$$

where G should be called propagator part of GF, and Λ should be called its terminal part. As in the Heisenberg model we split our equation (144) into two equations, one for Λ and one for G or, more exactly, for its self-energy part Σ:

$$\begin{aligned}\Lambda_{12}^{\sigma_1\sigma_2}&=\delta_{12}\left[\delta_{\sigma_1\sigma_2}(\hat{F}_1^{\sigma_1}\Phi)-\delta_{\bar{\sigma}_1\sigma_2}\frac{\delta\Phi}{\delta v_1^{\bar{\sigma}_1\sigma_1}}\right]+\\&+(tG)_{11'}^{\sigma_1\sigma_1'}(\hat{F}_1^{\sigma_1}\Lambda_{1'2}^{\sigma_1'\sigma_2})-(tG)_{11'}^{\bar{\sigma}_1\sigma_1'}\frac{\delta}{\delta v_1^{\bar{\sigma}_1\sigma_1}}\Lambda_{1'2}^{\sigma_1'\sigma_2},\end{aligned} \tag{147}$$

$$\begin{aligned}\Sigma_{12}^{\sigma_1\sigma_2}=(\hat{F}_1^{\sigma_1}\Phi)t_{12}^{\sigma_1\sigma_2}-\frac{\delta\Phi}{\delta v_1^{\bar{\sigma}_1\sigma_1}}t_{12}^{\bar{\sigma}_1\sigma_2}-\delta_{12}[\delta_{\sigma_1\sigma_2}(tG)_{11}^{\bar{\sigma}_1\bar{\sigma}_1}-\delta_{\bar{\sigma}_1\sigma_2}(tG)_{11}^{\sigma_1\bar{\sigma}_1}]+\\+(tG)_{11'}^{\sigma_1\sigma_1'}(\hat{F}_1^{\sigma_1}\Sigma_{1'2}^{\sigma_1'\sigma_2})-(tG)_{11'}^{\bar{\sigma}_1\sigma_1'}\frac{\delta}{\delta v_1^{\bar{\sigma}_1\sigma_1}}\Sigma_{1'2}^{\sigma_1'\sigma_2}.\end{aligned} \tag{148}$$

We have to add to the last equation the Dyson equation

$$(G^{-1})_{12}^{\sigma_1\sigma_2}=(G_{0V}^{-1})_{12}^{\sigma_1\sigma_2}-\Sigma_{12}^{\sigma_1\sigma_2},$$

in order for the system to be closed.

Here we introduced a convenient differential operator

$$\hat{F}_1^{\sigma_1}=-\left(\frac{\delta}{\delta v_1^0}+\frac{\delta}{\delta v_1^{\sigma\sigma}}\right),$$

then, according to definition (140),

$$(\hat{F}_1^{\sigma_1}\Phi) = \langle X_1^{00}\rangle + \langle X_1^{\sigma\sigma}\rangle \equiv \langle F_1^{\sigma}\rangle. \tag{149}$$

In equations (147) and (148) an expression $\frac{\delta}{\delta v_1^{\bar\sigma\sigma}}\Phi = -\langle X_1^{\bar\sigma\sigma}\rangle$ appears, which does not vanish when the fluctuating field $v_1^{\bar\sigma\sigma}$ switches on. The second derivative of that with respect to $v_2^{\sigma\bar\sigma}$ gives the magnon GF.

The iteration of equations for Λ and Σ generates a series of perturbation theory with respect to t for the terminal and self-energy parts of GF. When iterating, it is necessary to calculate derivatives of G over fluctuating fields. One can use the identities

$$\frac{\delta G_{12}^{\sigma_1\sigma_2}}{\delta v_3^0} = -G_{11'}^{\sigma_1\sigma_1'}\frac{\delta (G^{-1})_{1'2'}^{\sigma_1'\sigma_2'}}{\delta v_3^0}G_{2'2}^{\sigma_2'\sigma_2}, \tag{150}$$

$$\frac{\delta (G^{-1})_{12}^{\sigma_1\sigma_2}}{\delta v_3^0} = \frac{\delta (G_{0V}^{-1})_{12}^{\sigma_1\sigma_2}}{\delta v_3^0} - \frac{\delta}{\delta v_3^0}\Sigma_{12}^{\sigma_1\sigma_2}. \tag{151}$$

Taking the explicit expression (143) for G_{0V}^{-1} one finds

$$\frac{\delta (G_{0V}^{-1})_{12}^{\sigma_1\sigma_2}}{\delta v_3^0} = \delta_{12}\delta_{13}\delta_{\sigma_1\sigma_2},$$

$$\frac{\delta (G_{0V}^{-1})_{12}^{\sigma_1\sigma_2}}{\delta v_3^{\sigma_3\sigma_3'}} = -\delta_{31}\delta_{32}\delta_{\sigma_1\sigma_3'}\delta_{\sigma_2\sigma_3}.$$

In the iteration expansions of equations (147) and (148) any differentiation of G generates a product of two G's (with the addition of $\frac{\delta}{\delta v}\Sigma$), or increase the order of a derivative of G. As a result we obtain a series for Λ and Σ in powers of t. Coefficients in this series are products of the GF G and different derivatives of Φ.

Directly from equation (147) we obtain the zeroth approximation for the terminal part

$$(\Lambda_0)_{12}^{\sigma_1\sigma_2} = \delta_{12}\delta_{\sigma_1\sigma_2}(\hat{F}_2^{\sigma_2}\Phi_0) - \delta_{12}\delta_{\bar\sigma_1\sigma_2}\frac{\delta\Phi_0}{\delta v_2^{\sigma_2\bar\sigma_2}}.$$

The first approximation over t is given by an expression

$$\begin{aligned}(\Lambda_1)_{12}^{\sigma_1\sigma_2} &= (tG_0)_{11'}^{\sigma_1\sigma_1'}[\hat{F}_1^{\sigma_1}(\Lambda_0)_{1'2}^{\sigma_1'\sigma_2}] - (tG_0)_{11'}^{\bar\sigma_1\sigma_1'}\frac{\delta(\Lambda_0)_{1'2}^{\sigma_1'\sigma_2}}{\delta v_1^{\bar\sigma_1\sigma_1}} = \\ &= (tG_0)_{12}^{\sigma_1\sigma_2}(\hat{F}_1^{\sigma_1}\hat{F}_2^{\sigma_2}\Phi_0) + (tG_0)_{12}^{\bar\sigma_1\bar\sigma_2}\frac{\delta^2\Phi_0}{\delta v_2^{\sigma_2\bar\sigma_2}\delta v_1^{\bar\sigma_1\sigma_1}} - \\ &-(tG_0)_{12}^{\sigma_1\bar\sigma_2}(\hat{F}_1^{\sigma_1}\frac{\delta\Phi_0}{\delta v_2^{\sigma_2\bar\sigma_2}}) - (tG_0)_{12}^{\bar\sigma_1\sigma_2}(\hat{F}_2^{\sigma_2}\frac{\delta\Phi_0}{\delta v_1^{\bar\sigma_1\sigma_1}}).\end{aligned}$$

We can write now a general expression for the second order correction

$$(\Lambda_2)_{12}^{\sigma_1\sigma_2} = (tG_0)_{11'}^{\sigma_1\sigma_1'}[\hat{F}_1^{\sigma_1}(\Lambda_1)_{1'2}^{\sigma_1'\sigma_2}] - (tG_0)_{11'}^{\bar{\sigma}_1\sigma_1'}\frac{\delta(\Lambda_1)_{1'2}^{\sigma_1'\sigma_2}}{\delta v_1^{\bar{\sigma}_1\sigma_2}}.$$

We must use identities similar to (150):

$$\hat{F}_3^{\sigma_3}G_{12}^{\sigma_1\sigma_2} = -G_{11'}^{\sigma_1\sigma_1'}(\hat{F}_3^{\sigma_3}(G^{-1})_{1'2'}^{\sigma_1'\sigma_2'})G_{2'2}^{\sigma_2'\sigma_2},$$
$$\frac{\delta}{\delta v_3^{\sigma\bar{\sigma}}}G_{12}^{\sigma_1\sigma_2} = -G_{11'}^{\sigma_1\sigma_1'}\left(\frac{\delta}{\delta v_3^{\sigma\bar{\sigma}}}(G^{-1})_{1'2'}^{\sigma_1'\sigma_2'}\right)G_{2'2}^{\sigma_2'\sigma_2}.$$

Standard calculations lead to the result

$$\begin{aligned}
&(\Lambda_2)_{12}^{\sigma_1\sigma_2} = \\
&= (tG_0)_{11'}^{\sigma_1\sigma_1'}[(tG_0)_{1'2}^{\sigma_1'\sigma_2}(\hat{F}_{1'}^{\sigma_1'}\hat{F}_1^{\sigma_1}\hat{F}_2^{\sigma_2}\Phi_0) - (tG_0)_{1'2}^{\sigma_1'\bar{\sigma}_2}(\hat{F}_{1'}^{\sigma_1'}\hat{F}_1^{\sigma_1}\frac{\delta\Phi_0}{\delta v_2^{\sigma_2\bar{\sigma}_2}}) - \\
&-(tG_0)_{1'2}^{\bar{\sigma}_1'\sigma_2}(\hat{F}_1^{\sigma_1}\hat{F}_2^{\sigma_2}\frac{\delta\Phi_0}{\delta v_{1'}^{\bar{\sigma}_1'\sigma_1'}}) + (tG_0)_{1'2}^{\bar{\sigma}_1'\bar{\sigma}_2'}(\hat{F}_1^{\sigma_1}\frac{\delta^2\Phi_0}{\delta v_2^{\sigma_2\bar{\sigma}_2}\delta v_{1'}^{\bar{\sigma}_1'\sigma_1'}}) + \\
&+(tG_0)_{1'1}^{\sigma_1'\bar{\sigma}_1}(G_0)_{12}^{\bar{\sigma}_1\sigma_2}(\hat{F}_{1'}^{\sigma_1'}\hat{F}_2^{\sigma_2}\Phi_0) - (tG_0)_{1'1}^{\sigma_1'\bar{\sigma}_1}(G_0)_{12}^{\bar{\sigma}_1\bar{\sigma}_2}(\hat{F}_{1'}^{\sigma_1'}\frac{\delta\Phi_0}{\delta v_2^{\sigma_2\bar{\sigma}_2}}) - \\
&-(tG_0)_{1'1}^{\bar{\sigma}_1'\bar{\sigma}_1}(G_0)_{12}^{\bar{\sigma}_1\sigma_2}(\hat{F}_2^{\sigma_2}\frac{\delta\Phi_0}{\delta v_{1'}^{\bar{\sigma}_1'\sigma_1'}}) + (tG_0)_{1'1}^{\bar{\sigma}_1'\bar{\sigma}_1}(G_0)_{12}^{\bar{\sigma}_1\bar{\sigma}_2}\frac{\delta^2\Phi_0}{\delta v_2^{\sigma_2\bar{\sigma}_2}\delta v_{1'}^{\bar{\sigma}_1'\sigma_1'}}] - \\
&-(tG_0)_{11'}^{\bar{\sigma}_1\sigma_1'}[(tG_0)_{1'2}^{\sigma_1'\sigma_2}(\hat{F}_{1'}^{\sigma_1'}\hat{F}_2^{\sigma_2}\frac{\delta\Phi_0}{\delta v_1^{\bar{\sigma}_1\sigma_1}}) - (tG_0)_{1'2}^{\sigma_1'\bar{\sigma}_2}(\hat{F}_{1'}^{\sigma_1'}\frac{\delta^2\Phi_0}{\delta v_2^{\sigma_2\bar{\sigma}_2}\delta v_1^{\bar{\sigma}_1\sigma_1}}) - \\
&-(tG_0)_{1'2}^{\bar{\sigma}_1'\sigma_2}(\hat{F}_2^{\sigma_2}\frac{\delta^2\Phi_0}{\delta v_1^{\bar{\sigma}_1\sigma_1}\delta v_{1'}^{\bar{\sigma}_1'\sigma_1'}}) + (tG_0)_{1'2}^{\bar{\sigma}_1'\bar{\sigma}_2}\frac{\delta^3\Phi_0}{\delta v_{1'}^{\bar{\sigma}_1'\sigma_1'}\delta v_1^{\bar{\sigma}_1\sigma_1}\delta_2^{\sigma_2\bar{\sigma}_2}} + \\
&+(tG_0)_{1'1}^{\sigma_1'\bar{\sigma}_1}(G_0)_{12}^{\sigma_1\sigma_2}(\hat{F}_{1'}^{\sigma_1'}\hat{F}_2^{\sigma_2}\Phi_0) - (tG_0)_{1'1}^{\sigma_1'\bar{\sigma}_1}(G_0)_{12}^{\sigma_1\bar{\sigma}_2}(\hat{F}_{1'}^{\sigma_1'}\frac{\delta\Phi_0}{\delta v_2^{\sigma_2\bar{\sigma}_2}}) - \\
&-(tG_0)_{1'1}^{\bar{\sigma}_1'\bar{\sigma}_1}(G_0)_{12}^{\sigma_1\sigma_2}(\hat{F}_2^{\sigma_2}\frac{\delta\Phi_0}{\delta v_{1'}^{\bar{\sigma}_1'\sigma_1'}}) + (tG_0)_{1'1}^{\bar{\sigma}_1'\bar{\sigma}_1}(G_0)_{12}^{\sigma_1\bar{\sigma}_2}\frac{\delta^2\Phi_0}{\delta v_2^{\sigma_2\bar{\sigma}_2}\delta v_1^{\bar{\sigma}_1\sigma_1}}].
\end{aligned}$$

When fluctuating fields are absent

$$(G_0)_{12}^{\sigma_1\sigma_2} = \delta_{\sigma_1\sigma_2}(G_0)_{12}^{\sigma_1}, \quad \frac{\delta\Phi_0}{\delta v_1^{\sigma\bar{\sigma}}} = 0,\ldots,$$

and expressions for Λ_0, Λ_1 and Λ_2 are sinplified. Collecting together all these results we find $\Lambda_{12}^{\bar{\sigma}_1\sigma_2} = \delta_{\sigma_1\sigma_2}\Lambda_{12}^{\sigma_1}$, and

$$\begin{aligned}\Lambda_{12}^{\sigma} = \delta_{12}(\hat{F}_2^{\sigma}\Phi_0) + (t^{\sigma}G_0^{\sigma})_{12}(\hat{F}_1^{\sigma}\hat{F}_2^{\sigma}\Phi_0) + (t^{\bar{\sigma}}G_0^{\bar{\sigma}})_{12}\frac{\delta^2\Phi_0}{\delta v_2^{\sigma\bar{\sigma}}\delta v_1^{\bar{\sigma}\sigma}} + \\ +(t^{\sigma}G_0^{\sigma})_{11'}(G_0^{\bar{\sigma}})_{12}\frac{\delta^2\Phi_0}{\delta v_2^{\sigma\bar{\sigma}}\delta v_{1'}^{\bar{\sigma}\sigma}}(t^{\bar{\sigma}}G_0^{\bar{\sigma}})_{1'1} - (G_0^{\sigma})_{12}(\hat{F}_2^{\sigma}\hat{F}_{1'}^{\bar{\sigma}}\Phi_0)(t^{\bar{\sigma}}G_0^{\bar{\sigma}})_{1'1}(t^{\bar{\sigma}}G_0^{\bar{\sigma}})_{11'} + \\ +(t^{\sigma}G_0^{\sigma})_{11'}(t^{\sigma}G_0^{\sigma})_{1'2}(\hat{F}_{1'}^{\sigma}\hat{F}_1^{\sigma}\hat{F}_2^{\sigma}\Phi_0) + \\ +(t^{\sigma}G_0^{\sigma})_{11'}(t^{\bar{\sigma}}G_0^{\bar{\sigma}})_{1'2}(\hat{F}_1^{\sigma}\frac{\delta^2\Phi_0}{\delta v_2^{\sigma\bar{\sigma}}\delta v_{1'}^{\bar{\sigma}\sigma}}) + (t^{\bar{\sigma}}G_0^{\bar{\sigma}})_{11'}(t^{\sigma}G_0^{\sigma})_{1'2}(\hat{F}_2^{\sigma}\frac{\delta^2\Phi_0}{\delta v_{1'}^{\sigma\bar{\sigma}}\delta v_1^{\bar{\sigma}\sigma}}) + \\ +(t^{\bar{\sigma}}G_0^{\bar{\sigma}})_{11'}(t^{\bar{\sigma}}G_0^{\bar{\sigma}})_{1'2}(\hat{F}_{1'}^{\bar{\sigma}}\frac{\delta^2\Phi_0}{\delta v_2^{\sigma\bar{\sigma}}\delta v_1^{\bar{\sigma}\sigma}}) + \ldots,\end{aligned} \tag{152}$$

Here and in the sequel we formally keep the spin indexes for the matrix element t_{12}^{σ} to control the structure of terms. Notice that the quantity

$$(\hat{F}_1^{\sigma_1}\hat{F}_2^{\sigma_2}\Phi_0) = \langle T(F_1^{\sigma_1} - \langle F_1^{\sigma_1}\rangle)(F_2^{\sigma_2} - \langle F_2^{\sigma_2}\rangle)\rangle = \langle TF_1^{\sigma_1}F_2^{\sigma_2}\rangle_c \equiv D_{12}^{\sigma_1\sigma_2} \tag{153}$$

is a connected GF, constructed on operators $F_1^{\sigma_1} = X_1^{00} + X_1^{\sigma\sigma}$.

We rewrite the beginning of the series (152) for the terminal part in the form

$$\begin{aligned}\Lambda_{12}^{\sigma} = \delta_{12}\langle F_2^{\sigma_2}\rangle + (t^{\sigma}G_0^{\sigma})_{12}D_{12}^{\sigma\sigma} - (t^{\bar{\sigma}}G_0^{\bar{\sigma}})_{12}D_{21}^{\bar{\sigma}} - \\ -(t^{\sigma}G_0^{\sigma})_{11'}(t^{\bar{\sigma}}G_0^{\bar{\sigma}})_{1'1}(G_0^{\bar{\sigma}})_{12}D_{21'}^{\bar{\sigma}} - (t^{\bar{\sigma}}G_0^{\bar{\sigma}})_{11'}(t^{\bar{\sigma}}G_0^{\bar{\sigma}})_{1'1}(G_0^{\sigma})_{12}D_{21'}^{\sigma\bar{\sigma}} + \ldots\end{aligned} \tag{154}$$

We show this result graphically with the help of definitions in Fig. 40. The result (154)

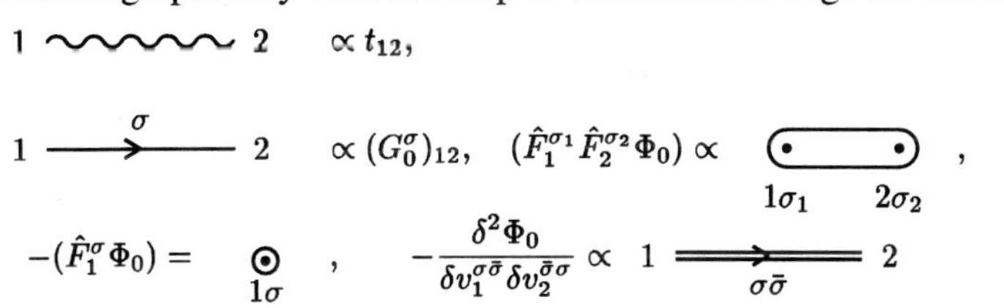

FIGURE 40. Graphic elements for the Hubbard model in the limit $U \to \infty$

is shown in Fig. 41. Graphic representation of second order terms takes additional cumulants of higher orders.

Similar iterations in equation (148) lead to the corrections for the self-energy

$$\begin{aligned}(\Sigma_1)_{12}^{\sigma_1\sigma_2} = (\hat{F}_1^{\sigma_1}\Phi_0)t_{12}^{\sigma_1\sigma_2} - \frac{\delta\Phi_0}{\delta v_1^{\bar{\sigma}_1\sigma_1}}t_{12}^{\bar{\sigma}_1\sigma_2} - \\ -\delta_{12}\delta_{\sigma_1\sigma_2}(tG_0)_{22}^{\bar{\sigma}_1\bar{\sigma}_1} + \delta_{12}\delta_{\bar{\sigma}_1\sigma_2}(tG_0)_{22}^{\sigma_1\bar{\sigma}_1}\end{aligned}$$

and

$$(\Sigma_2)_{12}^{\sigma_1\sigma_2} = (tG_0)_{11'}^{\sigma_1\sigma_1'}[\hat{F}_1^{\sigma_1}(\Sigma_1)_{1'2}^{\sigma_1'\sigma_2}] - (tG_0)_{11'}^{\bar{\sigma}_1\sigma_1'}\frac{\delta(\Sigma_1)_{1'2}^{\sigma_1'\sigma_2}}{\delta v_1^{\bar{\sigma}_1\sigma_1}}.$$

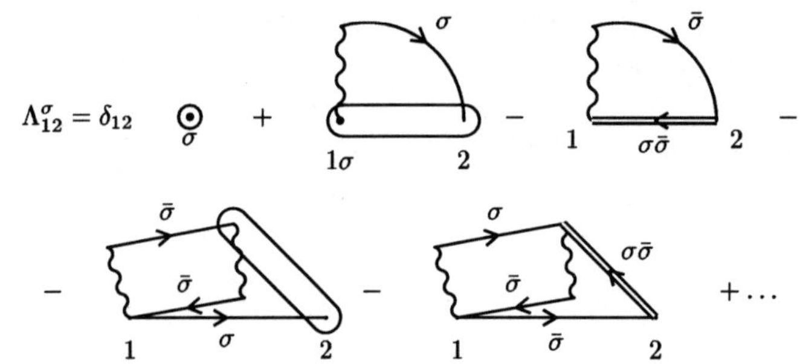

FIGURE 41. Graphic representation for the terminal part of electron GF

Direct calculations give

$$
\begin{aligned}
(\Sigma_1)_{12}^{\sigma_1\sigma_2} = {} & (tG_0)_{11'}^{\sigma_1\sigma_1'}(\hat{F}_{1'}^{\sigma_1'}\hat{F}_1^{\sigma_1}\Phi_0) - (tG_0)_{11'}^{\sigma_1\sigma_1'}(\hat{F}_1^{\sigma_1}\frac{\delta\Phi_0}{\delta v_1^{\bar{\sigma}_1'\sigma_1'}})t_{1'2}^{\bar{\sigma}_1'\sigma_2} - \\
& -(tG_0)_{12}^{\sigma_1\sigma_2}(tG_0)_{21}^{\bar{\sigma}_2\bar{\sigma}_1}G_{12}^{\bar{\sigma}_1\bar{\sigma}_2} + (tG_0)_{12}^{\sigma_1\bar{\sigma}_2}(tG_0)_{21}^{\bar{\sigma}_2\bar{\sigma}_1}G_{12}^{\bar{\sigma}_1\sigma_2} - \\
& -(tG_0)_{11'}^{\bar{\sigma}_1\sigma_1'}(\hat{F}_{1'}^{\sigma_1'}\frac{\delta\Phi_0}{\delta v_1^{\bar{\sigma}_1\sigma_1}})t_{1'2}^{\sigma_1'\sigma} + (tG_0)_{11'}^{\bar{\sigma}_1\sigma_1'}\frac{\delta^2\Phi_0}{\delta v_1^{\bar{\sigma}_1\sigma_1}\delta v_{1'}^{\sigma_1\bar{\sigma}_1}}t_{1'2}^{\bar{\sigma}_1'\sigma_2} + \\
& +(tG_0)_{12}^{\bar{\sigma}_1\sigma_2}(tG_0)_{21}^{\bar{\sigma}_2\bar{\sigma}_1}G_{12}^{\sigma_1\bar{\sigma}_2} - (tG_0)_{12}^{\bar{\sigma}_1\bar{\sigma}_2}(tG_0)_{21}^{\bar{\sigma}_2\bar{\sigma}_1}G_{12}^{\sigma_1\sigma_2}.
\end{aligned}
\tag{155}
$$

When fluctuating fields are taken to be zero, $\Sigma_{12}^{\sigma_1\sigma_2} = \delta_{12}\Sigma_{12}^{\sigma_1}$, and for $\Sigma_{12}^{\sigma_1}$ we obtain

$$
\begin{aligned}
\Sigma_{12}^{\sigma} = {} & (\hat{F}_1^{\sigma}\Phi_0)t_{12}^{\sigma} - \delta_{12}(t^{\bar{\sigma}}G_0^{\bar{\sigma}})_{11} + \\
& +(t^{\sigma}G_0^{\sigma})_{11'}t_{1'2}(\hat{F}_{1'}^{\sigma}\hat{F}_1^{\sigma}\Phi_0) + (t^{\bar{\sigma}}G_0^{\bar{\sigma}})_{11'}t_{1'2}^{\sigma}\frac{\delta^2\Phi_0}{\delta_{1'}^{\sigma\bar{\sigma}}\delta v_1^{\bar{\sigma}\sigma}} - \\
& -(t^{\sigma}G_0^{\sigma})_{12}(t^{\bar{\sigma}}G_0^{\bar{\sigma}})_{21}G_{12}^{\bar{\sigma}} - (t^{\bar{\sigma}}G_0^{\bar{\sigma}})_{12}(t^{\bar{\sigma}}G_0^{\bar{\sigma}})_{21}G_{12}^{\sigma} + \ldots
\end{aligned}
\tag{156}
$$

The graphic form is shown in Fig. 42.

Notice a deep similarity of the graph structure for the electron GF in the Hubbard model ($U = \infty$) and the magnon GF in the Heisenberg model. First, both have a multiplicative representation with propagator and terminal parts (compare expressions (146) and (62)). In both cases we have equations with variational derivatives over fluctuating fields, that is, field h for the Heisenberg model and fields v_1^0, $v_1^{0\sigma}$ and $v_1^{\sigma\bar{\sigma}}$ for the Hubbard model. As a consequence there are similar connections between terminal and self-energy parts. In the case of the Heisenberg model such a connection leads to the Larkin equation for magnon GF. A similar Larkin equation exists for the electron GF in the Hubbard model. This follows from the fact that Σ can be divided into two terms

$$\Sigma_{12}^{\sigma_1\sigma_2} = (\Sigma')_{12}^{\sigma_1\sigma_2} + (\Lambda t)_{12}^{\sigma_1\sigma_2}, \tag{157}$$

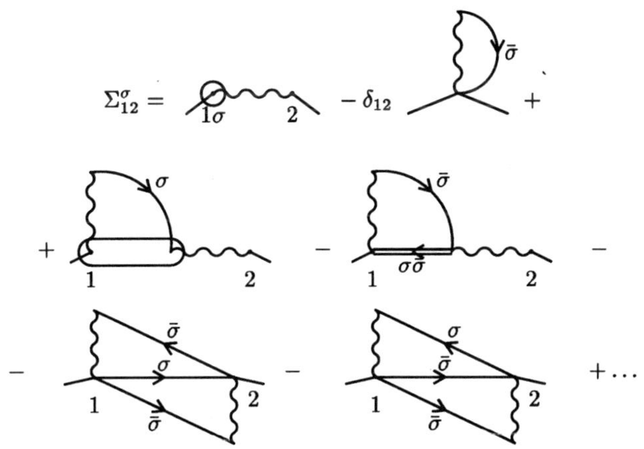

FIGURE 42. Graphic representation for self-energy part of electron GF

where Σ' is represented by graphs not-cut over a line t. From equation (148) and relation (157) one can obtain the equation for Σ':

$$(\Sigma')_{12}^{\sigma_1\sigma_2} = \delta_{12}[(t^{\sigma_1}G^{\sigma_1\sigma_2})_{11} - \delta_{\sigma_1\sigma_2}(t^{s'}G^{s's'})_{11}] - \\ -(t^{\sigma_2'}G^{\sigma_2'\sigma_1'})_{11'}\left(\delta_{\sigma_1\sigma_2'}\frac{\delta}{\delta v_1^0} + \frac{\delta}{\delta v_1^{\sigma_2'\sigma_1}}\right)(\Sigma')_{1'2}^{\sigma_1'\sigma_2}, \tag{158}$$

and indeed, upon iterations, it does not contain terms not-cut over a t-line. Using (157) it is easy to write an equation

$$\mathscr{G}_{12}^{\sigma_1\sigma_2} = (\Sigma')_{12}^{\sigma_1\sigma_2} + (\Sigma' t \mathscr{G})_{12}^{\sigma_1\sigma_2}, \tag{159}$$

which is just the Larkin equation for electron GF.

Equation for the two-particle vertex part. We shall try, by means of the method of the generating functional, to derive the Bethe-Salpether equation for the electron-electron vertex part corresponding to GRPA. To this purpose we follow the general Kadanoff-Baym approach [83], where such kind of equations are derived from the equation for the quantity $\delta\Sigma/\delta v$, with the self-energy part taken in the lowest approximation with respect to the fluctuating field. In the case of the Hubbard model we are interested in the quantity $\delta\Sigma_{12}^{\bar{\sigma}\sigma}/\delta v_3^{\bar{\sigma}\sigma}$.

We start from the basic equation (148) for the self-energy part. Let us take for it the first order approximation and write the expression for the off-diagonal matrix element.

$$\Sigma_{12}^{\bar{\sigma}\sigma} = -\frac{\delta\Phi}{\delta v_1^{\sigma\bar{\sigma}}}t_{12}^{\sigma} + \delta_{12}t_{11'}^{\bar{\sigma}}G_{1'2}^{\bar{\sigma}\sigma} + \ldots \tag{160}$$

GF $G_{12}^{\bar{\sigma}\sigma}$ vanishes when the fluctuating field $v_3^{\bar{\sigma}\sigma}$ is equal to 0, but the variational derivative is not zero. Below we show that $\frac{\delta\Phi}{\delta v_1^{\sigma\bar{\sigma}}}$ contains a contribution, proportional to the off-diagonal electron GF:

$$\frac{\delta\Phi}{\delta v_1^{\sigma\bar{\sigma}}} = -G_{11}^{\bar{\sigma}\sigma} + \ldots \tag{161}$$

So, instead of expression (160), we take the following expression for the off-diagonal matrix element of the self-energy

$$\Sigma_{12}^{\bar{\sigma}\sigma} = G_{11}^{\bar{\sigma}\sigma} t_{12}^{\sigma} + \delta_{12} t_{11'}^{\bar{\sigma}} G_{1'1}^{\bar{\sigma}\sigma}. \tag{162}$$

Taking the derivative with respect to $v_3^{\bar{\sigma}\sigma}$, with the help of (150) and (151), one obtains the equation

$$\frac{\delta\Sigma_{12}^{\bar{\sigma}\sigma}}{\delta v_3^{\bar{\sigma}\sigma}} = G_{13}^{\bar{\sigma}\bar{\sigma}} G_{31}^{\sigma\sigma} t_{12}^{\sigma} + \delta_{12}(tG)_{13}^{\bar{\sigma}\bar{\sigma}} G_{31}^{\sigma\sigma} +$$

$$+ G_{11'}^{\bar{\sigma}\sigma_1'} \frac{\delta\Sigma_{1'2'}^{\sigma_1'\sigma_2'}}{\delta v_3^{\bar{\sigma}\sigma}} G_{2'1}^{\sigma_2'\sigma} t_{12}^{\sigma} + \delta_{12}(t^{\bar{\sigma}} G^{\bar{\sigma}\sigma_1'})_{11'} \frac{\delta\Sigma_{1'2'}^{\sigma_1'\sigma_2'}}{\delta v_3^{\bar{\sigma}\sigma}} G_{2'2}^{\sigma_2'\sigma}.$$

We put $v_3^{\bar{\sigma}\sigma} = 0$ and take into account that in this case $G_{12}^{\sigma_1\sigma_2}$ is diagonal with respect to spin indexes. Then we come to a closed equation for $\frac{\delta\Sigma_{12}^{\bar{\sigma}\sigma}}{\delta v_3^{\bar{\sigma}\sigma}}$, which we write in the form

$$\frac{\delta\Sigma_{12}^{\bar{\sigma}\sigma}}{\delta v_3^{\bar{\sigma}\sigma}} = \Gamma_0 \begin{pmatrix} \bar{\sigma} & \sigma & ; & \sigma & \bar{\sigma} \\ 1 & 2 & ; & 3' & 4' \end{pmatrix} G_{4'3}^{\bar{\sigma}} G_{33'}^{\sigma} +$$

$$+ \Gamma_0 \begin{pmatrix} \bar{\sigma} & \sigma & ; & \sigma & \bar{\sigma} \\ 1 & 2 & ; & 3' & 4' \end{pmatrix} G_{4'1'}^{\bar{\sigma}} G_{2'3'}^{\sigma} \frac{\delta\Sigma_{1'2'}^{\bar{\sigma}\sigma}}{\delta v_3^{\bar{\sigma}\sigma}},$$

where

$$\Gamma_0 \begin{pmatrix} \bar{\sigma} & \sigma & ; & \sigma & \bar{\sigma} \\ 1 & 2 & ; & 3 & 4 \end{pmatrix} = \delta_{14}\delta_{43} t_{12}^{\sigma} + \delta_{12}\delta_{23} t_{34}^{\bar{\sigma}}. \tag{163}$$

Let us look for a solution in the form

$$\frac{\delta\Sigma_{12}^{\bar{\sigma}\sigma}}{\delta v_3^{\bar{\sigma}\sigma}} = \Gamma \begin{pmatrix} \bar{\sigma} & \sigma & ; & \sigma & \bar{\sigma} \\ 1 & 2 & ; & 3' & 4' \end{pmatrix} G_{4'3}^{\bar{\sigma}} G_{33'}^{\sigma},$$

then we come to the equation for Γ:

$$\Gamma \begin{pmatrix} \bar{\sigma} & \sigma & ; & \sigma & \bar{\sigma} \\ 1 & 2 & ; & 3 & 4 \end{pmatrix} = \Gamma_0 \begin{pmatrix} \bar{\sigma} & \sigma & ; & \sigma & \bar{\sigma} \\ 1 & 2 & ; & 3 & 4 \end{pmatrix} +$$

$$+ \Gamma_0 \begin{pmatrix} \bar{\sigma} & \sigma & ; & \sigma & \bar{\sigma} \\ 1 & 2 & ; & 3' & 4' \end{pmatrix} G_{4'1'}^{\bar{\sigma}} G_{2'3'}^{\sigma} \Gamma_0 \begin{pmatrix} \bar{\sigma} & \sigma & ; & \sigma & \bar{\sigma} \\ 1' & 2' & ; & 3 & 4 \end{pmatrix} \tag{164}$$

In its graphic representation, Fig. 43, one recognizes the Bethe-Salpether equation for the vertex part with antiparallel ladders, obtained earlier (see Section IIA). It lies in the base of GRPA.

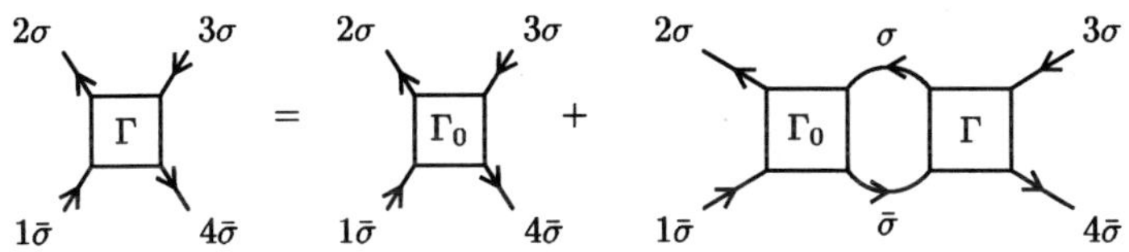

FIGURE 43. Equation for two-particle vertex part

An important feature, when deriving this equation, is the use of relation (161). To derive this relation we consider the Dyson equation

$$(G_{0V})_{12}^{\sigma_1\sigma_2} = (G_0)_{12}^{\sigma_1\sigma_2} + (G_0)_{11'}^{\sigma_1\sigma_1'} v_{1'}^{\sigma_1'\bar{\sigma}_1'} (G_{0V})_{1'2}^{\bar{\sigma}_1'\sigma_2},$$

where G_0 is determined by relation

$$(G_0^{-1})_{12}^{\sigma_1\sigma_2} = -\left(\frac{\partial}{\partial\tau_1} + \varepsilon_1^{\sigma_1}\right)\delta_{12}\delta_{\sigma_1\sigma_2}.$$

Write down separately equations for diagonal and off-diagonal matrix elements:

$$(G_{0V})_{12}^{\bar{\sigma}\sigma} = \left[(G_0^{\sigma\sigma})^{-1} - (v^{\sigma\bar{\sigma}}G_0^{\bar{\sigma}\bar{\sigma}}v^{\bar{\sigma}\sigma})\right]_{11'}^{-1} v_{1'}^{\bar{\sigma}\sigma}(G_0^{\bar{\sigma}\bar{\sigma}})_{1'2}.$$

$$(G_{0V})_{11}^{\bar{\sigma}\sigma} = \left[(G_0^{\sigma})^{-1} - (v^{\sigma\bar{\sigma}}G_0^{\bar{\sigma}}v^{\bar{\sigma}\sigma})\right]_{11'}^{-1} v_{1'}^{\bar{\sigma}\sigma}(G_0^{\bar{\sigma}})_{1'1} \approx (G_0^{\sigma}v^{\bar{\sigma}\sigma}G_0^{\bar{\sigma}})_{11}. \qquad (165)$$

Let us calculate the quantity $\delta\Phi/\delta v_1^{\sigma\bar{\sigma}}$. By standard methods, described in the Appendix, we write the equation for it:

$$(K_0^{\sigma\bar{\sigma}})_{11'}^{-1}\frac{\delta\Phi}{\delta v_{1'}^{\sigma\bar{\sigma}}} = -v_1^{\sigma\bar{\sigma}}\left(\frac{\delta\Phi}{\delta v_1^{\bar{\sigma}\bar{\sigma}}} - \frac{\delta\Phi}{\delta v_1^{\sigma\sigma}}\right) + t_{11'}^{\bar{\sigma}\sigma_1'}\frac{\delta\Phi}{\delta t_{1'1}^{\sigma_1'\sigma}} - \frac{\delta\Phi}{\delta t_{11'}^{\bar{\sigma}\sigma_1}}t_{1'1}^{\sigma_1'\sigma},$$

where

$$(K_0^{\sigma\bar{\sigma}})_{12}^{-1} = -\left(\frac{\partial}{\partial\tau_1} + \sigma h\right)\delta_{12}$$

is the propagator of the magnon GF of zeroth order. At $t = 0$ our equation can be rewritten in the form:

$$\frac{\delta\Phi_0}{\delta v_1^{\sigma\bar{\sigma}}} = (K_0^{\sigma\bar{\sigma}})_{11'}v_{1'}^{\sigma\bar{\sigma}}(\langle X_{1'}^{\bar{\sigma}\bar{\sigma}}\rangle_0 - \langle X_{1'}^{\sigma\sigma}\rangle_0). \qquad (166)$$

Now we compare the two relations (166) and (165) in momentum representation. From (166) we have

$$\frac{\delta\Phi_0}{\delta v_1^{\sigma\bar{\sigma}}} = T\sum_{\nu} \mathrm{e}^{-\mathrm{i}\omega_\nu\tau_1}\frac{v^{\sigma\bar{\sigma}}(\mathrm{i}\omega_\nu)}{\mathrm{i}\omega_\nu - \sigma h}(\langle X^{\bar{\sigma}\bar{\sigma}}\rangle_0 - \langle X^{\sigma\sigma}\rangle_0).$$

From (165) in turn we have

$$(G_{0V})_{11}^{\bar{\sigma}\sigma} = T^2\sum_{\nu n} \mathrm{e}^{-\mathrm{i}\omega_\nu\tau_1}G_0^{\bar{\sigma}}(\mathrm{i}\omega_n)G_0^{\sigma}(\mathrm{i}\omega_n - \mathrm{i}\omega_\nu)v^{\sigma\bar{\sigma}}(\mathrm{i}\omega_\nu).$$

Upon substitution of the propagator for the electron GF

$$G_0^\sigma(\mathrm{i}\omega_n) = \frac{1}{\mathrm{i}\omega_n - \varepsilon^\sigma},$$

we find

$$\begin{aligned}(G_{0V})_{11}^{\bar\sigma\sigma} &= T\sum_\nu \mathrm{e}^{-\mathrm{i}\omega_\nu\tau_1}\frac{v^{\sigma\bar\sigma}(\mathrm{i}\omega_\nu)}{\mathrm{i}\omega_\nu - \sigma h}[f(\varepsilon^\sigma) - f(\varepsilon^{\bar\sigma})] = \\ &= -(K_0^{\sigma\bar\sigma})_{11'}v_{1'}^{\sigma\bar\sigma}[(G_0^{\bar\sigma})_{1'1'} - (G_0^\sigma)_{1'1'}],\end{aligned} \tag{167}$$

here $f(\varepsilon)$ is the Fermi function appearing as a result of the summation over frequencies ω_n.

From comparison with expressions (166) and (167) we obtain the final relation between $\frac{\delta\Phi_0}{\delta v_1^{\sigma\bar\sigma}}$ and $(G_{0V})_{11}^{\bar\sigma\sigma}$:

$$\frac{\delta\Phi_0}{\delta v_1^{\sigma\bar\sigma}} = -(G_{0V})_{11}^{\bar\sigma\sigma} - (K_0^{\sigma\bar\sigma})_{11'}v_{1'}^{\sigma\bar\sigma}(\Delta_0^{\bar\sigma})_{1'1}, \tag{168}$$

where

$$(\Delta_0^{\bar\sigma})_{12} = \delta_{12}\{\langle F_2^\sigma\rangle_0 - \langle F_2^{\bar\sigma}\rangle_0 - (G_0^\sigma)_{22} + (G_0^{\bar\sigma})_{22}\} \equiv \Delta_0^{\bar\sigma}\delta_{12}. \tag{169}$$

Notice that instead of G_0^σ we could have taken the electron propagator in the Hartree-Fock approximation and our derivation would not have changed, therefore in (168) instead of G_{0V} we can take G_V, and relation (161) has been proved.

By differentiating the equality (168) over $v_2^{\bar\sigma\sigma}$, we obtain an important relation

$$-\frac{\delta^2\Phi_0}{\delta v_1^{\sigma\bar\sigma}\delta v_2^{\bar\sigma\sigma}} = \frac{\delta G_{11}^{\bar\sigma\sigma}}{\delta v_2^{\bar\sigma\sigma}} + (K_0^{\sigma\bar\sigma})_{12}\Delta_0^{\bar\sigma}.$$

Here in the left hand side stands the magnon GF of zeroth approximation $(D_0^{\sigma\bar\sigma})_{12}$. In the right hand side a derivative over $v_2^{\bar\sigma\sigma}$ gives a product $G_{12}^{\bar\sigma}G_{21}^\sigma$, therefore the obtained relation can be written in the form

$$(D_0^{\sigma\bar\sigma})_{12} = G_{12}^{\bar\sigma}G_{21}^\sigma + (K_0^{\sigma\bar\sigma})_{12}\Delta_0^{\bar\sigma}. \tag{170}$$

It will be used below during the calculation if the magnon GF.

Magnon Green's function in the limit $U = \infty$. Consider now the GF for transversal spin components

$$D_{12}^{\sigma\bar\sigma} = -\langle TX_1^{\sigma\bar\sigma}X_2^{\bar\sigma\sigma}\rangle = -\frac{\delta\Phi}{\delta v_1^{\sigma\bar\sigma}\delta v_2^{\bar\sigma\sigma}}.$$

We should start from the general definition of equation of motion given in the Appendix:

$$\begin{aligned}\frac{\partial}{\partial\tau_1}((TX_1^{\sigma\bar\sigma}X_2^{\bar\sigma\sigma}))_V &= ((T\{X_1^{\sigma\bar\sigma},X_2^{\bar\sigma\sigma}\}_-))_V + \\ &+ ((T\dot X_1^{\sigma\bar\sigma}X_2^{\bar\sigma\sigma}))_V - ((T\{X_1^{\sigma\bar\sigma},V\}X_2^{\bar\sigma\sigma}))_V.\end{aligned}$$

Here the permutation relations are

$$\{X_1^{\sigma\bar\sigma}, X_2^{\bar\sigma\sigma}\}_- = \delta_{12}(X_2^{\sigma\sigma} - X_2^{\bar\sigma\bar\sigma}),$$
$$\{X_1^{\sigma\bar\sigma}, V\} = X_1^{\sigma\sigma_2'} v_1^{\sigma_2'\bar\sigma} - v_1^{\sigma\bar\sigma_1'} X_1^{\sigma_1'\bar\sigma} + X_1^{\sigma 0} \ell_{11'}^{\bar\sigma\sigma_1'} X_{1'}^{0\sigma_1'} - X_{2'}^{\sigma_2' 0} \ell_{2'1}^{\sigma_2'\sigma} X_1^{0\bar\sigma},$$
$$\dot X_1^{\sigma\bar\sigma} = (\varepsilon^\sigma - \varepsilon^{\bar\sigma}) X_1^{\sigma\bar\sigma} - X_1^{\sigma 0} t_{11'}^{\bar\sigma\sigma_1'} X_{1'}^{0\sigma_1'} + X_{2'}^{\sigma_2' 0} t_{2'1}^{\sigma_2'\sigma} X_1^{0\bar\sigma}.$$

After substituting them in the equation, we get the equation of motion

$$(K_{0V}^{\sigma\bar\sigma})^{-1}_{11'}((TX_{1'}^{\sigma\bar\sigma} X_2^{\bar\sigma\sigma}))_V = \delta_{12}((X_2^{\sigma\sigma} - X_2^{\bar\sigma\bar\sigma}))_V -$$
$$- v_1^{\sigma\bar\sigma}((T(X_1^{\sigma\sigma} - X_1^{\bar\sigma\bar\sigma}) X_2^{\bar\sigma\sigma}))_V +$$
$$+ t_{11'}^{\bar\sigma\sigma_1'}((TX_{1'}^{0\sigma_1'} X_1^{\sigma 0} X_2^{\bar\sigma\sigma}))_V - ((TX_2^{\bar\sigma\sigma} X_1^{0\bar\sigma} X_{1'}^{\sigma_1' 0}))_V t_{1'1}^{\sigma_1'\sigma},$$

where propagator of zeroth order

$$(K_{0V}^{-1})_{12}^{\sigma\bar\sigma} = -\delta_{12}\left[\frac{\partial}{\partial\tau_1} + \sigma h + v_1^{\bar\sigma\bar\sigma} - v_1^{\sigma\sigma}\right]$$

(we used the relation $\varepsilon^\sigma - \varepsilon^{\bar\sigma} = -\sigma h$). In terms of variational derivatives the equation reads

$$-(K_{0V}^{-1})_{11'}^{\sigma\bar\sigma} \frac{\delta\Phi}{\delta v_{1'}^{\sigma\bar\sigma} \delta v_2^{\bar\sigma\sigma}} =$$
$$= -\delta_{12}\left(\frac{\delta\Phi}{\delta v_2^{\sigma\sigma}} - \frac{\delta\Phi}{\delta v_2^{\bar\sigma\bar\sigma}}\right) - v_1^{\sigma\bar\sigma}\left(\frac{\delta}{\delta v_1^{\sigma\sigma}} - \frac{\delta}{\delta v_1^{\bar\sigma\bar\sigma}}\right)\frac{\delta\Phi}{\delta v_2^{\bar\sigma\sigma}} - \tag{171}$$
$$- t_{11'}^{\bar\sigma\sigma_1'} \frac{\delta}{\delta v_2^{\bar\sigma\sigma}}\left(\frac{\delta\Phi}{\delta t_{1'1}^{\sigma_1'\sigma}}\right) + \frac{\delta}{\delta v_2^{\bar\sigma\sigma}}\left(\frac{\delta\Phi}{\delta t_{11'}^{\bar\sigma\sigma_1'}}\right) t_{1'1}^{\sigma_1'\sigma}.$$

This equation has not such a closed structure as the equation (144) for the electron GF, therefore it is necessary to add another equation for the mixture derivative over fields $v_1^{\bar\sigma}$ and $t_{12}^{\sigma_1\sigma_2}$. It will be more convenient to use some other approach, when we shall consider the tJ-model, but not the limit $U \to \infty$. Equation (171) is now useful because it allows to find the magnon GF of zeroth order approximation. Indeed, taking $t = 0$ in (171), we are looking for a solution of the form

$$\mathring{D}_{12}^{\sigma\bar\sigma} = -(K_{0V}^{\sigma\bar\sigma})_{12}(\hat F_2^\sigma - \hat F_2^{\bar\sigma})\Phi_0 - (K_{0V}^{\sigma\bar\sigma})_{11'} v_{1'}^{\sigma\bar\sigma} \frac{\delta}{\delta v_2^{\bar\sigma\sigma}}(\hat F_{1'}^\sigma - \hat F_{1'}^{\bar\sigma})\Phi_0.$$

Transform this expression with the help of the two relations

$$\hat F_2^\sigma \Phi_0 = \langle X_2^{00}\rangle_0 + \langle X_2^{\sigma\sigma}\rangle_0,$$
$$(K_{0V}^{-1})_{12}^{\sigma\bar\sigma}\left[(G_0^\sigma)_{22} - (G_0^{\bar\sigma})_{22}\right] = (G_0^{\bar\sigma})_{12}(G_0^\sigma)_{21},$$

(the last relation is proved by Fourier transformation of both sides). Now the expression for $\mathring{D}_{12}^{\sigma\bar\sigma}$ can be written in the form

$$\mathring{D}_{12}^{\sigma\bar\sigma} = (G_0^{\bar\sigma})_{12}(G_0^\sigma)_{21} + (K_{0V}^{\sigma\bar\sigma}\Delta_V^{\bar\sigma})_{12},$$

where

$$(\Delta_V^{\bar{\sigma}})_{12} = \delta_{12}\left\{\langle F_2^{\sigma}\rangle_0 - \langle F_2^{\bar{\sigma}}\rangle_0 - (G_0^{\sigma})_{22} + (G_0^{\bar{\sigma}})_{22} + v_1^{\sigma\bar{\sigma}}\frac{\delta}{\delta v_2^{\bar{\sigma}\sigma}}(\hat{F}_1^{\sigma} - \hat{F}_1^{\bar{\sigma}})\right\}.$$

When the fluctuating fields are absent, the last two relations reduce to the following ones

$$\mathring{D}_{12}^{\sigma\bar{\sigma}} = (G_0^{\bar{\sigma}})_{12}(G_0^{\sigma})_{21} + (K_0^{\sigma\bar{\sigma}})_{12}\Delta_{02}^{\bar{\sigma}}, \tag{172}$$

$$\Delta_{02}^{\bar{\sigma}} = \langle X_2^{\sigma\sigma}\rangle_0 - \langle X_2^{\bar{\sigma}\bar{\sigma}}\rangle_0 - (G_0^{\sigma})_{22} + (G_0^{\bar{\sigma}})_{22}. \tag{173}$$

Again we come to the relations (170) and (169), obtained at the end of previous part by means of another method.

Relation (172) has a simple graphic representation

$$-\left(\frac{\delta\Phi_0}{\delta v_1^{\sigma\bar{\sigma}}\delta v_2^{\bar{\sigma}\sigma}}\right)_0 = \mathring{D}_{12}^{\sigma\bar{\sigma}} =$$

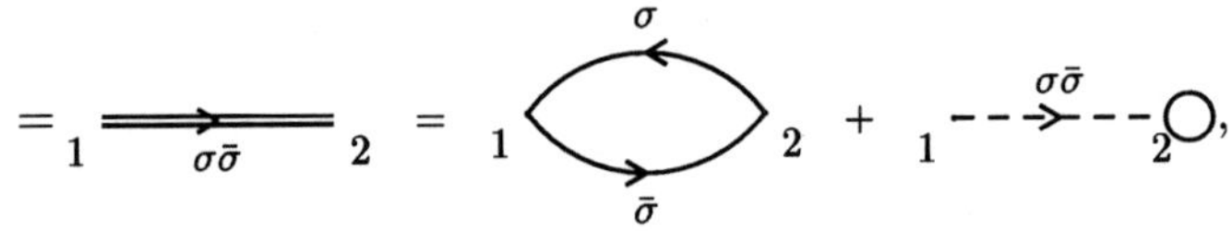

(174)

where a small circle denote the quantity $\Delta_2^{\bar{\sigma}}$, which we can write in a slightly different form

$$\Delta_2^{\bar{\sigma}} = [\langle F_2^{\sigma}\rangle_0 - (G_0^{\sigma})_{22}] - [\langle F_2^{\bar{\sigma}}\rangle_0 - (G_0^{\bar{\sigma}})_{22}] \equiv \tilde{F}_2^{\sigma} - \tilde{F}_2^{\bar{\sigma}}.$$

So, at the end of a dashed line in (174) there is the difference of two quantities

$\tilde{F}_2^{\sigma} =$ 

(175)

Earlier we called them noncomplete cumulants.

Relation (174) expresses a duality in the structure of the magnon GF, because it contains two contributions: one is purely electronic, but the other reminds the magnon GF in the Heisenberg model with localized spins. It reflects a dual behavior of strongly correlated systems as if they were simultaneously itinerant and localized magnets. A dual representation of magnon GF of zeroth order approximation leads to a dual character of the exact magnon GF, which can be seen from its graphic representation (Fig. 27). A detailed derivation of this representation has been done in a book [71].

We need the equation of motion for f- and b- operators

$$\begin{aligned}\dot{X}_1^{0\sigma_1} &= -\varepsilon^{\sigma_1}X_1^{0\sigma_1} - (X_1^{00}\delta_{\sigma_1\sigma_1'} + X_1^{\sigma_1'\sigma_1})t_{11'}X_{1'}^{0\sigma_1'} - \\ &-X_1^{0\bar{\sigma}_1}J_{11'}X_{1'}^{\bar{\sigma}_1\sigma_1} + X_1^{0\sigma_1}J_{11'}X_{1'}^{\bar{\sigma}_1\bar{\sigma}_1};\\ \dot{X}_1^{\sigma_1\bar{\sigma}_1} &= -(\varepsilon^{\bar{\sigma}_1} - \varepsilon_1^{\sigma})X_1^{\sigma_1\bar{\sigma}_1} + X_{1'}^{\sigma_1 0}t_{1'1}X_1^{0\bar{\sigma}_1} - X_1^{\sigma_1 0}t_{11'}X_{1'}^{0\bar{\sigma}_1} - \\ &-X_{1'}^{\sigma_1\bar{\sigma}_1}J_{1'1}(X_1^{\sigma_1\sigma_1} - X_1^{\bar{\sigma}_1\bar{\sigma}_1}) - X_1^{\sigma_1\bar{\sigma}_1}J_{11'}(X_{1'}^{\sigma_1\sigma_1} - X_{1'}^{\bar{\sigma}_1\bar{\sigma}_1}).\end{aligned}$$

The operator V, describing the interaction of a system with fluctuating fields, should be taken in the form

$$V = v^0_{1'} X^{00}_{1'} + v^{\sigma'_1\sigma'_2}_{1'} X^{\sigma'_2\sigma'_1}_{1'} + X^{\sigma'_2 0}_{2'} \ell^{\sigma'_2\sigma'_1}_{2'1'} X^{0\sigma'_1}_{1'} + \frac{1}{2} X^{\sigma'_1\bar{\sigma}'_1}_{2'} \mathscr{J}^{\bar{\sigma}'_1}_{2'1'} X^{\bar{\sigma}'_1\sigma'_1}_{1'}. \qquad (176)$$

So, comparing with the case $U \to \infty$, we added a quadratic form of b-like operators $X^{\sigma\bar{\sigma}}_1$ and $X^{\bar{\sigma}\sigma}_2$. The variational derivative over $\ell^{\sigma_1\sigma}_{12}$ should give the one-particle electron GF, while differentiation over $\mathscr{J}^{\sigma}_{12}$ gives the magnon GF. After differentiations we should put the fields ℓ and $\mathscr{J}$ equal to zero.

Introduce the electron GF in the fluctuating field V:

$$\mathscr{G}^{\sigma_1\sigma_2}_{12} = -\langle T X^{0\sigma_1}_1 X^{\sigma_2 0}_2 \mathrm{e}^{-V} \rangle$$

and write the equation of motion according to the Appendix:

$$\frac{\partial}{\partial \tau_1}((T X^{0\sigma_1}_1 X^{\sigma_2 0}_2 \mathrm{e}^{-V})) = ((T\{X^{0\sigma_1}_1, X^{\sigma_2 0}_2\}_+ \mathrm{e}^{-V})) + \\ + ((T\dot{X}^{0\sigma_1}_1 X^{\sigma_2 0}_2 \mathrm{e}^{-V})) - ((T\{X^{0\sigma_1}_1, V\} X^{\sigma_2 0}_2 \mathrm{e}^{-V})).$$

We substitute in it the expression for $\dot{X}^{0\sigma_1}_1$ and the two relations

$$\{X^{0\sigma_1}_1, X^{\sigma_2 0}_2\}_+ = X^{0\sigma_1}_1 X^{\sigma_2 0}_2 + X^{\sigma_2 0}_2 X^{0\sigma_1}_1,$$
$$\{X^{0\sigma_1}_1, V\} = -v^0_1 X^{0\sigma_1}_1 + X^{0\sigma'_1}_1 v^{\sigma'_1\sigma_1}_1 + \\ +[(X^{00}_1 + X^{\sigma_1\sigma_1}_1)\delta_{\sigma_1\sigma'_2} + X^{\bar{\sigma}_1\sigma_1}_1 \delta_{\bar{\sigma}_1\sigma'_2}] \ell^{\sigma'_2\sigma'_1}_{11'} X^{0\sigma'_1}_{1'} + X^{0\bar{\sigma}_1}_1 \mathscr{J}^{\bar{\sigma}}_{11'} X^{\bar{\sigma}_1\sigma_1}_{1'}.$$

In terms of variational derivatives one can write

$$\{(G^{\sigma}_{0V})^{-1}_{11'} + (\frac{\delta}{\delta v^0_1} + \frac{\delta}{\delta v^{\bar{\sigma}_1\bar{\sigma}_1}_1}) t^{\sigma_1\sigma'_1}_{11'} + \frac{\delta}{\delta v^{\bar{\sigma}_1\sigma_1}_1} t^{\bar{\sigma}_1\sigma'_1}_{11'} + \\ + \delta_{11'}\delta_{\bar{\sigma}_1\sigma'_1} J^{\bar{\sigma}}_{11'} \frac{\delta}{\delta v^{\bar{\sigma}_1\sigma_1}_1} + \delta_{11'}\delta_{\sigma_1\sigma'_1} J^z_{11'} (\frac{\delta}{\delta v^0_1} + \frac{\delta}{\delta v^{\bar{\sigma}_1\sigma_1}_1})\} \frac{\delta Z}{\delta t^{\sigma'_1\sigma_2}_{1'2}} = \\ = \delta_{12}\left[\delta_{\sigma_1\sigma_2}\left(\frac{\delta Z}{\delta v^0_1} + \frac{\delta Z}{\delta v^{\bar{\sigma}_1\bar{\sigma}_1}_1}\right) + \delta_{\bar{\sigma}_1\sigma_2} \frac{\delta Z}{\delta v^{\sigma_2\bar{\sigma}_2}_2}\right].$$

Here we introduced the electron propagator in the fluctuating field of zeroth order approximation with respect to V:

$$(G^{-1}_{0V})^{\sigma_1\sigma_2}_{12} = (G^{-1}_0)^{\sigma_1}_{12}\delta_{\sigma_1\sigma_2} - [(v^{\sigma_1\sigma_1}_1 - v^0_1)\delta_{\sigma_1\sigma_2} + v^{\bar{\sigma}_1\sigma_1}_1 \delta_{\bar{\sigma}_1\sigma_2}]\delta_{12}, \\ (G^{-1}_0)^{\sigma_1}_{12} = -\left(\frac{\delta}{\delta\tau_1} + \varepsilon^{\sigma_1}\right)\delta_{12}\delta_{\sigma_1\sigma_2}. \qquad (177)$$

Notice a renaming of the fields

$$t_{12}\delta_{\sigma_1\sigma_2} + \ell^{\sigma_1\sigma_2}_{12} \equiv t^{\sigma_1\sigma_2}_{12}; \quad J_{12} + \mathscr{J}^{\bar{\sigma}_1}_{12} \equiv J^{\bar{\sigma}_1}_{12}.$$

After taking the derivatives we should put the fields $t_{12}^{\sigma_1\sigma_2}$ and $J_{12}^{\bar{\sigma}}$ equal to the corresponding values of the Hamiltonian parameters (133).

Using the substitution $Z = \mathrm{e}^{\Phi}$, we obtain an equation for the electron GF in the form:

$$\left[(G_{1V}^{-1})_{11'}^{\sigma_1\sigma_1'} - \hat{\Sigma}_{11'}^{\sigma_1\sigma_1'}\right] \frac{\delta\Phi}{\delta t_{1'2}^{\sigma_1'\sigma_2}} = -\delta_{12}\left[\delta_{\sigma_1\sigma_2}(\hat{F}_2^{\sigma_2}\Phi) - \delta_{\bar{\sigma}_1\sigma_2}\frac{\delta\Phi}{\delta v_2^{\sigma_2\bar{\sigma}_2}}\right], \tag{178}$$

where we introduced the electron propagator G_{1V} in the Hatree-Fock approximation and the operator $\hat{\Sigma}_{12}^{\sigma_1\sigma_2}$ with variational derivatives

$$\begin{gathered}(G_{1V}^{-1})_{12}^{\sigma_1\sigma_2} = (G_{0V}^{-1})_{12}^{\sigma_1\sigma_2} - \\ -\gamma^z\begin{pmatrix}\sigma_1 & \sigma_2 & ; & \sigma_2' \\ 1 & 2 & ; & 2'\end{pmatrix}(\hat{F}_2^{\sigma_2'}\Phi) + \gamma\begin{pmatrix}\bar{\sigma}_1 & \sigma_2 & ; & \sigma_2' \\ 1 & 2 & ; & 2'\end{pmatrix}\frac{\delta\Phi}{\delta v_{2'}^{\sigma_2'\bar{\sigma}_2'}}; \\ \hat{\Sigma}_{12}^{\sigma_1\sigma_2} = \gamma^z\begin{pmatrix}\sigma_1 & \sigma_2 & ; & \sigma_2' \\ 1 & 2 & ; & 2'\end{pmatrix}\hat{F}_{2'}^{\sigma_2'} - \gamma\begin{pmatrix}\bar{\sigma}_1 & \sigma_2 & ; & \sigma_2' \\ 1 & 2 & ; & 2'\end{pmatrix}\frac{\delta}{\delta v_{2'}^{\sigma_2'\bar{\sigma}_2'}}.\end{gathered} \tag{179}$$

Here we used the following notation for bare amplitudes of interaction

$$\left.\begin{aligned}\gamma\begin{pmatrix}\bar{\sigma}_1 & \sigma_2 & ; & \sigma_3 \\ 1 & 2 & ; & 3\end{pmatrix} &= t_{12}^{\bar{\sigma}_1\sigma_2}\delta_{13}\delta_{\bar{\sigma}_1\sigma_3} + J_{23}^{\bar{\sigma}_1}\delta_{\bar{\sigma}_1\sigma_2}\delta_{\sigma_2\sigma_3}\delta_{12} \\ \gamma^z\begin{pmatrix}\sigma_1 & \sigma_2 & ; & \sigma_3 \\ 1 & 2 & ; & 3\end{pmatrix} &= t_{12}^{\sigma_1\sigma_2}\delta_{13}\delta_{\sigma_1\sigma_3} + J_{23}^{z}\delta_{\sigma_1\sigma_2}\delta_{\sigma_1\sigma_3}\delta_{12}\end{aligned}\right\} \tag{180}$$

As earlier we suggest a multiplicative representations for the GF

$$(\mathscr{G}_V)_{12}^{\sigma_1\sigma_2} = -\frac{\delta\Phi}{\delta t_{12}^{\sigma_1\sigma_2}} = (G\Lambda)_{12}^{\sigma_1\sigma_2}, \tag{181}$$

then the basic equation is decomposed into a couple of equations for G and Λ or for Σ and Λ:

$$\begin{gathered}\Lambda_{12}^{\sigma_1\sigma_2} = \delta_{12}\left[\delta_{\sigma_1\sigma_2}(\hat{F}_2^{\sigma_2}\Phi) - \delta_{\bar{\sigma}_1\sigma_2}\frac{\delta\Phi}{\delta v_2^{\sigma_2\bar{\sigma}_2}}\right] + \\ +G_{1'2'}^{\sigma_1'\sigma_2'}\gamma^z\begin{pmatrix}\sigma_1 & \sigma_1' & ; & \sigma_4' \\ 1 & 1' & ; & 4'\end{pmatrix}(\hat{F}_{4'}^{\sigma_4'}\Lambda_{2'2}^{\sigma_2'\sigma_2}) - G_{1'2'}^{\sigma_1'\sigma_2'}\gamma\begin{pmatrix}\bar{\sigma}_1 & \sigma_1' & ; & \sigma_4' \\ 1 & 1' & ; & 4'\end{pmatrix}\frac{\delta\Lambda_{2'2}^{\sigma_2'\sigma_2}}{\delta v_{4'}^{\sigma_4'\bar{\sigma}_4'}},\end{gathered} \tag{182}$$

$$\begin{gathered}\Sigma_{12}^{\sigma_1\sigma_2} = \\ = \gamma^z\begin{pmatrix}\sigma_1 & \sigma_2 & ; & \sigma_2' \\ 1 & 2 & ; & 2'\end{pmatrix}(\hat{F}_{2'}^{\sigma_2'}\Phi) + \gamma^z\begin{pmatrix}\sigma_1 & \sigma_1' & ; & \bar{\sigma}_2 \\ 1 & 1' & ; & 2\end{pmatrix}G_{1'2}^{\sigma_1'\sigma_2} - \\ -\gamma\begin{pmatrix}\bar{\sigma}_1 & \sigma_2 & ; & \sigma_2' \\ 1 & 2 & ; & 2'\end{pmatrix}\frac{\delta\Phi}{\delta v_{2'}^{\sigma_2'\bar{\sigma}_2'}} - \gamma\begin{pmatrix}\bar{\sigma}_1 & \sigma_1' & ; & \bar{\sigma}_2 \\ 1 & 1' & ; & 2\end{pmatrix}G_{1'2}^{\sigma_1'\bar{\sigma}_2} + \\ +G_{4'3'}^{\sigma_4'\sigma_3'}\gamma^z\begin{pmatrix}\sigma_1 & \sigma_4' & ; & \sigma_2' \\ 1 & 4' & ; & 2'\end{pmatrix}(\hat{F}_{2'}^{\sigma_2'}\Sigma_{3'2}^{\sigma_3'\sigma_2}) - \\ -G_{4'3'}^{\sigma_4'\sigma_3'}\gamma\begin{pmatrix}\bar{\sigma}_1 & \sigma_4' & ; & \sigma_2' \\ 1 & 4' & ; & 2'\end{pmatrix}\frac{\delta}{\delta v_{2'}^{\sigma_2'\bar{\sigma}_2'}}\Sigma_{3'2}^{\sigma_3'\sigma_2}.\end{gathered} \tag{183}$$

Equations (182) and (183) have the same structure as equations (147) and (148) for the limit case $U \to \infty$ ($J \to 0$). It is easy to prove that, if we put $J^{\bar{\sigma}}$ and J^{z} equal to zero, this equations reduce to equations (147) and (148). In the real situation fluctuating fields are $J^{\bar{\sigma}}_{12} = J^{z}_{12} = J_{12}$.

From equation (182) it follows that, when fluctuating fields are equal to zero, we find for the zero and first order approximation over t and J:

$$\begin{gathered}\Lambda^{\sigma_1\sigma_2}_{12} = \delta_{\sigma_1\sigma_2}\Lambda^{\sigma_1}_{12},\\ \Lambda^{\sigma_1}_{12} = \delta_{12}(\hat{F}^{\sigma_1}_{1}\Phi) +\\ +\gamma^{z}_{\sigma_1}(11';2')G^{\sigma_1}_{1'2}(\hat{F}^{\sigma_1}_{2}\hat{F}^{\sigma_1}_{2'}\Phi) - \gamma_{\bar{\sigma}_1}(11';2')G^{\bar{\sigma}_1}_{1'2}D^{\sigma\bar{\sigma}}_{22'},\end{gathered} \tag{184}$$

where

$$\left.\begin{aligned}\gamma^{z}_{\sigma}(12;3) &= t^{\sigma}_{12}\delta_{13} + \delta_{12}J^{z}_{13}\\ \gamma_{\bar{\sigma}}(12;3) &= t^{\bar{\sigma}}_{12}\delta_{13} + \delta_{12}J^{\bar{\sigma}}_{13}\end{aligned}\right\} \tag{185}$$

are the bare vertex parts (remind that t^{σ}_{12} and $J^{\bar{\sigma}}_{12}$ do not depend on σ and $J^{z}_{12} = J_{12}$).

In the unpacked form, expression (184) is written as

$$\begin{gathered}\Lambda^{\sigma}_{12} = \delta_{12}(\hat{F}^{\sigma}_{1}\Phi) + (t^{\sigma}G^{\sigma})_{12}(\hat{F}^{\sigma}_{2}\hat{F}^{\sigma}_{1}\Phi) - (t^{\bar{\sigma}}G^{\bar{\sigma}})_{12}D^{\sigma\bar{\sigma}}_{21} +\\ +G^{\sigma}_{12}(\hat{F}^{\sigma}_{2}\hat{F}^{\sigma}_{1'}\Phi)J_{1'1} - J_{11'}D^{\sigma\bar{\sigma}}_{21'}G^{\bar{\sigma}}_{12} + \dots\end{gathered} \tag{186}$$

In Fig. 44 we see the graphic representations of (184) and (186) for the terminal part of the electron GF.

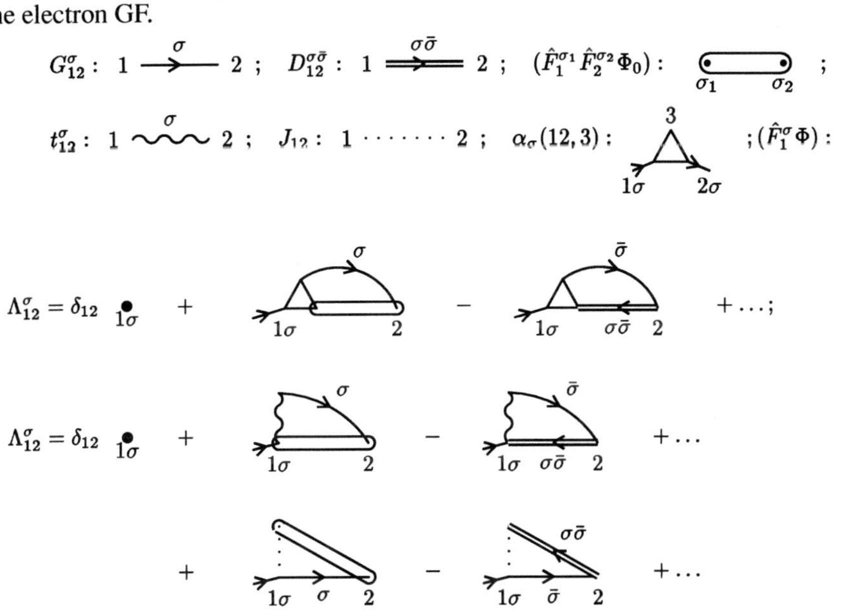

FIGURE 44. Graphic representations for the terminal part of the electron GF in compact and unpacked forms

Let us iterate equation (183) for the self-energy part. First order term $(\Sigma_1)^{\sigma_1\sigma_2}_{12}$ is given by the first four terms in equation (183). If we substitute this expression in the last two terms of equation (183), we find a correction $(\Sigma_2)^{\sigma_1\sigma_2}_{12}$ of second order. We can write the final result for zero fluctuating fields:

$$\begin{gathered}\Sigma^{\sigma_1\sigma_2}_{12} = \delta_{\sigma_1\sigma_2}\Sigma^{\sigma_1}_{12},\\ \Sigma^{\sigma}_{12} = \gamma^z_{\bar\sigma}(12;2')(\hat F^{\sigma}_{2'}\Phi) - \gamma_{\bar\sigma}(11';2)G^{\bar\sigma}_{1'2} +\\ +\gamma^z_{\bar\sigma}(14';2')G^{\sigma}_{4'3'}\gamma^z_{\bar\sigma}(3'2;1')(\hat F^{\sigma}_{1'}\hat F^{\sigma}_{2'}\Phi) - \gamma_{\bar\sigma}(14';2')G^{\bar\sigma}_{4'3'}\gamma_{\sigma}(3'2;1')D^{\sigma\bar\sigma}_{1'2'} -\\ -\gamma^z_{\bar\sigma}(14';2')G^{\sigma}_{4'3'}\gamma_{\bar\sigma}(3'1';2)G^{\bar\sigma}_{1'2'}G^{\bar\sigma}_{2'2} -\\ -\gamma_{\bar\sigma}(14';2')G^{\bar\sigma}_{4'3'}\gamma^z_{\bar\sigma}(3'1';2)G^{\bar\sigma}_{1'2'}G^{\sigma}_{2'2} + \ldots\end{gathered} \tag{187}$$

This expression is shown in Fig. 45.

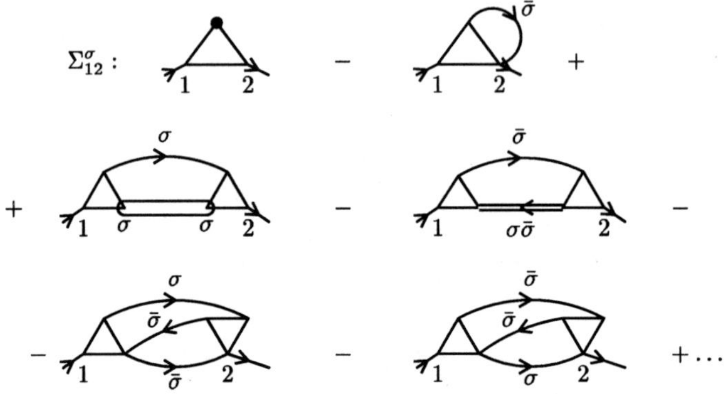

FIGURE 45. Self-energy part of electron GF

Now we write the first order correction for Σ_1 in the unpacked form and give the graphic representation

$$(\Sigma_1)^{\sigma}_{12} = (\hat F^{\sigma}_{1}\Phi)t^{\sigma}_{12} + \delta_{12}J_{12'}(\hat F^{\sigma}_{2'}\Phi) - (t^{\bar\sigma}G^{\bar\sigma})_{11}\delta_{12} - J_{12}G^{\bar\sigma}_{12} = \tag{188}$$

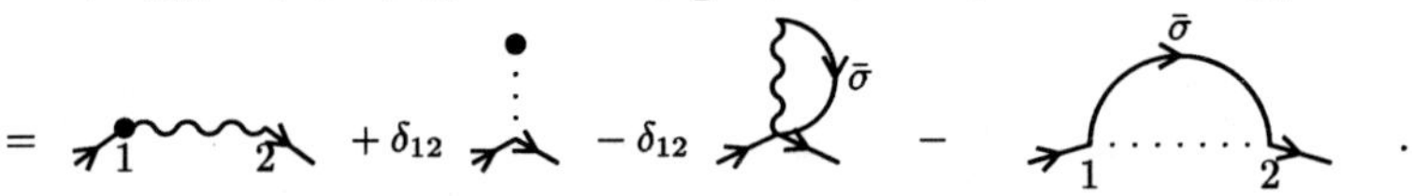

(189)

In the same way we shown terms Σ_2 of second order. The third and forth term in Fig. 45 with Bose-like GFs $(\hat F^{\sigma}_{1}\hat F^{\sigma}_{2}\Phi)$ and $D^{\sigma\bar\sigma}_{12}$ are made up of eight terms

$$\begin{gathered}(t^{\sigma}G^{\sigma})_{13'}t^{\sigma}_{3'2}(\hat F^{\sigma}_{3'}\hat F^{\sigma}_{1}\Phi) + (t^{\sigma}G^{\sigma})_{12}J_{23'}(\hat F^{\sigma}_{3'}\hat F^{\sigma}_{1}\Phi) +\\ +J_{12'}G^{\sigma}_{13'}t^{\sigma}_{3'2}(\hat F^{\sigma}_{3'}\hat F^{\sigma}_{2'}\Phi) + J_{12'}G^{\sigma}_{12}J_{23'}(\hat F^{\sigma}_{3'}\hat F^{\sigma}_{2'}\Phi) -\\ -(t^{\bar\sigma}G^{\bar\sigma})_{13'}t^{\bar\sigma}_{3'2}D^{\sigma\bar\sigma}_{3'1} - (t^{\bar\sigma}G^{\bar\sigma})_{12}J_{23'}D^{\sigma\bar\sigma}_{3'1} -\\ -J_{12'}G^{\bar\sigma}_{13'}t^{\bar\sigma}_{3'2}D^{\sigma\bar\sigma}_{3'2'} - J_{12'}G^{\bar\sigma}_{12}J_{23'}D^{\sigma\bar\sigma}_{3'2'}.\end{gathered} \tag{190}$$

shown in Fig. 46.

Comparing graphs in Fig. 44 for Λ with Fig. 41, obtained in the limit $U \to \infty$, shows how in the tJ-model graphs with exchange interaction lines appear. Comparing Fig. 46 with Fig. 42 shows how graphs are multiplied when we go from the case $U \to \infty$ to the tJ-model. We see that in the case $U \to \infty$, the Larkin equation for electron GF holds. Comparison with the tJ-model shows that Larkin equation also holds in this case. A detailed graphical analysis for electron GF in the tJ-model is given in monograph [71].

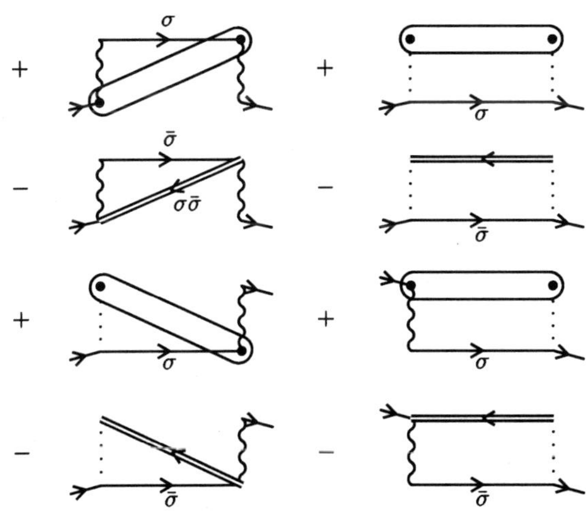

FIGURE 46. Result of unpacking the third and forth terms for Σ^{σ}_{12} in Fig. 45

tJ-model: magnon Green's function. Here we generalize the results for the GF $D^{\sigma\bar{\sigma}}_{12}$ obtained earlier for the limit $U \to \infty$. The general form of the equation of motion remains unchanged, and it is necessary to refine the expressions for operators $\dot{X}^{\sigma\bar{\sigma}}_1$ and V, when we move from the case $U = \infty$ to the tJ-model. In terms of variational derivatives the equation for $D^{\sigma\bar{\sigma}}_{12}$ can now be written in the form

$$
\begin{aligned}
&[(K_0^{-1})^{\sigma\bar{\sigma}}_{11'} - (v^{\bar{\sigma}\bar{\sigma}}_1 - v^{\sigma\sigma}_1)\delta_{11'} - (\alpha_{\bar{\sigma}}(\hat{F}^{\sigma} - \hat{F}^{\bar{\sigma}})\Phi)_{11'} - \\
&-(\alpha_{\bar{\sigma}}(\hat{F}^{\sigma} - \hat{F}^{\bar{\sigma}}))_{11'}]\frac{\delta^2\Phi}{\delta v^{\sigma\bar{\sigma}}_{1'}\delta v^{\bar{\sigma}\sigma}_2} = -\delta_{12}(\hat{F}^{\sigma}_1 - \hat{F}^{\bar{\sigma}}_1)\Phi - \\
&-v^{\sigma\bar{\sigma}}_1(\hat{F}^{\sigma}_1 - \hat{F}^{\bar{\sigma}}_1)\frac{\delta\Phi}{\delta v^{\bar{\sigma}\sigma}_2} + \frac{\delta\Phi}{\delta v^{\sigma\bar{\sigma}}_{1'}}(\alpha_{\bar{\sigma}}(\hat{F}^{\sigma} - \hat{F}^{\bar{\sigma}}))_{11'}\frac{\delta\Phi}{\delta v^{\bar{\sigma}\sigma}_2} + \\
&+\frac{\delta}{\delta v^{\bar{\sigma}\sigma}_2}(t^{\bar{\sigma}\sigma'_1}_{11'}\frac{\delta\Phi}{\delta t^{\sigma'_1\sigma}_{1'1}} - \frac{\delta\Phi}{\delta t^{\bar{\sigma}\sigma'_1}_{11'}}t^{\sigma'_1\sigma}_{1'1}).
\end{aligned}
\tag{191}
$$

Here the quantity

$$\alpha_{\bar{\sigma}}(12;3) = J^{\bar{\sigma}}_{12}\delta_{13} + J_{12}\delta_{12},$$

under the condition that in a real situation $J^{\bar{\sigma}}_{12} = J_{12}$, does not depend on spin, and coincides with the bare amplitude $\alpha_0(12;3)$ in the Heisenberg model, see definition (54).

When $t_{12} = 0$ this equation reduces to equation (57) of the Heisenberg model. We must notice that

$$(\hat{F}^{\sigma}_1 - \hat{F}^{\bar{\sigma}}_1)\Phi = \sigma\langle S^z_1\rangle.$$

Thus, equation (191) takes into account the dual nature of magnetic states of the system; from one hand they are itinerant but on the other, localized. The last term in (191) includes a mixed derivative $v^{\bar{\sigma}\sigma}_1$ and $t^{\bar{\sigma}\sigma}_1$. Its calculation shows that this term forms a part of the self-energy for the electron. To disentangle the mixed term we use a multiplicative representation for the electron GF:

$$\mathscr{G}_{12}^{\sigma_1\sigma_2} = -\frac{\delta\Phi}{\delta t_{12}^{\sigma_1\sigma_2}} = (G\Lambda)_{12}^{\sigma_1\sigma_2}. \tag{192}$$

By differentiating both factors in (192), we obtain a first expression for the calculation of the mixed term:

$$\frac{\delta}{\delta v_2^{\bar{\sigma}\sigma}}\left(\frac{\delta\Phi}{\delta t_{34}^{\bar{\sigma}\sigma}}\right) = -G_{32}^{\bar{\sigma}\bar{\sigma}}\mathscr{G}_{24}^{\sigma\sigma} - G_{31'}^{\bar{\sigma}\bar{\sigma}}\left[\frac{\delta\Sigma_{1'2'}^{\bar{\sigma}\sigma}}{\delta v_2^{\bar{\sigma}\sigma}} + \frac{\delta\Lambda_{1'3'}^{\bar{\sigma}\sigma}}{\delta v_2^{\bar{\sigma}\sigma}}\left(\mathscr{G}^{-1}\right)_{3'2'}^{\sigma\sigma}\right]\mathscr{G}_{2'4}^{\sigma\sigma}.$$

After calculating this expression in the limit of the first order over t we can show that equation (191) is decomposed into two equations: one for the self-energy $\mathscr{M}$ and one for the terminal part $\mathscr{P}$. The iteration of this equation up to the first order in t gives the expression for the self-energy

$$\begin{aligned}(\mathscr{M}_1)_{12}^{\sigma\bar{\sigma}} = {} & J_{12}(\hat{F}_1^{\sigma} - \hat{F}_1^{\bar{\sigma}})\Phi - \delta_{12}J_{11'}(\hat{F}_{1'}^{\sigma} - \hat{F}_{1'}^{\bar{\sigma}})\Phi + 2\sigma\delta_{12}(JK^{\bar{\sigma}\bar{\sigma}})_{11} - \\ & -2\sigma J_{12}K_{12}^{\bar{\sigma}\bar{\sigma}} + (t^{\bar{\sigma}}G^{\bar{\sigma}\bar{\sigma}})_{11}\delta_{12} + G_{12}^{\bar{\sigma}\bar{\sigma}}t_{21}^{\sigma},\end{aligned}$$

and for the terminal part

$$\mathscr{P}_{12}^{\sigma\bar{\sigma}} = \delta_{12}(\hat{F}_1^{\sigma} - \hat{F}_1^{\bar{\sigma}})\Phi + (t^{\bar{\sigma}}G^{\bar{\sigma}\bar{\sigma}})_{12}\mathscr{G}_{21}^{\sigma\sigma} - G_{12}^{\bar{\sigma}\bar{\sigma}}(\mathscr{G}^{\sigma\sigma}t^{\sigma})_{21} + \ldots$$

The iteration correction of second order to $\mathscr{M}_{12}^{\sigma\bar{\sigma}}$ is the following

$$(\mathscr{M}_2)_{12}^{\sigma\bar{\sigma}} = (t^{\bar{\sigma}}G^{\bar{\sigma}\bar{\sigma}})_{11'}\gamma^{\bar{\sigma}\sigma}(1'2';2)\mathscr{G}_{2'1}^{\sigma\sigma} - G_{11'}^{\bar{\sigma}\bar{\sigma}}\gamma^{\bar{\sigma}\sigma}(1'2';2)(\mathscr{G}^{\sigma\sigma}t^{\sigma})_{2'1}.$$

where

$$\gamma(12;3) = t_{12}\delta_{13} + J_{13}\delta_{12}.$$

Further, we must make an assumption concerning the ground state. It is known that in the tJ-model different types of magnetic ordering can occur. Now we suppose that the values of the parameter of the Hamiltonian are such that the ferromagnetic state is the ground state. Then, in momentum representation, the above expression can be rewritten in the form:

$$\begin{aligned}(\mathscr{M}_1)^{\sigma\bar{\sigma}}(q) = {} & -2\sigma m[J(0) - J(\boldsymbol{q})] + 2\sigma\sum_k[J(\boldsymbol{k}) - J(\boldsymbol{k}-\boldsymbol{q})]K^{\sigma\bar{\sigma}}(k) - \\ & -\sum_k[\varepsilon(\boldsymbol{k}) - \varepsilon(\boldsymbol{k}-\boldsymbol{q})]G^{\bar{\sigma}}(k),\end{aligned}$$

$$(\mathscr{M}_2)^{\sigma\bar{\sigma}}(q) = \sum_k[\varepsilon(\boldsymbol{k}) - \varepsilon(\boldsymbol{k}-\boldsymbol{q})]G^{\bar{\sigma}}(k)\mathscr{G}^{\sigma}(k-q)\gamma^{\bar{\sigma}\sigma}(k,k-q;q),$$

$$\mathscr{P}^{\sigma\bar{\sigma}}(q) = 2\sigma m + \sum_k[\varepsilon(\boldsymbol{k}) - \varepsilon(\boldsymbol{k}-\boldsymbol{q})]G^{\bar{\sigma}}(k)\mathscr{G}^{\sigma}(k-q).$$

Here we introduce the magnetization of electrons by the relation

$$\sigma m = \frac{1}{2}(\langle X^{\sigma\sigma}\rangle - \langle X^{\bar{\sigma}\bar{\sigma}}\rangle \equiv \frac{1}{2}(n_{\sigma} - n_{\bar{\sigma}}).$$

Let's substitute here the electron GF in the Hubbard-I approximation:

$$\mathscr{G}^{\sigma}(k) = G^{\sigma}(k)\langle F^{\sigma 0}\rangle; \quad G^{\sigma}(k) = \frac{1}{\mathrm{i}\omega_n - E_{\sigma}(\boldsymbol{k}) + \mu},$$
$$E_{\sigma}(\boldsymbol{k}) = \langle F^{\sigma 0}\rangle \varepsilon(\boldsymbol{k}) - \langle X^{\bar{\sigma}\bar{\sigma}}\rangle J(0) - \tfrac{1}{2}\sigma h.$$

Then after summation over frequencies we finally obtain

$$D^{\sigma\bar{\sigma}}(q) = \frac{2\sigma m + p_{\sigma}(q)}{\mathrm{i}\omega_{\nu} - M_{\sigma}(q)}, \quad q = \{\boldsymbol{q}, \mathrm{i}\omega_{\nu}\}, \tag{193}$$

where

$$\begin{aligned} M_{\sigma}(q) = &-2\sigma m[J(0) - J(\boldsymbol{q})] - \frac{1}{N}\sum_{\boldsymbol{k}}[\varepsilon(\boldsymbol{k}) - \varepsilon(\boldsymbol{k}-\boldsymbol{q})]f^{\bar{\sigma}}(\boldsymbol{k}) + \\ &+\langle F^{\sigma 0}\rangle \frac{1}{N}\sum_{\boldsymbol{k}}[\varepsilon(\boldsymbol{k}) - \varepsilon(\boldsymbol{k}-\boldsymbol{q})]\varepsilon(\boldsymbol{k})\frac{f^{\sigma}(\boldsymbol{k}-\boldsymbol{q}) - f^{\bar{\sigma}}(\boldsymbol{k})}{\mathrm{i}\omega_{\nu} + E_{\sigma}(\boldsymbol{k}-\boldsymbol{q}) - E_{\bar{\sigma}}(\boldsymbol{k})} + J(\boldsymbol{q})p_{\sigma}(q), \end{aligned} \tag{194}$$

$$p_{\sigma}(q) = \langle F^{\sigma 0}\rangle \frac{1}{N}\sum_{\boldsymbol{k}}[\varepsilon(\boldsymbol{k}) - \varepsilon(\boldsymbol{k}-\boldsymbol{q})]\frac{f^{\sigma}(\boldsymbol{k}-\boldsymbol{q}) - f^{\bar{\sigma}}(\boldsymbol{k})}{\mathrm{i}\omega_{\nu} + E_{\sigma}(\boldsymbol{k}-\boldsymbol{q}) - E_{\bar{\sigma}}(\boldsymbol{k})}. \tag{195}$$

Notice that $M_{\sigma}(q)$ includes terms directly depending only on $J(\boldsymbol{q})$, on $\varepsilon(\boldsymbol{q})$ and a mixed term $J(\boldsymbol{q})p_{\sigma}(q)$. The last term corresponds to contribution graphs in $M_{\sigma}(q)$, cut over an exchange interaction line. A similar term we had in the Heisenberg model. Poles of GF (193) determine, as one can see from expressions (193) – (195), a Goldstone mode.

The second and third terms in (194) are mainly compensating each others, due to the fact that, at $T \to T_C$ and $\mathrm{i}\omega_{\nu} \to 0$, $M_{\sigma}(q) \to 0$, and in the paramagnetic phase the expression (193) for magnon GF takes the form:

$$D^{\sigma\bar{\sigma}}(q) = \frac{\Pi(q)}{1 - Q(q) - J(\boldsymbol{q}\Pi(q)},$$

where $\Pi(q)$ and $Q(q)$ are determined by relations

$$\begin{pmatrix} \Pi(q) \\ Q(q) \end{pmatrix} = \frac{1}{N}\sum_{\boldsymbol{k}} \begin{pmatrix} 1 \\ \varepsilon(\boldsymbol{k}) \end{pmatrix} \frac{f(\boldsymbol{k}-\boldsymbol{q}) - f(\boldsymbol{k})}{\mathrm{i}\omega_{\nu} + E(\boldsymbol{k}-\boldsymbol{q}) - E(\boldsymbol{k})}.$$

The two terms in the denominator show a tendency of the system to magnetic ordering to account for electron hopping on the lattice and for indirect exchange interaction. From a comparison with formula (116), obtained in GRPA, we can see that our result is not complete now, because it does not contain two other terms $\Lambda(q)$ and $\Phi(q)$, which are of the same order $1/z$ as a loop. This demonstrate once more the importance of summing all loop types diagrams, when calculating the magnon GF in the tJ-model.

In conclusion of this Section we discuss a connection with other works. The Kadanoff-Baym approach [59] to the models, using X operators, was applied for the first time in the paper [60] for the limit case $U = \infty$. Later this approach was continued in the papers [61, 62, 84, 85, 86, 87, 88] and extended to the tJ-model. Equations with variational derivatives obtained in this Section for electron GFs are compatible with the corresponding equations of the work [61, 62, 84, 85, 86, 87, 88]. In the last ones, however, the generalized Hubbard model is studied (in the large U limit) for degenerate electron states with degree N. Comparison of the expressions for the effective electron-electron vertex part, obtained in [61, 62, 84, 85, 86, 87, 88] with the GRPA results, obtained by the diagrammatic technique for X operators [53], shows their total identity. By this comparison we demonstrate that GRPA results corresponds formally to the limit $N \to \infty$.

Such conclusion is a-priori understandable, because GRPA is based on the summation of loop types diagrams. But it is well known from many body-theory that RPA approach to usual Fermi systems is a zeroth order approximation with respect to the small parameter $1/N$, when loop types diagrams should be taken into account. Notice that all GRPA results for the tJ-model have been obtained using the diagrammatic technique with X operators in the paper [53], published before the papers [61, 62, 84, 85, 86, 87, 88], where they have been rediscovered.

IV. THE SD-MODEL

Generating functional of the model. The sd-model is a hybrid between the usual Fermi gas of electrons and the Heisenberg model with localized spins. The connection between these two systems is given by the sd-exchange interaction

$$\mathscr{H}_{\text{sd}} = -J_{\text{sd}} \sum_i (\boldsymbol{S}_i \boldsymbol{s}_i),$$

Here $\boldsymbol{S}_i$ is the operator of a spin localized on site i, and $\boldsymbol{s}_i$ is the electron spin operator on the same site. In the second quantization representation $\boldsymbol{s}_i$ is expressed by a quadratic form of Fermi operators

$$\boldsymbol{s}_i = \sum_{\sigma_1\sigma_2} \boldsymbol{s}_{\sigma_1\sigma_2} c^+_{i\sigma_1} c_{i\sigma_2},$$

where $\boldsymbol{s}$ is a vector of electron spin, represented by the Pauli matrices

$$s^+ = \frac{1}{\sqrt{2}} \begin{pmatrix} 0 & 1 \\ 0 & 0 \end{pmatrix}, \quad s^- = \frac{1}{\sqrt{2}} \begin{pmatrix} 0 & 0 \\ 1 & 0 \end{pmatrix}, \quad s^z = \frac{1}{2} \begin{pmatrix} 1 & 0 \\ 0 & -1 \end{pmatrix}.$$

The total Hamiltonian of the sd-model reads

$$\begin{aligned} \mathscr{H} &= \mathscr{H}_d + \mathscr{H}_s - J_{\text{sd}} \sum_{i\sigma_1\sigma_2} (\boldsymbol{S}_i \boldsymbol{s}_{\sigma_1\sigma_2}) c^+_{i\sigma_1} c_{i\sigma_2}, \\ \mathscr{H}_s &= \sum_{i\sigma} \varepsilon_\sigma c^+_{i\sigma} c_{i\sigma} + \sum_{ij\sigma} t_{ij} c^+_{i\sigma} c_{j\sigma}, \\ \mathscr{H}_d &= -h \sum_i S^z_i - \tfrac{1}{2} \sum_{ij} J_{ij} \boldsymbol{S}_i \boldsymbol{S}_j, \end{aligned} \tag{196}$$

where $\varepsilon_\sigma = \varepsilon - \sigma \frac{h}{2}$ is the atomic electron level in a magnetic field h. Thus the exchange term mixes electron states with different spin projections, and the amplitude of a mixture on a site depends on the mutual orientation of localized and electron spins. The equation of motion for electron and spin operators contains mixed terms as well

$$\begin{aligned} \dot{c}_{i\sigma} &= -\varepsilon^\sigma c_{i\sigma} - t_{ii'} c_{i'\sigma} + J_{\text{sd}} (\boldsymbol{S}_i \boldsymbol{s})_{\sigma\sigma'} c_{i\sigma'}, \\ -\dot{S}^\sigma_i &= \sigma h S^\sigma_i + \sigma S^\sigma_i J_{ii'} S^z_{i'} - \sigma S^z_i J_{ii'} S^\sigma_{i'} + \\ &\quad + \sigma J_{\text{sd}} c^+_{i\sigma'_1} c_{i\sigma'_2} (S^\sigma_i s^z_{\sigma'_1\sigma'_2} - S^z_i s^\sigma_{\sigma'_1\sigma'_2}). \end{aligned}$$

Consider formally the sd-interaction and the direct exchange interaction between localized spins as perturbations. A diagrammatic technique for this model should be a combination of the technique for usual Fermi system and the technique for spin operators. Such a technique was described in details in a book [54]. We will not reproduce it here, but we will develop the method of generating functional for the sd-model.

For the sd-model, the operator V can be chosen in the form [69]

$$\begin{aligned} V &= -v^z_{1'} S^z_{1'} - v^{\sigma'}_{1'} S^{\bar{\sigma}'}_{1'} - S^{\sigma'_1}_{1'} u^{\sigma'_1\sigma'_2}_{1'2'} S^{\sigma'_2}_{2'} + \\ &+ [w^z_{1'} (s^z_{1'})_{\sigma'_1\sigma'_2} + w^{\sigma'}_{1'} (s^{\bar{\sigma}'})_{\sigma'_1\sigma'_2}] c^+_{1'\sigma'_1} c_{1'\sigma'_2} + c^+_{1'\sigma'_1} \ell^{\sigma'_1\sigma'_2}_{1'2'} c_{2'\sigma'_2}, \end{aligned} \tag{197}$$

where summation over repeated prime indexes is implied. In particular, the second term in this expression means

$$v_{1'}^{\sigma'} S_{1'}^{\bar{\sigma}'} = v_{1'}^{+} S_{1'}^{-} + v_{1'}^{-} S_{1'}^{+}, \quad (\sigma' = +, -).$$

Here, one-point fields v_1^z, v_1^+, v_1^- and w_1^z, w_1^+, w_1^- are connected with spin and electron variables respectively. Two-point fields $u_{12}^{\sigma_1\sigma_2}$ and $\ell_{12}^{\sigma_1\sigma_2}$ connect spin and electron variables inside each group. The first row in (197) corresponds to expression (49) for V, which we used for the Heisenberg model. Here it is extended to account for transversal spin components. The second row in (197) refers to the electron system and is compatible with expression (137), which we used when we studied strongly correlated electrons.

As usual, we introduce a generating functional

$$Z[V] = ((T\mathrm{e}^{-V}))$$

and determine the electron and spin GF, depending on fluctuating fields[1].

$$\begin{aligned} G_{12}^{\sigma_1\sigma_2} &= -\langle T c_{1\sigma_1} c_{2\sigma_2}^{+} \mathrm{e}^{-V} \rangle, \\ D_{12}^{\sigma\bar{\sigma}} &= -\langle T S_1^{\sigma} S_2^{\bar{\sigma}} \mathrm{e}^{-V} \rangle, \end{aligned}$$

By differentiation of $Z[V]$ over a fluctuating field we obtain a set of intermediate expressions

$$\begin{aligned} &\frac{\delta Z[V]}{\delta \ell_{12}^{\sigma_1\sigma_2}} = ((T c_{1\sigma_1} c_{2\sigma_2}^{+} \mathrm{e}^{-V})), \\ &\frac{\delta Z[V]}{\delta u_{12}^{\sigma_1\sigma_2}} = ((T S_1^{\sigma_1} S_2^{\sigma_2} \mathrm{e}^{-V})), \\ &\frac{\delta Z[V]}{\delta v_1^z} = ((T S_1^z \mathrm{e}^{-V})), \quad \frac{\delta Z[V]}{\delta v_1^{\sigma}} = ((T S_1^{\bar{\sigma}} \mathrm{e}^{-V})), \\ &\frac{\delta^2 Z[V]}{\delta v_1^{\sigma} \delta v_2^z} = ((T S_1^{\bar{\sigma}} S_2^z \mathrm{e}^{-V})), \\ &\frac{\delta^2 Z[V]}{\delta t_{12}^{\sigma_1\sigma_2} \delta v_1^{\sigma}} = ((T S_1^{\bar{\sigma}} c_{1\sigma_1} c_{2\sigma_2}^{+} \mathrm{e}^{-V})), \\ &\ldots \end{aligned}$$

After the substitution $Z = \mathrm{e}^{\Phi}$ we can write the expressions for electron and magnon GFs

$$G_{12}^{\sigma_1\sigma_2} = -\frac{\delta \Phi}{\delta t_{12}^{\sigma_1\sigma_2}}, \tag{198}$$

---

[1] This definition includes both GF of transversal spin components: $\langle T S_1^{-} S_2^{+} \rangle$, which we used in Section II for the Heisenberg model, and GF $\langle T S_1^{+} S_2^{-} \rangle$. Notice that contrary definitions of Section II (formula (27)), we define GF with opposite sign. This corresponds more common definition of the spin wave GF in momentum representation.

$$D_{12}^{\sigma\bar{\sigma}} = -\frac{\delta\Phi}{\delta \mathscr{J}_{12}^{\sigma\bar{\sigma}}} = -\frac{\delta^2\Phi}{\delta v_1^{\bar{\sigma}} \delta v_2^{\sigma}}. \tag{199}$$

Here, instead of differentiation over variables $\ell_{12}^{\sigma_1\sigma_2}$ and u_{12}, we introduced new variables. After differentiation we have to put them equal to the matrix elements of hopping and exchange integrals respectively.

Electron Green's function. By standard way described in Appendix we write equation of motion for the average of a T-product of pair electron operators

$$\frac{\partial}{\partial\tau_1}((Tc_{1\sigma_1}c_{2\sigma_2}^+ \mathrm{e}^{-V})) = ((T[c_{1\sigma_1}, c_{2\sigma_2}^+]_+ \mathrm{e}^{-V}))+$$
$$+((T\dot{c}_{1\sigma_1}c_{2\sigma_2}^+ \mathrm{e}^{-V})) - ((T\{c_{1\sigma_1}, V\}c_{2\sigma_2}^+ \mathrm{e}^{-V})),$$

and

$$\{c_{1\sigma_1}, V\} = \ell_{11'}^{\sigma_1\sigma_1'} c_{1'\sigma_1'} + (\boldsymbol{ws})^{\sigma_1\sigma_1'} c_{1\sigma'}.$$

Then our equation is written

$$\left[\left(\frac{\partial}{\partial\tau_1} + \varepsilon_{\sigma_1}\right)\delta_{11'}\delta_{\sigma_1\sigma_1'} + t_{11'}^{\sigma_1\sigma_1'}\right]((Tc_{1'\sigma_1'}c_{2\sigma_2}^+ \mathrm{e}^{-V})) =$$
$$= \delta_{12}\delta_{\sigma_1\sigma_2}((T\mathrm{e}^{-V})) + J_{\mathrm{sd}}((T(\boldsymbol{S}_1\boldsymbol{s})^{\sigma_1\sigma_1'} c_{1\sigma_1'}c_{2\sigma_2}^+ \mathrm{e}^{-V})),$$

where we introduce the notation $t_{11'}^{\sigma_1\sigma_1'} = \ell_{11'}^{\sigma_1\sigma_1'} + \delta_{\sigma_1\sigma_1'} t_{11'}$.

Determine the electron GF of zeroth approximation

$$(G_{0V}^{-1})_{11'}^{\sigma_1\sigma_1'} = -\left(\frac{\partial}{\partial\tau_1} + \varepsilon_{\sigma_1}\right)\delta_{11'}\delta_{\sigma_1\sigma_1'} - t_{11'}^{\sigma_1\sigma_1'} - (\boldsymbol{w}_1\boldsymbol{s})^{\sigma_1\sigma_1'}\delta_{11'}. \tag{200}$$

Then our equation takes the form

$$(G_{0V}^{-1})_{11'}^{\sigma_1\sigma_1'} \frac{\delta Z[V]}{\delta t_{1'2}^{\sigma_1'\sigma_2}} = \delta_{12}\delta_{\sigma_1\sigma_2} Z - J_{\mathrm{sd}}(\boldsymbol{s}\hat{\boldsymbol{S}}_1)^{\sigma_1\sigma_1'} \frac{\delta Z[V]}{\delta t_{1'2}^{\sigma_1'\sigma_2}},$$

where we introduced a differential operator

$$(\boldsymbol{s}\hat{\boldsymbol{S}}_1) = s^z \frac{\delta}{\delta v_1^z} + s^+ \frac{\delta}{\delta v_1^+} + s^- \frac{\delta}{\delta v_1^-}. \tag{201}$$

Now transferring from Z to Φ, we write equation for electron GF in a final form

$$\left[(G_{0V}^{-1})_{11'}^{\sigma_1\sigma_1'} + J_{\mathrm{sd}}(\boldsymbol{s}\hat{\boldsymbol{S}}_1\Phi)^{\sigma_1\sigma_1'}\delta_{11'} + J_{\mathrm{sd}}(\boldsymbol{s}\hat{\boldsymbol{S}}_1)^{\sigma_1\sigma_1'}\delta_{11'}\right] G_{1'2}^{\sigma_1'\sigma_2} = \delta_{12}\delta_{\sigma_1\sigma_2}. \tag{202}$$

The last term in brackets is a differential operator over fluctuating fields v_1^z, v_1^+ and v_1^-. It acts directly on the electron GF, while the previous term acts on Φ and describe the exchange splitting of electron levels with respect to spin. The equation for the electron

GF in the sd-model is essentially different from the corresponding equation (144) in the tJ-model, because it does not contain in the right hand side some terms, which determine the terminal part of the GF in the tJ-model, so that in the sd-model it is just equal to 1, as for usual Fermi system.

From equation (202) one finds the equation for the self-energy part $\Sigma_{12}^{\sigma_1\sigma_2}$. To this purpose it is necessary to use an identity

$$(\boldsymbol{s}\hat{\boldsymbol{S}}_1)^{\sigma_1\sigma_1'}\delta_{11'}G_{1'2}^{\sigma_1'\sigma_2} = -G_{1'3'}^{\sigma_1'\sigma_3'}\left[(\boldsymbol{s}\hat{\boldsymbol{S}}_1)^{\sigma_1\sigma_1'}\delta_{11'}(G^{-1})_{3'4'}^{\sigma_3'\sigma_4'}\right]G_{4'2}^{\sigma_4'\sigma_2},$$

having in mind, that $\hat{\boldsymbol{S}}_1$ is a linear combination of derivatives. After substituting this equation in the initial equation (202) and using the Dyson equation $G^{-1} = (G^0)^{-1} - \Sigma$, determined Σ, we come to the final form of the equation for the self-energy part [69]:

$$\Sigma_{12}^{\sigma_1\sigma_2} = -J_{\mathrm{sd}}(\boldsymbol{s}\hat{\boldsymbol{S}}_1\Phi)^{\sigma_1\sigma_2}\delta_{12} - J_{\mathrm{sd}}G_{11'}^{\sigma_3'\sigma_1'}(\boldsymbol{s}\hat{\boldsymbol{S}}_1)^{\sigma_1\sigma_3'}\Sigma_{1'2}^{\sigma_1'\sigma_2}. \tag{203}$$

As in the case of the tJ-model, we obtained an equation for Σ with variational derivatives over fluctuating fields, acting on local spins. The first term in (203) gives the Hartree correction to the electron energy, describing spin splitting of an electron level, the second term describes all the remaining interactions. Iterations of this equation provide a series of perturbation theory in powers of the sd-exchange interactions. Thus for the first orders we have the simple expressions

$$(\Sigma_1)_{12}^{\sigma_1\sigma_2} = -J_{\mathrm{sd}}(\boldsymbol{s}\hat{\boldsymbol{S}}_1\Phi)^{\sigma_1\sigma_2}\delta_{12},$$
$$(\Sigma_2)_{12}^{\sigma_1\sigma_2} = J_{\mathrm{sd}}^2G_{12}^{\sigma_3'\sigma_1'}(\boldsymbol{s}\hat{\boldsymbol{S}}_1)^{\sigma_1\sigma_3'}(\boldsymbol{s}\hat{\boldsymbol{S}}_2\Phi)^{\sigma_1'\sigma_2}.$$

Using the relation (201) and the fact that $\langle S_1^{\pm}\rangle = 0$, we rewrite this expressions in the form

$$(\Sigma_1)_{12}^{\sigma_1\sigma_2} = -\frac{1}{2}J_{\mathrm{sd}}\langle S_1^z\rangle\sigma_1\delta_{\sigma_1\sigma_2}\delta_{12}, \tag{204}$$

$$(\Sigma_2)_{12}^{\sigma_1\sigma_2} = J_{\mathrm{sd}}^2G_{12}^{\sigma'}\{s_{\sigma_1\sigma'}^z s_{\sigma'\sigma_2}^z\frac{\delta^2\Phi}{\delta v_1^z\delta v_2^z} + s_{\sigma_1\sigma'}^{\sigma'}s_{\sigma'\sigma_2}^{\bar{\sigma}'}\frac{\delta^2\Phi}{\delta v_1^{\bar{\sigma}'}\delta v_2^{\sigma'}}\}. \tag{205}$$

The first order correction is referred to as the electron GF of "zeroth" order, and, in momentum representation, it is written as

$$G_0^{\sigma}(k) = \frac{1}{\mathrm{i}\omega_n - \varepsilon_\sigma(\boldsymbol{k})},$$

where

$$\varepsilon_\sigma(\boldsymbol{k}) = \varepsilon(\boldsymbol{k}) - \frac{1}{2}\sigma(h + \langle S_1^z\rangle J_{\mathrm{sd}})$$

describes the exchange splitting.

It is easy to see that the second order correction is diagonal with respect to spin indexes, and the matrix element (205) is equal to

$$(\Sigma_2^{\sigma})_{12} = \frac{1}{4}J_{\mathrm{sd}}^2G_{12}^{\sigma}D_{21}^z - \frac{1}{2}J_{\mathrm{sd}}^2G_{12}^{\bar{\sigma}}D_{21}^{\sigma\bar{\sigma}}. \tag{206}$$

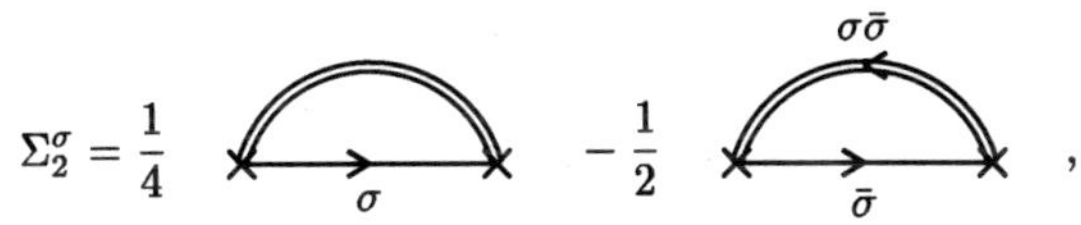

FIGURE 47. Second order correction to the electron GF in the sd-model. Slanting cross denotes value J_{sd}

So, Σ_2^σ is expressed through the electron GF and spin GFs of longitudinal and transversal components. The result (206) corresponds to the self consistent Born approximation (SCBA), because the electron GF includes the self-energy part, and in momentum representation is equal to

$$G_0^\sigma(k) = \frac{1}{\mathrm{i}\omega_n - \varepsilon_\sigma(\boldsymbol{k}) - \Sigma_2^\sigma(\boldsymbol{k})}, \tag{207}$$

Fig. 47. It gives correction to electron energy due to interaction with magnons and longitudinal spin fluctuations.

Now we refuse from iterations and study the exact equation (203) for the self-energy part of electron GF. With zero fluctuating field $\Sigma_{12}^{\sigma\sigma} \equiv \Sigma_{12}^\sigma$, and the equation for the matrix element is written as

$$\Sigma_{12}^{\sigma\sigma} = -\frac{1}{2}\sigma J_{\mathrm{sd}}\frac{\delta\Phi}{\delta v_1^z}\delta_{12} - \frac{1}{2}\sigma J_{\mathrm{sd}} G_{11'}^{\sigma\sigma}\frac{\delta\Sigma_{1'2}^{\sigma\sigma}}{\delta v_1^z} - \frac{1}{\sqrt{2}} J_{\mathrm{sd}} G_{11'}^{\bar\sigma\bar\sigma}\frac{\delta\Sigma_{1'2}^{\bar\sigma\sigma}}{\delta v_1^\sigma}. \tag{208}$$

For the derivatives of matrix elements $\Sigma^{\sigma\sigma}$ and $\Sigma^{\bar\sigma\sigma}$ we may write the equations by differentiating the basic equation (203) over v_1^z and v_1^σ. From the structure of this equation it follows the representation for derivatives of $\Sigma^{\sigma\sigma}$ and $\Sigma^{\bar\sigma\sigma}$

$$\frac{\delta\Sigma_{12}^{\bar\sigma\sigma}}{\delta v_3^\sigma} = \frac{1}{\sqrt{2}}\Delta^{\bar\sigma\sigma}(12;3')D_{3'3}^{\sigma\bar\sigma} - \frac{1}{\sqrt{2}}R\begin{pmatrix} \bar\sigma & \sigma & ; & \sigma & \bar\sigma \\ 1 & 2 & ; & 3' & 4' \end{pmatrix} G_{4'3}^{\bar\sigma} G_{33'}^{\sigma}, \tag{209}$$

$$\frac{\delta\Sigma_{12}^{\sigma\sigma}}{\delta v_3^z} = -\frac{\sigma}{2}\Delta^{\sigma\sigma}(12;3')D_{3'3}^{z} - \frac{\sigma}{\sqrt{2}}R\begin{pmatrix} \sigma & \sigma & ; & \sigma' & \sigma' \\ 1 & 2 & ; & 3' & 4' \end{pmatrix} G_{4'3}^{\sigma'} G_{33'}^{\sigma'}, \tag{210}$$

Here 3-point quantities Δ represent the vertex parts of the interaction of an electron with a spin wave or a longitudinal spin fluctuation. The 4-point vertex parts R give the effective electron-electron interactions. If we substitute relations (209) and (210) in (208), we obtain the following representation for the self-energy

$$\begin{aligned} \Sigma_{12}^\sigma &= -\tfrac{1}{2}\sigma J_{\mathrm{sd}}\delta_{12} - \\ &-\tfrac{1}{4}J_{\mathrm{sd}}G_{11'}^{\sigma}\left[\Delta^{\sigma\sigma}(1'2;3')D_{3'1}^{z} + R\begin{pmatrix} \sigma & \sigma & ; & \sigma' & \sigma' \\ 1' & 2 & ; & 3' & 4' \end{pmatrix} G_{4'1}^{\sigma'}G_{13'}^{\sigma'}\right] - \\ &-\tfrac{1}{2}J_{\mathrm{sd}}G_{11'}^{\bar\sigma}\left[\Delta^{\bar\sigma\sigma}(1'2;3')D_{3'1}^{\sigma\bar\sigma} + R\begin{pmatrix} \bar\sigma & \sigma & ; & \sigma & \bar\sigma \\ 1' & 2 & ; & 3' & 4' \end{pmatrix} G_{4'1}^{\bar\sigma}G_{13'}^{\sigma}\right]. \end{aligned} \tag{211}$$

From Fig. 48 it can be grasped the interpretation of Δ and R as vertex parts.

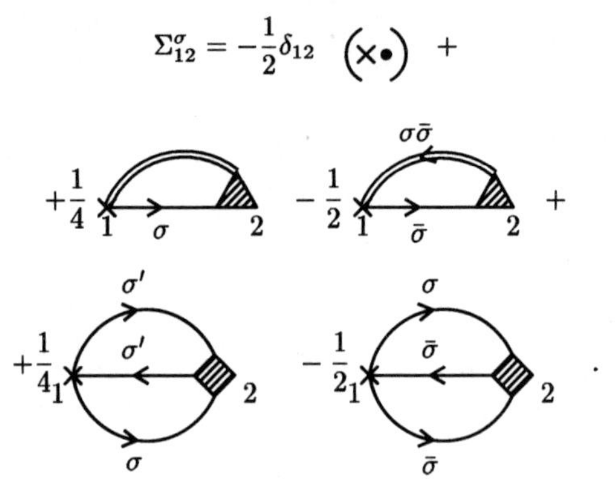

FIGURE 48. Structure of self-energy part for electron GF in the sd-model

In the lowest approximation, tree-point vertex parts are equal to

$$\Delta_0^{\bar{\sigma}\sigma}(12;3) = J_{\text{sd}}\delta_{12}\delta_{23}, \quad \Delta_0^{\sigma\sigma}(12;3) = J_{\text{sd}}\delta_{12}\delta_{23}. \tag{212}$$

If we substitute them in Fig. 48 and neglect four-point vertex part, we come to result, Fig. 47, corresponding to SCBA.

Now we explicitly calculate the second order correction (206). We take the electron GF in the approximation (207) with $\Sigma = 0$, and the magnon GF in the form corresponding to the one of the Heisenberg model

$$D^{\sigma\bar{\sigma}}(k) = \frac{\sigma\langle S_1^z\rangle}{\mathrm{i}\omega_n - \sigma\omega(\boldsymbol{k})} \equiv \sigma\langle S_1^z\rangle K^{\sigma\bar{\sigma}}(k), \tag{213}$$

where $\omega(\boldsymbol{k}) = \sigma\langle S_1^z\rangle[J(0) - J(\boldsymbol{k})]$ is the spin wave energy. Neglecting the contribution of longitudinal fluctuations, we write the expression (206), in the form

$$\Sigma^{\sigma}(\boldsymbol{k},\mathrm{i}\omega_n) = -\sigma\frac{1}{2}J_{\text{sd}}\langle S_1^z\rangle - \frac{1}{2}J_{\text{sd}}^2\langle S_1^z\rangle\sigma\sum_q G^{\bar{\sigma}}(k+q)K^{\sigma\bar{\sigma}}(q).$$

After summation over frequencies (that is, the fourth component of the 4-momentum $q = (\boldsymbol{q},\mathrm{i}\omega_q)$) we obtain

$$\Sigma^{\sigma}(\boldsymbol{k},\mathrm{i}\omega_n) = -\sigma\frac{1}{2}J_{\text{sd}}\langle S_1^z\rangle + \frac{1}{2}\langle S_1^z\rangle\sigma J_{\text{sd}}^2\frac{1}{N}\sum_{\boldsymbol{q}}\frac{f[\varepsilon_{\bar{\sigma}}(\boldsymbol{k}+\boldsymbol{q})] - n[\sigma\omega(\boldsymbol{q})]}{\mathrm{i}\omega_n - \varepsilon_{\bar{\sigma}}(\boldsymbol{k}+\boldsymbol{q}) + \sigma\omega(\boldsymbol{q})}.$$

Doing the analytical continuation $\mathrm{i}\omega_n \to E + \mathrm{i}\delta$ we obtain a correction to the electron energy taking into account the interaction with spin waves

$$\mathrm{Re}\Sigma^{\uparrow}(\boldsymbol{k},E) = -\frac{1}{2}J_{\text{sd}}\langle S_1^z\rangle + \frac{1}{2}\langle S_1^z\rangle J_{\text{sd}}^2\frac{1}{N}\sum_{\boldsymbol{q}}\frac{f^{\downarrow}(\boldsymbol{k}+\boldsymbol{q}) - n(\boldsymbol{q})}{E - \varepsilon_{\downarrow}(\boldsymbol{k}+\boldsymbol{q}) + \omega(\boldsymbol{q})},$$

$$\mathrm{Re}\Sigma^{\downarrow}(\boldsymbol{k},E) = \frac{1}{2}J_{\mathrm{sd}}\langle S_1^z\rangle + \frac{1}{2}\langle S_1^z\rangle J_{\mathrm{sd}}^2 \frac{1}{N}\sum_{\boldsymbol{q}} \frac{1-f^{\uparrow}(\boldsymbol{k}+\boldsymbol{q})+n(\boldsymbol{q})}{E-\varepsilon_{\uparrow}(\boldsymbol{k}+\boldsymbol{q})-\omega(\boldsymbol{q})},$$

Here we denote by $f[\varepsilon_{\bar{\sigma}}(\boldsymbol{k}+\boldsymbol{q})] \equiv f^{\bar{\sigma}}(\boldsymbol{k}+\boldsymbol{q})$ the Fermi function and by $n[\omega(\boldsymbol{q})] \equiv n(\boldsymbol{q})$ the Bose-function, and used a relation $n(-\omega) = -1 - n(\omega)$.

Magnon Green's function. Consider now the GF of transversal spin components

$$D_{12}^{\sigma\bar{\sigma}} = -\langle TS_1^{\sigma}S_2^{\bar{\sigma}}\rangle = -\frac{\delta^2\Phi}{\delta v_1^{\bar{\sigma}}\delta v_2^{\sigma}}. \tag{214}$$

It is convenient first to write an equation for the transversal spin component

$$\langle S_1^{\sigma}\mathrm{e}^{-V}\rangle = \frac{\delta\Phi}{\delta v_1^{\bar{\sigma}}},$$

and then by differentiation over the field v^{σ} obtain an equation for GF (214).

We start from an equation of a general form

$$\frac{\partial}{\partial\tau_1}((TS_1^{\sigma}\mathrm{e}^{-V})) = ((T\dot{S}_1^{\sigma}\mathrm{e}^{-V})) - ((T\{S_1^{\sigma},V\}\mathrm{e}^{-V}))$$

and substitute in it the expressions for $\dot{S}_1^{\sigma}$ and $\{S_1^{\sigma},V\}$:

$$-\dot{S}_1^{\sigma} = \sigma hS_1^{\sigma} + \sigma S_1^{\sigma}J_{11'}S_{1'}^z - \sigma S_1^z J_{11'}S_{1'}^{\sigma} + \sigma J_{\mathrm{sd}}c_{1\sigma_1'}^{+}c_{1\sigma_2'}(S_1^{\sigma}s_{\sigma_1'\sigma_2'}^z - S_1^z s_{\sigma_1'\sigma_2'}^{\sigma}),$$

$$\{S_1^{\sigma},V\} = -v_1^z S_1^{\sigma} + v_1^{\sigma}S_1^z + S_1^z u_{11'}S_{1'}^{\sigma}.$$

We obtain an equation

$$\begin{aligned} -\left(\frac{\partial}{\partial\tau_1} - \sigma h_1\right)\delta_{11'}((TS_{1'}^{\sigma}\mathrm{e}^{-V})) &= \\ = -\sigma v_1^{\sigma}((TS_1^z\mathrm{e}^{-V})) - \sigma\alpha_0(11';2')((TS_{1'}^{\sigma}S_{2'}^z\mathrm{e}^{-V})) &- \\ -\sigma J_{\mathrm{sd}}s_{\sigma_1'\sigma_2'}^z((TS_1^{\sigma}c_{1\sigma_2'}c_{1\sigma_1'}^{+}\mathrm{e}^{-V})) + \sigma J_{\mathrm{sd}}s_{\sigma_1'\sigma_2'}^{\sigma}((TS_1^z c_{1\sigma_1'}c_{1\sigma_2'}^{+}\mathrm{e}^{-V})), \end{aligned}$$

where we used the following definition for the amplitude of spin interactions in the Heisenberg model

$$\alpha_0(12;3) = J_{12}\delta_{13} - J_{13}^z\delta_{12}. \tag{215}$$

We express the two-particle GFs in this equation as a variational derivative

$$((TS_1^{\sigma}S_2^z\mathrm{e}^{-V})) = \frac{\delta}{\delta v_1^z}\frac{\delta Z}{\delta v_1^{\bar{\sigma}}},$$

$$s_{\sigma_2'\sigma_1'}^{\sigma}((TS_1^z c_{1\sigma_1'}c_{1\sigma_2'}^{+}\mathrm{e}^{-V})) = \frac{\delta}{\delta v_1^z}\frac{\delta Z}{\delta w_1^{\bar{\sigma}}},$$

$$s^z_{\sigma_2'\sigma_1'}\langle\langle T S_1^\sigma c_{1\sigma_1'} c^+_{1\sigma_2'} \mathrm{e}^{-V}\rangle\rangle = \frac{\delta}{\delta w_1^z}\frac{\delta Z}{\delta v_1^{\bar\sigma}}.$$

As a result we come to the following equation for the transversal spin component

$$\begin{gathered}\Bigg\{(K_{0V}^{-1})^{\sigma\bar\sigma}_{11'}+\\ +\sigma\left[\left(\alpha_0\frac{\delta\Phi}{\delta v^z}\right)_{11'}+J_{\mathrm{sd}}\frac{\delta\Phi}{\delta w_1^z}\delta_{11'}+\left(\alpha_0\frac{\delta}{\delta v_1^z}\right)_{11'}+J_{\mathrm{sd}}\frac{\delta}{\delta w_1^z}\delta_{11'}\right]\Bigg\}\frac{\delta\Phi}{\delta v_{1'}^{\bar\sigma}}=\\ =-\sigma\left[v_1^\sigma\frac{\delta\Phi}{\delta v_1^z}-J_{\mathrm{sd}}\left(\frac{\delta\Phi}{\delta w_1^{\bar\sigma}}\frac{\delta\Phi}{\delta v_1^z}+\frac{\delta^2\Phi}{\delta w_1^{\bar\sigma}\delta v_1^z}\right)\right],\end{gathered} \tag{216}$$

where a propagator of zeroth order approximation is introduced

$$(K_{0V}^{-1})^{\sigma\bar\sigma}_{12} = -\left[\frac{\partial}{\partial\tau_1}+\sigma(h+v_1^z)\right]\delta_{12}. \tag{217}$$

Differentiate this equation over the field v_2^σ, then we obtain an equation, which may be treated as the equation of motion for the magnon GF. Write it in the form [69]:

$$\begin{gathered}\Bigg\{(K_{0V}^{-1})^{\sigma\bar\sigma}_{11'}+\\ +\sigma\left[\left(\alpha_0\frac{\delta\Phi}{\delta v^z}\right)_{11'}+J_{\mathrm{sd}}\frac{\delta\Phi}{\delta w_1^z}\delta_{11'}+\left(\alpha_0\frac{\delta}{\delta v_1^z}\right)_{11'}+J_{\mathrm{sd}}\frac{\delta}{\delta w_1^z}\delta_{11'}\right]\Bigg\}\frac{\delta^2\Phi}{\delta v_{1'}^{\bar\sigma}\delta v_2^\sigma}=\\ =-\sigma\left[\delta_{12}\frac{\delta\Phi}{\delta v_1^z}+v_1^\sigma\frac{\delta^2\Phi}{\delta v_1^z\delta v_2^\sigma}-J_{\mathrm{sd}}\left(\frac{\delta^2\Phi}{\delta w_1^{\bar\sigma}\delta v_2^\sigma}\frac{\delta\Phi}{\delta v_1^z}+\frac{\delta\Phi}{\delta w_1^{\bar\sigma}}\frac{\delta^2\Phi}{\delta v_1^z\delta v_2^\sigma}+\right.\right.\\ \left.\left.+\frac{\delta}{\delta v_1^z}\frac{\delta^2\Phi}{\delta w_1^{\bar\sigma}\delta v_2^\sigma}\right)+\alpha_0(11';2')\frac{\delta^2\Phi}{\delta v_{2'}^z\delta v_2^\sigma}\frac{\delta\Phi}{\delta v_{1'}^{\bar\sigma}}+J_{\mathrm{sd}}\frac{\delta^2\Phi}{\delta w_1^z\delta v_2^\sigma}\frac{\delta\Phi}{\delta v_1^{\bar\sigma}}\right],\end{gathered} \tag{218}$$

By transformation of separate terms in the right hand side, we can get terms proportional to the magnon GF. It is clear that such terms should be replaced in the left hand side, and then they will form the self-energy part for the magnon GF. The remaining terms in the right hand side should form the terminal part. For example, a mixed term is best expressed first through the electron GF, and then differentiated with the help of identity (150), that is

$$\frac{\delta^2\Phi}{\delta w_1^{\bar\sigma}\delta v_2^\sigma} = \frac{\delta}{\delta v_2^\sigma}\left(-s^\sigma_{\sigma_1'\sigma_2'}G^{\sigma_2'\sigma_1'}_{11}\right) = -\frac{1}{\sqrt{2}}G^{\bar\sigma\bar\sigma}_{11'}\frac{\delta\Sigma^{\bar\sigma\sigma}_{1'2'}}{\delta v_2^\sigma}G^{\sigma\sigma}_{2'1}.$$

In the expression (209) for $\delta\Sigma^{\bar\sigma\sigma}_{12}/\delta v_3^\sigma$ we take only the first term, including the magnon GF. As a result, a mixed derivative is represented in the form

$$\frac{\delta^2\Phi}{\delta w_1^{\bar\sigma}\delta v_2^\sigma} = -\frac{1}{2}G^{\bar\sigma\bar\sigma}_{11'}G^{\sigma\sigma}_{2'1}\Delta^{\bar\sigma\sigma}(1'2';3')D^{\sigma\bar\sigma}_{3'2}. \tag{219}$$

We do not plan the differentiation equation (218) over v^σ, therefore we can drop all anomalous terms proportional to the averaged value of transversal spin component. In this case, equation (218) is written in the form [69]

$$\Big\{(K_{0V}^{-1})_{11'}^{\sigma\bar{\sigma}} + +\sigma\Big[\Big(\tilde{\alpha}_0^\sigma \frac{\delta\Phi}{\delta v^z}\Big)_{11'} + \Big(\tilde{\alpha}_0^\sigma \frac{\delta}{\delta v^z}\Big)_{11'} + J_{\rm sd}\frac{\delta\Phi}{\delta w_1^z}\delta_{11'} + J_{\rm sd}\frac{\delta}{\delta w_1^z}\delta_{11'} +$$

$$+\tfrac{1}{2}J_{\rm sd}\frac{\delta}{\delta v_1^z}\left(G_{13'}^{\bar{\sigma}\bar{\sigma}}\Delta^{\bar{\sigma}\sigma}(3'4';1')G_{4'1}^{\sigma\sigma}\right)\Big]\Big\}\frac{\delta^2\Phi}{\delta v_{1'}^{\bar{\sigma}}\delta v_2^{\sigma}} = -\sigma\delta_{12}\frac{\delta\Phi}{\delta v_2^z} \qquad (220)$$

Here we introduced the definition for the renormalized amplitude of interactions in the Heisenberg model

$$\tilde{\alpha}_0^\sigma(12;3) = \tilde{J}_{12}^{\sigma\bar{\sigma}}\delta_{13} - J_{13}^z\delta_{12}, \qquad (221)$$

$$\tilde{J}_{12}^{\sigma\bar{\sigma}} = J_{12} + \tfrac{1}{2}J_{\rm sd}G_{13'}^{\bar{\sigma}\bar{\sigma}}\Delta^{\bar{\sigma}\sigma}(3'4';2)G_{4'1}^{\sigma\sigma}. \qquad (222)$$

Equation (220) contains the unknown value $\Delta^{\bar{\sigma}\sigma}(12;3)$. For its calculation we must know the derivative $\delta\Sigma^{\bar{\sigma}\sigma}/\delta v_1^\sigma$. Therefore the magnon and electron GFs are tightly connected through $\Delta^{\bar{\sigma}\sigma}$. By calculating $\Delta^{\bar{\sigma}\sigma}$ in some approximation, we can obtain a closed equation for the magnon GF.

Like in the Heisenberg model we suggest a multiplicative representation for the magnon GF

$$D_{12}^{\sigma\bar{\sigma}} = K_{11'}^{\sigma\bar{\sigma}}P_{1'2}^{\sigma\bar{\sigma}} \qquad (223)$$

and obtain equations for the propagator $K^{\sigma\bar{\sigma}}$ and the terminal part $P^{\sigma\bar{\sigma}}$. To this purpose we have to transform some terms in the basic equation (221) similarly as we did it in the Heisenberg model.

First of all we introduce the self-energy part $M^{\sigma\bar{\sigma}}$ for the magnon GF from the Dyson equation

$$(K^{-1})_{12}^{\sigma\bar{\sigma}} = (K_{0V}^{-1})_{12}^{\sigma\bar{\sigma}} - \sigma M_{12}^{\sigma\bar{\sigma}}.$$

Using the rules of differentiation for the GF, we obtain two relations

$$\Big(\tilde{\alpha}_0^\sigma\frac{\delta}{\delta v^z}\Big)_{11'}D_{1'2}^{\sigma\bar{\sigma}} = \sigma K_{2'1'}^{\sigma\bar{\sigma}}\tilde{\alpha}_0^\sigma(12';1')D_{1'2}^{\sigma\bar{\sigma}} +$$

$$+\sigma K_{4'3'}^{\sigma\bar{\sigma}}\Big[\Big(\tilde{\alpha}_0^\sigma\frac{\delta}{\delta v^z}\Big)_{14'}M_{3'1'}^{\sigma\bar{\sigma}}\Big]D_{1'2}^{\sigma\bar{\sigma}} + K_{2'1'}^{\sigma\bar{\sigma}}\Big(\tilde{\alpha}_0^\sigma\frac{\delta}{\delta v^z}\Big)_{12'}P_{1'2}^{\sigma\bar{\sigma}},$$

$$-\frac{\delta}{\delta w_1^z}\left(\frac{\delta^2\Phi}{\delta v_1^{\bar{\sigma}}\delta v_2^{\sigma}}\right) = \frac{\delta}{\delta w_1^z}D_{12}^{\sigma\bar{\sigma}} = \sigma K_{13'}^{\sigma\bar{\sigma}}\frac{\delta M_{3'4'}^{\sigma\bar{\sigma}}}{\delta w_1^z}D_{4'2}^{\sigma\bar{\sigma}} + K_{11'}^{\sigma\bar{\sigma}}\frac{\delta P_{1'2}^{\sigma\bar{\sigma}}}{\delta w_1^z}.$$

With the help of them we rewrite the basic equation (220) in the following form

$$\Big\{(K_{0V}^{-1})_{11'}^{\sigma\bar{\sigma}} + \sigma\Big[\Big(\tilde{\alpha}_0^\sigma\frac{\delta\Phi}{\delta v^z}\Big)_{11'} + K_{2'1'}^{\sigma\bar{\sigma}}\tilde{\alpha}_0^\sigma(12';1') + K_{4'3'}^{\sigma\bar{\sigma}}\Big(\tilde{\alpha}_0^\sigma\frac{\delta}{\delta v^z}\Big)_{14'}M_{3'1'}^{\sigma\bar{\sigma}} +$$

$$+J_{\rm sd}\frac{\delta\Phi}{\delta w_1^z}\delta_{11'} + J_{\rm sd}K_{12'}^{\sigma\bar{\sigma}}\frac{\delta M_{2'1'}^{\sigma\bar{\sigma}}}{\delta w_1^z} + \tfrac{1}{2}J_{\rm sd}\frac{\delta}{\delta v_1^z}\left(G_{13'}^{\bar{\sigma}\bar{\sigma}}\Delta^{\bar{\sigma}\sigma}(3'4';1')G_{4'1}^{\sigma\sigma}\right)\Big]\Big\}D_{1'2}^{\sigma\bar{\sigma}} =$$

$$= -\sigma\left[\delta_{12}\frac{\delta\Phi}{\delta v_2^z} + K_{2'1'}^{\sigma\bar{\sigma}}\Big(\tilde{\alpha}_0^\sigma\frac{\delta}{\delta v^z}\Big)_{12'}P_{1'2}^{\sigma\bar{\sigma}} + K_{11'}^{\sigma\bar{\sigma}}\frac{\delta P_{1'2}^{\sigma\bar{\sigma}}}{\delta w_1^z}\right]. \qquad (224)$$

We can split equation (224) into two equations, one for $M^{\sigma\bar{\sigma}}$ and one for $P^{\sigma\bar{\sigma}}$:

$$M_{12}^{\sigma\bar{\sigma}} = -\left(\tilde{\alpha}_0^{\sigma}\frac{\delta\Phi}{\delta v^z}\right)_{12} - K_{1'2}^{\sigma\bar{\sigma}}\tilde{\alpha}_0^{\sigma}(11';2) - K_{4'3'}^{\sigma\bar{\sigma}}\left(\tilde{\alpha}_0^{\sigma}\frac{\delta}{\delta v^z}\right)_{14'} M_{3'2}^{\sigma\bar{\sigma}}$$
$$-J_{\rm sd}\frac{\delta\Phi}{\delta w_1^z}\delta_{12} - J_{\rm sd}K_{11'}^{\sigma\bar{\sigma}}\frac{\delta M_{1'2}^{\sigma\bar{\sigma}}}{\delta w_1^z} - \tfrac{1}{2}J_{\rm sd}\frac{\delta}{\delta v_1^z}\left(G_{13'}^{\bar{\sigma}\bar{\sigma}}\Delta^{\bar{\sigma}\sigma}(3'4';2)G_{4'1}^{\sigma\sigma}\right), \quad (225)$$

$$P_{12}^{\sigma\bar{\sigma}} = -\sigma\left[\delta_{12}\frac{\delta\Phi}{\delta v_2^z} + K_{2'1'}^{\sigma\bar{\sigma}}\left(\tilde{\alpha}_0^{\sigma}\frac{\delta}{\delta v^z}\right)_{12'} P_{1'2}^{\sigma\bar{\sigma}} + K_{11'}^{\sigma\bar{\sigma}}\frac{\delta P_{1'2}^{\sigma\bar{\sigma}}}{\delta w_1^z}\right]. \quad (226)$$

When $J_{\rm sd} = 0$ (if we put $\sigma = -$) these equations reduce to the corresponding equations (66) and (64) for the GF $\langle TS_1^- S_2^+\rangle$ of the Heisenberg model. Sd-interaction leads to the renormalization of the amplitude of the direct exchange interaction to account for the indirect exchange interaction of localized spins via electrons. Moreover, in equations (225) and (226) additional terms with variational derivatives over w_1^z appear. Finally, in the equation for the self-energy a special term with a derivative of $\Delta^{\bar{\sigma}\sigma}$ over field v^z appears.

We consider first a simple situation, where the system of localized spins is taken in the mean-field approximation with spin GF (213). We find a correction to the magnon spectrum $\omega(\boldsymbol{k})$ due to the sd-interaction. By iteration of equation (225) up to the second order we find the following expression for the magnon self-energy

$$M_{12}^{\sigma\bar{\sigma}} = \langle S^z\rangle[(\sum_{1'}J_{11'})\delta_{12} - J_{12}] + \tfrac{1}{2}J_{\rm sd}\sigma' G_{11}^{\sigma'\sigma'}\delta_{12} + \tfrac{1}{2}J_{\rm sd}^2\langle S^z\rangle G_{12}^{\bar{\sigma}}G_{21}^{\sigma} +$$
$$+\tfrac{1}{2}J_{\rm sd}^2 G_{11'}^{\bar{\sigma}}G_{1'1}^{\sigma}K_{1'1}^{\sigma\bar{\sigma}}\delta_{12} - \tfrac{1}{4}J_{\rm sd}^2 G_{12}^{\sigma'}G_{21}^{\sigma'}K_{12}^{\sigma\bar{\sigma}}. \quad (227)$$

Here we take into account expressions (221) and (222) for the renormalized amplitude $\tilde{\alpha}_0^{\sigma}$, and the value for $\Delta^{\bar{\sigma}\sigma}$ in lowest approximation:

$$\Delta^{\bar{\sigma}\sigma}(12;3) = J_{\rm sd}\delta_{12}\delta_{23}.$$

Let us take only the first row in expression (227), so the self-energy for the magnon is

$$M_{12}^{\sigma\bar{\sigma}} = \langle S^z\rangle\left[\left(\sum_{1'}J_{11'}\right)\delta_{12} - J_{12}\right] + \frac{1}{2}J_{\rm sd}\sigma(G_{11}^{\sigma} - G_{11}^{\bar{\sigma}})\delta_{12} + \frac{1}{2}J_{\rm sd}^2\langle S^z\rangle G_{12}^{\bar{\sigma}}G_{21}^{\sigma}. \quad (228)$$

In momentum representation, after summation over frequencies, we find

$$M^{\sigma\bar{\sigma}}(q) = \langle S^z\rangle[J(0) - J(\boldsymbol{k})] +$$
$$+\frac{J_{\rm sd}}{2N}\sum_{\boldsymbol{k}}\sigma\left[f^{\sigma}(\boldsymbol{k}) - f^{\bar{\sigma}}(\boldsymbol{k})\right] + \tfrac{1}{2}J_{\rm sd}^2\langle S^z\rangle\frac{1}{N}\sum_{\boldsymbol{k}}\frac{f^{\sigma}(\boldsymbol{k}-\boldsymbol{q}) - f^{\bar{\sigma}}(\boldsymbol{k})}{\mathrm{i}\omega_q + \varepsilon_{\sigma}(\boldsymbol{k}-\boldsymbol{q}) - \varepsilon_{\bar{\sigma}}(\boldsymbol{k})}. \quad (229)$$

In such a form this result was obtained by Vonsovskii and Izyumov [89]. This expression can be presented in another form

$$M^{\sigma\bar{\sigma}}(q) = \langle S^z\rangle[J(0) - J(\boldsymbol{k})] +$$
$$+\frac{J_{\rm sd}}{2N}\sum_{\boldsymbol{k}}\left[f^{\sigma}(\boldsymbol{k}-\boldsymbol{q}) - f^{\bar{\sigma}}(\boldsymbol{k})\right]\left\{\sigma + \frac{J_{\rm sd}\langle S^z\rangle}{\mathrm{i}\omega_q + \varepsilon(\boldsymbol{k}-\boldsymbol{q}) - \varepsilon(\boldsymbol{k}) - \sigma J_{\rm sd}\langle S^z\rangle}\right\}. \quad (230)$$

Notice that at $\boldsymbol{q}=0$ and $\omega_q=0$ $M^{\sigma\bar{\sigma}}(q)=0$, and Goldstone's theorem is fulfilled. When calculating expression (230) we used the second order correction, and we considered the dimensionless parameter J_{sd}/W as small. However in the numerator of (230) the quantity $\varepsilon(\boldsymbol{k}-\boldsymbol{q})-\varepsilon(\boldsymbol{k})$ at small $\boldsymbol{q}$ may be small when compared to $J_{\mathrm{sd}}\langle S^z\rangle$, in spite of a wide band width $W \gg J_{\mathrm{sd}}$. Region of q's, where the effect of spin splitting in (230) must be strong, is determined by the inequality

$$q < q_0 = \frac{J_{\mathrm{sd}}\langle S^z\rangle}{\hbar v_F},$$

where v_F is the velocity at the Fermi surface.

The magnon GF for a ferromagnetic metal is

$$D^{+-}(\boldsymbol{q},\omega) \approx \frac{\langle S^z\rangle}{\omega-\tilde{\omega}(q)+\mathrm{i}\gamma(q)}, \tag{231}$$

where $\tilde{\omega}(q)$ is the spin wave energy and $\gamma(q)$ is the attenuation. $\tilde{\omega}(q)$ is determined by the equation

$$\begin{gathered}\tilde{\omega}(\boldsymbol{q})=\omega(\boldsymbol{q})+\\ +\frac{J_{\mathrm{sd}}}{2N}\sum_{\boldsymbol{k}}\left[f^{\sigma}(\boldsymbol{k}-\boldsymbol{q})-f^{\bar{\sigma}}(\boldsymbol{k})\right]\left\{\sigma+\frac{J_{\mathrm{sd}}\langle S^z\rangle}{\tilde{\omega}(q)+\varepsilon(\boldsymbol{k}-\boldsymbol{q})-\varepsilon(\boldsymbol{k})-\sigma J_{\mathrm{sd}}\langle S^z\rangle}\right\},\end{gathered} \tag{232}$$

where $\omega(\boldsymbol{q})=\langle S^z\rangle[J(0)-J(\boldsymbol{k})]$ and $\gamma(\boldsymbol{q})$ is [89]

$$\gamma(\boldsymbol{q})=\frac{\pi}{2}J_{\mathrm{sd}}\frac{1}{N}\sum_{\boldsymbol{k}}\left[f^{+}(\boldsymbol{k}-\boldsymbol{q})-f^{-}(\boldsymbol{k})\right]\delta\left[\tilde{\omega}(\boldsymbol{q})+\varepsilon_{+}(\boldsymbol{k}-\boldsymbol{q})-\varepsilon_{-}(\boldsymbol{k})\right], \tag{233}$$

It follows from here that

$$\frac{\gamma(q)}{\tilde{\omega}(q)}=\frac{3\pi}{8}\langle S^z\rangle n\left(\frac{J_{\mathrm{sd}}}{\varepsilon_F}\right)^2\frac{k_F}{q}\begin{cases}1, & q>q_0,\\ 0, & q<q_0.\end{cases}$$

Thus, spin waves with small momentum do not attenuate in a ferromagnetic metal to account for the scattering on electrons. When $q=q_0$ the relative damping of a spin wave has the value

$$\frac{\gamma(q)}{\tilde{\omega}(q)}\sim n\frac{J_{\mathrm{sd}}}{\varepsilon_F}\ll 1.$$

Notice that equation (232) has a second solution, corresponding to an optical mode. In fact, if we put $q=0$ in it, we obtain, together with $\tilde{\omega}(0)=0$, another solution

$$\tilde{\omega}(0)=J_{\mathrm{sd}}(\langle S^z\rangle+s),$$

where

$$s=\frac{1}{2N}\sum_{\boldsymbol{k}}\left[f^{+}(k)-f^{-}(k)\right]$$

is the spontaneous moment of conduction electrons. An optical mode should always appear in the spin wave spectrum of a ferromagnet if sublattices exist in it. In the sd-model two sublattices are formed by localized spins and conduction electrons.

Results (231) and (232) were obtained much later also by the technique of decoupling double-time GFs in the spirit of RPA [31]. Since RPA is not associated with small interactions (in our case the parameter $|J_{sd}| \ll W$), formulas (231)–(233), corresponding to RPA results, may be considered also for large $|J_{sd}| \gg W$ [31]. As it was shown in [31], the strong coupling character of the spin wave spectrum corresponds to that of the Heisenberg model with exchange interactions only between nearest neighbors. Such a spectrum is used for the explanation of experimental data on inelastic neutron scattering in manganites (see also new works [90, 31, 11]).

In this Section we follow the paper [69], where it was shown also how to construct a theory of the magnetic polaron in the sd-model, and also how to describe the principal properties of the quasiparticle spectrum and spin waves in the double exchange model.

V. CONCLUSION

We discussed two specific approaches for the study of the basic models of magnetism. One is based on the use of the Wick theorem for operators expressing Hamiltonians of models, the other one is based on the generating functional of fluctuating fields. In both cases the goal is the same: to construct a perturbation theory for the temperature GFs, describing these models. It is remarkable that for different models some common features in the structures of GFs are observed, and such similarities are revealed in the structure of corresponding GFs in both mentioned approaches. Comparison of these two approaches and elucidation of common features for the different models is the main subject of our Conclusion. First of all we resume the initial positions of the each of two approaches.

Approach based on the generalized Wick theorem. In the standard theory of perturbation for usual Fermi systems one deals with the calculation of averages over a statistical ensemble of a T-product of Fermi operators c and c^+, taken in the interaction representation

$$\langle T \ldots c_1 \ldots c_2^+ \ldots \rangle. \tag{234}$$

According to Wick's theorem, this average is decomposed into a sum of terms with products of pair averages of two operators c and c^+:

$$\overbrace{c_1 c_2^+} = G_{12}^0 [c_1, c_2^+]_+ = G_{12}^0 \delta_{12}, \tag{235}$$

where G_{12}^0 is one-particle Fermi function of zeroth approximation.

In the Heisenberg model it is necessary to calculate averages of a T-product of the spin operators S^-, S^+ and S^z

$$\langle T \ldots S_1^- \ldots S_2^+ \ldots S_3^z \ldots \rangle. \tag{236}$$

The Wick theorem for spin operators states that such an average is decomposed into a set of terms with all possible pairings of spin operators according to the scheme:

$$\left. \begin{aligned} \overbrace{S_1^- S_2^+} &= K_{12}^0 [S_2^+, S_1^-]_- = K_{12}^0 \delta_{12} S_2^z \\ \overbrace{S_1^- S_2^z} &= K_{12}^0 [S_2^z, S_1^-]_- = -K_{12}^0 \delta_{12} S_2^- \end{aligned} \right\}, \tag{237}$$

where K_{12}^0 is a Bose GF of zeroth approximation.

For the third basic model – the tJ-model – the Hamiltonian is expressed by a quadratic form of X operators, among them there are Fermi-like operators $X^{\sigma\bar{\sigma}}$ and Bose-like operators $X^{\sigma\bar{\sigma}}$ and $X^{\bar{\sigma}\sigma}$. In the general case we deal with an average

$$\langle T \ldots X_1^{0\sigma} \ldots X_2^{\sigma 0} \ldots X_3^{\sigma\bar{\sigma}} \ldots X_4^{\bar{\sigma}\sigma} \ldots \rangle. \tag{238}$$

The Wick theorem for X operators states that this average is decomposed into a set of terms with pairings

$$\left.\begin{array}{l}
\overrightarrow{X_1^{0\sigma}X_2^{\sigma 0}} = G_{12}^0[X_2^{\sigma 0},X_1^{0\sigma}]_+ = G_{12}^0\delta_{12}(X_2^{00}+X_2^{\sigma\sigma}) \\
\overrightarrow{X_1^{\sigma\bar{\sigma}}X_2^{\bar{\sigma}\sigma}} = K_{12}^0[X_2^{\bar{\sigma}\sigma},X_1^{\sigma\bar{\sigma}}]_- = K_{12}^0\delta_{12}(X_2^{\bar{\sigma}\bar{\sigma}}-X_2^{\sigma\sigma}) \\
\overrightarrow{X_1^{0\sigma}X_2^{\sigma\bar{\sigma}}} = G_{12}^0[X_2^{\sigma\bar{\sigma}},X_1^{0\sigma}]_- = -G_{12}^0\delta_{12}X_2^{0\bar{\sigma}} \\
\ldots
\end{array}\right\}, \tag{239}$$

In this case a pairing leads to the appearance of either a Fermi GF G_{12}^0 or a Bose GF K_{12}^0 of zeroth approximation.

Comparison of expressions (235), (237) and (239) shows that, upon pairing of Fermi operators, the number of the remaining operators decreases by two, while in the case of spin or X operators a pairing lowers the total number in the T-product only by one. Because of that, the reduction of an average of a T-product of S^α or X operators goes more slowly with respect to the case of usual Fermi operators. However this process is finishing after the disappearance of (S^-, S^+) operators in the Heisenberg model and of $(X^{0\sigma}, X^{\sigma 0})$ and also $(X^{\sigma\bar{\sigma}}, X^{\bar{\sigma}\sigma})$ operators in the tJ-model. As a result of the total system of pairings in the Fermi system an initial average is represented by a sum of terms, each of which is a product of G_{12}^0 GFs. However in the Heisenberg model, together with the GFs K_{12}^0, an average of S^z operators appears as a factor. In the tJ-model, together with products of the GFs G_{12}^0 and K_{12}^0, appears an average of diagonal X operators X^{00}, $X^{\sigma\sigma}$, $X^{\bar{\sigma}\bar{\sigma}}$.

In the diagrammatic elements for the Fermi system only Green's lines are present, but in the Heisenberg and the tJ-model, in addition to Green's lines, appear the cumulants, which are averages of diagonal operators. There is an additional complexity for these two models. Since as a result of the pairing of two S^α or X operators some third operator appears, and this can participate in further pairings, complicate vertices with many Green's lines appear.

Nevertheless in the Heisenberg model a unique graphic representation of perturbation series takes place as in the case of Fermi system. However in the tJ-model, operating with X operators, the situation is much more complicated. The reason for that is the existence of two operators of "annihilation", $X^{0\sigma}$ and $X^{\sigma\bar{\sigma}}$, and we can start pairing either from the first of them or from the second. Graphic representation will be different in the two cases. We tried to use a system of priority among such operators. This ambiguity makes it desirable to develop some other approach not based on the Wick theorem for X operators. Such an alternative approach is just the method of generating functional, which was widely used in the theory of usual Fermi system.

Method of generating functional. The main idea of the method is to include in the system under consideration the interaction V with fluctuating fields, and to determine the GFs as variational derivatives of this functional. The generating functional is a generalization of a partition function of the system

$$Z[V] = \langle T\mathrm{e}^{-V} \rangle. \tag{240}$$

It is convenient to shift from the functional $Z[V]$ to $\Phi[V]$ by a substitution

$$Z[V] = \mathrm{e}^{\Phi[V]}. \tag{241}$$

For a given model the functional V is chosen in accordance with the Hamiltonian. For example, for the Heisenberg model, we introduced the operator

$$V = \sum_{1} h_1 S_1^z + \sum_{11'} \mathscr{J}_{11'} S_1^- S_{1'}^+ \tag{242}$$

with fluctuating fields h_1 and $\mathscr{J}_{11'}$, depending on one and two four-dimensional points. Then the transversal and longitudinal spin GFs are expressed through variational derivatives over these fields:

$$\mathscr{K}_{12}^{\perp} = \langle T S_1^- S_2^+ \rangle = \frac{\delta \Phi}{\delta \mathscr{J}_{12}}, \tag{243}$$

$$\mathscr{K}_{12}^{z} = \langle T(S_1^z - \langle S_1^z \rangle)(S_2^z - \langle S_2^z \rangle) \rangle = \frac{\delta^2 \Phi}{\delta h_1 \delta h_2}. \tag{244}$$

For the tJ-model, V is chosen in the form

$$V = \sum_{1} \left[v_1 X_1^{00} + \sum_{\sigma} (v_1^{\sigma\sigma} X_1^{\sigma\sigma} + v_1^{\sigma\bar{\sigma}} X_1^{\sigma\bar{\sigma}}) \right] + \sum_{11'} t_{11'}^{\sigma} X_1^{\sigma 0} X_{1'}^{0\sigma} \tag{245}$$

with one-point fields v_1, $v_1^{\sigma\sigma}$, $v_1^{\sigma\bar{\sigma}}$ and the two-point field $t_{11'}^{\sigma}$. The variational derivative over the field $t_{11'}^{\sigma}$ gives the electron GF, and over one-point fields gives the Bose-like GFs, in particular the magnon GF:

$$\mathscr{G}_{12} = \langle T X_1^{0\sigma} X_2^{\sigma 0} \rangle = \frac{\delta \Phi}{\delta t_{12}^{\sigma}}, \tag{246}$$

$$\mathscr{D}_{12}^{\perp} = \langle T X_1^{\sigma\bar{\sigma}} X_2^{\bar{\sigma}\sigma} \rangle = \frac{\delta^2 \Phi}{\delta v_1^{\sigma\bar{\sigma}} \delta v_2^{\bar{\sigma}\sigma}}, \tag{247}$$

and so on. Of course after differentiation in expressions (243), (244) and (246), (247) the fluctuating fields should be put equal to zero.

For the Heisenberg and tJ-model one can obtain the equations of motion in terms of variational derivatives over fluctuating fields. It is remarkable that GF (243) in the Heisenberg model and (246) in the tJ-model, which are derivatives over two-point fields, obey to equations of motion with derivatives over the one-point field h for the Heisenberg model and fields v_1, $v_1^{\sigma\sigma}$, $v_1^{\sigma\bar{\sigma}}$ for the tJ-model. From these equations one can see that such GFs have a multiplicative structure, that is, they can be represented in the form

$$\mathscr{K}_{12}^{\perp} = (K^{\perp} \mathscr{P}^{\perp})_{12}, \tag{248}$$

$$\mathscr{G}_{12}^{\sigma} = (G^{\sigma} \Lambda^{\sigma})_{12}. \tag{249}$$

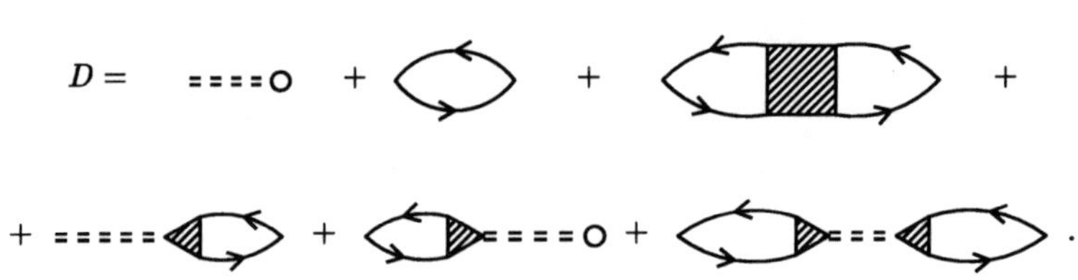

FIGURE 49. Structure of Bose-like GF of type (244) or (247)

Factors $K^{\perp}$ and G^{σ} are propagators, and $\mathscr{P}^{\perp}$ and Λ^{σ} are terminal parts of GFs. Propagators obey to Dyson equations with bare propagators $K_0^{\perp}$ and G_0^{σ} and self-energy parts $\Sigma^{\perp}$ and Σ^{σ}, respectively:

$$(K^{\perp})^{-1} = (K_0^{\perp})^{-1} - \Sigma^{\perp}, \tag{250}$$

$$(G^{\sigma})^{-1} = (G_0^{\sigma})^{-1} - \Sigma^{\sigma}. \tag{251}$$

For each GF (248) and (249) one can derive a couple of equations for self-energy and terminal parts, which includes derivatives over field h_1 in the first case and over fields v_1, $v_1^{\sigma\sigma}$, $v_1^{\sigma\bar{\sigma}}$ in the second case. The structure of these equations for the transversal GF in the Heisenberg model and the electron GF in the tJ-model is similar. Iteration of the equation generates a series of perturbation theory for the self energy and the terminal part. For the Heisenberg model these series exactly corresponds to the diagrammatic technique for spin operators. In case of the tJ-model the situation is more complicated because of the not-unique graphic representations of the series. The theory of perturbation based on equations with variational derivatives turns out to be more consistent and not contradictory.

General structure of one-particle Green's functions. It is important to note that, although the structure of GFs (248) and (249), expressed by second derivatives over one-point fields, is different, the equations for them in both models are similar. Thus, the longitudinal GF in the Heisenberg model is isomorph to the transversal GF in the tJ-model. Summation of loop diagrams for GFs of type (244) or (247) leads to the graphic representation for these quantities of Fig. 49. This result corresponds to GRPA. Solid lines with arrows are propagators for magnons in the Heisenberg model with GF (244), or propagators for electrons in the tJ-model with GF (247). Double dashed line corresponds to GF of collective modes in a system: fluctuations of longitudinal spin components in the Heisenberg model or magnons in the tJ-model.

Fig. 49 takes into account for the GF $\mathscr{D}$ all the loop diagrams, consisting of four types of loops, Fig. 50. Here a wavy line denotes the exchange interaction in the Heisenberg model or the hopping matrix element in the tJ-model. GF $\mathscr{D}$ in GRPA has the following form

$$\mathscr{D}(q) = \frac{\pi(q) + \Pi(q)}{[1-\Lambda(q)][1-Q(q)] - [\pi(q)+\Pi(q)][\Phi(q)+J(\boldsymbol{q})]}, \tag{252}$$

where $q = (\boldsymbol{q}, \mathrm{i}\omega_n)$ is the 4-momentum and $J(\boldsymbol{q})$ is the Fourier component of the exchange interaction. In the case of the tJ-model $J(\boldsymbol{q})$ should be change for $t(\boldsymbol{q})$), and $\pi(q) \sim \delta_{\omega_n,0}$ is a singular contribution.

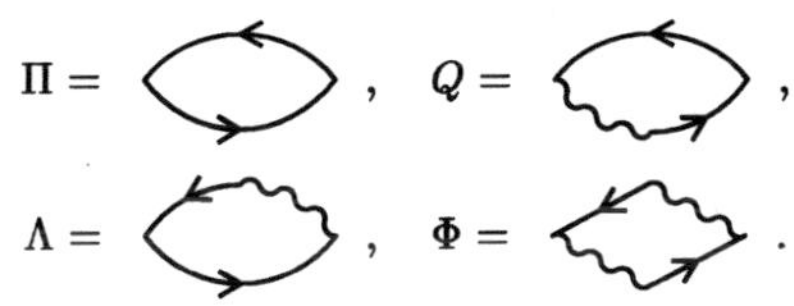

FIGURE 50. Loop diagrams of GRPA approximation

It is remarkable that the structure of a GF of type (244) (longitudinal GF for the Heisenberg model) and of type (247) (Bose-like GF of transversal spin component for the tJ-model) is similar in the GRPA-approximation. Such isomorphism for so different models comes from the similarity of the permutation relations for spin S operators and $X^{\sigma\bar{\sigma}}$ operators.

We discuss now the structure of the one-particle GF of type (243) and (246), constructed on the basic operators in each model. We mean by this termini operators S^- and S^+ for the Heisenberg model and Fermi-like operators for the tJ-model. For these GFs we have multiplicative representations (248) and (249) and the propagator and the terminal parts obey to equations with derivatives over one-point fields.

Along with multiplicative form these GFs have another general property: they obey to the Larkin equations, from which the special structure of self-energy parts follows: for the transversal GF in the Heisenberg model (see relations (248) – (251))

$$\Sigma^{\perp} = \Sigma^{\perp}_{\text{ir}} + P^{\perp}J \tag{253}$$

and the electron GF in the tJ-model

$$\Sigma^{\sigma} = \Sigma^{\sigma}_{\text{ir}} + \Lambda^{\sigma}t. \tag{254}$$

Thus, in both cases Σ is given by two terms: not-cut over the interaction line Σ_{ir} and cut over a line. Terms $P^{\perp}J$ and $\Lambda^{\sigma}t$ are products of the terminal part and the interaction.

Combining, for example, expressions (249), (251) and (254), we can write the electron GF in the form

$$\mathscr{G}^{\sigma}(k) = \frac{1}{F^{\sigma}(k) - \varepsilon(\boldsymbol{k})}, \tag{255}$$

where

$$F^{\sigma}(k) = (\Lambda^{\sigma}(k))^{-1}[(G^{\sigma}_{0}(k))^{-1} - \Sigma^{\sigma}_{\text{ir}}(k)]. \tag{256}$$

In the method of decoupling of double-time GFs often electron GFs are obtained in the form (255). Such representation can be treated as exact, if the quantity $F^{\sigma}(k)$ is exact, which is expression (256).

VI. APPENDIX

Equations of motion for Green's function in fluctuating fields

Consider an average of T-product of operators $X_1, X_2, X_3, \ldots$, depending on thermodynamic times $\tau_1, \tau_2, \tau_3, \ldots$:

$$((TX_1X_2X_3\ldots e^V)) \equiv \mathrm{Tr}[e^{-\beta\mathscr{H}}TX_1X_2X_3\ldots e^V],$$

where $\mathscr{H}$ Hamiltonian, and V operator of interaction of a system with fluctuating fields. If operators $X_1, X_2, X_3\ldots$ commute with each other then an identity takes place [65, 70, 71]:

$$\begin{gathered}\frac{\partial}{\partial\tau_1}((TX_1X_2X_3\ldots e^V)) = ((T\dot{X}_1X_2X_3\ldots e^V))+\\ +((T\{X_1,X_2\}X_3\ldots e^V)) + ((T\{X_1,X_3\}X_2\ldots e^V)) + \ldots -\\ -((T\{X_1,V\}X_2X_3\ldots e^V)).\end{gathered} \tag{257}$$

Here $\dot{X}_1$ is determined by equation of motion for operator X_1:

$$\dot{X}_1 = \frac{\partial X_1}{\partial\tau} = -[X_1,\mathscr{H}]_-, \quad X_i = e^{\tau\mathscr{H}}X_ie^{-\tau\mathscr{H}}.$$

Symbol with curl brackets for operators X_1 and X_2 means a commutator multiplied by δ-function

$$\{X_1,X_2\} = \delta(\tau_1-\tau_2)[X_1,X_2]_-. \tag{258}$$

Notice that if operators $X_1, X_2, X_3, \ldots$ anticommute, in equation (257) relation (258) should be changed for a similar relation with anticommutators

$$\{X_1,X_2\} = \delta(\tau_1-\tau_2)[X_1,X_2]_+, \tag{259}$$

It is also necessary to write a sign in terms of equation (257) dependently on parity of f-like operators.

ACKNOWLEDGMENTS

I should like to thank the students to whom I lectured this course, for their patience and their questions, many of which helped sharpen the notes. I am grateful to Mr. Renzo Rubele for improving the English text. Also I should like thank Prof. Ferdinando Mancini for hospitality during my stay in Vietri and Salerno. This work was supported by the Russian Fond of Support of Scientific Schools, grant 747.2003.2.

REFERENCES

1. Heisenberg W., *Zs. Physik*, **49**, 619 (1928).
2. Hubbard J., *Proc. Roy. Soc. A*, **276**, 238 (1963).
3. Hubbard J., *Proc. Roy. Soc. A*, **281**, 401 (1964).
4. Anderson P. W., *Phys. Rev.*, **124**, 41 (1961).
5. Izuyama T., Kim D., and Kubo R., *J. Phys. Soc. Japan*, **18**, 1025 (1963).
6. Chao K. A., Spalek J., and Oles A. M., *J. Phys. C*, **10**, L271 (1972).
7. Anderson P. W., *Phys. Rev.*, **115**, 2 (1959).
8. Anderson P. W. and Hasegawa H., *Phys. Rev.*, **100**, 675 (1955).
9. de Gennes P. G., *Phys. Rev.*, **118**, 141 (1960).
10. Zener C., *Phys. Rev*, **82**, 403 (1951).
11. Izyumov Yu. A. and Skryabin Yu. N., *Physics-Uspekhi*, **44**, 109 (2001).
12. Edwards D. M., *Advances in Physics*, **51**, 1259 (2002).
13. Coey J. M. D., Viret M., and von Molnar S., *Advances in Physics*, **48**, 167 (1999).
14. Georges A., Kotliar G., Krauth W., and Rozenberg M. J., *Rev. Mod. Phys.*, **68**, 13 (1996).
15. Barnes S. E., *J. Phys. F*, **6**, 1375 (1976).
16. Barnes S. E., *J. Phys. F*, **7**, 2637 (1977).
17. Coleman P., *Phys. Rev. B*, **29**, 3035 (1984).
18. Read N. and Newns D. M., *J. Phys. C*, **16**, 3273 (1983).
19. Kotliar G. and Ruckenstein A. E., *Phys. Rev. Lett.*, **57**, 1362 (1986).
20. Lee P. A. and Nagaosa N., *Phys. Rev. B*, **46**, 5621 (1988).
21. Zou Z. and Anderson P. W., *Phys. Rev. B*, **37**, 627 (1988).
22. Mancini F., Marra S., and Matsumoto H., *Physica C*, **244**, 49 (1995).
23. Mancini F., Marra S., and Matsumoto H., *Physica C*, **250**, 184 (1995).
24. Mancini F., Marra S., and Matsumoto H., *Physica C*, **252**, 361 (1995).
25. Matsumoto H. and Mancini F., *Phys. Rev. B*, **55**, 2095 (1997).
26. Avella A., Mancini F., Villani D., Siurakshina L., and Yushankhai V.Yu., *Int. J. Mod. Phys. B*, **12**, 81 (1998).
27. Mancini F. and Avella A. (2002), cond-mat/0206377.
28. Richard J. L. and Yushankhai V. Yu., *Phys. Rev. B*, **47**, 1103 (1993).
29. Wang Y. R. and Rice M. J., *Phys. Rev. B*, **49**, 4360 (1994).
30. Khaliullin G., *Pis'ma v ZhETF*, **52**, 999 (1990).
31. Wang X., *Phys. Rev. B*, **57**, 7427 (1998).
32. Loors J., *Phys. Rev. B*, **53**, 12556 (1996).
33. Roth L. M., *Phys. Rev.*, **184**, 451 (1969).
34. Roth L. M., *Phys. Rev.*, **186**, 428 (1969).
35. Bickers N. F., *Rev. Mod. Phys.*, **59**, 845 (1987).
36. Plakida N. M., Yushankhai V. Yu., and Stasyuk I. V., *Physica C*, **160**, 80 (1989).
37. Goryachev E. G., Kuzmin E. V., and Ovchinnikov S. G., *J. Phys. C*, **15**, 1481 (1982).
38. Nolting W. and Borguel W., *Phys. Rev. B*, **39**, 6962 (1089).
39. Kane C. L., Lee P. A., Read N., *Phys. Rev. B*, **39**, 6880 (1989).
40. Yushankhai V. Yu., Oudovenko V. S., and Hayn R., *Phys. Rev. B*, **55**, 15562 (1997).
41. Zarubin A. V. and Irkhin V. Yu. , *Phys. Solid State*, **41**, 963 (1999).
42. Irkhin V. Yu. and Zarubin A. V., *Eur. Phys. J. B*, **16**, 463 (2000).

43. Irkhin V. Yu. and Zarubin A. V. (2001), cond-mat/0103054.
44. Vaks V. G., Larkin A. I., and Pikin S. A., *Sov. Phys. JETP*, **26**, 188 (1968).
45. Vaks V. G., Larkin A. I., and Pikin S. A., *Sov. Phys. JETP*, **26**, 647 (1968).
46. Izyumov Yu. A. and Kassan-Ogly F. A., *Fiz. Met. Metalloved.*, **26**, 385 (1968).
47. Izyumov Yu. A. and Kassan-Ogly F. A., *Fiz. Met. Metalloved.*, **30**, 225 (1970).
48. Izyumov Yu. A. and Kassan-Ogly F. A., *Fiz. Met. Metalloved.*, **30**, 449 (1970).
49. Hubbard J., *Proc. Roy. Soc. A*, **285**, 542 (1965).
50. Westwanski B., *Phys. Lett. A*, **44**, 27 (1973).
51. Slobodyan P. M. and Stasyuk I. V., *Teor. Mat. Fiz.*, **19**, 423 (1974).
52. Zaitsev R. O., *Zh. Eksp. Teor. Fiz.*, **70**, 1100 (1976).
53. Izyumov Yu. A. and Letfulov B. M., *J. Phys.: Cond. Mat.*, **2**, 8905 (1990).
54. Izyumov Y. A., Kassan-Ogly F.A., and Skryabin Y. N., *Field Methods in Ferromagnetic Theory*, Nauka, Moscow, 1974, in Russsian.
55. Izyumov Y. A., Katsnelson M.I., and Skryabin Y. N., *Magnetism of Itenerant Electrons*, Nauka, Moscow, 1994, in Russsian.
56. Izyumov Y. A. and Skryabin Y. N., *Statistical Mechanics of Magnetically Ordered Systems*, Consultant Bureau, New York and London, 1988.
57. Schwinger J., *Proc. Natl. Acad. Sci. U. S.*, **37**, 452 (1951).
58. Baym G. and Kadanov L. P., *Phys. Rev.*, **124**, 287 (1961).
59. Kadanoff L. P. and Baym G., *Quantum Statistical Mechanics*, Benjamin, New York, 1962.
60. Ruckenstein A.E. and Schmitt-Rink S., *Phys. Rev B*, **38**, 7188 (1988).
61. Kulic M. L. and Zeyher R., *Phys. Rev. B*, **49**, 4395 (1994).
62. Zeyher R. and Kulic M. L., *Phys. Rev. B*, **51**, 1234 (1995).
63. Sandalov I., Johanson B., and Eriksson O. (2001), cond-mat/0111259.
64. Sandalov I., Lundin U., and Eriksson O. (2001), cond-mat/0111260.
65. Izyumov Yu. A. and Chaschin N. I., *The Physics of Metalls and Metallography*, **92**, no 5, 451 (2001).
66. Izyumov Yu. A. and Chaschin N. I., *The Physics of Metalls and Metallography*, **92**, no 6, 531 (2001).
67. Izyumov Yu. A. and Chaschin N. I., *The Physics of Metalls and Metallography*, **93**, no 1, 18 (2002).
68. Izyumov Yu. A. and Chaschin N. I., *The Physics of Metalls and Metallography*, **94**, no 6, 527 (2002).
69. Izyumov Yu. A. and Chaschin N. I., *The Physics of Metalls and Metallography*, **94**, no 6, 539 (2002).
70. Izyumov Yu. A., Chaschin N. I., and Yushankhai V. Yu., *Phys. Rev. B*, **65**, 214425 (2002).
71. Izyumov Y. A. and Skryabin Y. N., *Basic Models in Quantum Theory of Magnetism*, Ural Division of the Russian Academy of Sciencies, Ekaterinburg, 2002, in Russsian.
72. Abrikosov A. A., Gor'kov L. P., and Dzyaloshinskii I. E., *Methods of Quantum Field Theory in Statistical Physics*, Dover, New York, 1975.
73. Bogolyubov N. N. and Tyablikov S. V., *Sov. Phys. Doklady*, **4**, 589 (1959).
74. Wilcox R. N., *Phys. Rev.*, **174**, 624 (1968).
75. Kwok P. C. and Schultz T. D., *J. Phys. C: Solid State Phys.*, **2**, 1196 (1969).
76. Forster D., *Hydrodynamic Fluctuations, Broken Symmetry, and Correlation Functions*, W. A. Benjamin, Inc.(Reading, MA), London, 1975.
77. Mitchell P. W., Cowley R. A., and Penn R., *J. Phys. C: Solid State Phys.*, **17**, L875 (1984).
78. Bunker A. and Landau D. P., *Phys. Rev. Lett.*, **85**, 2601 (2000).
79. Sibberglitt R. and Harris A. B., *Phys. Rev.*, **174**, 640 (1968).
80. Vedyaev A. V. and Nikolaev M. Yu., *Teor. Mat. Fiz.*, **59**, 293 (1984).
81. Izyumov Yu. A., Letfulov B. M., and Shipitsyn E. V., *J. Phys.: Cond. Mat.*, **4**, 9955 (1992).
82. Izyumov Yu. A. and Henderson J. A., *Int. J. Mod. Phys. B*, **8**, 1877 (1994).
83. Baym G. and Kadanoff L. P., *Phys. Rev.*, **124**, 287 (1961).
84. Gehlhoff L. and Zeyler R., *Phys. Rev. B*, **52**, 4635 (1995).
85. Zeyher R., Kulic M. L., *Phys. Rev. B*, **53**, 1 (1996).
86. Zeyher R. and Kulic M. L., *Phys. Rev. B*, **54**, 8985 (1996).
87. Kulic M. L. and Zeyher R., *Mod. Phys. Lett.*, **11**, 333 (1997).
88. Cappelluti E. and Zeyher R., *Int. J. Mod. Phys. B*, **13**, 2607 (1999).
89. Vonsovsky S. V. and Izyumov Yu. A., *Fiz. Met. Metalloved.*, **10**, 321 (1960).
90. Furukawa N., *J. Phys. Soc. Jpn.*, **65**, 1174 (1996).

PART II

PARTICIPANT CONTRIBUTIONS

How to detect unconventional density waves?

Balázs Dóra*, Attila Virosztek†** and Kazumi Maki‡

*The Abdus Salam ICTP, Strada Costiera 11, I-34014, Trieste, Italy
†Department of Physics, Budapest University of Technology and Economics, H-1521 Budapest, Hungary
**Research Institute for Solid State Physics and Optics, P.O.Box 49, H-1525 Budapest, Hungary
‡Department of Physics and Astronomy, University of Southern California, Los Angeles CA 90089-0484, USA

Abstract. We report about our study on unconventional density waves (UDW) (i.e. density waves with wavevector dependent gap) in quasi-one dimensional systems. Due to the zero average of the gap over the Fermi surface, these systems are not characterized by the periodic modulation of either the charge or spin density, and are referred to as systems with hidden-order parameter. The quantities signaling the phase transition are determined. Depending on the explicit wavevector dependence of the gap, optical absorption is possible at low frequencies. In the presence of magnetic field, the different phases exhibit distinct behaviour, and the nonlinear electric response is strongly influenced by the applied field.

INTRODUCTION

In the quasi-one dimensional and quasi-two dimensional systems the normal Fermi liquid state is destroyed due to the quasiparticle interaction at low temperature and enters into one of four canonical states with energy gap: singlet superconductor, triplet superconductor, charge density wave (CDW) and spin density wave (SDW) [1, 2].

However since the discovery of heavy fermion superconductors, organic superconductors, high T_c superconductors and Sr_2RuO_4, this simple picture has to be necessarily modified. Interestingly most of these new superconductors are unconventional or nodal, the quasiparticle spectrum has no energy gap[3, 4, 5].

From this point it looks natural to consider condensates formed by electron hole pairs (for example density waves) with a wavevector dependent gap[6]. Recently considerable attention has been focused on investigating unconventional density waves (UDW) under various circumstances[7, 8, 9, 10, 11, 12, 13, 14]. The common feature of these systems is the zero average of the gap over the Fermi surface, resulting in the absence of any spatially periodic modulation of the charge or spin density. This property makes UDW a very likely candidate for those systems in which clear thermodynamic signals of a phase transition are seen without any order parameter, which could be detected by conventional methods (i.e. x-ray or NMR). From this the notion "hidden-order" follows naturally[15]. Furthermore due to the zeros of gap at the Fermi surface, these systems are gapless similarly to most of unconventional superconductors, and as a result exhibit fascinating physical properties.

In this paper we report about our study on different properties of UDW. First the basic

CP678, *Lectures on the Physics of Highly Correlated Electron Systems VII: Seventh Training Course*, edited by A. Avella and F. Mancini

properties such as the excitation energy, the density of states or the order parameters are determined. A few possible materials, whose ground state can be some kind of UDW, are introduced. The quasiparticle velocity can couple directly to the wavevector dependence of the gap, hence the optical conductivity exhibits characteristic properties depending on the explicit form of the gap. The magnetic field acts differently on UDW states with respect to the spin symmetry of the gap, and produces diverse phase diagrams. Finally the Goldstone mode of incommensurate density waves, the phason mode is discussed through the threshold electric field.

ORDER PARAMETER

The Hamiltonian of a system of quasi-one dimensional interacting electrons is given by:

$$H = \sum_{\mathbf{k},\sigma} \varepsilon(\mathbf{k}) a^+_{\mathbf{k},\sigma} a_{\mathbf{k},\sigma} + \frac{1}{2V} \sum_{\substack{\mathbf{k},\mathbf{k}',\mathbf{q} \\ \sigma,\sigma'}} \tilde{V}(\mathbf{k},\mathbf{k}',\mathbf{q}) a^+_{\mathbf{k}+\mathbf{q},\sigma} a_{\mathbf{k},\sigma} a^+_{\mathbf{k}'-\mathbf{q},\sigma'} a_{\mathbf{k}',\sigma'}. \tag{1}$$

The system is based on an orthorhombic lattice, but the probability of electron transport is much larger in one direction (chain direction) than in the two other spatial directions. Beyond the usual form of the interaction, it depends not only on the momentum transfer of electrons ($\mathbf{q}$), but also the momenta of the incoming electrons $\mathbf{k}$ and $\mathbf{k}'$, allowing us to consider wavevector dependent gap. To obtain such interaction, we considered interaction between at most nearest neighbour electrons: on site and direct Coulomb, exchange, pair-hopping and bond charge terms[12]. In the DW phase, the new energy spectrum is given by

$$E_\pm(\mathbf{k},\sigma) = \frac{\varepsilon(\mathbf{k}) + \varepsilon(\mathbf{k}-\mathbf{Q})}{2} \pm \sqrt{\left(\frac{\varepsilon(\mathbf{k}) - \varepsilon(\mathbf{k}-\mathbf{Q})}{2}\right)^2 + |\Delta(\mathbf{k},\sigma)|^2}, \tag{2}$$

where $\mathbf{Q}$ is the nesting vector, $\Delta(\mathbf{k},\sigma)$ plays the role of the energy gap. Its wavevector dependence can be determined by solving the gap-equation:

$$\Delta(\mathbf{l}) = \sum_{\mathbf{k}}{}' \frac{1}{V} P(\mathbf{k},\mathbf{l}) \frac{\Delta(\mathbf{k}) \tanh\left(\frac{\beta}{2} E(\mathbf{k})\right)}{2E(\mathbf{k})}, \tag{3}$$

where the kernel $P(\mathbf{k},\mathbf{l})$ is the linear combination of the different matrix elements of the interaction, and the $\mathbf{k}$ and $\mathbf{l}$ dependence comes from the dependence of the interaction on the momentum of incoming electrons. From this, the gap reads as[12]:

$$\Delta(\mathbf{l}) = \Delta_0 + \Delta_1 \cos(l_y b) + \Delta_2 \sin(l_y b) + \Delta_3 \cos(l_z c) + \Delta_4 \sin(l_z c) \tag{4}$$

From now on, we assume that only one kind of gap, whose T_c is the highest, opens and persists all the way down to zero temperature. If it is unconventional ($\Delta_{i\neq 0}$), the expectation value of the spin and charge density is zero: $\langle m(\mathbf{Q})\rangle = 0$, $\langle n(\mathbf{Q})\rangle = 0$. In this case, the energy spectrum is shown in Fig. 1: gapless excitations can occur around the zeros of the gap. According to the spin symmetry of the gap, we call our system unconventional charge density wave for $\Delta(\sigma) = \Delta(-\sigma)$ and unconventional spin density wave for $\Delta(\sigma) = -\Delta(-\sigma)$.

Since the gap vanishes on certain subsets of the Fermi surface as it is shown in Fig. 1, the density of states (DOS) is finite except at the Fermi energy. Also DOS shows a weak logarithmic divergence at the gap maximum, as it can be seen in Fig. 1. Since the DOS is very close to those of a d-wave superconductor (SC), the thermodynamics was also found to be similar to the one in d-wave SC. Due to the gapless excitations, we expect that power-law temperature dependence in the physical quantities at low temperatures replace the exponential freeze-out of the fully gapped conventional density wave.

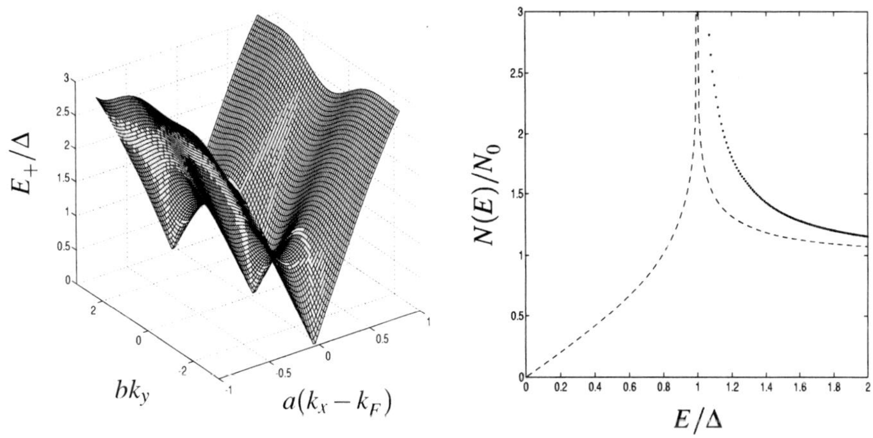

FIGURE 1. The quasiparticle spectrum is shown (left panel) above the Fermi energy at half filling for $\Delta(\mathbf{k}) = \Delta\sin(bk_y)$. In the right panel, the density of states is plotted in the conventional (dotted line) and unconventional (dashed line) case.

Now it is clear, that those materials are successful candidates for UDW ground state, which show phase transition without any obvious long range order (without superconductivity or spin or charge ordering). Possible materials with UDW ground state:

- the micromagnetic phase of URu_2Si_2: the extremely small amplitude of antiferromagnetic ordering (0.02-0.04 μ_B) below $T \sim 10$K is orders of magnitude smaller than what is expected from the transition temperature, and implies that AF may not be the dominant order parameter, but can arise from interaction of impurities and USDW ground state[13, 16, 17].
- the mysterious low temperature phase of $\alpha-$(BEDT-TTF)$_2$KHg(SCN)$_4$: its quasi-one dimensional Fermi surface and the lack of any spin and charge ordering in the low temperature phase (below 10K) makes this material an ideal candidate for UDW. Furthermore the destruction of this phase suggests a kind of CDW. Indeed we described many of its properties including the threshold electric field and angular dependent magnetoresistance assuming UCDW ground state[18, 19, 20, 21].

- 2H-$TaSe_2$: among the fascinating properties of this material, which possesses a two dimensional, sixfold symmetric Fermi surface, it enters into a CDW phase, whose gap was found to depend on the wavevector by angle resolved photoemission spectroscopy, and the theory of f-wave UCDW was worked out by Castro-Neto[22].
- pseudogap phases: $(TaSe_4)_2I$[23], high temperature superconductors[7, 24], transition metal oxides ($SrRuO_3$, $BaRuO_3$)[25], among which the most well known is the one in HTSC, for which the d-DW phase was proposed, which is UCDW with a gap of $d_{x^2-y^2}$ symmetry. This phase consists of circulating currents around the plaquettes of the lattice inducing orbital magnetic moment in their centers.

In our quasi-one dimensional picture, we identified the following quantities as order parameter for the different gap functions:

phase	gap	order parameter: the **Q**-th Fourier component of
UCDW	$\Delta\cos(bk_y)$	electric current density
UCDW	$\Delta\sin(bk_y)$	kinetic energy density
USDW	$\Delta\cos(bk_y)$	spin current density
USDW	$\Delta\sin(bk_y)$	spin kinetic energy density

These phases were already mentionned in the context of the two dimensional Hubbard model as orbital antiferromagnet[8] or staggered-flux phase[26], bond order wave[27], spin nematic state[9] and spin bond-order wave[27], respectively. In Fig. 2, the different patterns of ground state are shown for the half-filled case for simplicity.

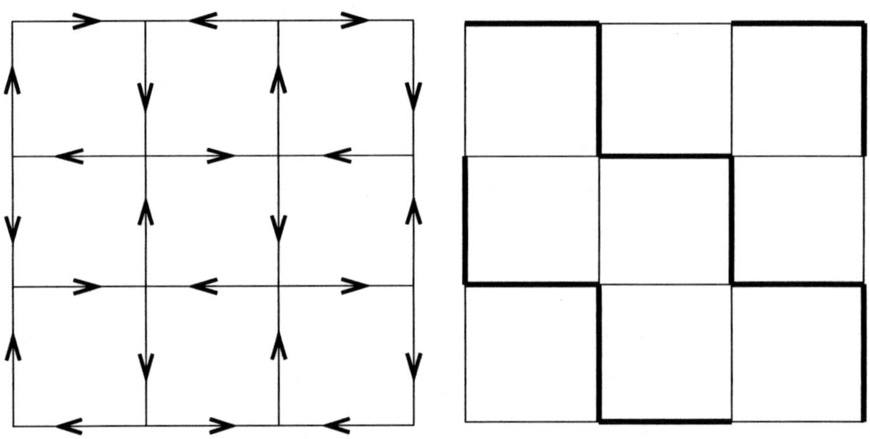

FIGURE 2. The pattern of the ground state is shown for electric (spin) currents in the left panel in the half-filled case. In the right panel, the (spin) bond order phase can be seen. Thick (thin) lines indicate higher (lower) hopping probability. In the spin bond order phase, on thick lines up spin electrons hop more often than down spin electrons, on thin lines vice versa.

The differentiate between the phases according to their spin symmetry, the magnetic response is decisive: USDW, in spite of the lack of periodic modulation of the spin density, exhibit antiferromagnetic (or SDW) like spin susceptibility: it freezes out in one direction as the temperature is lowered while remains unchanged in the two other directions[12], while the spin susceptibility of UCDW remains isotropic[28]. Consequently, an applied magnetic field is expected to destroy UCDW, while small fields do not influence the USDW phase.

As to the wavevector symmetry, one has to options: performing an experiment, which either
1. is sensitive to whether the ground state breaks or conserves time reversal symmetry. In this respect, the first and fourth phases are time reversal symmetry breaking, the other two are conserving,
or
2. couples directly to the wavevector dependence of the gap. Such a quantity is the velocity ($v_y \sim \sin(bk_y)$), and by measuring the optical conductivity, distinct curves from the different gaps are expected, as will be shown in the followings.

OPTICAL CONDUCTIVITY IN THE PRESENCE OF IMPURITIES

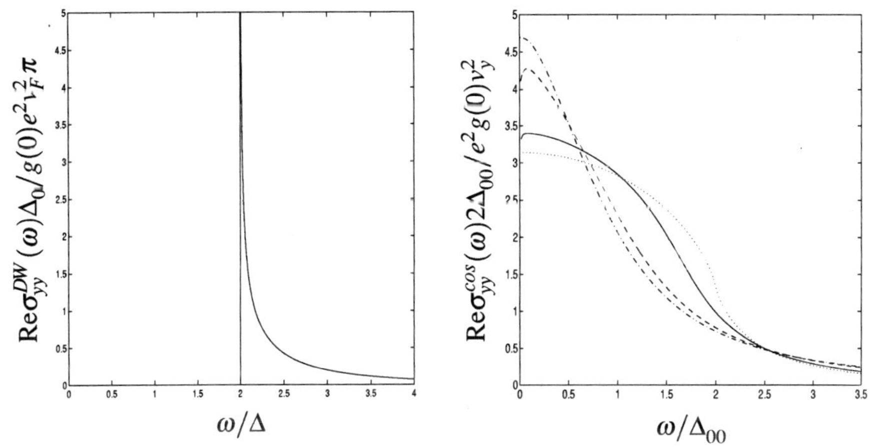

FIGURE 3. The optical conductivity in the conventional case is shown in the left panel in the perpendicular directions. In the right panel, $\mathrm{Re}\sigma_{yy}(\omega)$ is plotted for a gap with $\cos(bk_y)$. The dotted line denotes the pure case, the others stands for the impure system. The larger the DC response, the stronger the impurities are.

The optical conductivity contains relevant informations about the possible excitations of a systems, and allows one to get insight in the nature of the ground state. In the chain or quasi-one dimensional direction, the phason mode is present in the incommensurate case as in conventional DW, hence collective modes are important, and change the quasiparticle part significantly. As opposed to this, in the perpendicular direction, only Fermi liquid type renormalization might occur[28], and the one bubble contribution describes the optical response. In Fig. 3, we show $\mathrm{Re}\sigma(\omega)$ in the conventional case for comparison. For $\omega < 2\Delta$, there is no absorption and at twice the gap maximum a square root divergence from the DOS (see Fig. 1) shows up[2]. Compared to this, a clear distinction can be made in the unconventional case, where assuming a gap ($\Delta(\mathbf{k}) \sim \cos(bk_y)$), considerable absorption is present at low frequencies, but the divergent peak is suppressed because of the zero velocity of quasiparticles at the gap maximum[12].

On the other hand, if we have a gap with $\sin(bk_y)$ (Fig. 4), the low frequency part is chopped due to matching of the wavevector dependence of the velocity and the gap, but a weak logarithmic divergence reminiscent to that in the DOS (Fig. 1) is present. In an electric field applied in the z direction, both features, namely absorption at low frequencies and divergence at 2Δ, can be seen, and there is no difference with respect to the gap function. In the presence of impurities[29], the curves are smoother, the peaks are broadened and more and more spectral weight is transferred to lower frequencies, transforming the conductivity curves into a Lorentzian. By measuring the optical conductivity in the perpendicular directions, one can clearly distinguish between the possible gap functions.

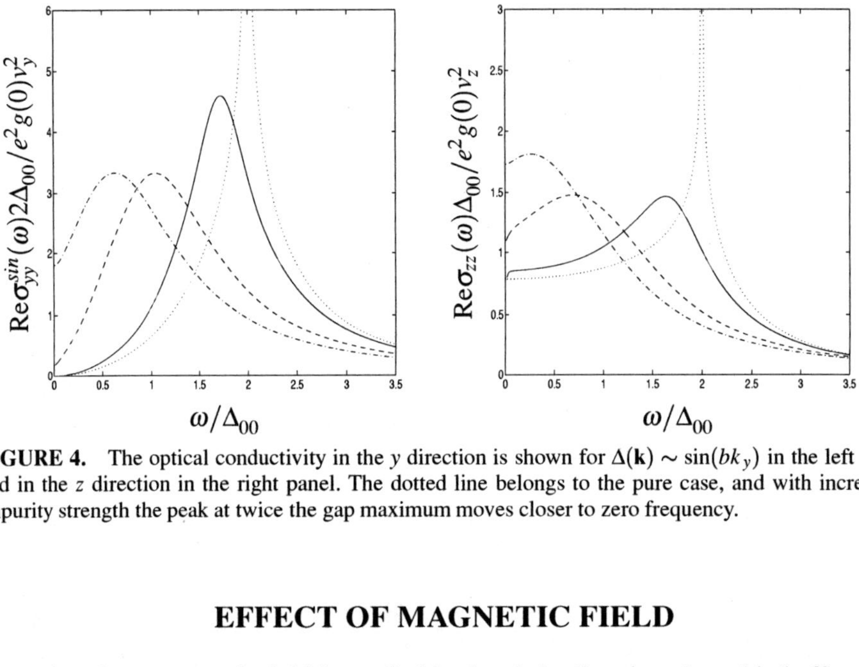

FIGURE 4. The optical conductivity in the y direction is shown for $\Delta(\mathbf{k}) \sim \sin(bk_y)$ in the left panel, and in the z direction in the right panel. The dotted line belongs to the pure case, and with increasing impurity strength the peak at twice the gap maximum moves closer to zero frequency.

EFFECT OF MAGNETIC FIELD

Assuming that a magnetic field is applied in the chain direction, the orbital effects can be neglected, and it suffices to study the effect of an applied field through the Zeeman term. In USDW, similarly to conventional SDW or antiferromagnets, this has no effect due to the spin-flop mechanism: small fields leave this phase unaffected. On the other, UCDW is strongly affected by the magnetic field[19, 30], which can readily be seen in the phase diagram in Fig. 5. In small fields, the transition temperature starts decreasing, and the order parameter (Δ) goes continuously to zero along the second order phase boundary. But at the bicritical point given by $\mu_B H \approx 1.91 k_B T$, the order of the transition changes from second to first, and the single second order boundary splits into two first order ones, leading to hysteresis. Further increasing the field, the analog of the FFLO state[31, 32, 33], some kind of hybrid UDW looks possible, when the nesting vector starts depending on the magnetic field, but this part of the phase diagram has not been explored yet.

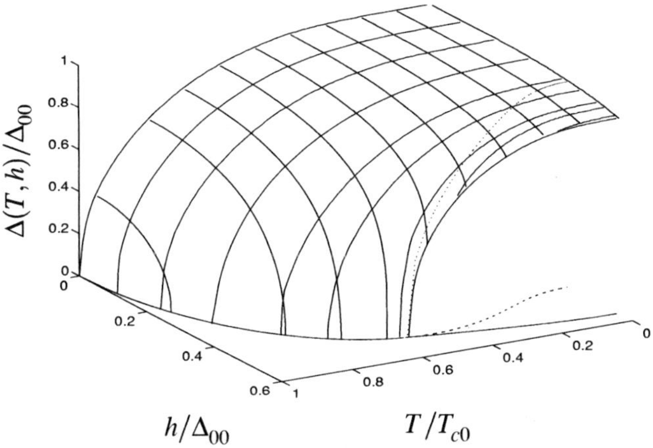

FIGURE 5. The phase diagram of UCDW in magnetic field is shown. The solid line in the $h-T$ plane denotes the second order phase boundary, and it splits into two first order ones (dashed lines). Δ_{00} is the order parameter at $h=0$ and $T=0$.

Among the transport properties, we concentrate on the threshold electric field. In quasi-one dimensional systems density wave can be stable at arbitrary (incommensurate) fillings, and generally the phase of the order parameter is free. If impurities are present, the DW adjusts its phase to the impurities in order to minimize the energy loss due to impurities. On the other hand, if an electric field is applied in the chain direction, it couples to the phase and wants to move the condensate as a whole[34, 35, 36]. The equilibrium of these two competing forces defines the threshold electric field. In fact this is the Goldstone mode of DW corresponding to the freedom of the phase, and it is absent for example in the d-DW phase, where only discrete symmetry is broken due the commensurate nesting vector and the restricted phase at half filling ($\phi = \pm\pi/2$).

Assuming point-like impurities, which are localized as a Dirac-delta function in real space, the threshold electric field (E_T) is zero, these impurities are not able to couple to the wavevector dependence of the gap[18]. Extended impurities[37] with some real space tail are given in the form:

$$U(\mathbf{Q}+\mathbf{q}) = V_0 + \sum_{i=y,z} V_i \cos(q_i \delta_i), \tag{5}$$

where V_0 corresponds to the Dirac delta like feature of the impurity, and the wavevector dependent terms ($V_i \ll V_0$) describe the real space extension. The threshold electric field turned out to be finite, and strongly magnetic field dependent, and even diverges at the bicritical point as

$$\frac{E_T^S(T,h)}{E_T^S(0,0)} = \frac{1.12}{1.91-h/T} \tag{6}$$

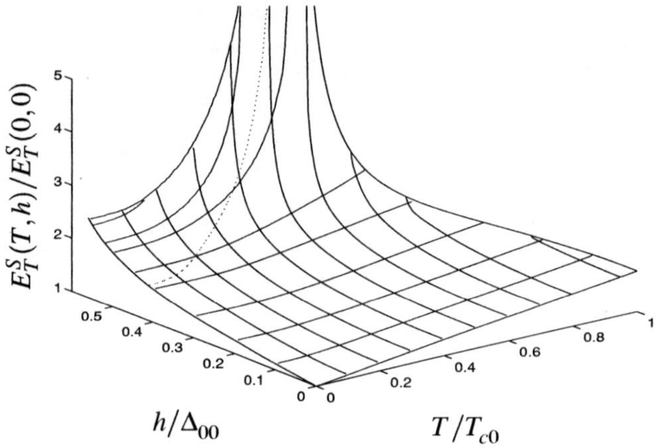

FIGURE 6. Stereograph of the threshold electric field in the strong pinning limit as a function of the reduced temperature and field. The dashed line is the threshold field belonging to the metastability line.

For arbitrary magnetic fields E_T is shown in Fig. 6. The threshold field is a monotonically increasing function of both the magnetic field and the temperature, and it should be accessible experimentally.

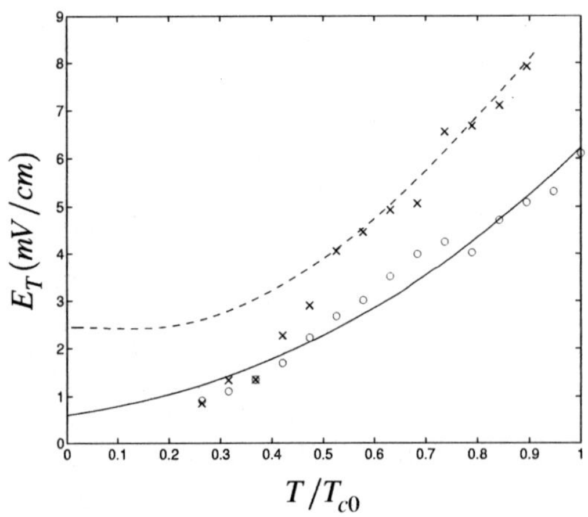

FIGURE 7. The theoretical and experimental threshold electric fields are plotted as a function of the reduced temperature. The measured E_T in the α-(BEDT-TTF)$_2$KHg(SCN)$_4$ salt[38] is shown for $H = 0$T (open circles) and $H = 1$T (crosses). The solid (dashed) line represents the theoretical curve at $h/\Delta_{00} = 0$ ($h = 0.3\Delta_{00}$).

In general two distinct pinning regions can be realized: strong and weak pinning[34, 35]. The former applies when the impurities are strong but the concentration is dilute. The latter requires weak but dense impurities. The strong pinning limit applies when impurities are introduced by X-ray irradiation or some violent means. In this case one single impurity is enough to pin the DW locally. For high quality crystals, however, the weak pinning limit is more appropriate. Then for the 3 dimensional weak pinning limit we obtain

$$\frac{E_T^W(T,h)}{E_T^W(0,0)} = \left(\frac{E_T^S(T,h)}{E_T^S(0,0)}\right)^4 . \tag{7}$$

As a comparison with experiment, we show E_T in the 3 dimensional weak pinning limit together with the experimental data on α-(BEDT-TTF)$_2$KHg(SCN)$_4$ salt at $H = 0$T és $H = 1$T in Fig. 7. In zero field the agreement is remarkable, but at 1T below half T_c some deviations are visible. Imperfect nesting may account for this discrepancy[20].

CONCLUSIONS

In this paper we summarized the basic properties of unconventional density waves (density waves with wavevector dependent gap) in quasi-one dimensional systems. The gap vanishes on lines at the Fermi surface, resulting in gapless behaviour at low temperatures. Due to this, the transition to UDW is metal to metal instead of metal to semiconductor as is in the case of constant gap. Moreover, due to the zero average of the gap over the Fermi surface, UDW do not exhibit periodic modulation of either the charge or spin density, and are referred to as systems with hidden order parameter. Four order parameters were found corresponding to the different spin and wavevector dependent gap functions: orbital antiferromagnet, bond order wave, spin nematic phase and spin bond order wave.

Materials with UDW ground state are expected to show robust features of a phase transition without superconductivity or without any long range spin or charge ordering. From angle resolved photoemission experiments, the nodes of the gap should be accessible, and the power law temperature dependent quantities can provide further evidence of a gapless DW phase transition.

To differentiate between the spin symmetry of the various ground states, we proposed to study the effect of magnetic field on the stability of the phases. UCDW are destroyed in an external field similarly to their conventional version, but USDW remains unaffected by small fields. Beyond the phase diagram, the threshold electric field is also modified in UCDW, and the experimental observation of the predicted changes can lead to decisive conclusion with regard to the UCDW nature of the low temperature phase.

The optical conductivity was shown to be a useful tool to test the wavevector dependence of the gap. In certain cases, depending on the interference of the velocity and the gap function, significant optical absorption is present at low frequencies below the gap maximum, and the divergent peak at twice the gap maximum might also be absent.

ACKNOWLEDGMENTS

We are benefitted from discussions with Hyekyung Won, Peter Thalmeier, Bojana Korin-Hamzić and Mark Kartsovnik. This work was supported by the Hungarian National Research Fund under grant numbers OTKA T032162 and TS040878.

REFERENCES

1. Sólyom, J., *Adv. Phys.*, **28**, 201 (1979).
2. Grüner, G., *Density waves in solids*, Addison-Wesley, Reading, 1994.
3. Harlingen, D. J. V., *Rev. Mod. Phys.*, **67**, 515 (1995).
4. Won, H., and Maki, K., "Day of Unconventional Superconductors," in *Symmmetry and Pairing in Superconductors*, edited by M. Ausloos and S. Kruchinin, Kluwer, Dordrecht, 1999.
5. Tsuei, C. C., and Kirtley, J. R., *Rev. Mod.Phys.*, **72**, 969 (2000).
6. Halperin, B. I., and Rice, T. M., "The Excitonic State at the Semiconductor-Semimetal Transition," in *Solid State Physics*, edited by F. Seitz, D. Turnbull, and H. Ehrenreich, Academic Press, New York, 1968, vol. 21, p. 115.
7. Chakravarty, S., Laughlin, R. B., Morr, D. K., and Nayak, C., *Phys. Rev. B*, **63**, 094503 (2001).
8. Nersesyan, A. A., and Vachnadze, G. E., *J. Low T. Phys.*, **77**, 293 (1989).
9. Nersesyan, A. A., Japaridze, G. I., and Kimeridze, I. G., *J. Phys. Cond. Mat.*, **3**, 3353 (1991).
10. Schulz, H. J., *Phys. Rev. B*, **39**, 2940 (1989).
11. Benfatto, L., Caprara, S., and Di Castro, C., *Eur. Phys. J. B*, **17**, 95 (2000).
12. Dóra, B., and Virosztek, A., *Eur. Phys. J. B*, **22**, 167 (2001).
13. Ikeda, H., and Ohashi, Y., *Phys. Rev. Lett.*, **81**, 3723 (1998).
14. Gulácsi, Z., and Gulácsi, M., *Phys. Rev. B*, **36**, 699 (1987).
15. Nayak, C., *Phys. Pev. B*, **62**, 4880 (2000).
16. Ohashi, Y., *Phys. Rev. B*, **60**, 15388 (1999).
17. Virosztek, A., Maki, K., and Dóra, B., *Int. J. Mod. Phys. B*, **16**, 1667 (2002).
18. Dóra, B., Virosztek, A., and Maki, K., *Phys. Rev. B*, **64**, 041101(R) (2001).
19. Dóra, B., Virosztek, A., and Maki, K., *Phys. Rev. B*, **65**, 155119 (2002).
20. Dóra, B., Maki, K., and Virosztek, A., *Phys. Rev. B*, **66**, 165116 (2002).
21. Dóra, B., Maki, K., Korin-Hamzić, B., Basletić, M., Virosztek, A., Kartsovnik, M. V., and Müller, H., *Europhys. Lett.*, **60**, 737 (2002).
22. Castro-Neto, A. H., *Phys. Rev. Lett.*, **86**, 4382 (2001).
23. Németh, L., Matus, P., Kriza, G., and Alavi, B., *Synth. Metals*, **120**, 1007 (2001).
24. Kim, W., and Carbotte, J. P., *Phys. Rev. B*, **66**, 033104 (2002).
25. Schroeter, D. F., and Doniach, S., *Phys. Rev. B*, **66**, 075120 (2002).
26. Affleck, I., and Marston, J. B., *Phys. Rev. B*, **37**, 3774 (1988).
27. Ozaki, M., *Int. J. Quantum Chem.*, **42**, 55 (1992).
28. Dóra, B., and Virosztek, A. (2002), cond-mat/0211316, to be published in Europhys. Lett.
29. Dóra, B., Virosztek, A., and Maki, K., *Phys. Rev. B*, **66**, 115112 (2002).
30. Dóra, B., Virosztek, A., and Maki, K., *Curr. Appl. Phys.*, **1**, 313 (2001).
31. Fulde, P., and Ferrell, R. A., *Phys. Rev.*, **35**, A550 (1964).
32. Larkin, A. I., and Ovchinnikov, N., *Sov. Phys. JETP*, **20**, 762 (1965).
33. Maki, K., and Won, H., *Czech. J. Phys.*, **46**, S2, 1035 (1996).
34. Fukuyama, H., and Lee, P. A., *Phys. Rev. B*, **17**, 535 (1978).
35. Lee, P. A., and Rice, T. M., *Phys. Rev. B*, **19**, 3970 (1979).
36. Maki, K., and Virosztek, A., *Phys. Rev. B*, **42**, 655 (1990).
37. Haran, G., and Nagi, A. D. S., *Phys. Rev. B*, **45**, 15463 (1996).
38. Fujita, T., Sasaki, T., Yoneyama, N., Kobayashi, N., and Fukase, T., *Synth. Met.*, **120**, 1077 (2001).

Mott domains of bosons and fermions confined in optical lattices

M. Rigol*, A. Muramatsu*, G. G. Batrouni†, V. Rousseau†, R. T. Scalettar**, P. J. H. Denteneer‡ and M. Troyer§

*Institut für Theoretische Physik III, Universität Stuttgart, Pfaffenwaldring 57, D-70550 Stuttgart, Germany
†Institut Non-Linéaire de Nice, Université de Nice–Sophia Antipolis, 1361 route des Lucioles, 06560 Valbonne, France
**Physics Department, University of California, Davis, California 95616, USA
‡Lorentz Institute, Leiden University, P.O. Box 9506, 2300 RA Leiden, The Netherlands
§Theoretische Physik, Eidgenössische Technische Hochschule Zürich, CH-8093 Zürich, Switzerland

Abstract.
We study the ground state properties of the 1-D boson and fermion Hubbard models in harmonic traps with an underlying lattice. The physics of these systems is found to be fundamentally different from the homogeneous case. Global quantities lose their meaning and Mott domains appear for a continuous range of incommensurate fillings. A local order parameter is defined to characterize the Mott insulating phases that coexist with the metallic ones. Finally, the phase diagrams show richer structures than in the homogeneous models.

INTRODUCTION

Since the experimental achievement of Bose-Einstein condensation (BEC) of trapped atomic gases in 1995 [1, 2, 3], leading to the 2001 Nobel Prize being awarded to Eric Cornell, Carl Wieman and Wolfang Ketterle, this subject has been the focus of intense experimental and theoretical activity [4, 5]. Recently, a new feature has been added to the experiments: The trapped Bosons are placed on optical lattices generated by interfering laser beams.

The atomic Fermi gases have only been investigated experimentally in the last three years [6, 7, 8, 9, 10]. Unlike bosons, fermions cannot be directly evaporatively cooled because they are unable to undergo the collisions necessary to rethermalize the gas during evaporation. This problem has been circumvented in various ways, for example, by simultaneously trapping and evaporatively cooling two different spin states of the same fermionic atom [6, 9] or introducing mixed gases of bosons and fermions in which bosons enable fermions to rethermalize through their elastic interactions [7, 8, 10].

These experiments with bosons and fermions allow the study of a great variety of experimental regimes from the non-interacting to the strongly correlated case, with good control of the parameters involved. With such control, it is possible to study the superfluid-Mott insulator (for bosons) or the metal-Mott insulator (for fermions) transitions when the particles have repulsive interactions; and the BCS-like fermionic

CP678, *Lectures on the Physics of Highly Correlated Electron Systems VII: Seventh Training Course*, edited by A. Avella and F. Mancini

superfluidity when the fermions are able to form Cooper pairs.

On the other hand, a significant amount of theoretical and numerical work has been done in the last decades to determine the ground state phase diagram of correlated boson (fermions) on a lattice described by the boson (fermion) Hubbard Hamiltonian [11, 12, 13, 14, 15] ([16]). It has been found that on site repulsion can produce Mott insulating phases at commensurate fillings with quantum phase transitions to superfluid (bosons) or metal (fermions) phases when the on-site interaction is weakened or the density changed.

The presence of the confining potential, in addition to the regular lattice, in the boson (fermion) Hubbard Hamiltonian, leads to new fundamental questions within this model: 1) Does the confining potential preclude the formation of local Mott insulator regions coexisting with superfluid (bosons) or metal (fermions) regions? 2) If a Mott insulating phase exists, how can it be characterized. 3) What are the quantitative values of the trap curvature and on-site repulsion that support these Mott phases?

In the present work we will address the above questions for boson and fermion systems. In the case of bosons we summarize our paper [17] in which we reported the first quantum Monte Carlo (QMC) simulation of the one-dimensional boson-Hubbard model in a confining quadratic potential. For fermions we will discuss new results.

In general we find the confining potential to be a fundamental issue in the physics of the trapped systems. For example, Mott regions form at incommensurate fillings, something that never occurs in the homogeneous system. Similarly the global compressibility of the trapped boson system [17] is non-zero for all densities, including the commensurate ones. In this context global properties like density (defined as the total number of particles divided by the system size) and the global compressibility lose their meaning and local quantities have to be defined to characterize the system.

The paper is organized as follows: In next section we review the trapped boson system, the section after that is devoted to the trapped fermion system and finally conclusions are given in last section.

CONFINED BOSONS ON OPTICAL LATTICES

It is well known that the homogeneous d-dimensional boson-Hubbard model with only on-site repulsion has two phases: A Mott insulator at commensurate fillings and sufficiently strong repulsion, and a superfluid phase elsewhere [11, 12]. The critical behavior is of two types: Mean field for transitions induced by tuning the density, and of the $(d+1)$ dimensional XY universality class when the interaction strength is modified at fixed commensurate fillings.

In the case of the Hubbard model for a confined system, we study the Hamiltonian

$$H = -t\sum_i (a_i^\dagger a_{i+1} + a_{i+1}^\dagger a_i) + V_0 \sum_i n_i(n_i - 1) + V\sum_i (i - N/2)^2\, n_i, \tag{1}$$

where t measures the boson kinetic energy, V_0 the on-site interaction, V describes the curvature of the quadratic confining potential and N is the number of lattice sites considered. N is chosen such that all the particles are confined in the trap, and do not see the edges. The simulations were done with the world-line QMC algorithm, in the canonical ensemble [12, 20]. The chemical potential $\mu = \frac{\partial E}{\partial N}$ is obtained by differentiating numerically the energy with respect to the total number of particles [21] and the limit of zero temperature was taken.

To our knowledge the only previous (to Ref. [17]) non-mean field work on this inhomogeneous problem is Ref. [18] in which very small systems (five particles) are analyzed; and in a recent work [19] relations between the local density and the momentum distribution function in a three dimensional system with up to 10^3 bosons were studied.

The first quantity we studied was the local boson density $n_i = \langle a_i^\dagger a_i \rangle$, which indicates the formation of Mott insulator regions for integer values of the local density. In Fig. 1 we show the evolution of the local boson density with increasing total occupancy of the trap. It can be clearly seen that at low fillings the density profile is smooth and looks like an inverted parabola. Above a certain filling (number of bosons $N_b = 30$, for the selected parameters) a plateau with local filling of one boson per site starts to develop. Increasing the filling, the Mott plateau widens spatially until a certain critical filling for which the Mott insulator in the middle of the trap ceases to exist because a second superfluid phase starts to form there. With increasing filling, this superfluid phase spreads spatially replacing the Mott phase surrounding it. Increasing further the filling (up to $N_b = 116$ here) a second Mott phase appears in the middle of the trap with a local filling of two bosons per site. Finally, increasing even further the number of particles in the trap, the previous scheme repeats and new Mott regions with higher integer local density can appear in the middle. The Mott regions previously formed are squeezed farther into the boundaries of the system and may be totally destroyed.

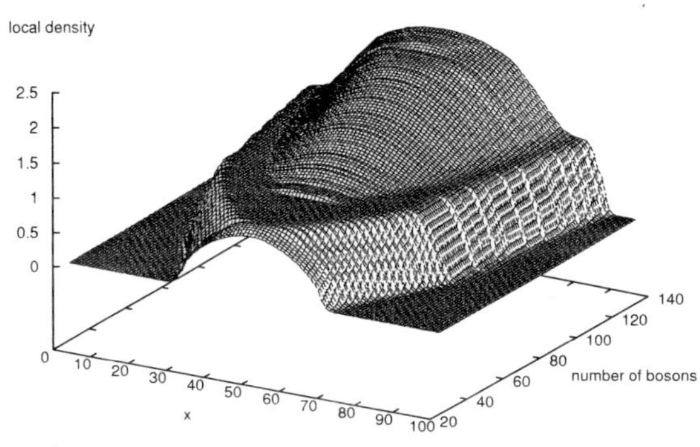

FIGURE 1. Evolution of the local density as a function of the position x and increasing total number of bosons. The parameters are $V_0/t = 4.0$, $V/t = 0.008$ and $N = 100$.

A central feature of the Mott phase transition in the unconfined boson-Hubbard model is that the global compressibility is zero. A charge gap opens and the density remains constant at integer values for a range of values of the chemical potential μ, as it is shown inset in Fig. 2. However in the confined case, even if locally incompressible regions appear in the system, the global compressibility is never zero, which is clearly seen in Fig. 2.

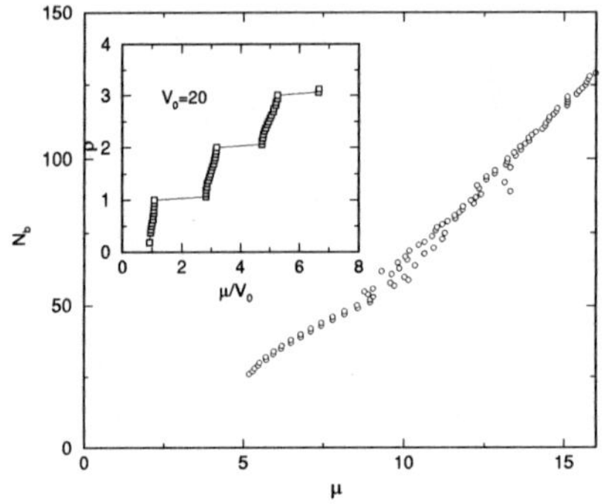

FIGURE 2. Number of bosons as a function of chemical potential for $V_0/t = 4.5$. Inset shows the unconfined case.

In order to characterize the local Mott region, we study the behavior of the local density variance as a first approximation for the local compressibility [17]: $\Delta_i = \langle n_i^2 \rangle - \langle n_i \rangle^2$ (we will discuss a definition of the local compressibility, and its relation with the local variance, in next section). The behavior of this quantity is shown in Fig. 1 for four profiles from Fig. 3. The figure shows that in the local Mott regions Δ_i is strongly suppressed, going to very small non-zero values. In the limit $V_0/t \to \infty$, Δ_i goes to zero. The arrows in the figure show the site at which a Mott domain is entered or exited.

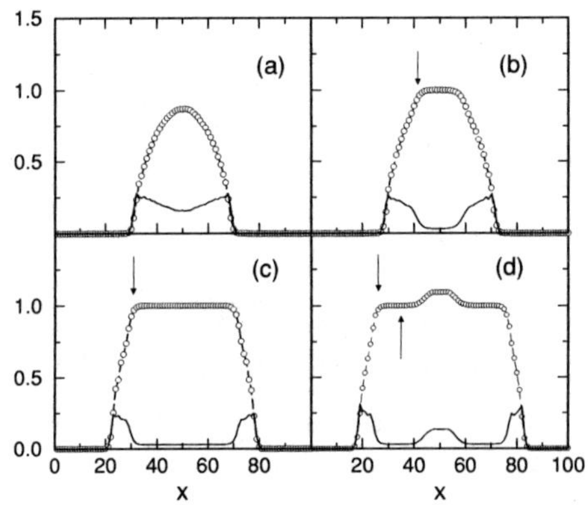

FIGURE 3. Four density profiles of Fig. 1 and their variance. The fillings are $N_b = 25$ (a), 33 (b), 50 (c), 60 (d).

An important difference between the behavior of the local variance Δ_i and the global compressibility in the unconfined case is that the latter diverges [12] as the Mott lobe is approached, $\kappa \propto |\mu - \mu_c|^{-1/2}$, where μ_c is the critical chemical potential at which the Mott insulator is formed. κ vanishes inside the Mott region. Instead, the local variance in the confined case decreases continuously to its minimum non-zero value for the Mott region: No divergence is evident. The origin of the above mentioned difference is that the global compressibility probes density-density correlations to all length scales, which means that in the unconfined case, unlike in the confined trapped system, the establishment of the Mott phase occurs collectively through the whole system and the correlation length diverges.

Different runs like the ones presented in Fig. 1 allow us to determine the phase diagram as a function of the total boson filling and the interaction strength, for a given trap curvature. Results are shown in Fig. 4. Region A admits locally incompressible domains with $n_i = 1$ in the center of the trap (like in Figs. 3(b) and 3(c)). Region B has $n_i > 1$ in the trap center surrounded by incompressible $n_i = 1$ regions (like in Fig. 3(d)). In region C the central part of the trap has an $n_i = 2$ Mott insulating phase inside superfluid and Mott ($n_i = 1$) regions respectively. Region D is where the center of the system is compressible with $n_i > 2$ surrounded by Mott ($n_i = 2$), superfluid and Mott ($n_i = 1$) regions. Region E has no locally incompressible regions (like in Fig. 3(a)). We do not refer the superfluid region at the edges because it is always there for all possible configurations of the system.

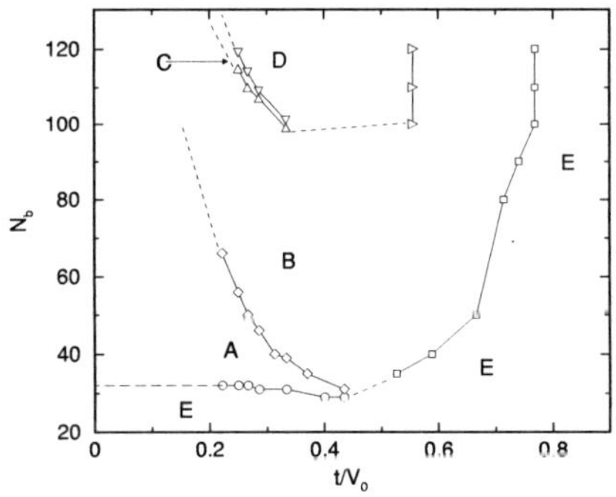

FIGURE 4. Phase diagram of confined bosons in optical lattices. Solid lines are to guide the eye and dashed lines are extrapolations.

Two interesting things to note in this phase diagram are: 1) The values of V_0/t at which the A and C phases are develop are of the same order as those of the first two true Mott lobes in the non-confined case [12], which seems to be in agreement with the experimental results on three dimensional optical lattices [22], where the theoretical values expected in the homogeneous case are obtained. 2) The narrowness of region C could help to understand why the experiments [22] have not shown signs of the $n_i = 2$ Mott region, even when they reach up to $n_i = 2.5$ particles in the core of the system.

CONFINED FERMIONS ON OPTICAL LATTICES

The phase diagram of the single band fermionic Hubbard Model for a homogeneous system has a phase transition from a Mott insulator to a metal at half filling and finite U/t (except for the case of perfect nesting in which the transition occurs at $U/t = 0$). There are two routes to this metal insulator transition (MIT), the filling-controlled MIT (FC-MIT) and the bandwidth controlled MIT (BC-MIT). In the neighborhood of the Mott insulating phase, the metallic region is under the strong influence of the MIT and anomalous features like mass enhancement or carrier-number reduction have been observed [16]. For a non-degenerate band $n = 0$ and $n = 2$ fillings correspond to band insulators.

We want to study the model for fermions confined on a lattice and described by the Hubbard Hamiltonian

$$H = -t\sum_{i\sigma}\left(c^{\dagger}_{i\sigma}c_{i+1\sigma} + c^{\dagger}_{i+1\sigma}c_{i\sigma}\right) + U\sum_{i}\left(\hat{n}_{i\uparrow} - \frac{1}{2}\right)\left(\hat{n}_{i\downarrow} - \frac{1}{2}\right) + V\sum_{i}\left(i - \frac{N}{2}\right)^{2}\hat{n}_{i} \qquad (2)$$

where t is the electron hopping parameter, U the on-site repulsion (we consider only the case $U > 0$), V is the curvature of the confining potential and N is the number of lattice sites. The number of lattice sites was selected in such a way that all the fermions were confined within the trap and do not feel the effects of the borders. The simulations were done using the zero temperature projector method [23, 24] adapted from the Quantum Monte Carlo determinantal algorithm by Blankenbecler, Scalapino and Sugar (BSS) [25, 26, 27]. The discrete Hubbard-Stratonovich transformation by Hirsch [28] was used.

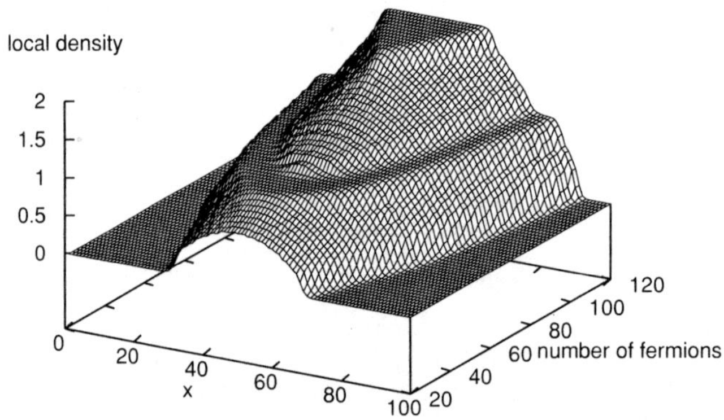

FIGURE 5. Evolution of the local density as a function of the position x and increasing total number of fermions. The parameters are $U/t = 6.0$, $V/t = 0.006$ and $N = 100$.

As in the boson, case we are interested in the evolution of the local density $n_i = \langle \hat{n}_{i\uparrow} + \hat{n}_{i\downarrow} \rangle$ when the total filling of the trap is increased. Results are shown in Fig. 5, and are similar to those for bosons (Fig. 1). The difference with bosons is that the allowed local integer fillings for fermions are the trivial ones $n_i = 0$, $n_i = 2$ and $n_i = 1$, because n_i must be always smaller or equal to two. This fact reduces the number of possible configurations for fermions compared to bosons. For the lowest fillings in Fig. 5 there are only metallic phases, for which the density profiles look like inverted parabolas. After a certain filling (number of fermions $N_f = 36$) a Mott insulating phase starts to develop in the middle of the trap and when the total filling is $N_f = 48$ this phase is destroyed in the center and a new metallic phase starts to develop there (in addition to the metallic phase always present at the borders). Increasing the filling, the metallic phase in the center of the trap widens and can completely destroy the Mott insulating phase surrounding it, depending on the strength of the on-site repulsion (as will be shown later in the phase diagram). Finally for the highest fillings a band insulator with $n_i = 2$ develops in the middle of the trap (in the present case after $N_f = 98$).

Four characteristic density profiles from Fig. 5 are shown in Fig. 6 (left) together with profiles of their variance (right). We see that, like for the bosonic case, the variance of the density is strongly suppressed in the Mott phases. In addition, as expected, the band insulator with $n_i = 2$ has a variance that vanishes completely, like for the $n_i = 0$ insulator.

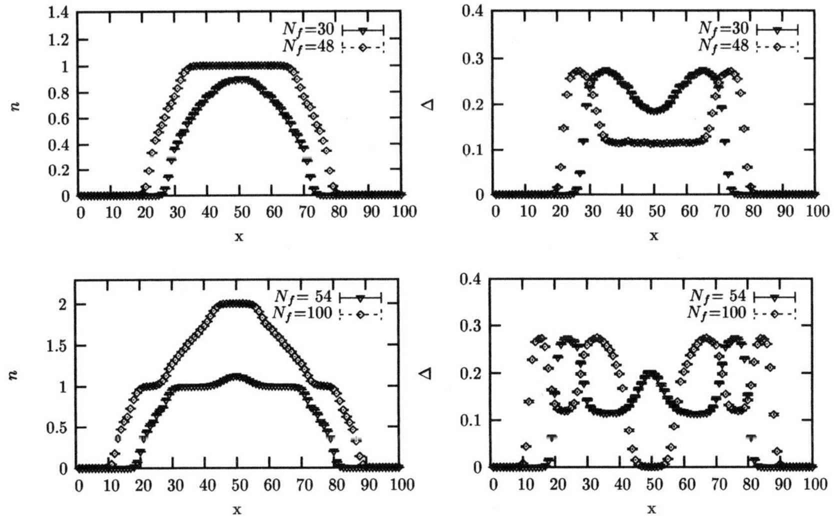

FIGURE 6. Four density profiles from Fig 5 (left) and their variances (right).

Another interesting question to address is how the density and variance profiles evolve when the on-site repulsion is changed; results for other simulations are presented in Fig. 6. It is easy to see in this figure that the variance decreases in the Mott phase as the on-site repulsion is increased (and will vanish for $U/t \to \infty$). This property of the local variance (which is small but not zero for finite U/t) lead us to look for a local quantity that vanishes when the Mott insulator is formed, this new quantity we call local compressibility and is defined as

$$\kappa_i^l = \sum_{j=-l}^{l} \chi_{i,i+j} \tag{3}$$

where $\chi_{i,j} = \left\langle n_i n_j \right\rangle - \langle n_i \rangle \left\langle n_j \right\rangle$ is the density-density correlation function and l is a U/t dependent length. We chose this length to be greater than the density-density correlation length for the homogeneous Mott phase at the U/t value under consideration. That means that we studied the density-density correlations of the Mott insulators in the homogeneous system (at half filling) and for a given U/t. These correlations decay exponentially as a consequence of the charge gap $\chi_{(x)} \propto \exp^{-\frac{x}{\xi(U/t)}}$ which gives the correlation length ξ as function of U/t. We used this in the confined case and chose l in Eq. 3 to be about ten times $\xi(U/t)$. Using this definition of the local compressibility in the homogeneous case we obtained for the Mott insulator a value $\kappa^l = 0$ and for a metal a non-vanishing value.

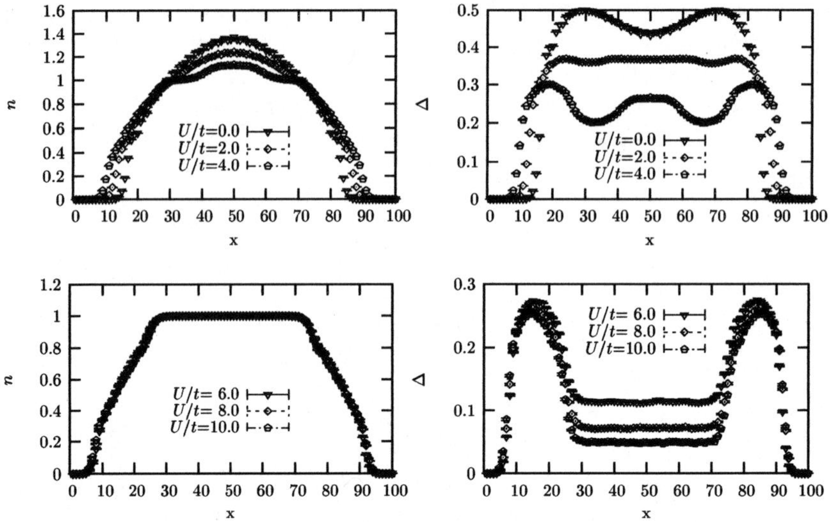

FIGURE 7. Evolution of local density profiles (left) and their variances (right) when the on-site repulsion is increased. The parameters of the trap are $N = 100$, $N_f = 70$ and $V/t = 0.0025$.

With this definition, the local compressibility in the trapped system of Fig. 7 is shown in Fig. 8. We see that whenever the Mott insulator is formed and independently of the value of the on-site repulsion, the local compressibility is indeed zero. This quantity, therefore, appears to be a good local order parameter to characterize the Mott insulator phases in the trap.

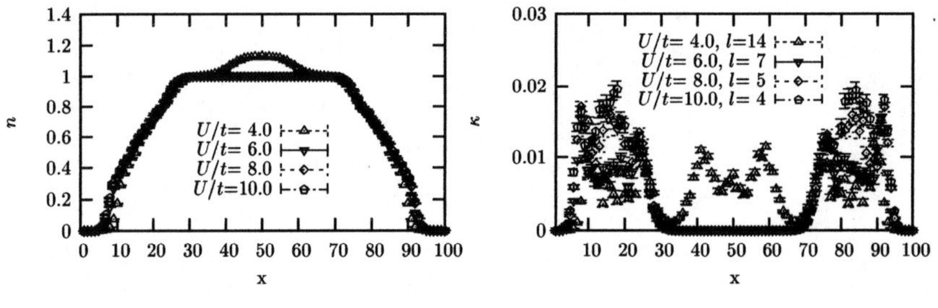

FIGURE 8. Evolution of four density profiles of Fig. 7 (left) and their local compressibilities (right) when the on-site repulsion is increased.

To end our analysis of the confined fermionic system, we present in Fig. 9 the phase diagram obtained for a trap with $N = 50$ sites. The y axis shows the quantity $N_f\sqrt{V/t}$ which has the dimensions of density while the x axis shows the dimensionless quantity U/t. We have verified that using these magnitudes the phase diagram scales quite well when the confining potential and the number of lattice sites are changed.

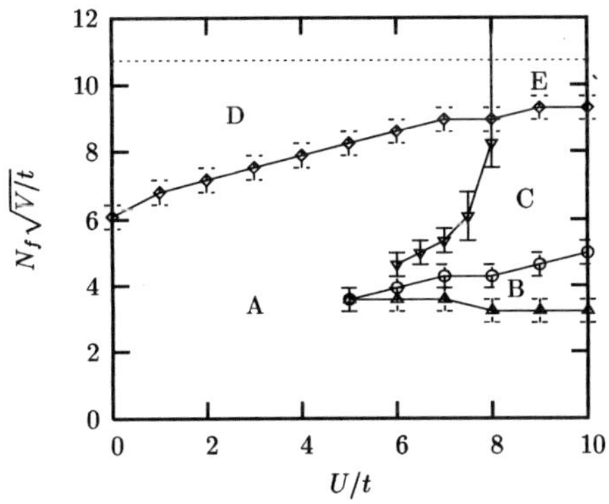

FIGURE 9. Phase Diagram of confined fermions on optical lattices. The dashed line (horizontal in the top of the figure) indicates the highest density at which we studied the system. The traps had $N = 50$ and $V/t = 0.032$

The phases present in Fig. 9 are as follows. In region A, only the metallic phase is present (like $N_f = 30$ in Fig. 6). Region B has the Mott insulating phase in the middle of the trap ($N_f = 48$ in Fig. 6). Region C has a metallic phase in the center of the surrounded by Mott insulator zones with $n_l = 1$ ($N_f = 54$ in Fig. 6). In region D the middle of the trap has a band insulator with $n_i = 2$ surrounded by a pure metallic phase and in region E the band insulator in the middle is surrounded by a metallic phase that is inside a Mott insulating phase and all of them surrounded by the usual metallic phase at the borders.

CONCLUSIONS

We have presented the results of the first QMC simulations for the 1-D boson and fermion Hubbard models in the presence of a confining parabolic potential. To our knowledge these are the first non-mean field studies of the problem (in the case of bosons excepting Refs. [18, 19]).

We showed that the physics in the traps is fundamentally different from the unconfined case. Mott domains appear for a continuous range of incommensurate fillings and the global compressibility of the system was always found to be non-vanishing even when local Mott phases are present. Global quantities, like the total density ($N_{particles}/N_{sites}$) and global compressibility, no longer provide the best way to characterize the system.

In general we find that local phases appear in the system, so that there is coexistence between metallic and insulating phases. In particular we have studied the local incompressible Mott regions and we have defined a local order parameter to characterize them. This local order parameter (local compressibility) is always zero in the Mott domains for independently of the values of the parameters. The variance of the density was found to be strongly suppressed in the Mott phases and vanished in the limit $U/t \to \infty$.

Finally, we determined the phase diagrams quantitatively for both bosonic and fermionic trapped systems. We found that they are much richer than the ones for the periodic lattices, showing how different the physics of the confined case is compared with the homogeneous case. These phase diagrams can also be useful for experimentalists to understand future results for fermions and some of result obtained for bosonic systems, like the coincidence between the values of V_0/t at which Mott phases develop in the confined case and those of the true Mott lobes in the non-confined case; and the narrowness of the Mott region with $n_i = 2$, discussed in the boson section.

We gratefully acknowledge financial support from the LFSP Nanomaterialien, a PROCOPE grant (France-Germany), NSF DMR 9985978, a CNRS-NSF co-operative grant (France-USA), the Stichting FOM and the Swiss National Science Foundation. We thank M. Feldbacher, F. F. Assaad, R. Kaiser, C. Miniatura and T. Pfau for useful conversations. The calculations for fermions were carried out on the HITACHI SR8000 of the HLRS (Stuttgart), M. Rigol thanks B. Krischok for his support there.

REFERENCES

1. M. H. Anderson, J. R. Ensher, M. R. Matthews, C. E. Wieman and E. A. Cornell, Science **269**, 198 (1995).
2. C. C. Bradley, C. A. Sackett, J. J. Tollett and R. G. Hulet, Phys. Rev. Lett. **75**, 1687 (1995).
3. K. B. Davis, M.-O. Mewes, M. R. Andrews, N. J. van Druten, D. S. Durfee, D. M. Kurn and W. Ketterle, Phys. Rev. Lett. **75**, 3969 (1995).
4. *Bose-Einstein Condensation in Atomic Gases*, Proccedings of the International School of Physics "Enrico Fermi", edt. by M. Inguscio, S. Stringari and C. E. Wieman (IOS Press, Amsterdam 1999).
5. F. Dalfovo, S. Giorgini, L. P. Pitaevskii and S. Stringari, Rev. Mod. Phys. **71**, 463 (1999).
6. B. DeMarco and D. S. Jin, Science **285**, 1703 (1999).
7. A. G. Truscott, K. E. Strecker, W. I. McAlexander, G. B. Partridge and R. G. Hulet, Science **291**, 2570 (2001).
8. F. Schreck, L. Khaykovich, K. L. Corwin, G. Ferrari, T. Bourdel, J. Cubizolles and C. Salomon, Phys. Rev. Lett. **87**, 080403 (2001).
9. S. R. Granade, M. Gehm, K. M. O'Hora and J. E. Thomas, Phys. Rev. Lett. **88**, 120405 (2002).
10. Z. Hadzibabic, C. A. Stan, K. Dieckmann, S. Gupta, M. W. Zwierlein, A. Görlitz and W. Ketterle, Phys. Rev. Lett. **88**, 160401 (2002).
11. M. P. A. Fisher, P. B. Wiechman, G. Grinstein and D. S. Fisher, Phys. Rev. B **40**, 546 (1989).
12. G. G. Batrouni, R. T. Scalettar and G. T. Zimanyi, Phys. Rev. Lett. **65**, 1765 (1990).
13. W. Krauth and N. Trivedi, Europhys. Lett. **14**, 627 (1991).
14. M. Cha, M. P. A. Fisher, S. M. Girvin, M. Wallin and A. P. Young, Phys. Rev. B **44**, 6883 (1991).
15. J. K. Freericks and H. Monien, Phys. Rev. B **53**, 2691 (1996).
16. M. Imada, A. Fujimori and Y. Tokura, Rev. Mod. Phys. **70**, 1039 (1998).
17. G. G. Batrouni, V. Rousseau, R. T. Scalettar, M. Rigol, A. Muramatsu, P. J. H. Denteneer and M. Troyer, Phys. Rev. Lett. **89**, 117203 (2002).
18. D. Jaksch, C. Bruder, J. I. Cirac, C. W. Gardiner and P. Zoller, Phys. Rev. Lett. **81**, 3108 (1998).
19. V. A. Kaskurnikov, N. V. Prokof'ev and B. V. Svistunov, Phys. Rev. A **66**, 031601(R) (2002).
20. J. E. Hirsch, R. L. Sugar, D. J. Scalapino and R. Blankenbecler, Phys. Rev. B **26**, 5033 (1982).
21. G. G. Batrouni and R. T. Scalettar, Phys. Rev. B **46**, 9051 (1992).
22. M. Greiner, O. Mandel, T. Esslinger, T. W. Hänsch and I. Bloch, Nature (London) **415**, 39 (2002).
23. G. Sugiyama and S. Koonin, Annals of Physics **168**, 1 (1986).
24. S. Sorella, E. Tosatti, S. Baroni, R. Car and M. Parrinello, Int. Jour. Mod. Phys. B **1**, 993 (1988).
25. D. J. Scalapino and R. L. Sugar, Phys. Rev. Lett. **46**, 519 (1981).
26. R. Blankenbecler, D. J. Scalapino and R. L. Sugar, Phys. Rev. D **24**, 2278 (1981).
27. D. J. Scalapino and R. L. Sugar, Phys. Rev. B **24**, 4295 (1981).
28. J. E. Hirsch, Phys. Rev. B **28**, 4059 (1983).
29. F. Gleisberg, W. Wonneberger, V. Schlöder and C. Zimmermann, Phys. Rev. A **62**, 063602 (2000).

Competition between Vertical and Diagonal Stripes in the Hartree-Fock Approximation

Marcin Raczkowski and Andrzej M. Oleś

Marian Smoluchowski Institute of Physics, Jagellonian University, Reymonta 4, PL-30059 Kraków, Poland

Abstract. Using the Hartree-Fock approximation we analyze the properties and stability of the filled (one doped hole per stripe site) vertical stripes (VS) and diagonal stripes (DS), found in the two-dimensional Hubbard model at two representative doping levels $x = 1/8$ and $x = 1/6$, by varying the ratio of the on-site Coulomb repulsion U to the hopping element t. In the weak coupling regime of $U \sim 3.5t$ the stability of VS is best explained by the solitonic mechanism which leads to the kinetic energy gain due to the hopping perpendicular to the stripes, while DS have a lower potential energy which explains their stability in the regime of large U. The results obtained beyond the Hubbard model are shortly reviewed as well.

INTRODUCTION

The instability towards a novel type of charge and magnetically ordered state, the so-called stripe phases, has been the subject of numerous theoretical investigations in the past few years supported by the experimental observations of striped domain walls (DW) in several doped transition metal oxides, including nickelates, cuprates, and manganites. Such states result from the competition between the superexchange interactions, which stabilize an antiferromagnetic (AF) long-range order in the undoped two-dimensional (2D) system, and the kinetic energy of holes, which are introduced to this AF state.

On one hand, the magnetic energy is gained when the electrons occupy the neighboring sites with the spin order as in the Néel state, while on the other hand, the kinetic energy is gained when the holes can move and the AF order is suppressed, at least locally. At large but finite Coulomb interaction U a compromise between these two energy gains has to be reached. This gives therefore a phase separation which manifests itself in the formation of nonmagnetic lines of holes, one-dimensional DW or just stripes, which separate AF domains of opposite phases.

The stripe instability was predicted, on the basis of mean-field calculations, well before their experimental confirmation [1]. The most stable structures obtained in the Hartree-Fock (HF) approximation have the density of one doped hole per one atom in DW and correspond to the so-called *filled stripes*. Indeed, such stripes were observed in the insulating nickelates: $La_2NiO_{4+\delta}$ [2, 3] and $La_{2-x}Sr_xNiO_4$ [4]. On the contrary, it is clear from variety of experiments that magnetic states within doped CuO_2 planes of the cuprates, $La_{1.6-x}Nd_{0.4}Sr_xCuO_4$ [5–7] and $La_{2-x}Sr_xCuO_4$ [8, 9], are the *half-filled stripes*, with the density of one doped hole per two atoms along DW. Such solutions could only be obtained as locally (but not globally) stable in the HF calculations [10].

CP678, *Lectures on the Physics of Highly Correlated Electron Systems VII: Seventh Training Course,* edited by A. Avella and F. Mancini

The above discrepancy between the theory and experiments indicates that it is necessary to go beyond the HF approach by including explicitly local electron correlations [11]. In general, in the strongly correlated regime $U \geq 5t$ the mean-field approximation works well only for polarized states within AF domains, in which the correlation corrections are small. A quite different situation is found on nonmagnetic DW - here the correlation corrections are large and one has to go *beyond* the HF approximation to investigate the relative stability of different phases. Taking into account that filled stripes have smaller electron densities on the atoms which belong to DW, it is obvious that more correlation energy can be gained, by reducing double occupancies, in the half-filled than in filled stripes [11].

The question of filling is not the only difference between the nickelate and cuprate stripes. Neutron diffraction measurements performed on $La_{1.6-x}Nd_{0.4}Sr_xCuO_4$ revealed that magnetic peaks are displaced from the AF maximum at $\mathbf{Q}_{AF} = (\pi,\pi)$ to the points $\mathbf{Q} = [(1 \pm 2\eta)\pi, \pi]$ and $\mathbf{Q} = [\pi, (1 \pm 2\eta)\pi]$ and the shift η depends linearly on hole doping $\eta \simeq x$ for $x < 1/8$ while it is almost constant at higher doping [6, 7]. Such a crossover behavior was obtained within the dynamical mean field theory (DMFT) [12]. Note, that the used samples with $x = 0.12$, 0.15, and 0.20 are superconducting with $T_c \simeq 4$, 11, and $15K$ [13]. The position of the peaks indicates that the magnetic structure is modulated either along the x or along the y axis, to which we refer as horizontal stripes (HS), or as vertical stripes (VS). The essentially identical modulation and doping dependence of η was observed in superconducting crystals of $La_{2-x}Sr_xCuO_4$ with $x > 0.05$ [8]. This suggests a local coexistence of superconductivity and static VS.

On the contrary, experiments performed in the insulating regime of $x \leq 0.05$ demonstrate unambiguously that spin modulation runs diagonally within CuO_2 planes (i.e., along the $\langle 110 \rangle$ direction) [9]. Such stripes are called diagonal stripes (DS) and were obtained as the most stable solutions in the DMFT calculations [12]. In fact, a diagonal stripe pattern identical (but with a hole concentration of 1 hole/Ni ion) to that in $La_{1.95}Sr_{0.05}CuO_4$, was established in the mentioned nickelates as well [2–4].

MECHANISM BEHIND THE STRIPE STABILITY

In what follows we describe the essential mechanism which favors the stripes at finite doping, by considering only a small cluster consisting of three atoms filled by two electrons and one hole (with respect to half-filling with the electron density $n = 1$ per site). For simplicity we assume that the electrons are confined to the considered cluster owing to large Coulomb interaction $U \gg t$, and we do not take into account any interactions with the AF background. There are two possibilities of the ground state, the first one corresponds to a hole added to a single AF domain in which, if we suppose that a $\downarrow$-spin electron is replaced by a hole, the two remaining $\uparrow$-spin electrons can be found in one of three allowed configurations: $\{\uparrow,0,\uparrow\}$, $\{\uparrow,\uparrow,0\}$, and $\{0,\uparrow,\uparrow\}$ (the other configurations are excluded by the Pauli principle). Hence, this polaronic state gives the total energy,

$$E_P = -\sqrt{2}t, \tag{1}$$

and the Coulomb interaction U does not contribute. A different situation is obtained when a hole occupies instead a domain wall separating two AF domains. Delocalization leads to the similar three configurations to those obtained above: $\{\uparrow,0,\downarrow\}$, $\{\uparrow,\downarrow,0\}$, and $\{0,\uparrow,\downarrow\}$, but in addition, three configurations with one doubly occupied site $\{\uparrow\downarrow,0,0\}$, $\{0,\uparrow\downarrow,0\}$, and $\{0,0,\uparrow\downarrow\}$, can be reached now because the Pauli principle is absent. Moreover, three other configurations with the interchanged $\uparrow$- and $\downarrow$- spins are also accesible via the double occupancies: $\{\downarrow,0,\uparrow\}$, $\{\downarrow,\uparrow,0\}$, and $\{0,\downarrow,\uparrow\}$. In the regime of large U the total energy can be found in a perturbative way and as a result one obtains,

$$E_S = -\sqrt{2}t - \frac{4t^2}{U}. \tag{2}$$

Therefore, the Hilbert space for the latter solitonic solution is larger and this solution is always more stable than a polaronic one [10]. The argument applies also to 2D systems, where DW are more stable than lines of polarons in an AF background.

HAMILTONIAN AND THE HF APPROXIMATION

In spite of difficulty to stabilize the ground state with half-filled DW, the HF framework can provide a generic microscopic description of filled inhomogeneous structures. In order to demonstrate and investigate the effect of the solitonic mechanism on the relative stability of VS and DS, we use the effective single band Hubbard model, which can be justified by taking the realistic parameters [14],

$$H = -t\sum_{\langle ij\rangle\sigma}\left(c^{\dagger}_{i\sigma}c_{j\sigma} + c^{\dagger}_{j\sigma}c_{i\sigma}\right) + U\sum_{i} n_{i\uparrow}n_{i\downarrow}, \tag{3}$$

where $n_{i\sigma} = c^{\dagger}_{i\sigma}c_{i\sigma}$, and $n_i = c^{\dagger}_{i\uparrow}c_{i\uparrow} + c^{\dagger}_{i\downarrow}c_{i\downarrow}$ gives the electron density at site i. The first sum runs over the bonds $\langle ij\rangle$ between nearest neighbors. The model is solved self-consistently within the most straightforward version of the HF method, with wave function being a product of two separate Slater determinants for up and down spins,

$$n_{i\uparrow}n_{i\downarrow} \simeq n_{i\uparrow}\langle n_{i\downarrow}\rangle + \langle n_{i\uparrow}\rangle n_{i\downarrow} - \langle n_{i\uparrow}\rangle\langle n_{i\downarrow}\rangle. \tag{4}$$

Calculations were performed on 12×12 (16×16) clusters with AF domains of width five (seven) atoms separated by DW, adequate for doping $x=1/6$ ($x=1/8$), respectively, using periodic boundary conditions. Such clusters give the most stable solutions for the selected doping. Furthermore, they are large enough to reproduce the results obtained by using the zero temperature HF equations in the thermodynamic limit for the total energy $E(\bar{m},n)$ and its derative,

$$E(\bar{m},n) = \frac{Un}{2} - \frac{1}{2}\int\frac{d^2\mathbf{k}}{(2\pi)^2}\sqrt{(U\bar{m})^2 + (2\varepsilon_{\mathbf{k}})^2} - \frac{U}{4}(n^2 - \bar{m}^2), \tag{5}$$

$$\frac{\partial E(\bar{m},n)}{\partial\bar{m}} = \bar{m} - \int\frac{d^2\mathbf{k}}{(2\pi)^2}\frac{1}{\sqrt{1+\left(\frac{2\varepsilon_{\mathbf{k}}}{\bar{m}U}\right)^2}} = 0, \tag{6}$$

TABLE 1. Magnetization density $\bar{m} = |\langle n_{i\uparrow} - n_{i\downarrow}\rangle|$ and site-normalized ground state energy $E(\bar{m},n)$ for an undoped AF system ($n = 1$) at $U/t = 5$ as a function of cluster size.

cluster	4×4	6×6	8×8	10×10	12×12	16×16	$\sqrt{N}\times\sqrt{N}$*
$\bar{m}$	0.7783	0.7745	0.77237	0.77241	0.77226	0.77241	0.77241
$E(\bar{m},n)/t$	-0.67365	-0.68126	-0.681970	-0.681973	-0.681972	-0.681976	-0.681970

* Calculated from Eqs. (5) and (6) in the limit of large N.

where $\varepsilon_{\mathbf{k}} = -2t(\cos k_x + \cos k_y)$ is the 2D free electron dispersion. The energy $E(\bar{m},n)$ is determined by summing over the occupied lower Hubbard band. The latter Eq. (6) can be easily solved self-consistently yielding the value of magnetization $\bar{m} = |\langle n_{i\uparrow} - n_{i\downarrow}\rangle|$ for any filling n and interaction U, which gives, after inserting into the former Eq. (5), the site-normalized total energy $E(\bar{m},n)$. In Table 1 we have compared the values of $\bar{m}$ and $E(\bar{m},n)$ for chosen half-filled finite clusters ($n = 1$) at $U/t = 5$ with the corresponding values obtained from Eqs. (5) and (6).

NUMERICAL RESULTS

The effective Hamiltonian (3) solved in the HF approach (4) give the striped solutions presented in real space (reciprocal space) in Fig. 1 (Fig. 3) with doping $x = 1/8$ and in Fig. 2 (Fig. 4) with $x = 1/6$, respectively, for intermediate interaction $U/t = 5$. In order to describe the charge and spin distribution functions we calculated the Fourier transforms of the static hole-hole and spin-spin correlation functions, which give the charge and magnetic structure factor,

$$C(\mathbf{k}) = \frac{1}{N}\sum_{ij} e^{i\mathbf{k}(\mathbf{R}_i-\mathbf{R}_j)}\langle n_{\mathrm{h}i}\rangle\langle n_{\mathrm{h}j}\rangle, \tag{7}$$

$$S^z(\mathbf{k}) = \frac{1}{N}\sum_{ij} e^{i\mathbf{k}(\mathbf{R}_i-\mathbf{R}_j)}\langle S^z_i\rangle\langle S^z_j\rangle, \tag{8}$$

where the summations include all pairs $\{ij\}$ in a considered cluster with N sites. The obtained strong maxima in the magnetic structure factor $S^z(\mathbf{k})$, displaced slightly from the *AF* value of $\mathbf{k} = (\pi,\pi)$, are accompanied by some weaker satellities (Figs. 3 and 4). The appearence of similar satellities were reported in Ref. [11]. The local hole densities $\langle n_{\mathrm{h}i}\rangle = 1 - \langle n_{i\uparrow} + n_{i\downarrow}\rangle$ and the local magnetization densities $\langle S^z_i\rangle = \frac{1}{2}|\langle n_{i\uparrow} - n_{i\downarrow}\rangle|$ of nonequivalent atoms of both stripe phases used in Eqs. (7) and (8) are given in Tables 2 ($x = 1/8$) and 3 ($x = 1/6$). They are all characterized by decreasing hole density along the x-direction towards the center of an AF domain, beginning at a nonmagnetic domain wall. The values of $\langle n_{\mathrm{h}i}\rangle$ and $\langle S^z_i\rangle$, calculated in the weak coupling regime $U/t = 3.5$ (in parenthesis) show more delocalized DW and stronger doped AF domains. On the contrary, increasing U leads to the reinforcement of stripe order best seen in the increasing hole densities within DW, and the decreasing hole densities within AF domains. Hence, VS and DS become apparently narrower.

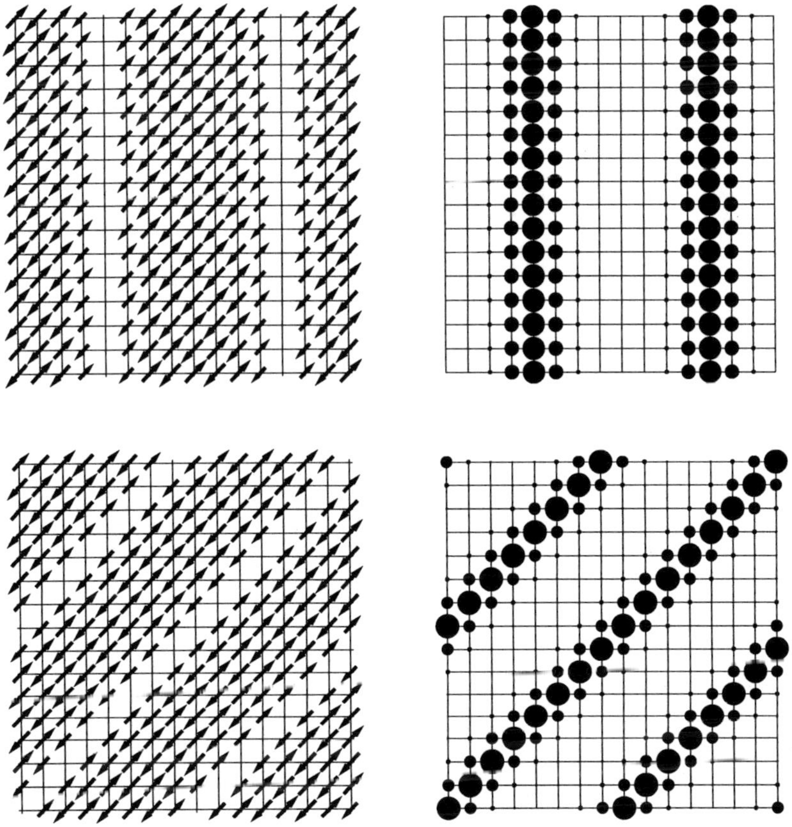

FIGURE 1. VS and DS as found for $U/t = 5$ at hole doping $x = 1/8$. The length of the arrows is proportional to the magnetization $\langle S_i^z \rangle$ and the hole density $\langle n_{hi} \rangle$ is scaled by the diameter of the black circles.

TABLE 2. Local charge density $\langle n_{hi} \rangle$ and local magnetization density $\langle S_i^z \rangle$, all labeled by decreasing doped hole density in the x-direction at $x = 1/8$ for $U/t = 5$ and $U/t = 3.5$ (in parenthesis).

	i	1	2	3	4	5
VS	$\langle n_{hi} \rangle$	0.364	0.234	0.067	0.014	0.006
		(0.258)	(0.206)	(0.107)	(0.044)	(0.027)
	$\langle S_i^z \rangle$	0.000	0.222	0.348	0.381	0.384
		(0.000)	(0.135)	(0.241)	(0.292)	(0.305)
DS	$\langle n_{hi} \rangle$	0.388	0.193	0.070	0.032	0.020
		(0.283)	(0.175)	(0.096)	(0.062)	(0.051)
	$\langle S_i^z \rangle$	0.000	0.262	0.352	0.373	0.380
		(0.000)	(0.184)	(0.262)	(0.289)	(0.299)

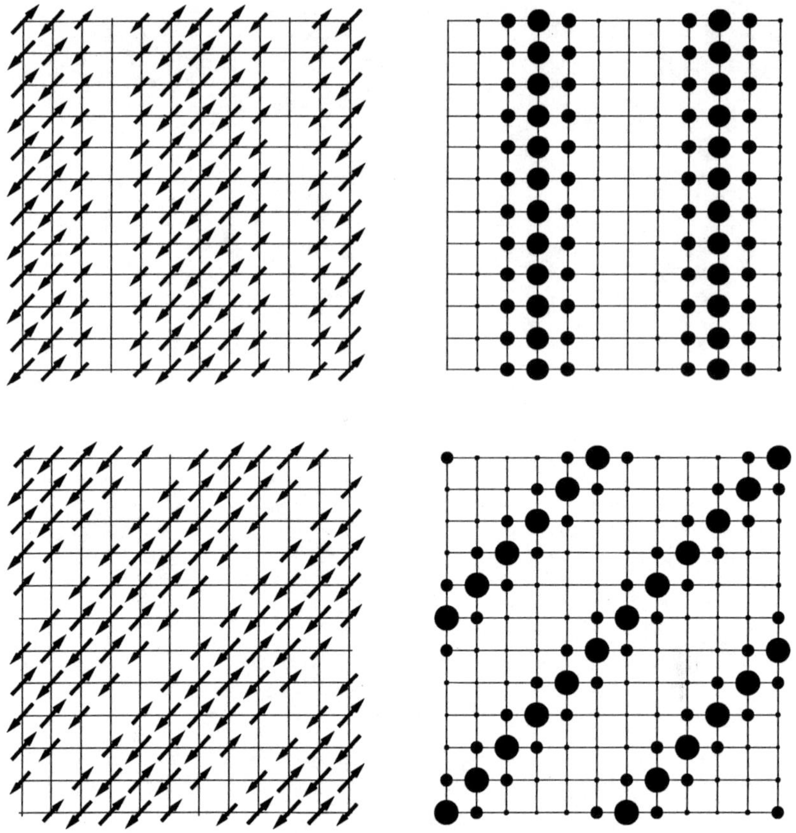

FIGURE 2. VS and DS as found for $U/t = 5$ at hole doping $x = 1/6$. The length of the arrows is proportional to the magnetization $\langle S_i^z \rangle$ and the hole density $\langle n_{hi} \rangle$ is scaled by the diameter of the black circles.

TABLE 3. Local charge and magnetization density, all labeled by decreasing doped hole density in the x-direction at $x = 1/6$ for $U/t = 5$ and $U/t = 3.5$ (in parenthesis).

	i	1	2	3	4
VS	$\langle n_{hi} \rangle$	0.365	0.235	0.069	0.026
		(0.262)	(0.211)	(0.119)	(0.079)
	$\langle S_i^z \rangle$	0.000	0.223	0.348	0.376
		(0.000)	(0.134)	(0.236)	(0.271)
DS	$\langle n_{hi} \rangle$	0.396	0.198	0.078	0.053
		(0.306)	(0.188)	(0.113)	(0.093)
	$\langle S_i^z \rangle$	0.000	0.264	0.354	0.367
		(0.000)	(0.191)	(0.267)	(0.289)

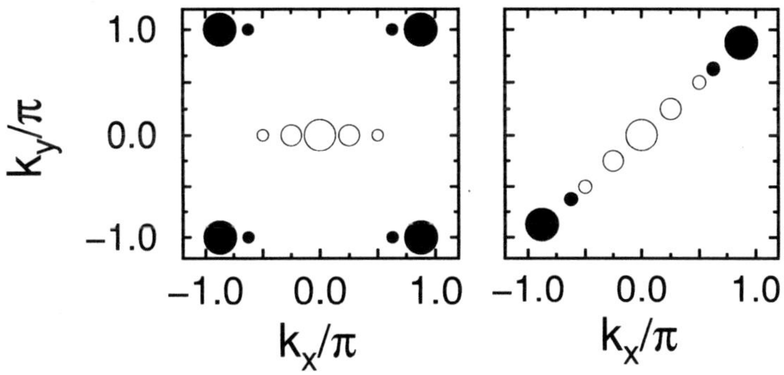

FIGURE 3. Reciprocal space pattern of VS (left panel) and DS (right panel) shown in Fig. 1. Filled (open) circles correspond to the magnetic (charge) peaks, respectively, at hole doping $x = 1/8$.

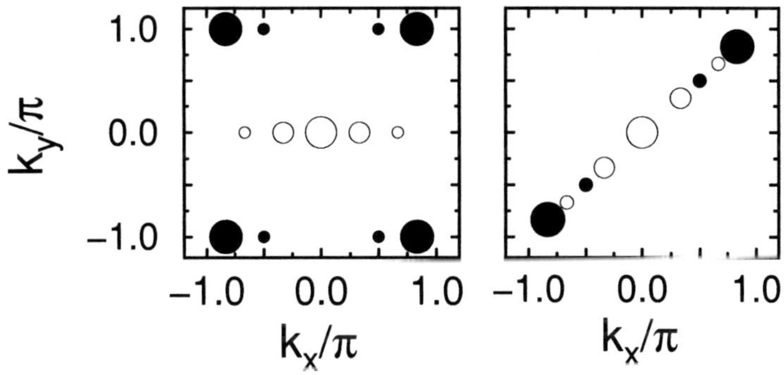

FIGURE 4. The same as in Fig. 3 but for doping $x = 1/6$.

TABLE 4. Site-normalized ground-state energy E_{tot}, kinetic energy ($E^x_{\text{kin}}, E^y_{\text{kin}}$), and potential energy E_U components of VS and DS at $x = 1/8$ for $U/t = 5$ and $U/t = 3.5$ (in parenthesis).

VS				DS			
E^x_{kin}/t	E^y_{kin}/t	E_U/t	E_{tot}/t	E^x_{kin}/t	E^y_{kin}/t	E_U/t	E_{tot}/t
-0.6753	-0.6147	0.4900	-0.8000	-0.6368	-0.6368	0.4696	-0.8040
(-0.7564)	(-0.7147)	(0.4937)	(-0.9774)	(-0.7236)	(-0.7236)	(0.4728)	(-0.9744)

TABLE 5. The same as in Table 4 but for doping $x = 1/6$.

VS				DS			
E^x_{kin}/t	E^y_{kin}/t	E_U/t	E_{tot}/t	E^x_{kin}/t	E^y_{kin}/t	E_U/t	E_{tot}/t
-0.7010	-0.6222	0.4841	-0.8391	-0.6455	-0.6455	0.4489	-0.8421
(-0.7719)	(-0.7238)	(0.4822)	(-1.0135)	(-0.7219)	(-0.7219)	(0.4396)	(-1.0042)

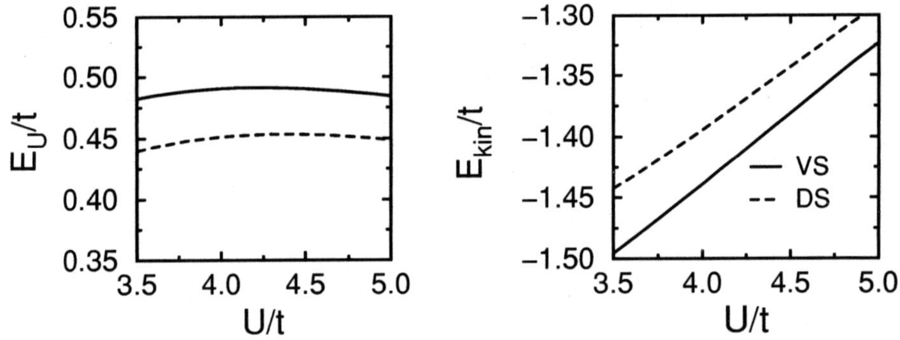

FIGURE 5. Site-normalized potential energy E_U and kinetic energy E_{kin} of VS and DS as a function of increasing interaction U at $x = 1/6$. Note the same energy scale of the both panels.

Looking into these tables one can notice two important features. First of all, for a given value of U hole density along DW is higher in the case of DS. Second, magnetization densities of nonequivalent atoms within AF domains are larger in the case of DS. Both these features indicate that DS have lower probabilities of double occupancy and consequently - a more favorable potential energy E_U (see Tables 4 and 5).

An inspection of Tables 4 and 5 with the values of site-normalized ground state energy E_{tot}, kinetic energies E^x_{kin} and E^y_{kin}, calculated separately for both main directions, and potential energy E_U, leads us to conclude that DS have indeed better potential energy E_U which explains their stability in the strong coupling regime $U/t \sim 5$. On the contrary, VS are stabilized in weak coupling regime and their stability follows from larger kinetic energy gain E_{kin}.

The above features are better seen in Fig. 5 with a behavior of E_U as a function of increasing U (left panel) together with a variation of E_{kin} (right panel) at $x = 1/6$. For

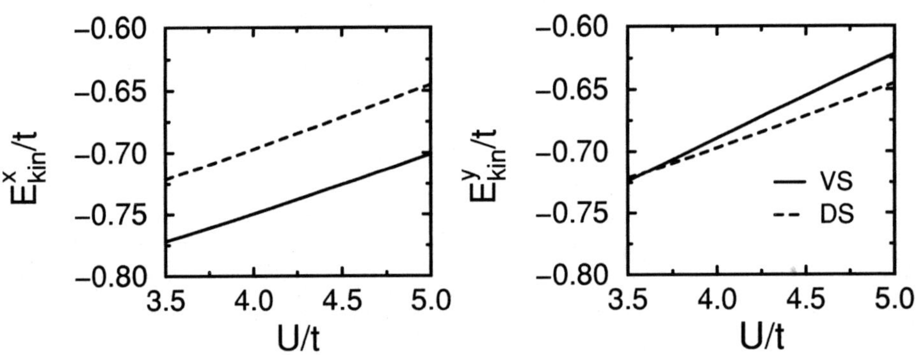

FIGURE 6. Site-normalized kinetic energy along x-direction E^x_{kin} and along y-direction E^y_{kin} of VS and DS as a function of increasing interaction U at $x = 1/6$.

convenience, we keep the same energy scale in both panels, so one can easily recognize that hole dynamics is responsible for the stability of VS in the regime of small $U/t \sim 3.5$, while in the regime of large $U/t \sim 5$ strong enough interaction stabilizes DS. Therefore, a generic crossover from VS to DS occurs with increasing U. Indeed, the critical value of the interaction $U/t = 4.1$ is found at doping level $x = 1/8$, and it is shifted to $U/t = 4.6$ at higher doping $x = 1/6$. This increase can be explained by a lower electron density at a higher value of x, which reduces the number of double occupancies, so the potential energy becomes relevant only at a higher value of U.

Furthermore, a relative better kinetic energy E_{kin} of VS appears due to the hopping perpendicular to the stripes while the hopping parallel to the stripes is less important for their stability as shown in Fig. 6 with $x = 1/6$. The large anisotropy in the values of energies E^x_{kin} and E^y_{kin} follows from the mentioned solitonic mechanism. Therefore, it might be expected that the above mechanism will always tune the direction of the DW along a weaker hopping element in the anisotropic model. Indeed, this expectation was confirmed by recent studies [15, 16]. On the contrary, the anisotropy disappears in the case of DS (see Fig. 6).

BEYOND THE HUBBARD MODEL

Here, we describe briefly the results of our calculations performed beyond the Hubbard model. First, we investigated the stripes within the extended Hubbard model with the second neighbor hopping t' and the nearest-neighbor Coulomb interaction V. It has been found that both repulsive V $(V > 0)$ and negative t' $(t'/t < 0)$, which correspond to the realistic parameters of the superconducting cuprates, enhance the relative stability of DS [16]. We attribute it to the large difference between charge densities of DW atoms and their nearest-neighbor sites, which is avoided by VS (HS) along DW (see Figs. 1 and 3). Furthermore, negative t' gives a positive kinetic energy contribution, which is much more readily minimized by the DS charge configuration. Therefore, the VS are destabilized when $t'/t \simeq -0.3$, as in $YBa_2Cu_3O_{6+x}$ [12].

Next, we studied the Hubbard model with a static Peierls electron-lattice coupling,

$$H = -t_0 \sum_{\langle ij\rangle\sigma} (1+\alpha u_{ij}) c^\dagger_{i\sigma} c_{j\sigma} + U \sum_i n_{i\uparrow} n_{i\downarrow} + \frac{1}{2} K \sum_{\langle ij\rangle} u^2_{ij}, \tag{9}$$

where K stands for the elastic energy, and u_{ij} is a fractional change of the bondlength for the bond $\langle ij\rangle$ between two nearest-neighbor sites i and j. Unlike the realistic values of t' and V, it has been established that the electron-phonon interaction stabilizes VS by the bond contractions, especially pronounced on the bonds which connect DW atoms with their neighbors in the transverse direction. They lead to a stronger relative enhancement of the local hopping elements $t_{ij} = t_0(1 + \alpha u_{ij})$, and consequently to larger kinetic energy gain [17].

SUMMARY

In summary, we have investigated the tendencies of doped hole towards self-organization into vertical (horizontal) and diagonal striped pattern. Our studies confirm that the HF method is capable of finding the relevant ground state of the doped AF system, and allows to draw qualitative conclusions about the nature of possible phases. In general, we have found that the DS are more stable at large values of the on-site Coulomb repulsion U. We believe that the present results do not explain sufficiently why the DS are more stable in the nickelates, even though it seems that the ratio U/t is larger in the nickelates than in cuprates [18]. This question could be answered only by investigating a degenerate Hubbard model with two e_g orbitals filled by $2-x$ electrons per site. Further, we have showed that even such a simple model as the single-band Hubbard Hamiltonian is sufficient to demonstrate the solitonic mechanism which primarily stabilizes stripes owing to the hopping perpendicular to the direction of DW.

ACKNOWLEDGMENTS

This work was supported by the Polish State Committee for Scientific Research (KBN), Project no. 5 P03B 036 21.

REFERENCES

1. Zaanen, J. and Gunnarsson, O., *Phys. Rev. B* **40**, 7391 (1989).
2. Tranquada, J. M., Buttrey, D. J., Sachan, V., and Lorenzo, J. E., *Phys. Rev. Lett.* **73**, 1003 (1994).
3. Tranquada, J. M., Lorenzo, J. E., Buttrey, D. J., and Sachan, V., *Phys. Rev. B* **52**, 3581 (1995).
4. Sachan, V., Buttrey, D. J., Tranquada, J. M., Lorenzo, J. E., and Shirane, G., *Phys. Rev. B* **51**, 12 742 (1995).
5. Tranquada, J. M., Sternlieb, B. J., Axe, J. D., Nakamura, Y., and Uschida, S., *Nature (London)* **375**, 561 (1995).
6. Tranquada, J. M., Axe, J. D., Ichikawa, N., Nakamura, Y., Uschida, S., and Nachumi, B., *Phys. Rev. B* **54**, 7489 (1996).
7. Tranquada, J. M., Axe, J. D., Ichikawa, N., Moodenbaugh, A. R, Nakamura, Y., and Uschida, S., *Phys. Rev. Lett.* **78**, 338 (1997).
8. Yamada, K., Lee, C. H., Kurahashi, K., Wada, J., Wakimoto, S., Ueki, S. Kimura, H., Endoh, Y., Hosoya, S., Shirane, G., Birgeneau, R. J., Greven, M., Kastner, M. A., and Kim, Y. J., *Phys. Rev. B* **57**, 6165 (1998).
9. Wakimoto, S., Birgeneau, R. J., Kastner, M. A., Lee, Y. S., Erwin, R., Gehring, P. M., Lee, S. H., Fujita, M., Yamada, K., Endoh, Y., Hirota, K., and Shirane, G., *Phys. Rev. B* **57**, 3699 (2000).
10. Zaanen, J. and Oleś, A. M., *Ann. Phys. (Leipzig)* **5**, 224 (1996).
11. Góra, D., Rościszewski, K., and Oleś, A. M., *Phys. Rev. B* **60**, 7429 (1999).
12. Fleck, M., Lichtenstein, A. I., Pavarini, E., and Oleś, A. M., *Phys. Rev. Lett.* **84**, 4962 (2000).
13. Kastner M. A., Birgeneau, R. J., Shirane, G., and Endoh, Y., *Rev. Mod. Phys.* **70**, 897 (1998).
14. Inui, M. and Littlewood, P. B., *Phys. Rev. B* **44**, 4415 (1991).
15. Normand, B. and Kampf, A. P., *Phys. Rev. B* **64**, 024521 (2001).
16. Raczkowski, M., Normand, B., and Oleś, A. M., to appear in *Phys. Stat. Sol. (a)*.
17. Raczkowski, M. and Oleś, A. M., to appear in *Physica C*.
18. Zaanen, J. and Littlewood, P. B., *Phys. Rev. B* **50**, 7222 (1994).

Pseudogaps from classical fluctuations of the order parameter

A. Posazhennikova* and P. Coleman†

*Institut für Theorie der Kondensierten Materie Universität Karlsruhe , Germany
†Center for Materials Theory, Department of Physics and Astronomy, Rutgers University, Piscataway, NJ 08854, USA

Abstract. The problem of pseudo-gap formation in an electronic system, induced by the fluctuations of the order parameter is revisited. We make the observation that a large class of current theories are theoretically equivalent to averaging the Free energy of the pseudo-gap system over quenched-disordered distribution of the order parameter. To illustrate how our approach works we examine the interplay of pseudogap, superconductivity and disorder.

INTRODUCTION

The interplay between fluctuation and order phenomena has been a subject of great scientific interest for already three decades. It was intriguing prediction, made by Lee, Rice and Anderson [1] that in 1D systems, which undergo the Peierls transition, the order parameter fluctuations near a charge density wave instability induce a suppression of the density of states (DOS). Other theoretical papers, addressing the problem of fluctuation effect above the transition temperature have also resulted in challenging predictions [2]. First experimental investigation of the problem was performed on quasi-1D KCP-solts [3]. The observation of a marked suppression of the optical conductivity at high temperatures in those materials was impossible to explain within theoretical models with a true gap. It was then when era of pseudogap models made its start.

According to Mott the term pseudogap describes a region with a suppressed DOS within the energy interval corresponding to a forbidden band in an ideal crystal, representing the reminiscence of this band gap, which is conserved after strong disordering [7]. Since first experimental study of Brüesch et al the manifestation of pseudogap has been observed in a tremendous number of various experiments and numerous low-dimensional strongly-correlated systems. The material classes displaying pseudogap behavior include transition metal dichalcogenides, organic layered superconductors [4], colossal magneto-resistance compounds [5], ruthenades [6] and high-temperature superconductors (HTSC) [7, 8].

In our research we confine ourselfs to a study of a particular class of models, originated from a pioneering paper of Lee, Rice and Anderson and describing the drop of single particle density of states due to the fluctuations of the order parameter within so-called fluctuation gap model (FGM). The key ideas of FGM is that the fluctuations are considered to be classical and time-independent. Importantly, this basic assumption threads through most of subsequent developments during past 30 years. One of the

CP678, *Lectures on the Physics of Highly Correlated Electron Systems VII. Seventh Training Course,* edited by A. Avella and F. Mancini first successors of LRA was M. Sadovskii, who has shown how the problem of nested electrons moving in a critical fluctuating order parameter field could be solved exactly [9, 10]. With the discovery of HTSC the interest to the models, descibing pseudogap behavior was resurrected. The mechanism of superconductivity in high-temperature compounds still remains a topic of great debate. Most researchers consider the important keys to unlocking the secrets of the mechanism for SC to be found by understanding the nature of the normal state, where the pseudogap manifests itself in anomalous behaviour of the majority of the physical properties, observed in the underdoped region of a phase diagram of cuprates. Therefore most of the efforts have been put in experimental and theoretical study of the normal state of HTSC. Many researches [11, 12] (for review see [7]) have been trying to elaborate the original ideas of 1D FGM with respect to HTSC by incorporating the spin degrees of freedom into consideration and by extending the 1D model to higher dimensions or even by taking into account the non -Gaussian character of the fluctuations [13].

In our paper we revisit the underlying concepts of FGM by making the observation that the whole class of current theories is theoretically equivalent to averaging the Free energy of the pseudo-gap system over a *quenched-disordered* distribution of the order parameter. The intuitive equivalence between the slow critical fluctuations responsible for pseudo-gap formation and a quenched random potential was certainly known to several authors in the past[1, 14]. In this paper, we vocalize this equivalence and set it in a formal framework. Using the idea we show how the Sadovskii model of pseudo-gap formation can be formulated in terms of a quenched average free energy. To illustrate the utility of this method, we develop a toy model for the interplay of pseudo-gap formation and weak impurity scattering with d-wave superconductivity.

We also discuss the shortcomings of this kind of theoretical approach and the perspectives in applying quenched -disorder pseudogap model to HTSC and other experimental systems.

QUENCHED AVERAGE OVER CLASSICAL FLUCTUATIONS

One of the central assumptions of a large class of pseudo-gap models, is that the order-parameter fluctuations which scatter electrons to form the pseudo-gap are so slow that they can be considered to be infinitely retarded. In other words, the characteristic relaxation timescale of the fluctuations τ_{OP} is not actually infinite, but may be treated as such because it is much greater than the inelastic scattering rate of the electrons τ_e ($\tau_{OP} >> \tau_e$). The result of this condition, is that electrons perceive the order parameter as a frozen degree of freedom, that may be treated as a classical variable.

We now need to consider the interactions between the electrons induced by these classical order parameter fluctuations. In a Feynman diagram, the net interaction induced by these fluctuations has a number of important features (Fig. 1.):

- It transfers no energy, and is thus infinitely retarded. (Fig. 1 (a))
- Polarization bubble insertions into the effective interaction are absent, since these have already been taken into account in forming the effective interaction (Fig. 1(b)).

- Interactions between the classical fluctuations may be ignored. This assumption permits the summation of a large class of diagrams. Unfortunately, it can not really be justified near a critical point, where non-linear interactions between classical fluctuations are almost always relevant. This is a major weakness of the current class of theories, and one we shall return to in our final discussion.

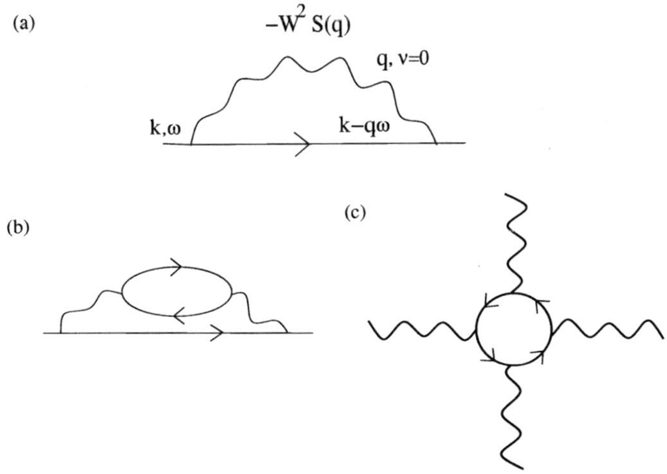

FIGURE 1. (a) Classical fluctuations induce an interaction $V_{eff} = -W^2 S(\vec{q})$ on a single line. (b) Fermion loop insertions into the interaction line are neglected which means that the average over the classical fluctuations is a quenched average. (c) By carrying out a quenched average, we also eliminate interactions between fluctuations of the form shown here.

The first feature can be incorporated into a model by introducing a coupling into the Hamiltonian

$$H[\phi] = H_o + W \sum_{\vec{q}} \rho_{\vec{q}} \phi_{-\vec{q}}. \tag{1}$$

In this paper we will imagine that H_0 is a Hamiltonian in which effects of interaction have been taken into account by a mean-field approximation- so that for instance, H_0 might be a BCS Hamiltonian with explicit pairing terms. In (1) $\rho_{\vec{q}} = \sum_{\vec{k}} c^\dagger_{\vec{k}-\vec{q}} c_{\vec{k}}$ is the density of electrons at momentum $\vec{q}$ and $\phi_{\vec{q}}$ is a purely classical field representing the order parameter fluctuations. This quantity is to be averaged over the distribution

$$P[\phi] = exp\left[-\beta \sum \frac{|\phi_{\vec{q}}|^2}{2S(\vec{q})}\right]$$

and it thus gives rise to an effective, and infinitely retarded (attractive) interaction of the form

$$V_{eff}(\vec{q}) = -W^2 S(\vec{q})$$

At first sight, it might appear sufficient average the partition function over $P[\phi]$, i.e

$$
\begin{aligned}
Z &= \int P[\phi] Z[\phi] \\
Z[\phi] &= \mathrm{Tr}\left[e^{-\beta H[\phi]}\right].
\end{aligned}
\tag{2}
$$

This is an annealed average. However, such a scheme allows fermion loop renormalizations and non-linear interactions to develop within the interaction lines as illustrated in Fig. 1 (b). These terms are to be dropped, under the assumptions mentioned above. To ensure this, we must carry out a quenched average over the classical field, which means that we average the Free energy, rather than the partition function

$$F = -T \int P[\phi] \log Z[\phi] \tag{3}$$

This procedure is implicit to the approach of Sadovskii, and subsequent developments, by the selection of Feynman diagrams containing a single Fermion line. One way to implement the quenched average is to use the $N \to 0$ trick [15],

$$F = -T \lim_{N\to 0} \int P[\phi] \frac{1}{N} \left(Z[\phi]^N - 1\right). \tag{4}$$

The term containing $Z[\phi]^N$ can be considered as the partition function of N identical replicas in the frozen classical field ϕ. Diagrammatic contributions to $Z[\phi]^N$ containing n fermion loops, scale as $O(N^n)$, so that in the $N \to 0$ limit, the only terms surviving in the diagrammatic expansion of the Free energy, are those with a single fermion loop, so that

$\Delta F = -T$ 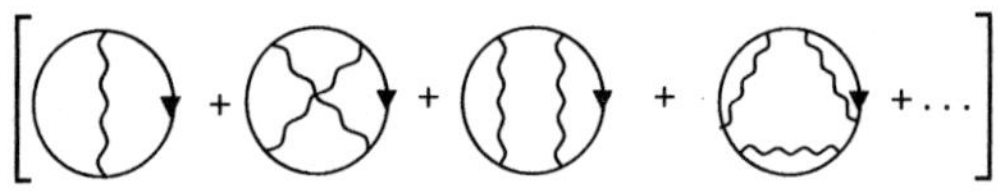 (5)

where $\Delta F = F_I - F_0$ is the change in the Free energy due to turning on the interactions. The advantage of this approach is that we can now start to formulate a theory of the pseudo-gap in the language of an effective Free energy. This becomes particularly useful when considering the case of a superconductor with a pseudo-gap.

EFFECTIVE ACTION APPROACH TO THE PSEUDOGAP PROBLEM

To illustrate the above approach, we consider a two-dimensional electronic system with a nested Fermi surface spanned by the commensurate wavevector $\vec{Q} = (\pi, \pi)$ as illustrated in Fig. 2. Fluctuations of the underlying order parameter are governed by a correlation function of the form

$$S(\vec{q}) = \frac{1}{\pi^2} \frac{\xi^{-1}}{(q_x - Q_x)^2 + \xi^{-2}} \frac{\xi^{-1}}{(q_y - Q_y)^2 + \xi^{-2}} \tag{6}$$

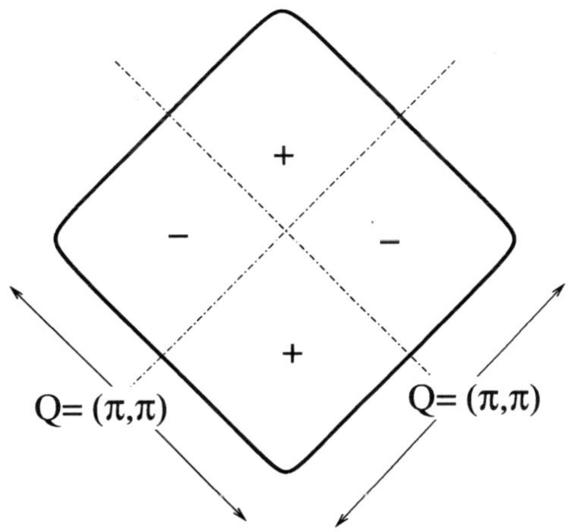

FIGURE 2. Nested Fermi surface with commensurate fluctuations considered in the toy model of this paper.

fluctuations couple the nested Fermi surfaces via the effective interaction $V_{eff} = -W^2 S(\vec{q})$. It was shown [12] that in the limit $\xi \to \infty$ the model under consideration can be solved exactly and the effective interaction takes the oversimplified form

$$V_{eff}(\vec{q}) = -W^2\{\delta(q_x - \pi)\delta(q_y - \pi)\}. \tag{7}$$

The class of models we shall discuss take the form (1). Let us start with a simple model where

$$H_0 = \sum_{\vec{k}} \varepsilon_{\vec{k}} c^{\dagger}_{\vec{k}\sigma} c_{\vec{k}\sigma}. \tag{8}$$

In this case the diagrams for the free energy may be completely summed. It turns out for example that each term of a given order gives rise to the same contribution and the only thing one should be careful about is the combinatorics. Fir instance, in incommensurate case the diagrams with crossing lines can be omitted, so that we have less diagrams to sum in comparison with commensurate case. In commensurate case the number of diagrams of the order W^{2n} is shown to be $(2n-1)!!$ [**?**], which can be expressed via Γ-function and thus absorbed into an integral over a Gaussian distribution of gap sizes as follows

$$W^{2n}(2n-1)!! = \frac{1}{\sqrt{2\pi}} \int_{-\infty}^{\infty} d\zeta e^{-\zeta^2/2} (\zeta W)^{2n} \tag{9}$$

Then, taking into account the expression for the free energy of "non-interacting" system $F_o = -T\sum_{\vec{k},n} \ln[-G_o^{-1}(\vec{k}, i\omega_n)]$ we obtain finally

$$F = \frac{1}{\sqrt{2\pi}} \int_{-\infty}^{\infty} d\zeta e^{-\zeta^2/2} F[\zeta, W] \tag{10}$$

where

$$F[\zeta, W] = -T \sum_{n,\, \vec{k} \in \frac{1}{2}\text{BZ}} \text{Tr} \ln \begin{pmatrix} -G_o^{-1}(\vec{k} - \vec{Q}/2, i\omega_n) & \zeta W \\ \zeta W & -G_o^{-1}(\vec{k} + \vec{Q}/2, i\omega_n) \end{pmatrix} \tag{11}$$

is the Free energy for a gap size ζW. The most important feature of our derivation above, is the conservation of the momentum associated with each nested Fermi surface. In what follows we discuss how our derivation can be easily generalized to include pairing (which conserves momentum) and a restricted class of disorder scattering which scatters between states on the same nested side of the Fermi surface.

Pairing in the presence of a pseudo-gap

Now we can incorporate pairing in our model in order to demonstrate how our approach actually works. Let us take

$$H_0 = \sum_{\vec{k}} \varepsilon_{\vec{k}} c^{\dagger}_{\vec{k}\sigma} c_{\vec{k}\sigma} + \sum_{\vec{k}} \left[\Delta \gamma_{\vec{k}} c^{\dagger}_{\vec{k}\uparrow} c^{\dagger}_{-\vec{k}\downarrow} + \text{H.c.} \right] + \frac{|\Delta|^2}{g}. \tag{12}$$

where $\gamma_{\vec{k}} = \cos(k_x) - \cos(k_y)$ is a d-wave order parameter whose nodes bisect the nested Fermi surfaces. We have included an additional term $|\Delta|^2/g$ to take account of the BCS decoupling of the pairing interaction. To make the model more interesting, we shall consider a pseudo-gap with d-wave symmetry,

$$H_I = W \sum_{\vec{k},\vec{q}} \gamma_{\vec{k}} c^{\dagger}_{\vec{k}+\vec{q}} c_{\vec{k}} \phi_{\vec{q}},$$

where $\gamma_{\vec{k}} = \cos(k_x) - \cos(k_y)$ as before. Since the pairing does not change the momentum of particles, the reasoning for a normal state is quickly generalized by using the appropriate Nambu electron propagators.

The expression for the quenched-averaged Free energy takes the form

$$F = \frac{1}{\sqrt{2\pi}} \int_{-\infty}^{\infty} d\zeta e^{-\zeta^2/2} \left\{ -2T \sum_{n,\, k \in \frac{1}{2}\text{BZ}} \ln[\omega_n^2 + \varepsilon^2_{k+Q/2} + \Delta^2_{k+Q/2} + (\zeta W_k)^2] + \frac{\Delta^2}{g} \right\}. \tag{13}$$

The gap-equation is provided by the condition that the derivative of the free energy is equal to zero, $\frac{\partial F}{\partial \Delta} = 0$, and leads to a T_c-equation under condition $\Delta = 0$, i.e.

$$\frac{1}{g} = \frac{1}{\sqrt{2\pi}} \int_{-\infty}^{\infty} d\zeta e^{-\zeta^2/2} \left\{ 2T \sum_{n,\, k \in \frac{1}{2}\text{BZ}} \frac{\gamma_k^2}{\omega_n^2 + \varepsilon_k^2 + \zeta^2 W^2_{k-Q/2}} \right\}. \tag{14}$$

In this way, we have successfully eluded the diagrammatic derivation of the two-particle Green's function[12] and derived the main result in a compact and more general way.

Impurity effect

As a further application of these methods, let us now introduce disorder into the paired pseudo-gap model of the last section. The interplay of disorder and pseudo-gap formation on pairing is of particular interest in the context of underdoped cuprate superconductors.

In the presence of impurity scattering we have to include the additional terms in our Hamiltonian of the form

$$H' = \sum_{j,kk'\sigma} e^{-i(k-k')\cdot\vec{R}_j} V_{kk'} c^{\dagger}_{k\sigma} c_{k'\sigma}$$

In order to apply the methods of the last section, we make the important assumption that impurity potential $V_{kk'}$ does not scatter between different sides of the Fermi surface. Such scattering can still transfer electrons across the node, and thus remains severely pair-breaking. With this simplified assumption, the impurity scattering does not renormalize the strength of the pseudo-gap potential.

We can write now

$$F = \frac{1}{\sqrt{2\pi}} \int_{-\infty}^{\infty} d\zeta e^{-\zeta^2/2} F[\zeta, V_{kk'}]. \tag{15}$$

In this oversimplified model the effect of impurities reduces just to replacing of thin lines to there "thick", impurity dressed equivalents in the graphic representation for the two-particle Green's function in our model.

Using this result, we can generalize equation (14) to obtain

$$\frac{1}{g} = T_c \sum_n \int_{-\infty}^{\infty} \frac{d\zeta}{\sqrt{2\pi}} e^{-\frac{\zeta^2}{2}} \int_0^{\infty} \frac{d^2k}{(2\pi)^2} \frac{\gamma^2(\phi)}{\tilde{\omega}_n^2 + \varepsilon_k^2 + \zeta^2 W^2(\phi)}, \tag{16}$$

where $\tilde{\omega}_n$ is the usual renormalized Matsubara frequency $\tilde{\omega}_n = \omega_n + \Gamma sign\omega_n$, $\Gamma = \pi\rho N(0) V_{imp}^2$, and ρ is the impurity concentration.

For simplicity we assume the d-wave symmetry of the pseudogap as well as of the superconducting gap function

$$W(\phi) = W\gamma(\phi), \tag{17}$$

where $\gamma(\phi) = \sqrt{2}\cos(2\phi)$ is the angular dependence of the gap around the Fermi surface.

After integrating over energy and carrying out the Matsubara sum, we obtain the final equation for the critical temperature:

$$\ln\frac{T_c}{T_{c0}} = \int_0^{2\pi} \frac{d\phi}{2\pi} \gamma^2(\phi) \int_{-\infty}^{\infty} dx \Phi(x) \left\{ \Psi\left(\frac{1}{2}\right) - \Psi\left(\frac{1}{2} + \frac{\Gamma + ixW(\phi)}{2\pi T_c}\right) \right\}, \tag{18}$$

where the normalized distribution function $\Phi(x)$ can be written in terms of the modified Bessel function $K_0(x)$ as

$$\Phi(x) = \frac{1}{\sqrt{2\pi^3}} e^{-x^2/4} K_0\left(\frac{x^2}{4}\right).$$

Note, that impurity scattering produces pair breaking, by scattering particles across the nodes of the order parameter, while the pseudogapby contrast suppresses superconductivity by removing states from around the Fermi energy.

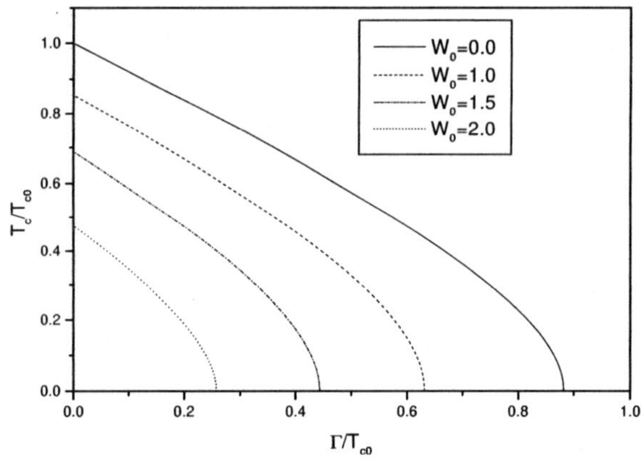

FIGURE 3. Critical temperature of a superconductor, normalized by the critical temperature of a pure superconductor without pseudogap T_{c0}, vs impurity scattering Γ in the presence of the pseudogap (W is an effective pseudogap width).

Fig.3 shows the variation of the transition temperature with impurity scattering rate for a set of different values of the pseudogap. In the limit where the pseudogap vanishes, Eq.(18) naturally reverts to the standard Abrikosov-Gorkov mean-field theory.

For illustrative purposes we now develop a phenomenology for the cuprate phase diagram (Fig.4). We consider a model, in which the bare superconducting transition temperature is a linear function of doping x, vanishing at $x = 0.3$ and the pseudogap is a rapidly reducing function of x ($W = W_{max}(1 - x/x_1)$), which vanishes at $x_1 \simeq 0.19$. The effect of the pseudogap is to suppress superconductivity at low doping, leading to a maximum in the transition temperature in the vicinity of where the pseudogap goes to zero. When disorder is introduced the superconducting region is pushed to higher doping.

DISCUSSION AND CONCLUSIONS

Our central idea was to point out that a large class of treatments of the pseudogap problem may be formulated in terms of a quenched average over a distribution of static classical order parameter fluctuations. The static nature of the fluctuations was stated by Lee, Rice and Anderson [1], so that the dynamics was forbidden to develop from the very beginning. If such an assumption makes sense or not is not a subject of our research. We address the problem of the type of field theoretical description of the basic concepts of FGM. The thrilling answer we get is that time-independent and non-

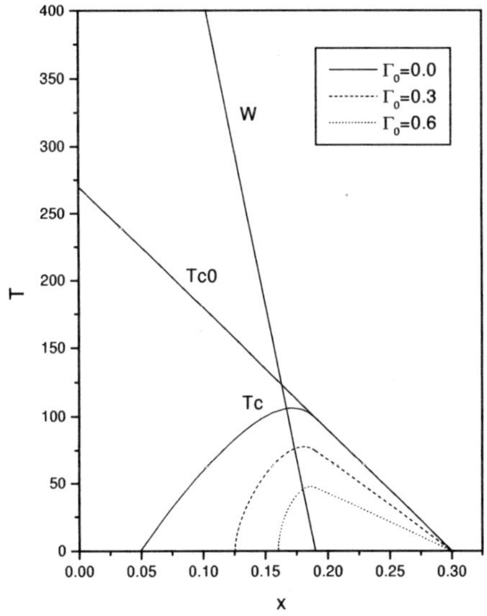

FIGURE 4. Toy phase diagram of cuprates. Here W is the effective energy scale of the pseudogap, for simplicity we assume that this quantity follows a linear dependence on doping, while we let the critical temperature of a pure superconductor T_{c0} vanish at $x \simeq 0.3$, since in most experiments superconductivity is not observed for doping value exceeding $0.4 \div 0.5$. It is clearly seen now how the superconducting area shrinks while the scattering by normal impurities increases (Γ is impurity scattering rate (34), normalized by T_{c0}). This phase diagram is surprisingly similar to the figure 5 of experimental paper [16].

interacting fluctuations of the order parameter are described by quenched disorder field theory.

The microscopic justification of a static approximation if of course much more complicated issue. Clearly, the fluctuations can be considered as static close to classical criticality while in the vicinity of quantum criticality the interactions are to be taken into account. Some derivation of the static limit in the spin-fluctuation model is provided by Chubukov et al [17]. But still the problem is far from being exactly solved and fully understood.

The main advantage of the method we develop is that it permits us to begin with the Free energy functional, from which all mean-field equation can be derived, rather than working directly with Green functions. Thus we are able to introduce superconducting pairing into the formalism without having to reconsider a whole class of two-particle correlation functions[12].

The disadvantages of the method is that we ignore the frequency dependence in the fluctuations and that the classical modes we are dealing with are strictly non-interacting. Both weaknesses are clearly issues of time scales. So long as the characteristic timescale of the electrons is longer than the characteristic time scale of the fluctuations, then it is reasonable to treat them in the way we have outlined. If by contrast, we are interested in situations where the characteristic electron time scales become very large, such as the vicinity to a quantum critical point, then we might expect the methods used here to become invalid.

Nevertheless, there are clearly a wide number of applications for this approach. One of the possible applications are transport properties of a pseudogap system; another interesting line of research could be to use the ideas of quenched- disorder to model the fluctuations of the density of states in the normal state of HTSC. The fluctuations of DOS could provide vast amount of information about the underlying quasiparticle physics and correlations, responsible for pseudogap formation.

ACKNOWLEDGMENTS

Discussions with M. V. Sadovskii, P. Chandra , T. Giamarchi, E. Kuchinskii, A. Chubukov are gratefully acknowledged. This work is supported in part by the National Science Foundation grant NSF-DMR 9983156 (PC). The hospitality of LVSM, KULeuven and the MPICPS, Dresden, Germany where parts of this work were carried out, is highly appreciated.

REFERENCES

1. Lee P., Rice M. and Anderson P. W.,*Phys. Rev. Lett.* **31**, 462 (1973).
2. Patton B. R. et al, *Phys. Rev. Lett.* **31**, 631 (1973); Fukuyama H. et al, *Phys. Rev. Lett.* **33**, 305 (1974); Luther A. et al, *Phys. Rev. Lett.* **33**, 589 (1974).
3. Brüesch P. et al, *Phys. Rev B* **12**, 219 (1975).
4. Grüner G., *Density Waves in Solids*, Addison Wesley (1994); Klemm R. A. "Origin of the pseudogap in HTSC", *lecture notes Nathiagali Summer college* July, 2000, to be published in Kalus' book.
5. Dessau D. S. et al.,*J. Electron Spectrosc. Rel. Phenom.* **117-118**, 265 (2001).
6. Lee Y. S. et al.,*Phys. Rev. B* **66**, 041104(R) (2002), Noh T. W. et al.,*Physica C* **364-365**, 480 (2001).
7. Sadovskii M. V., *Usp. Fiz. Nauk* **171**, 539 (2001); (*Phys. Uspekhi* **44**, 515 (2001)).
8. Timusk T., Statt B., *Rep. Prog. Phys.* **62**, 61 (1999).
9. Sadovskii M. V., *Zh. Eksp. Teor. Fiz.* **66**, 1720 (1974) [*Sov. Phys. JETP* **39**, 845 (1974)].
10. Sadovskii M. V., *Zh. Eksp. Teor. Fiz.* **77**, 2070 (1979) [*Sov. Phys. JETP* bf 50, 989 (1979)].
11. Kampf A. P., Schrieffer J. R.,*Phys. Rev. B* **41**, 6399 (1990), **42**, 7967(1990); Barzykin V., Pines D., *Phys. Rev. B* **52**,13585(1995); Chubukov A., Pines D., Stojkovic B., *J. Phys. Condensed Matter* **8**, 10017 (1996).
12. Posazhennikova A. I. , Sadovskii M. V., *JETP* **115**, 632 (1999).
13. Monien H., *Phys. Rev. Lett.* **87**, 126402 (2001).
14. De Gennes P. G. and Friedel J., *J. Phys. Chem Solids* **4**, 71 (1958); Fisher M. E. and Langer J. S., *Phys. Rev. Lett.*, **20**, 665 (1968); Chandra P., *J. Phys. Condens Matter* **50**, 10067-10080, (1989); Monthoux P., *Phil. Mag B.* **79**, 15 (1999).
15. Edwards S. F., Anderson P. W., *J. Phys.* **F5**, 965 (1975), also reference book Chowdhury D. *Spin glasses and other frustrated systems*, World Scientific, 1986, Singapore.
16. Naqib S. H., Cooper J. R., Tallon J. L., and Panagopoulos C., cond-mat/0209457.
17. Chubukov A. V., Pines D., Schmalian J., "A Spin fluctuation model of d-wave superconductivity", Review Chapter to appear in *The Physics of Conventional and Unconventional Superconductors* edited by K.H. Bennemann and J.B. Ketterson (Springer-Verlag).

Properties of a correlated nanoscopic chain close to the metal-insulator transition

Adam Rycerz* and Jozef Spałek*

*Marian Smoluchowski Institute of Physics, Jagiellonian University, Reymonta 4, 30-059 Kraków, Poland

Abstract. We review our recent results concerning the electronic properties of a nanoscopic system with inclusion of *long-range* Coulomb interactions that are obtained by combining *exact diagonalization* in the Fock space with an *ab initio* method (EDABI). The system evolution is discussed as a function of lattice parameter R. A transformation from a nanometal to localized spin system is observed. With the help of a *finite-size scaling* we determine the discontinuity of the momentum distribution at the Fermi points. Our results imply *Fermi-liquid-like behavior* for $R \lesssim 3a_0$ (a_0 is the Bohr radius). The quasiparticle mass is almost divergent at the localization threshold, where the particle distribution of the Fermi-Dirac type gets smeared out. The charge-gap evolution is compared with that of the Slater gap. Also, the single-particle spectral function and the corresponding density of states are presented. The analysis is performed using $1s$-like Gaussian-type STO-3G basis.

INTRODUCTION

Theoretical understanding of the electrical and mechanical properties of nanometer-scale systems containing $N = 1 \div 100$ atoms is crucial from the two principal prospectives. *First,* as the size of microelectronic devices continues to shrink, interest is focused on the nature of electron transport through essentially one-dimensional devices, such as nanowires and quantum point contacts, with properties that are qualitatively different from those of larger size. For instance, it has recently been discovered that the gold nanowires evolve spontaneously into chains of single atoms [1], which are surprisingly stable, have nearly ideal quantum values of the conductance $2e^2/h$ ($\approx 12.9\ \mathrm{k}\Omega^{-1}$), and are able to sustain enormous current densities. *Second,* it is important to determine the electronic structure of such small systems to see the evolution from the collection of atoms to a piece of a solid with the delocalized (*Bloch-type* or *Fermi-liquid-like*) states. The latter aspect opens up few separate problems for further investigations such as the presence of specific one-dimensional excitations (*holons* and *spinons*), the *Peierls distortion*, and the *Tomonaga-Luttinger liquid* states [2].

In this paper we apply a novel method of approach (EDABI) developed in our group recently [3, 4], in which we combine the *exact diagonalization* providing us with rigorous treatment of interparticle interaction, with an *ab-initio* optimization of a single-particle wave function in the interacting-system ground state. In this manner, the interaction and the single-particle terms in the Hamiltonian are treated on the same footing. Hence, the method is particularly useful for strongly correlated systems, when the kinetic (band) energy of electrons propagating throughout the system is at best comparable to the interaction among those particles.

CP678, *Lectures on the Physics of Highly Correlated Electron Systems VII: Seventh Training Course,* edited by A. Avella and F. Mancini

The major part of the paper contains a numerical study of correlated atomic chains containing $N = 6 \div 14$ atoms, with periodic boundary conditions, including *long-range* Coulomb interactions. Such nanoscopic systems are interesting objects of study for few reasons. First of all, the role of long-range Coulomb interaction in 1D is crucial, as the charge screening becomes less effective. The exact solutions of the parametrized models with inclusion of intersite interactions [5] prove the existence of the metal-insulator transition (MIT) for the half-filled-band case, in contradistinction to the Hubbard-model solution [6]. Separate questions concern the appearance of the Tomonaga-Luttinger liquid behavior [7] in the metallic state, and the so-called *dielectric catastrophe* [8] in the insulating state. *Last but not least*, the questions of the metallicity appearance in a correlated nanoscopic atomic chain or ring and that concerning the applicability of the concept of *Mott insulator* in the situation when the separation between the quasiparticle states (in a finite system) is of the same order of magnitude as the Hubbard gap must be addressed separately. Similar problems are encountered in the weak-coupling renormalization group analysis [9], which predicts that *umklapp* processes lead any half-filled 1D system with two Fermi points to an insulating state. This is the reason for discussing in detail here a rigorous numerical evidence for the transition from a nanometal to the localized spin system.

THE EXACT DIAGONALIZATION AB-INITIO (EDABI) METHOD

The idea of EDABI is illustrated on the block–diagram exhibited in Figure.1. We start from choosing the initial Wannier basis set $\{w_i(\mathbf{r})\}$, composed of atomic-like wave functions with their radii a_i. Next, we write down the system Hamiltonian in the second-

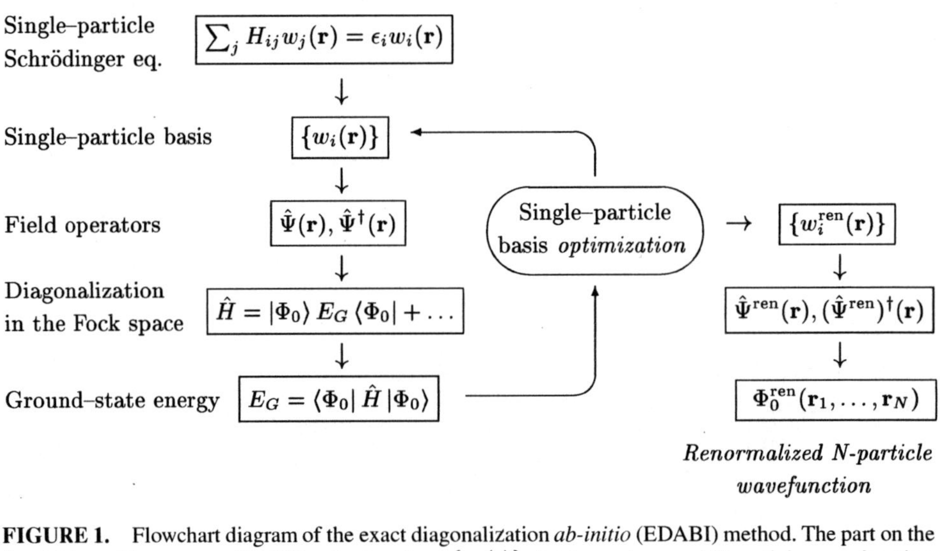

FIGURE 1. Flowchart diagram of the exact diagonalization *ab-initio* (EDABI) method. The part on the far right provides *renormalized* Wannier functions $\{w_i(\mathbf{r})\}$, field operators, and N-particle wave function.

quantization form and determine the ground-state energy E_G and the corresponding state $|\Phi_0\rangle$ by e.g. Lanczos procedure. Than, the Wannier basis set $\{w_i(\mathbf{r})\}$ is optimized with respect to the atomic radii a_i, contained in the atomic wave functions composing $w_i(\mathbf{r})$, until the minimal ground-state energy E_G is reached [3] for given lattice parameter R.

ONE-DIMENSIONAL NANOCHAIN

Model Hamiltonian and the optimized ground state energy

We now consider a nanoscopic linear chain of $N = 6 \div 14$ atoms, each containing a single valence electron (hydrogenic-like atoms), including *all* long-range Coulomb interactions (3– and 4–site terms are treated exactly in the Gaussian STO-3G basis). We focus here on the so-called *Extended Hubbard Model* with Hamiltonian of the form

$$H = \varepsilon_a^{\text{eff}} \sum_i n_i + t \sum_{i\sigma} \left(a_{i\sigma}^\dagger a_{i+1\sigma} + \text{h.c.} \right) + U \sum_i n_{i\uparrow} n_{i\downarrow} + \frac{1}{2} \sum_{ij}{}' K_{ij} \delta n_i \delta n_j, \quad (1)$$

where $\delta n_i \equiv n_i - 1$, $\varepsilon_a^{\text{eff}} = \varepsilon_a + N^{-1} \sum_{i<j} (2/R_{ij} + K_{ij})$ (in Ry) is the effective atomic level defined including the mean–field part of the Coulomb interaction, R_{ij} is the distance between the atoms i and j, t is the nearest-neighbor hopping integral, U and $K_{ij} = K_{|i-j|}$ are the intra– and inter–site Coulomb repulsion amplitudes, respectively. The model parameters, corresponding to the lattice spacing $R/a_0 = 1.5 \div 10$ (a_0 is the Bohr radius) and the optimal values of the inverse orbital size α_{min} (see below), are presented in Table 1. We also provided values of the *correlated hopping V* and the *Heisenberg-exchange integral J* to show one could disregard the corresponding terms in (1).

The Hamiltonian (1) is diagonalized in the Fock space with the help of Lanczos technique. As the microscopic parameters $\varepsilon_a^{\text{eff}}$, t, U, and K_{ij} are calculated numerically in the Gaussian STO-3G basis, the inverse orbital size α of the $1s$-like state is subsequently optimized to obtain the ground state energy E_G as a function of the interatomic distance R. We have already shown [4] that such a combined exact diagonalization - *ab initio* study of the one dimensional system provides precise values of the localization threshold, the electron–lattice couplings, and the dimerization magnitude. However, the convergence

TABLE 1. Microscopic parameters (in Ry) of the nanochain, calculated in the Gaussian STO-3G basis. Corresponding values of the optimal inverse orbital size α_{min} are also provided.

R/a_0	$\alpha_{\text{min}} a_0$	$\varepsilon_a^{\text{eff}}$	t	U	$V/10^{-3}$	$J/10^{-3}$	K_1	K_2	K_3
1.5	1.363	0.100	-0.8309	2.054	-43.93	30.92	1.165	0.667	0.447
2.0	1.220	-0.550	-0.4423	1.733	-23.81	21.06	0.911	0.501	0.334
2.5	1.122	-0.797	-0.2644	1.531	-14.95	15.13	0.750	0.401	0.267
3.0	1.062	-0.902	-0.1708	1.407	-10.99	10.91	0.639	0.334	0.222
4.0	1.013	-0.971	-0.0796	1.291	-8.74	4.93	0.493	0.250	0.167
5.0	1.004	-0.987	-0.0374	1.258	-7.07	1.57	0.399	0.200	0.133
8.0	1.000	-0.992	-0.0027	1.247	-0.70	$5\cdot10^{-3}$	0.250	0.125	0.083
10.0	1.000	-0.992	$-2.5\cdot10^{-3}$	1.247	-0.05	$2\cdot10^{-5}$	0.200	0.100	0.067

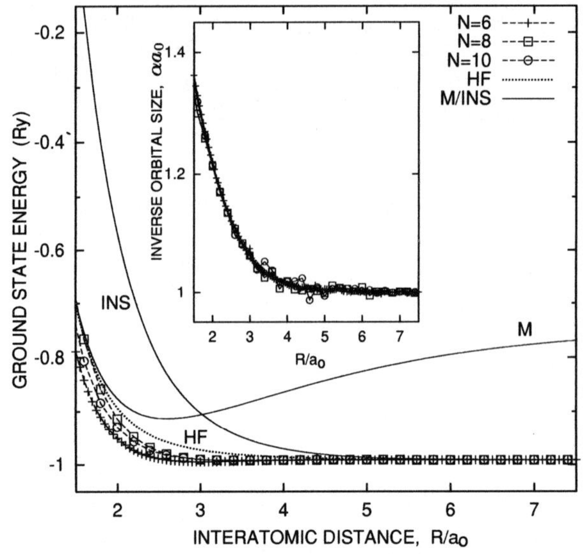

FIGURE 2. The ground state energy per atom for the linear chain of $N = 6 \div 10$ atoms with periodic boundary conditions. The Gaussian-type orbitals (STO-3G basis) have been used. The energies of the *ideal* metallic (M), *ideal* insulating (INS), and Hartree-Fock (HF) solutions for an *infinite* system are shown for comparison. The *inset* provides the optimal inverse orbital size $\alpha_{\min}$.

of the results obtained with the Slater-type orbitals is not sufficient to be able to perform a finite–size scaling with the lattice size $N \to \infty$. This is because, when calculating the microscopic parameters in the single–particle (Wannier) basis, one ignores the three– and the four–site interaction terms, that represents an uncontrolled approximation.

In contrast, for the Gaussian-type orbitals we can treat the three– and four–site terms exactly. Their effects on the convergence of the results for the ground-state energy E_G and the optimal inverse orbital size $\alpha_{\min}$ are shown in Fig.2 for $N = 6 \div 10$ atoms. These results were used to extrapolate the value of the variational parameter $\alpha_{\min}$ to larger N to speed up the computations. Fig.2 illustrates also the *Hubbard localization criterion*. Namely, for the interatomic distance $R \approx 3a_0$ the energy of the *ideal metallic* state (M), determined as $E_G^{\mathrm{M}} = \varepsilon_a^{\mathrm{eff}} - 4|t|/\pi + N^{-1}\sum_{i<j} K_{ij} \langle \delta n_i \delta n_j \rangle$, where $\langle \delta n_i \delta n_j \rangle = -2\sin^2(\pi|i-j|/2)/(\pi|i-j|)^2$, crosses over to that representing the *Mott insulating* state (INS), with $E_G^{\mathrm{INS}} = \varepsilon_a^{\mathrm{eff}}$. The critical value of R is very close to obtained for the $1s$ Slater-type orbitals [4]. The validity of the Hubbard criterion for this one-dimensional system is controversial, as the energy of the antiferromagnetic (so *Slater-type*) Hartree-Fock solution (HF) is lower than those of M and INS.

Spin and charge correlation functions

In Fig.3 we present the set of basic correlation functions for the nanochain described by Hamiltonian (1), obtained with the EDABI method. The spin-spin correlation function $\langle \mathbf{S}_i \cdot \mathbf{S}_j \rangle$ illustrates the *antiferromagnetic* nature of the system ground state, particularly for $R/a_0 \gtrsim 3$ (*cf.* Fig.3a). The *long-range* character of this order is confirmed by the nearest neighbor spin-spin correlation function $|\langle \mathbf{S}_i \cdot \mathbf{S}_{i+1} \rangle|$, which is almost independent of R, whereas the single-particle correlation function $| < a_i^\dagger a_{i+1} > |$ and the average double occupancy $\langle n_{i\uparrow} n_{i\downarrow} \rangle$, decrease exponentially with the increasing R (*cf.* Fig.3b, *note* the logarithmic scale of the y-axis). The analysis of the density-density correlation function $\langle \Delta n_i \Delta n_j \rangle$, (where $\Delta n_i \equiv n_i - \bar{n}$ and $\bar{n} = N_e/N$ is the band filling) provides the evidence that the system gradually transforms from a nanoscopic metal to an insulator with growing R. In the half-filed band case (*cf.* Fig.3c), $\langle \Delta n_i \Delta n_j \rangle$ diminishes significantly faster with the distance $|i-j|$ for larger R, whereas for the system containing 2 holes (*cf.* Fig.3d) an additional maximum appears at $|i-j| = N/2$ for $R/a_0 \gtrsim 3$; this can be interpret as follows: in the metallic-like regime the added holes are distributed almost uniformly throughout the system to minimize their band energy, whereas in the insulating limit (large R) the Coulomb repulsion dominates, so the minimal energy respects the holes located at the opposite sides of the system (i.e. at $|i-j| = N/2$ for the periodic boundary conditions).

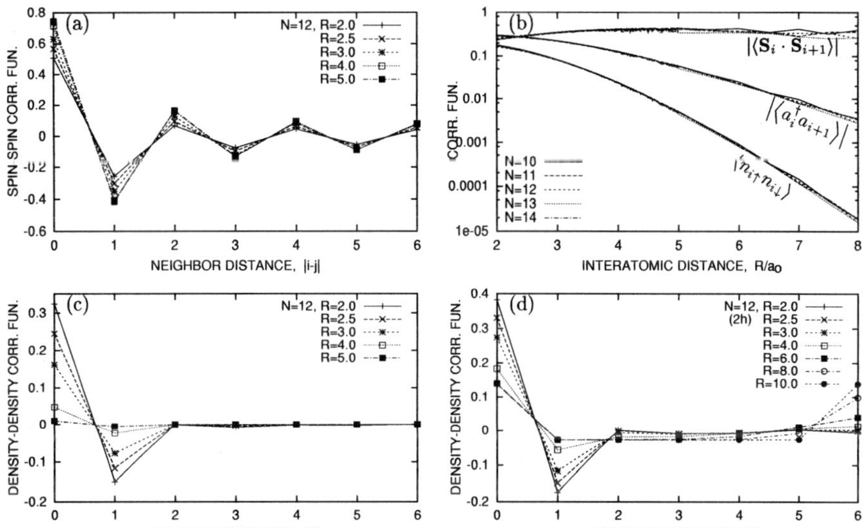

FIGURE 3. Basic correlation functions for nanochain: (a) spin-spin correlation function $\langle \mathbf{S}_i \cdot \mathbf{S}_j \rangle$ vs. distance $|i-j|$; (b) absolute values of the nearest neighbor spin-spin $|\langle \mathbf{S}_i \cdot \mathbf{S}_{i+1} \rangle|$, single-particle (hopping) correlation function $| < a_i^\dagger a_{i+1} > |$, and average double occupancy per site $\langle n_{i\uparrow} n_{i\downarrow} \rangle$, all vs. interatomic distance R/a_0; (c) density-density fluctuation correlation function $\langle \Delta n_i \Delta n_j \rangle$ vs. distance $|i-j|$ for the half-filled band case, and (d) for the system containing two holes ($N_e = N-2$).

Quantum critical behavior of quasiparticle states

The the electron momentum distribution $n_{\mathbf{k}\sigma} \equiv < a^{\dagger}_{\mathbf{k}\sigma} a_{\mathbf{k}\sigma} >$ is shown in Fig.4 for $N = 6 \div 14$ atoms. The continuous lines represent the scaling formula

$$n_{\mathbf{k}\sigma} = \frac{1}{2} + \mathrm{sgn}(k - k_F)[\alpha |k - k_F|^2 + \beta |k - k_F| - C], \qquad (2)$$

with fitted parameters α, β, and C. We obtain the critical distance R_{crit} depending on value of N (*cf.* Table 2), at which the scaled Fermi-ridge discontinuity $\Delta n_F = 2C$ disappears, signaling the electron localization. No Luttinger-liquid effects were detected in the system with one-electron per atom. A clear metallic behavior is evidenced by the presence of the Fermi ridge, since the energy-level spread produces remarkably smaller variations of $n_{\mathbf{k}\sigma}$ than Δn_F for $R < R_{\mathrm{crit}}$.

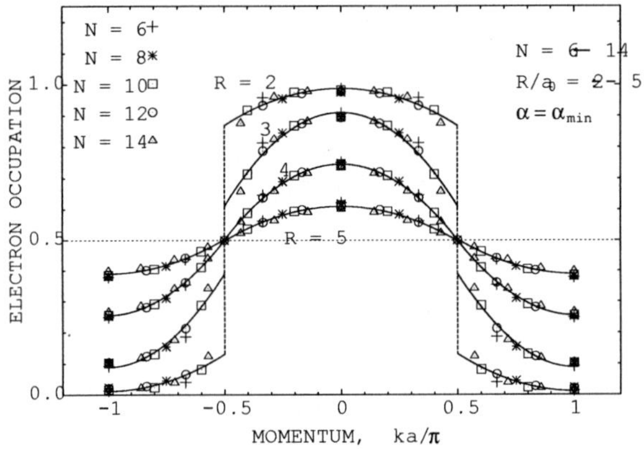

FIGURE 4. Momentum distribution $n_{\mathbf{k}\sigma}$ for electrons on a linear chain of $N = 6 \div 14$ atoms; the interatomic distance R is specified in units of Bohr radius a_0. The continuous line represents the parabolic interpolation (2), which is of the same type for both $k > k_F$ and $k < k_F$ regimes.

With the help of the interpolation formula (2) we can determine the quasiparticle mass enhancement $m^{\star}/m_B$ by equating it with $(\Delta n_F)^{-1}$. This means that we relate the Fermi-liquid wave-function renormalization factor Z to Δn_F. The parameters resulting from the postulated critical behavior $m^{\star}/m_B = A|R - R_{\mathrm{crit}}|^{-\gamma}$ are listed in Table 2. The values of critical exponent are in agreement with those obtained for Slater-type orbitals [4]. It would be very important to test experimentally this result, as it represents the critical exponent behavior of the Fermi discontinuity $\Delta n_F \sim |R - R_{\mathrm{crit}}|^{\gamma}$ near the localization border approached from the metallic side.

TABLE 2. The fitted parameters of the critical exponent formula for the quasiparticle mass scaling $m^{\star}/m_B \equiv (\Delta n_F)^{-1} = A|R - R_{\mathrm{crit}}|^{-\gamma}$, m_B is effective band mass.

N	A	a_{crit}	γ	N	A	a_{crit}	γ
8	3.5 ± 2.6	4.18(1)	1.39(14)	12	7.0 ± 1.1	4.030(5)	1.29(10)
10	6.5 ± 2.3	3.900(3)	1.32(7)	14	7.4 ± 0.4	3.56(2)	1.58(7)

Finite-size-scaling estimate of MIT

We discuss now finite-size scaling to determine whether the system shows Fermi- or non-Fermi-liquid-like behavior at the localization threshold. But first, we briefly summarize the basic features of the homogeneous conductors, following [10].

In a Fermi liquid quasiparticles are well defined, which implies: *i*) a nonzero quasiparticle pole strength Z in a single-particle propagator

$$G(\mathbf{k},\omega) = \frac{Z}{\omega - (\varepsilon_{\mathbf{k}} - \varepsilon_F) + i\delta \mathrm{sgn}(k - k_F)} + G_{\mathrm{incoh}}(\mathbf{k},\omega) \quad (3)$$

(with the standard notation), and *ii*) a finite discontinuity Δn_F of the momentum distribution $n_{\mathbf{k}} = < a_{\mathbf{k}}^{\dagger} a_{\mathbf{k}} >$ at the Fermi ridge ($k = k_F$). Moreover, in a Fermi-liquid the discontinuity $\Delta n_{\mathbf{k}}$ is equal to the quasiparticle mass renormalization Z.

In a system of finite size N, however, $\Delta n_{\mathbf{k}}(N)$ is always nonzero due to the finite distance between neighboring momenta in the first Brillouin zone. Nevertheless, its asymptotic behavior for $N \to \infty$ can belongs to one of the three categories: *a*) $\Delta n_{\mathbf{k}}(N) \to A^{-N}$ *(insulator)*, *b*) $\Delta n_{\mathbf{k}}(N) \to N^{-\theta}$ *(marginal conductor)*, and *c*) $\Delta n_{\mathbf{k}}(N) \to$ const *(normal conductor)*. Although, in principle, it is straightforward to justify whether a correlated electron liquid is a Fermi liquid, provided finite-size scaling can be performed on $\Delta n_{\mathbf{k}}(N)$, one should carefully check the condition *i*), concerning the quasiparticle pole strength Z in Eq. (3). This part of analysis can also be done with the help of Lanczos diagonalization technique [11], details will be presented elsewhere.

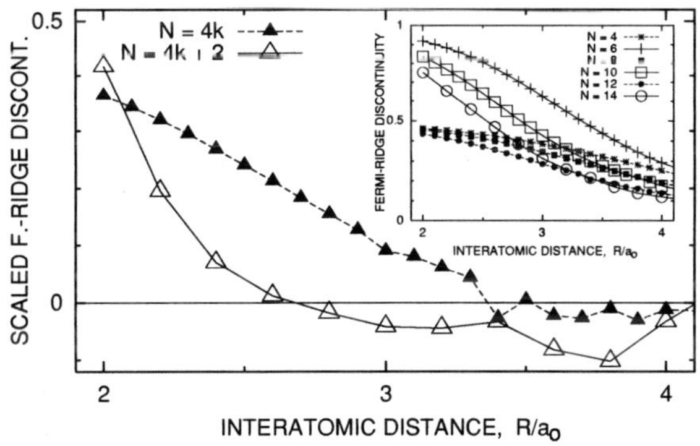

FIGURE 5. Fermi-ridge discontinuity $\Delta n_{\mathbf{k}}$ obtained through the finite-size scaling with $1/N \to 0$, and the original data *(inset)* for $N = 4 \div 14$.

In the Fermi-liquid regime *c*), one can expand $\Delta n_{\mathbf{k}}(N)$ in powers of $1/N$, to obtain the following extrapolation formula for large lattice size N:

$$\Delta n_{\mathbf{k}}(N) = \Delta n_{\mathbf{k}} + a(1/N) + b(1/N)^2 + \ldots. \quad (4)$$

The extrapolated values of $\Delta n_{\mathbf{k}}$ for $1/N \to 0$, are presented in Fig.5 as a function of the lattice parameter R. The source data for $N = 4 \div 14$, used for finite-size scaling, are shown in the *inset*. Both groups of curves for $N = 4k$ and $N = 4k + 2$ show a clear transition-like behavior for $R_{\mathrm{crit}} \approx 3a_0$, in contradistinction to the corresponding numerical results for *pure* Hubbard model [10]. The presented value of R_{crit} is in agreement with that obtained from the Hubbard localization criterion illustrated in Fig.2.

Similar extrapolation with $1/N \to 0$ were performed for the charge-gap defined as

$$\Delta E_C(N) = E_G^{N+1} + E_G^{N-1} - 2E_G^N, \tag{5}$$

where $E_G^{N_e}$ is the ground-state energy of the system containing N_e electrons. The corresponding numerical results are shown in Fig.6. Although the value of ΔE_C is always nonzero (for any R), it reaches the order of extrapolation error for the lowest examined value of $R = 1.5a_0$. It also becomes significantly smaller than the corresponding Hartree-Fock (HF) value for $R \lesssim 4.5a_0$, that suggests a dielectric catastrophe discussed for parametrized models [8]. This hypothesis is substantiated further by a nonzero conductivity of the system and will be discussed elsewhere.

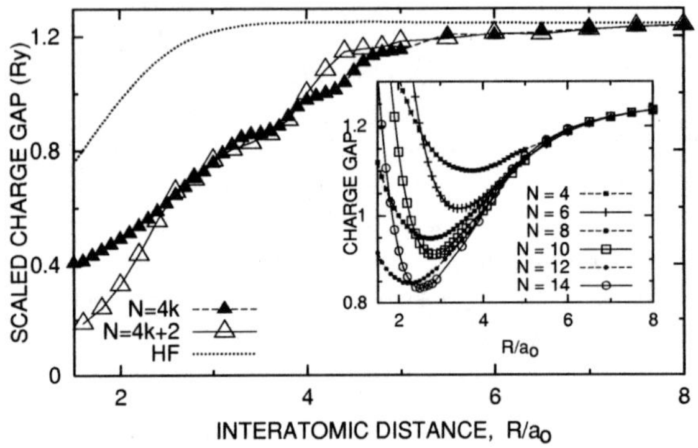

FIGURE 6. Charge energy gap obtained through finite-size scaling for the chains of $N = 4 \div 14$ atoms. The corresponding Hartree-Fock value (the Slater gap) for an infinite system is also drawn (*dotted* line). *Inset* provides an original data used for the scaling.

Spectrum of single-particle excitations

The evolution of the single-particle density of states $N(\omega) = \sum_{\mathbf{k}} A(\mathbf{k}, \omega)$, where $A(\mathbf{k}, \omega)$ is the spectral function, with the increasing R is shown in Fig.7. The spectral functions are defined in the standard manner

$$A(\mathbf{k}, \omega) = \sum_n \left| \langle \Phi_n^{N\pm1} | a_{\mathbf{k}\sigma}^\dagger | \Phi_0^N \rangle \right|^2 \delta\left[\omega - \left(E_n^{N\pm1} - E_0^N\right)\right], \tag{6}$$

where the upper (lower) sign correspond to $\omega > \mu$ ($\omega < \mu$), respectively, $|\Phi_n^N\rangle$ is the n-th eigenstate of the system containing N particles, E_n^N is the corresponding eigenenergy, $\langle \Phi_n^{N\pm1} | a_{\mathbf{k}\sigma}^\dagger | \Phi_0^N \rangle$ terms are calculated within the Lanczos technique set up by Dagotto [11]. For plotting purposes we have used analytical representation of Diracs delta $\delta(x) \to (1/\pi)\varepsilon/(x^2 + \varepsilon^2)$ with $\varepsilon = 0.01$ Ry. In the metallic range ($R/a_0 \lesssim 2$) the quasiparticles

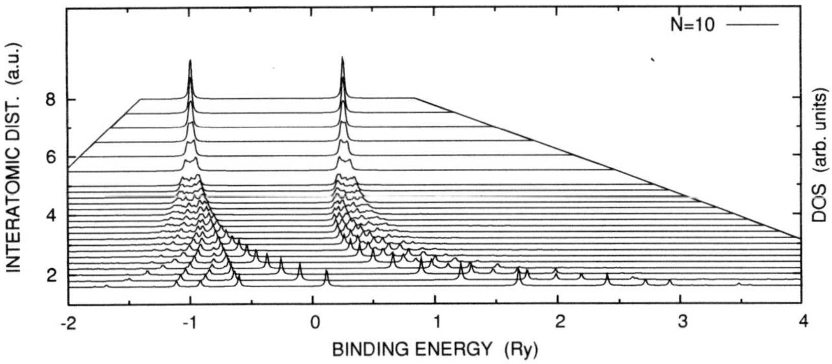

FIGURE 7. Density of states $N(\omega) = \sum_{\mathbf{k}} A(\mathbf{k}, \omega)$ for the linear chain of $N = 10$ atoms.

are well defined, but incoherent part of the spectra grows systematically with raising R. In effect, at the intermediate region of $a/a_0 = 3 \div 4$ the lower- and the upper Hubbard bands are formed, and then they continuously evolve to discrete atomic levels located at the positions $\omega = \varepsilon_a$ and $\omega = \varepsilon_a + U$ as $R \to \infty$. Those limiting peak positions correspond to the ground H^0 and excited H^- atomic states. Probably the most interesting feature of this spectrum is its incoherent nature for $R \sim 3a_0$, where the band and the interaction energies are comparable.

Additional illustration of the spectrum evolution is provided by the spectral functions plotted in Fig.8. The chain of $N = 10$ atoms (or generally $N = 4k + 2$), for which there are no quasiparticle states at the Fermi level, could be regarded as a nanoscopic metal

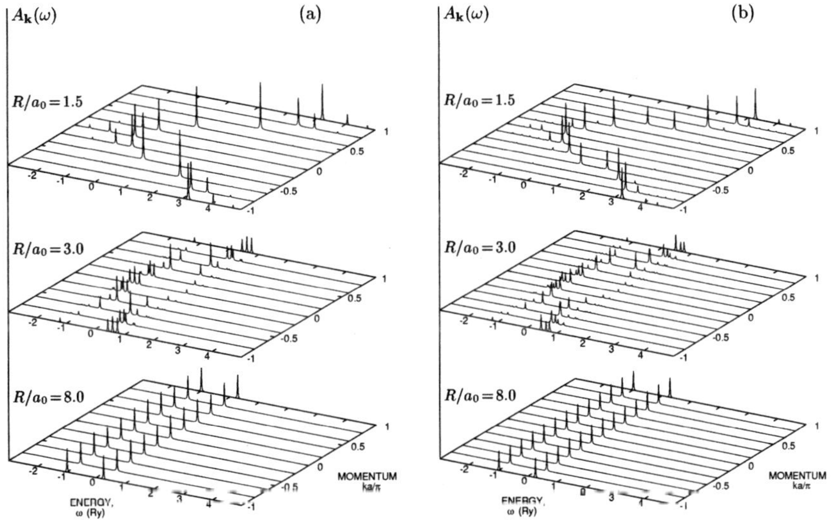

FIGURE 8. Spectral functions $A(\mathbf{k}, \omega)$ for the chains of $N = 10$ (a) and $N = 12$ (b) atoms. The values of the lattice parameter a are specified in the units of Bohr radius a_0.

in the small R limit, since the separation between the single-particle levels are mainly determined by the geometrical quantization of their momenta (cf. *top panel* in Fig.8a, where a clear band dispersion can be seen). For $N = 12$ ($N = 4k$), however, the situation is slightly different, as the states at the Fermi level are split (cf. *top panel* in Fig.8b). This phenomenon looks like the presence of Slater-split states at the Fermi level, but may also appear as a signature of the charge–spin separation in 1D system (*cf.* Voit, Ref. [7]).

SUMMARY

We have calculated ground-state correlation functions and spectral properties of a nanochain within the novel method (EDABI) of approach to the strongly correlated electron systems. The proposed microscopic criterion for the *crossover* transition from the nanometal to the localized spin system in one-dimension has been verified and the spectral properties were also obtained with the help of the Lanczos technique for Gaussian-type basis set. A quantum critical behavior at the metal-insulator boundary is indicated.

ACKNOWLEDGMENT

One of the authors (A.R.) thanks to Profs. F.F. Assaad and Yu.A. Izyumov for informative discussions during VII Training Course in Vietri. Discussions with our colleagues, Krzysztof Rościszewski, Marcin Mierzejewski, and Maciek Maśka are appreciated. The work was supported by the Committee for Scientific Research (KBN) of Poland, Grant Nos. 2P03B 050 23 and 2 P03B 064 22.

REFERENCES

1. Yanson, A.I., *et al.*, *Nature (London)* **395**, 783 (1998); Ohnishi, H., Kondo, Y., and Takayanagi, K., *ibid.* 780. For the wide selection of metals and density functional calculations, see: Smit, R.H.M, *et al.*, *Phys. Rev. Lett.* **87**, 266102 (2001); Bahn, S.R., and Jacobsen, K.W., *ibid.* 266101.
2. Fisher, M.P.A., and Glazman, L.I., in *Mesoscopic Electron Transport*, eds Sohn, L.L., Kouwenhoven, L.P., and Schön, G., NATO ASI Ser. E, Vol. 345, Kluwer Academic, Dortrecht, p.331 (1997).
3. Spałek, J., *et al.*, *Phys. Rev. B* **61**, 15676 (2000); Rycerz, A., and Spałek, J., *ibid.* **63**, 073101 (2001). For a brief review, see: Spałek, J., *et. al.*, *Acta. Phys. Pol. B* **31**, 2879 (2000); *ibid.* **32**, 3189 (2001).
4. Spałek, J., Rycerz, A., *Phys. Rev. B* **64**, R161105 (2001); Rycerz, A., Spałek, J., Podsiadły, R., and Wójcik, W., in *Lectures on the Physics of Highly Correlated Electron Systems VI*, edited by F. Mancini, AIP Conf. Proc. Vol. 629, p.213 (2002). For a treatment of both electronic and lattice degrees of freedom see: Rycerz, A., Spałek, J., *Phys. Rev. B* **65**, 035110 (2002).
5. Strack, R., and Vollhardt, D., *Phys. Rev. Lett.* **70**, 2637 (1993); Arachea, L., and Aligia, A.A., *ibid.* **73**, 2240 (1994); Michielsen, K., Raedt, H.De, and Schneider, T., *ibid.* **68**, 1410 (1992).
6. Lieb, E.H., and Wu, F.Y., *Phys. Rev. Lett.* **20**, 1445 (1968); see also Andrei, N., *these proceedings*.
7. Schulz, H.J., *Phys. Rev. Lett.* **71**, 1864 (1993); Voit, J., *Rep. Prog. Phys.* **57**, 977 (1995).
8. Resta, R., Sorella, S., *Phys. Rev. Lett.* **82**, 370 (1999); Aebisher, C., *et al.*, *ibid.* **86**, 468 (2001).
9. Fabrizio, M., *Phys. Rev. B* **54**, 10054 (1996); see also Voit, Ref. [7].
10. Sorella, S., *et al.*, *Europhys. Lett.* **12**, 721 (1990); see also Voit, Ref. [7].
11. Dagotto, E., *Rev. Mod. Phys.* **66**, 763 (1994); Prelovšek, P., and Zotos, X., in *Lectures on the Physics of Highly Correlated Electron Systems VI*, ed. by F. Mancini, AIP Conf. Proc. Vol. 629, p.161 (2002).

Scaling, Entanglement, and Quantum Phase Transitions

A. Osterloh*, L. Amico*, G. Falci* and R. Fazio†

*NEST-INFM & Dipartimento di Metodologie Fisiche e Chimiche (DMFCI), viale A. Doria 6, I-95125 Catania, Italy.
†NEST-INFM & Scuola Normale Superiore, I-56127 Pisa, Italy.

Abstract. We study the entanglement near a quantum phase transition in a class of exactly solvable one-dimensional quantum magnets, the anisotropic XY models. The critical behaviour (scaling) is reflected by the derivatives of the bipartite entanglent, quantified by the concurrence. However, we found a profound difference between classical correlation and non-local quantum correlation: The range of the concurrence is found to behave in a non-universal way, being extremely short-ranged for the Ising model. This range tends to infinity towards the isotropic XY model.

INTRODUCTION

A quantum mechanical system possesses correlations that do not have a classical counterpart. This phenomenon, called entanglement, is one of the most astonishing features of quantum mechanics. It is the resource for quantum speed-up in computation and communication. The role of the entanglement at a phase transition goes beyond the knowledge of the statistical mechanics of the system. A complete classification of the critical many-body state requires the introduction of concepts from quantum information theory. Here we connect the theory of critical phenomena with quantum information by exploring the entangling resources of a system close to its quantum critical point. Classical phase transitions occur when a physical system below a critical temperature reaches a state characterized by a macroscopic order [1]. Driven by quantum fluctuations, phase transitions can occur also at zero temperature where they are induced by the change of an external parameter or coupling constant [2]. Examples are transitions in quantum Hall systems [3], localization in Si-MOSFETs [4] or the superconductor-insulator transition in two-dimensional systems [5, 6]. Close to the quantum critical point the system is characterized by long-range correlations which manifest in the singular behavior of different physical observables. Both classical and quantum critical points are governed by a diverging correlation length. A quantum mechanical system possesses additional correlations that do not have a classical counterpart. This phenomenon, called entanglement [7], is one of the most astonishing features of quantum mechanics. It is the resource for quantum speed-up in computation and communication [8]. The role of the entanglement at a phase transition goes beyond the knowledge of the statistical mechanics of the system. A complete classification of the critical many-body state requires the introduction of concepts from quantum information theory [10]. The aim of our work [11] is to connect the theory of critical phenomena with quantum information by exploring the

CP678, *Lectures on the Physics of Highly Correlated Electron Systems VII: Seventh Training Course*, edited by A. Avella and F. Mancini

entangling resources of a system close to its quantum critical point. There are various questions that emerge in the study of this problem. Since the ground state wave-function undergoes qualitative changes at a quantum phase transition, it is important to understand how its genuine quantum aspects evolve throughout the transition. Will entanglement between distant subsystems be extended over macroscopic regions, as correlations are? Will it carry distinct features of the transition itself and show scaling behavior? Answering these questions is important for a deeper understanding of quantum phase transitions and from the perspective of quantum information theory as well. Therefore results that bridge between these two areas of research are of great relevance. We study a set of localized spins coupled through exchange interaction and subject to an external magnetic field (we consider only spin-1/2 particles), a paradigm model central both to condensed matter and information theory and subject to intense study. O'Connors and Wootters showed that in the Heisenberg chain the maximization of the entanglement at zero temperature is related to the energy minimization [12]. It is known that Werner states [13] can be generated in a one dimensional XY model [14] and that temperature and magnetic field can increase the entanglement of the systems as shown for the Ising and Heisenberg model in [15, 16]. Finally we mention the study on the role of the entanglement in the density matrix renormalization group flow and the introduction of entanglement-preserving renormalization schemes [17]. In this Letter we address the problem of the relation between macroscopic order, classical correlations and quantum correlations. Therefore we analyze the entanglement near the critical point of the XY model in a transverse field. Due to the universality principle, meaning that the critical behavior depends only on the dimension of the system and the symmetry of the order parameter, our results have much broader validity. The concurrence, a measure for the resource entanglement [22], is calculated exactly for generic number of spins. We find that in the vicinity of a quantum phase transition the entanglement can be classified in the framework of scaling theory. On the other hand this analysis provides a clear distinction between the role of entanglement and correlations in quantum systems close to a critical point.

THE MODELS

The system under consideration is a spin-1/2 ferromagnetic chain with an exchange coupling J in a transverse magnetic field of strength h. The Hamiltonian is

$$H = -\frac{J}{2}\sum_{i=1}^{N}(1-\gamma)\sigma_i^x\sigma_{i+1}^x + (1+\gamma)\sigma_i^y\sigma_{i+1}^y - h\sum_{i=1}^{N}\sigma_i^z \tag{1}$$

where σ^a are the Pauli matrices ($a = x,y,z$) and N is the number of sites. We assume periodic boundary conditions. It is convenient, for later purposes, to define the dimensionless coupling constant $\lambda = J/2h$. If the anisotropy parameter γ continuously connects the quantum Ising model for $\gamma = 1$ with the isotropic XY model for $\gamma = 0$. For all the interval $0 < \gamma \leq 1$ the models belong to the Ising universality class and for $N = \infty$ they undergo a quantum phase transition at the critical value $\lambda_c = 1$. The magnetization $\langle\sigma^x\rangle$

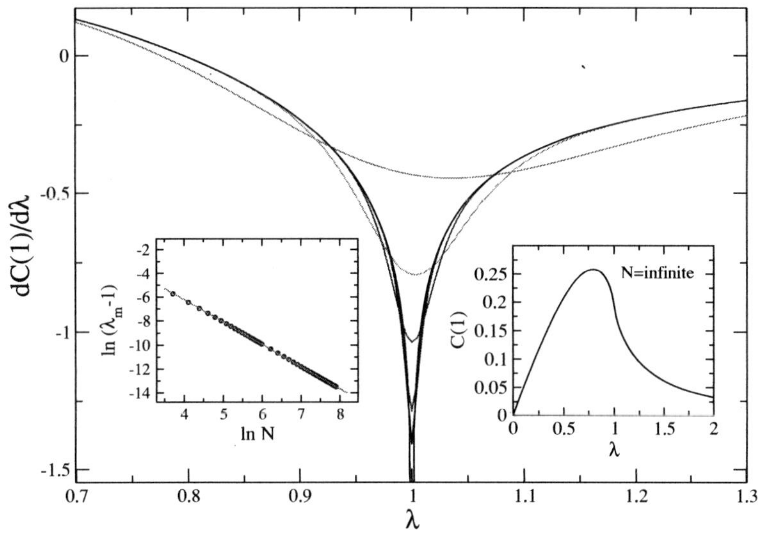

FIGURE 1. The change in the ground state wave-function in the critical region is analyzed considering $\partial_\lambda C(1)$ as a function of the reduced coupling strength λ. The curves correspond to different lattice sizes $N = 11, 41, 101, 251, 401, \infty$. We choose N odd to avoid the subtleties connected with boundary phases. From [11], with kind permission from Nature.

is different from zero for $\lambda > 1$ and it vanishes at the transition. On the contrary the magnetization along the z-direction $\langle\sigma^z\rangle$ is different from zero for any value of λ. At the phase transition the correlation length ξ diverges as $\xi \sim |\lambda - \lambda_c|^\nu$ with $\nu = 1$ [19, 20]. We confine our interest to the entanglement between two spins on position i and j in the chain. All the information needed is contained in the reduced density matrix $\rho(i,j)$ obtained from the ground state wave-function after all the spins except those at positions i and j have been traced out. The resulting $\rho(i,j)$ represents a mixed state of a bipartite system for which a good deal of work has been devoted to quantify its entanglement [21, 22]. We use the concurrence between sites i and j, related to the *entanglement of formation*, and defined as [22]

$$C(i,j) = \max\{0, r_1(i,j) - r_2(i,j) - r_3(i,j) - r_4(i,j)\} \tag{2}$$

In Eq.(2) $r_\alpha(i,j)$ are the square roots of the eigenvalues of the product matrix $R = \rho(i,j)\tilde{\rho}(i,j)$ in descending order; the spin flipped matrix $\tilde{\rho}$ is defined as $\tilde{\rho} \doteq \sigma^y \otimes \sigma^y \rho^* \sigma^y \otimes \sigma^y$. In the previous definition, the eigenstates of $\sigma^z \in \{|\uparrow\rangle |\downarrow\rangle\}$ should be used. Translation invariance implies that $C(i,j) = C(|i-j|)$. The concurrence will be evaluated as a function of the relative position $|i-j|$ between the spins and the distance $|\lambda - \lambda_c|$ from the critical point. The structure of the reduced density matrix is obtained by exploiting symmetries of the model. These are the transalational invariance, the parity symmetry (i.e., whether the number of flipped spins is odd or even is unchanged by the

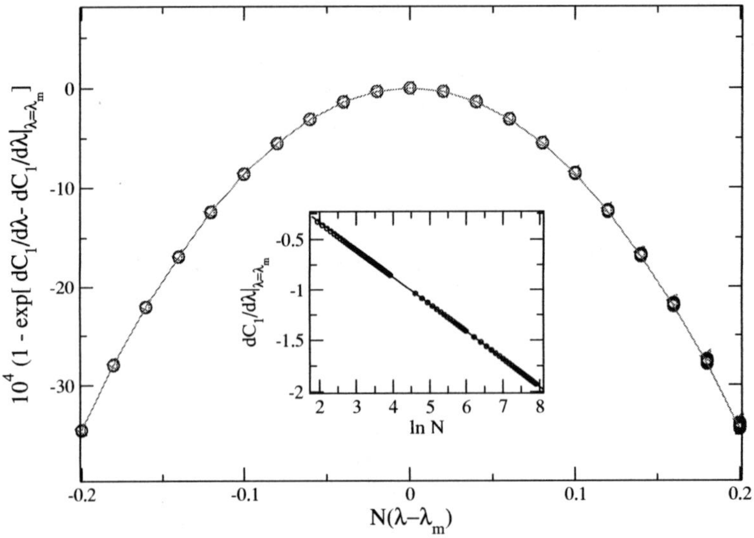

FIGURE 2. The finite size scaling is performed for the case of logarithmic divergences. All the data from $N = 41$ up to $N = 2701$ collapse on a single curve, such that the concurrence is a function of $N^{1/\nu}(\lambda - \lambda_m)$. The critical exponent is $\nu = 1$ as one expects for the Ising model. The inset shows the divergence of the value at the minimum as the system size increases. From [11], with kind permission from Nature.

Hamiltonian) and the reality of the Hamiltonian. They make ρ to be real symmetric with $\rho_{11}, \rho_{22} = \rho_{33}, \rho_{23}, \rho_{14}, \rho_{44}$ as the only non zero entries

$$\rho = \begin{pmatrix} \rho_{11} & 0 & 0 & \rho_{14} \\ 0 & \rho_{22} & \rho_{23} & 0 \\ 0 & \rho_{23} & \rho_{33} & 0 \\ \rho_{14} & 0 & 0 & \rho_{44} \end{pmatrix} \tag{3}$$

They can be related to the various correlation functions by means of $\left\langle \sigma_i^a \sigma_j^b \right\rangle = Z^{-1}\mathrm{Tr}\sigma_i^a \sigma_j^b \rho$. We point out that $\left\langle \sigma_i^a \sigma_j^b \right\rangle = 0$ for $a \neq b$. From (3) the concurrence of the ground state is evaluated exactly starting from the pfaffian expression of the correlation functions [18, 19, 20].

ISING MODEL ($\gamma = 1$)

The first question we consider is the range of the entanglement ξ_E, i.e. the maximum distance between two spins at which the concurrence is different from zero. The result is quite surprising: even at the critical point, where spin-spin correlations extend over a long range (the correlation length is diverging for an infinite system), the concurrence vanishes unless the two sites are at most next-nearest neighbors. That is: the truly non-local quantum part of the two-point correlations is nonetheless very short ranged. In order to quantify the change of the many-body wave-function when the system crosses the critical point, we look at the derivatives of the concurrence as a function of λ. In this case we need to consider only the nearest neighbor and next nearest neighbor concurrence. We first discuss the behavior of the nearest neighbor concurrence. The results for systems of different size (including the thermodynamic limit) are presented in Fig.1. For the infinite chain $\partial_\lambda C(1)$ diverges on approaching the critical value as

$$\partial_\lambda C(1) = \frac{8}{3\pi^2} \ln|\lambda - \lambda_c| + const. \tag{4}$$

Eq.(4) quantifies non-local correlations in the critical region. An aspect, particularly relevant for quantum information is the study of the precursors of the critical behavior in finite samples. This study goes under the rubric of finite size scaling [23]. In the scaling ansatz, the concurrence, a function of the system size and the coupling constant, is considered to depend only on the combination $N^{1/\nu}(\lambda - \lambda_m)$ of both with critical exponent ν. In the case of log divergence it behaves as

$$\exp \frac{\partial_\lambda C(1)(N,\lambda)}{\partial_\lambda C(1)(N,\lambda_0)} \sim \exp \frac{Q[N^{1/\nu}(\lambda - \lambda_m)]}{Q[N^{1/\nu}(\lambda_0 - \lambda_m)]}$$

where λ_0 is some non critical value and $Q(x) \sim Q(\infty)\ln x$ (for large x). In Fig.1 the derivative of $C(1)$ respect to λ is considered for different system sizes. As expected, there is no divergence for finite N, but there are clear anomalies. The position of the minimum λ_m scales as $\lambda_m \sim \lambda_c + N^{-1.86}$ and its value diverges logarithmically with increasing system size as

$$\partial_\lambda C(1)\big|_{\lambda_m} = -0.2702 \ln N + const. \tag{5}$$

According to the scaling ansatz [23] in the case of logarithmic singularities, the ratio between the two prefactors of the logarithm in Eqs.(4) and (5) is the exponent which governs the divergence of the correlation length ξ [9].In this case ($8/3\pi^2 \approx 0.2702$) it follows that $\nu = 1$, as it is known from the solution of the Ising model [19]. By proper scaling [23] and taking into account the distance of the minimum of $C(1)$ from the critical point, it is possible to make all the data for different N collapse onto a single curve (shown in Fig.2). It contains the data for the lattice size ranging from $N = 41$ up to $N = 2701$. **These results show that all the key ingredients of the finite size scaling are present in the concurrence.** It is worthwhile to note that finite size scaling scaling is fulfilled in a very broad range of values of N which are of interest in several protocols in quantum information.

A similar analysis can be carried on for the next nearest neighbor concurrence $C(2)$. The result is presented in Fig.3. In the inset a peculiarity of the Ising model is shown: $C(2)$ is maximal exactly at the critical point. Note that this maximum decreases rapidly

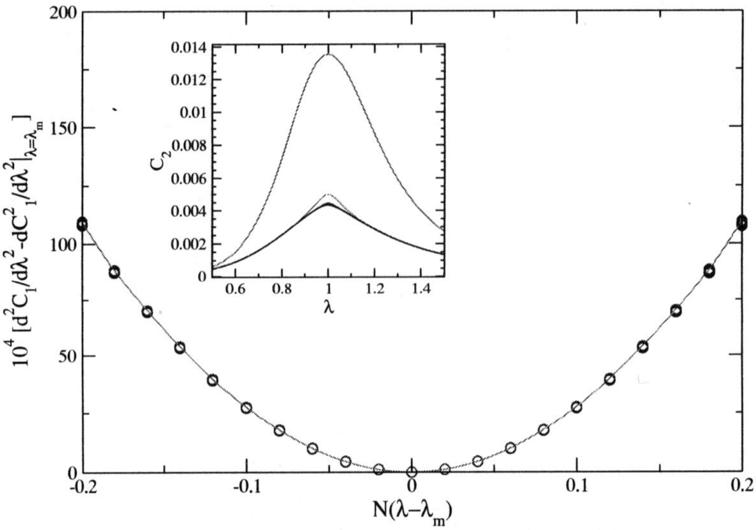

FIGURE 3. As in the case of the nearest neighbor concurrence, data collapse is also obtained for the next nearest neighbor concurrence C(2) and for $\nu = 1$. In the figure data for system size from $N = 41$ to $N = 401$ are plotted. The inset shows a peculiarity of the Ising model: $C(2)$ has its maximum precisely at the critical point for arbitrary system size. From [11], with kind permission from Nature.

as the system size increases and that $C(2)$ is two orders of magnitude smaller than $C(1)$. Since $\partial_\lambda C(2)|_{\lambda_c} = 0$ for all N, the logarithmic singularity here appears first in the second derivative respect to λ, which we use to perform the scaling analysis. We want to emphasize that for $\gamma \neq 1$ $\partial_\lambda C(2)|_{\lambda_c} \neq 0$ and the critical behavior can be extracted from the first derivative as is the case for $C(1)$. Furthermore we find that only for the smallest system sizes the concurrence is different from zero for $|i-j| = 3$ and $\lambda > 1.05$ (i.e. for $N = 7$; for $N \geq 9$, $C(3) = 0$ for all λ). For sufficiently large Ising chains $C(n) = 0$ for $n \geq 3$. In contrast the correlation functions are long-ranged at the critical point. In the thermodynamic limit $\partial_\lambda^2 C(2) = 0.108 \ln|\lambda - \lambda_c| + const$. Also in this case data collapse and finite size scaling, presented in Fig.3, agree with the expected scaling behavior, $\nu = 1$. This completes the analysis of entanglement for the one-dimensional Ising model.

UNIVERSALITY: ANISOTROPIC XY MODEL, $0 < \gamma < 1$

A cornerstone in the theory of critical phenomena is the concept of universality, i.e. the critical properties depend only on the dimensionality of the system and the broken

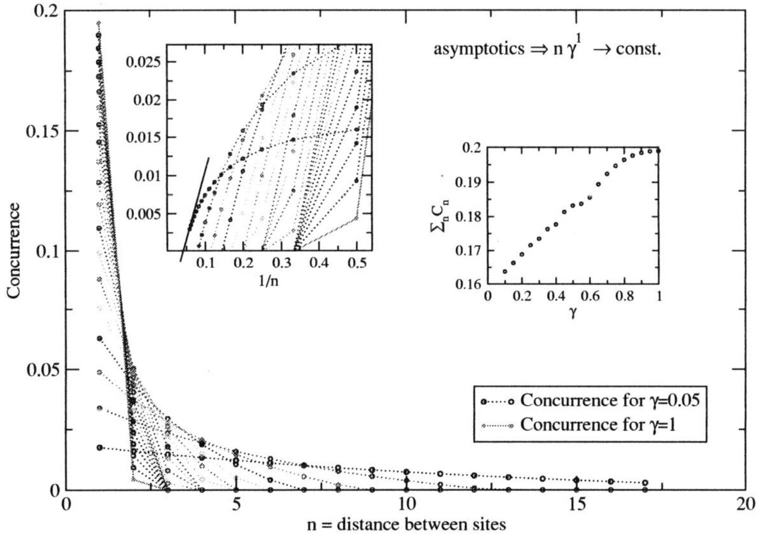

FIGURE 4. Range of the concurrence for different values of γ. The rightmost inset shows that the concurrence approachs its largest value in a universal way; the other inset shows how the "integrated" concurrence is bounded.

symmetry in the ordered phase. Universality in the critical properties of entanglement was verified considering the properties of the family of models defined in Eq.(1) with $\gamma = 1/2$. The range of entanglement ξ_E is not universal. The maximum possible distance between entangled pairs increases and tends to infinity as γ tends to zero. From the asymptotic behavior of the reduced density matrix we extract that ξ_E goes as $\gamma^{-5/9}$. This however has no dramatic consequences, the total concurrence $\sum_n C(n)$ stored in the chain is a weakly increasing function of γ (for $0 < \gamma \leq 1$, $0.15 < \sum_n C(n) < 0.2$). These facts are summarized in Fig. 4. More interesting is the critical behavior of the concurrence. To be specific we consider $C(1)$ in the case $\gamma = 0.5$, shown in Fig.4. As it was obtained for the Ising model, scaling is fulfilled with the critical exponent $\nu = 1$ in agreement with the universality hypothesis. From this we obtain that the anomaly of the concurrence at the quantum critical point is of critical origin.

CONCLUSIONS

The analysis of the resource entanglement for a condensed matter system close to a quantum critical point allows to characterize both quantitatively and qualitatively the change in the wave-function of the ground state on passing the phase transition. A striking feature which emerges is that though the entanglement itself is not an indicator of the phase transition, an intimate connection between entanglement, scaling and universality exists. In a way this analysis allows to dig out what is genuinely quantum in a zero temperature phase transition.

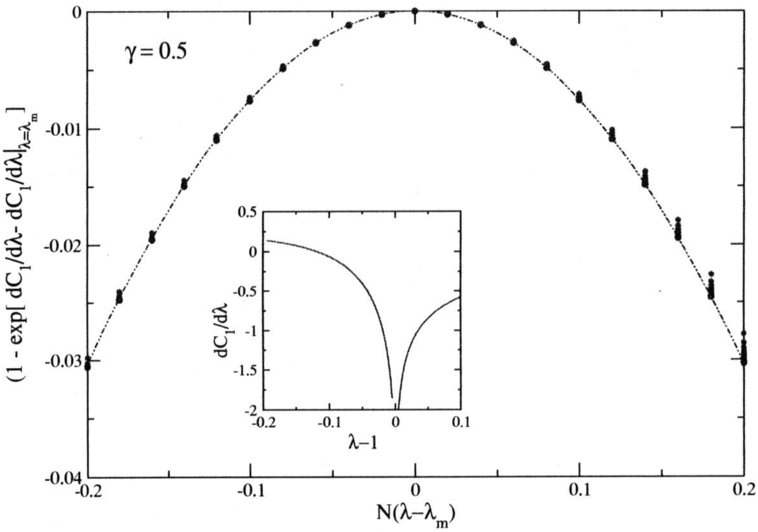

FIGURE 5. The universality hypothesis for the entanglement is checked considering the model Hamiltonian, defined in Eq.(1), for a different value of γ. In this case we chose $\gamma = 0.5$ and N ranging from 41 up to 401. Data collapse, shown here for $C(1)$, is obtained for $\nu = 1$ consistent with the model being in the universality class of the Ising model. In the inset the divergence at the critical point for the infinite system. From [11], with kind permission from Nature.

The results presented here might be tested by measuring different correlation functions, for example with neutron scattering, and to extract from these the entanglement properties of the ground state close to the critical point [15]. We finally discuss the perspectives of this work towards quantum computation. First of all, system sizes considered here ($\sim 10^3$) can be those of a typical/realistic quantum computer. Quantum computation close to a critical point may offer great advantages: it may be more robust against both errors due to imperfections and against decoherence due to long range correlations and diverging dephasing time respectively. Moreover, our study shows how the resource entanglement is shared among the qubits and how at the phase transition the entanglement 'switches' with the external field. Development of our work is to analyse the dynamics of maximally entangled states released on the vacuum and on the ground state [24]. Certain non-equilibrium correlation functions are calculated exactly and permit an access to the concurrence. We see the concurrence and the one-site entanglement with the system, imported by the singlet, propagating with roughly the velocity of the isotropic XY model. Though the concurrence propagation is strongly suppressed at the critical point, this damping is not of critical but rather of dynamical origin. total entanglement stored in the system beyond the two-qubit entanglement. In particular a preliminar analysis suggests that the entanglement seen by the concurrence should only constitute a small fraction of the total entanglement stored in the system.

ACKNOWLEDGMENTS

The authors would like to thank G.M. Palma, F. Plastina, and J. Siewert for helpful discussions. This work was supported by the European Community (IST-SQUIBIT), and by INFM-PRA-SSQI.

REFERENCES

1. N. Goldenfeld , *Lectures on phase transitions and the renormalization group*, (Addison Wesley, New York, 1992).
2. S. Sachdev, *Quantum Phase Transitions*, (Cambridge University Press, Cambridge, 2000).
3. S.I. Sondhi , S.M. Girvin, J.P. Carini, D. Shahar, Rev. Mod. Phys. **69**, 315-333 (1997).
4. S.V. Kravchenko, G.V. Kravchenko, J.E. Furneaux, V.M. Pudalov, M. D'Iorio, Phys. Rev. B **50**, 8039-8042 (1994).
5. D.B. Haviland, Y. Liu, and A.M. Goldman, Phys. Rev.Lett. **62**, 2180-2183 (1989).
6. H.S.J. van der Zant, F.C. Fritschy, W.E. Elion, L.J. Geerligs, and J.E. Mooij, Phys. Rev.Lett. **69**, 2971-2974 (1992).
7. J.S. Bell, *Speakable and unspeakable in Quantum Mechanics*, (Cambridge University Press, Cambridge , 1987).
8. M. Nielsen and I. Chuang, *Quantum Computation and Quantum Communication*, (Cambridge University Press, Cambridge, 2000).
9. C. J. Hamer and M. N. Barber, J. Phys. A **14**, 241 (1981); M. E. Fisher and M. N. Barber, Phys. Rev. Lett. **28**, 1516 (1972).
10. J. Preskill, J. Mod. Optics **47**, 127-137 (2000).
11. A. Osterloh, L. Amico, G. Falci, and R. Fazio, Nature **416** 608 (2002).
12. K.M. O'Connors and W.K. Wootters, Phys. Rev. A **63**, 052302 (2001).
13. W. Dür, G. Vidal, and J.I. Cirac, Phys. Rev. A **62** 062314 (2000).
14. X. Wang, Phys. Rev. A **64**, 012313 (2001).
15. M.C. Arnesen, S. Bose, and V. Vedral, Phys. Rev. Lett. **87**, 017901 (2001).
16. D. Gunlycke, S. Bose, V.M. Kendon, and V. Vedral, Phys. Rev. A **64**, 042302 (2001).
17. T.J. Osborne and M.A. Nielsen, *quant-ph/0109024*.
18. E. Lieb, T. Schultz, D. Mattis, Ann. Phys. **60**, 407 (1961).
19. P. Pfeuty, Ann. Phys. (N.Y.), Ann. Phys. **57**, 79-90 (1970).
20. E. Barouch and B.M. McCoy, Phys. Rev. A. **3**, 786-804 (1971).
21. C.J. Bennett, H.J. Bernstein, S. Popescu, and B. Schumacher, Phys. Rev. A **53**, 2046-2052 (1996). V. Vedral, M.B. Plenio, M.A. Rippin, and P. Knight, Phys. Rev. Lett. **78**, 2275-2279 (1997). C.H. Bennett, D.P. DiVincenzo, J. Smolin, and W.K. Wootters, Phys. Rev. A **54**, 3824-3851 (1996).
22. W.K. Wootters, Phys. Rev. Lett. **80**, 2245-2248 (1998).
23. M.N. Barber, *Finite size scaling, in Phase Transitions and critical phenomena*, Domb C. and Leibovitz J.L. Eds., Vol. 8 146-259 (Academic Press, London,1983).
24. L. Amico, R. Fazio, A. Osterloh, G.M. Palma, and F. Plastina preprint.

Structural, Transport and Magnetic Properties of the Doped Magnetic Superconductor Ru-1212

A. Hassen*, A. Krimmel*, J. Hemberger*, P. Mandal† and A. Loidl*

*Institut für Physik, Elektronische Korrelationen und Magnetismus, Universität Augsburg, D-86159 Augsburg, Germany
†Saha Institute of nuclear Physics, 1/AF Bidhannagar, Calcutta 7000064, India

Abstract. Single phase materials of $RuSr_2Gd_{0.5}Eu_{0.5}Cu_2O_8$ (Ru-1212$Gd_{0.5}Eu_{0.5}$) and $Ru(Sr_{1-x}Na_x)_2GdCu_2O_8$ ($0 \leq x \leq 0.10$) have been successfully synthesized in oxygen at 1065 ^oC. DC-resistivity and magnetic susceptibility measurements show a decreases of the superconducting transition temperature T_c from 46 K in the pure Gd compound to 41 K in Ru-1212$Gd_{0.5}Eu_{0.5}$, whereas the magnetic ordering temperature T_M does not change. The influence of the monovalent substitution Na^+ for Sr^{2+} on the properties of Ru-1212Gd is investigated and compared with previous work. Finally, the effects of Cu substituted $RuSr_2Gd(Cu_{1-x}M_x)_2O_8$ compounds, where M is Co, Ni and Ga for $0 \leq x \leq 0.03$ are reported and compared with similar doping experiments in other high-T_c superconductors.

INTRODUCTION

The antagonistic nature of superconductivity and ferromagnetism was studied theoretically by Ginzburg[1] in 1956, while experimental progress in the field began after the discovery of the so-called Chevrel phases $REMo_6X_8$ (RE=rare earth, X=S, Se) or in $RERh_4B_4$ (see for example Ref. [2]). In many of these compounds, superconductivity with a critical temperature T_c coexists with antiferromagnetic (AFM) order below the Néel temperature T_N where $T_N \ll T_c$. On the other hand, singlet superconductivity and ferromagnetism can not coexist in bulk samples with realistic physical parameters. However, under certain conditions the ferromagnetic order is transformed, in the presence of superconductivity, into a spiral or domain-like structure depending on the type and strength of magnetic anisotropy in the system [3]. The internal field of a ferromagnetically ordered state leads to a Zeeman splitting of the Fermi surface that excludes electron pairing of opposite spin and momentum. Allowing for finite angular momentum of electron pairs led to the theory of Fulde-Ferrel-Larkin-Ovchinnikov (FFLO) superconductivity [4], where a suitable spatial modulation of the superconducting or ferromagnetic order parameter (or both) leads to their mutual coexistence.

Recently, a new class of magnetic superconductors among the layered perovskite ruthenocuprates $RuSr_2RECu_2O_8$ (Ru-1212RE, RE = Gd, Eu, Y and Sm) has been synthesized [5]. In particular, the Ru-1212Gd system has attracted great interest since it exhibits a ferromagnetic transition at $T_M \approx 135$ K and bulk superconductivity below $T_c \approx 46$ K, depending on the sample preparation conditions [6, 7, 8]. For the isostructural Ru-1212Eu compound, these transitions are shifted to 133 K and 32 K, respectively.

The tetragonal crystal structure of both Gd-and Eu-based Ru-1212 is very similar to

CP678, *Lectures on the Physics of Highly Correlated Electron Systems VII: Seventh Training Course*, edited by A. Avella and F. Mancini

the prototype high-T_c cuprate superconductors $REBa_2Cu_3O_{7-\delta}$ (RE-123) [6, 9, 10]. The structure of Ru-1212RE contains CuO_2 bilayers carrying superconductivity and RuO_2 monolayers substituting the CuO chains of the RE-123 compounds.

Neutron powder diffraction measurements [10, 11] revealed a G-type AFM structure of the Ru sublattice with an ordered magnetic moment of the order of $1\mu_B$. The ferromagnetic ordering observed in magnetic experiments can be accounted for by a canting of the Ru moments. However, the origin of the FM component is still unclear. Both neutron powder diffraction studies lead to upper limits of any net FM component of $0.3\mu_B$ [11] or $0.1\mu_B$ [10], respectively. The mixed valent state of the Ru ions, as evidenced by XANES [12] and NMR [13] spectroscopy enables a double exchange interaction between different Ru ions and the proportion of Ru^{4+}/Ru^{5+} ions may change with cation substitution in Ru-1212RE. Therefore, the hole concentration in the superconducting CuO_2 planes may be changed in a controlled way by doping [12]. In case of the cuprate superconductors for example, hole carriers can be induced either by cation substitution (as in $La_{2x-1}Sr_xCuO_4$) or by oxygen intercalation (as in $YBa_2Cu_3O_{7-\delta}$) or by a combination of them. Unlike the RE-123 compounds, the oxygen content of Ru-1212RE can virtually not be changed. Thermogravimetric experiments [14, 15], as well as the thermoelectric power (TEP) [16] indicate that the oxygen content is always very close to 8. In order to change the hole carrier concentration in Ru-1212RE, one therefore has to rely on cation substitution [17, 18]. The effect of substituting cations on the electrical and magnetic properties of Ru-1212Gd was reported by different groups [12, 18, 19, 20, 21, 22]. It is generally observed that both magnetism and superconductivity depend on the type of cation and the substitutional site. It was noticed that Ti and Rh doping reduced the magnetic ordering temperatures [22] while Nb and V enhanced it [18, 20]. T_c decreases for all cations investigated so far, except in case of $Ru_{1-x}Sr_2GdCu_{2+x}O_8$, where it reaches 70 K for $x = 0.7$ [19].

In this work, we first complete the comparison between pure Gd and Eu-based compounds by investigating Ru-1212$Gd_{0.5}Eu_{0.5}$. Then we compare pervious results of doped Ru-1212Gd compounds [21, 22] with a new type of substitution $Ru(Sr_{1-x}Na_x)_2GdCu_2O_8$ for $0 \leq x \leq 0.10$. Finally, the T_c-suppression by copper substituting cations $RuSr_2Gd(Cu_{1-x}M_x)_2O_8$, (M = Co, Ni and Ga) for $0 \leq x \leq 0.03$ is studied and compared with Zn-doping in Ru-1212Gd [23] and other high-T_c superconductors.

EXPERIMENTAL DETAILS

Polycrystalline samples have been synthesized by a solid-state reaction method [23] using high purity RuO_2, $SrCO_3$, Gd_2O_3, CuO, NiO, Ga_2O_3, CoO and Na_2O powders. At the final stage, the samples were annealed for 6 days at 1065 ^{o}C in flowing oxygen followed by slow cooling down to room temperature. Powder x-ray diffraction (PXD) data at room temperature were collected on a Stoe x-ray diffractometer using the Cu-K_α radiation (λ = 1.5406 Å) in the range of $20^o \leq 2\theta \leq 80^o$ at an increment of 0.02^o. The data were analyzed by standard Rietveld refinement. Magnetic measurements were performed employing a SQUID magnetometer (Quantum Designs, MPMS). The resistivity was measured by a conventional four-probe technique.

RESULTS AND ANALYSIS

Crystal Structure

Based on Rietveld analyses of XRD patterns, all samples are found to be single phase except for the pure Eu-based compounds which exhibit an additional very weak intensity close to $2\theta \approx 31.5^o$. This spurious reflection is attributed to residues of $SrRuO_3$ or Gd_2CuO_4 [24] and its occurrence is well-known for the Ru-1212RE system. The tetragonal lattices of the pure Gd- and Eu-based samples are in good agreement with previous reports [5, 6, 18, 25]. The lattice parameters of Ru-1212$Gd_{0.5}Eu_{0.5}$ are $a = b = 3.842$ Å and $c = 11.573$ Å which are close to those of the pure compounds. Moreover, all diffraction patterns of the $x = 0$, 0.03 and 0.10 Na-doped compounds could be successfully indexed on the basis of the usual tetragonal unit cell with space group $P4/mmm$, as shown in Fig. 1. The crystallographic structures remain almost unchanged within the present range of doping.

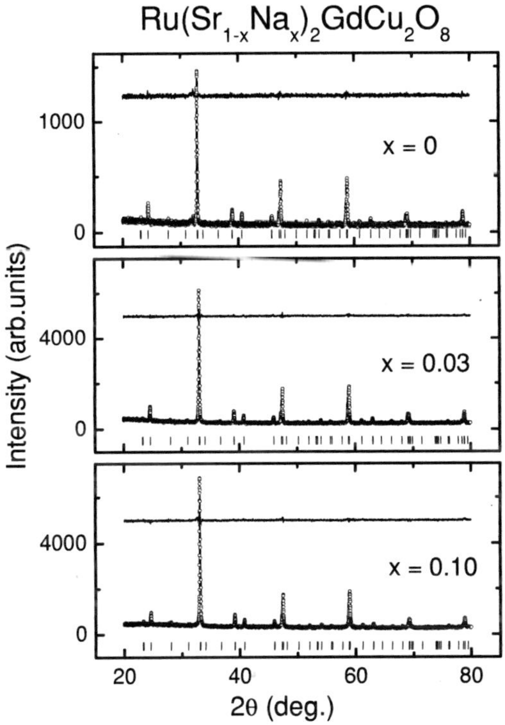

FIGURE 1. Room temperature x-ray diffraction patterns of $Ru(Sr_{1-x}Na_x)_2GdCu_2O_8$ for $x = 0$, 0.03 and 0.10. The vertical bars denote the peak positions as indexed by space group $P4/mmm$. The intensities as calculated on the basis of the Rietveld refinement are indicated as solid lines. The difference between observed and calculated intensities are shown at the top of each pattern.

Magnetic Properties

Fig. 2 shows the temperature dependence on dc magnetic susceptibility $\chi(T) = M/H$ at an external field of $H = 1$ kOe. As expected, $\chi(T)$ of Ru-1212$Gd_{0.5}Eu_{0.5}$ is located between the pure Gd-and Eu-based compounds. This implies that only the Gd moments contribute to the total magnetization of the doped sample whereas Eu is nonmagnetic. There is no change in T_M for the mixed compound since the RuO_2 layers are responsible for the magnetic ordering in Ru-1212RE. Similarly, Bernhard et al. [26] reported that the ferromagnetic transition is not significantly affected by the partial substitution of non-magnetic Y^{3+} for the magnetic Gd^{3+} ions. The inverse magnetic susceptibility, $\chi^{-1}(T)$ of both Ru-1212Gd and Ru-1212$Gd_{0.5}Eu_{0.5}$ were fitted using two independent Curie-Weiss functions, where the parameters of Gd were kept fixed at θ_{Gd} = -4 K and μ_{eff} = 7.94 μ_B [21]. The $\chi^{-1}(T)$ data of Ru-1212Eu were fitted at $T \geq 200$ K by using a susceptibility χ^{-1} (T) = $(1/(T+\theta)+\chi_0)^{-1}$, where χ_0 is the temperature independent part. Its value of $\chi_0 = 3.34 \times 10^{-3}$ emu/mol is in agreement with earlier reports [27]. It is noticed that the value of the magnetic parameters significantly depend on the rare earth element.

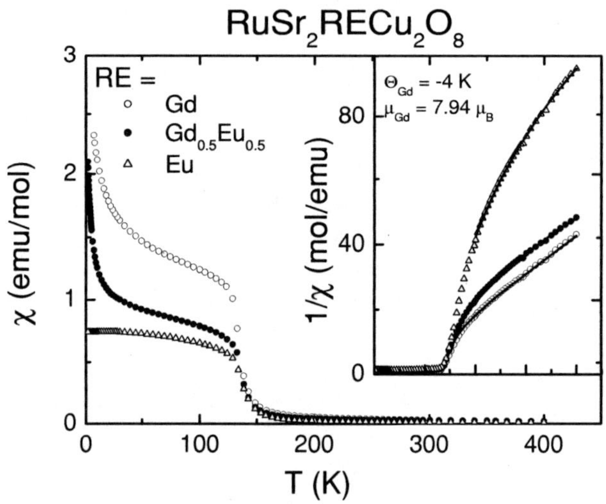

FIGURE 2. The temperature dependence of dc susceptibilities $\chi_{dc}(T)$ at 1 kOe for $RuSr_2Gd_{0.5}Eu_{0.5}Cu_2O_8$, as well as for the pure Gd- and Eu-based compounds. The inset shows the corresponding inverse susceptibilities $\chi^{-1}(T)$ including the Curie-Weiss fitting functions (solid lines)

The substitution of monovalent Na^+ for Sr^{2+} in Ru-1212Gd is now compared with La-doping [21]. Fig. 3 shows $\chi(T)$(upper frame) for x =0, 0.03 and 0.10 Na-doping at an external field 1 kOe. Similar to La-doping, the magnetic ordering temperature T_M is enhanced while at low temperatures the magnetic moments are decreased with increasing Na concentration x due to an increase of the canting angle of the Ru moments. The inset of the figure shows FC/ZFC curves of the sample x = 0.03. A splitting at the magnetic ordering temperature is observed like for the pure and La-doped Ru-1212Gd compounds. There is no clear evidence about the absence of a Meissner effect down to

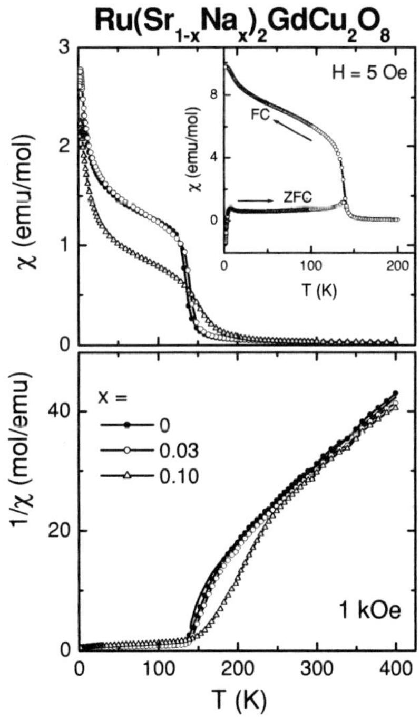

FIGURE 3. Temperature dependence of the dc susceptibility (upper frame) of $Ru(Sr_{1-x}Na_x)_2GdCu_2O_8$ for $x = 0$, 0.03 and 0.10 at an external field of 1 kOe and the corresponding inverse susceptibility (lower frame). The inset shows FC/ZFC measurements for $x = 0.03$ Na at $H = 5$ Oe.

2 K. The sample $x = 0.03$ clearly is a superconductor while for the same La concentration superconductivity is already completely suppressed [21]. The lower frame of the figure shows $\chi^{-1}(T)$ of some Na-doped samples and the data can be fitted as for pure and La-doped Ru-1212Gd. It is evident that any change in the SrO layers caused by either Na or La substitution, respectively enhances the magnetic ordering temperature and destroys superconductivity in Ru-1212Gd. However, the physical origin must be different for the two doping series since the substitution of Na^+ for Sr^{2+} corresponds to hole doping, whereas the substitution of La^{3+} for Sr^{2+} corresponds to electron doping.

The dependence of the real part of the ac susceptibility χ' on magnetic field at constant temperatures for $x = 0.03$ Na is shown in Fig. 4. As in case of the pure and La-doped Ru-1212Gd compounds, the data were taken after zero-field cooling at 1 Hz with an ac field of 1 Oe from above and below T_M. Maxima in the field dependence of χ' can be detected and signal a metamagnetic transition. The corresponding phase diagram, where the solid points separate a ferromagnetic (FM) phase at high fields from a canted antiferromagnetic (CA) phase at low fields is illustrated in the inset of Fig. 4. The $H-T$ phase diagram of the Na-doped compounds is similar to the La-doped one, where the ferromagnetic phase transition can be induced by fields smaller than 1kOe [21].

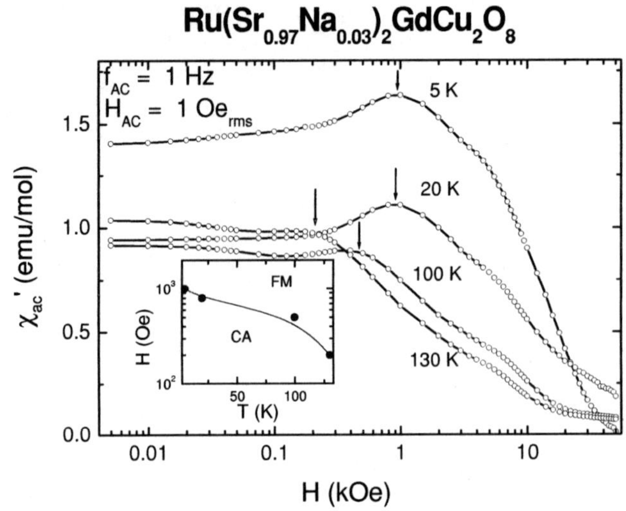

FIGURE 4. Field dependence of the real part of the ac-susceptibility for the $x = 0.03$ Na doped sample at different temperatures below T_M. The arrows indicate a characteristic temperature where a metamagnetic transition occurs. The inset shows a schematic (H,T) phase diagram separating a canted antiferromagnetic (CA) from a purely ferromagnetic (FM) phase.

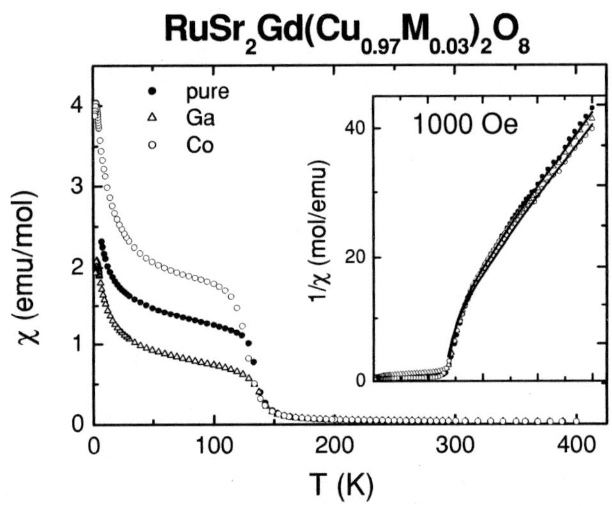

FIGURE 5. Temperature dependence of the dc-susceptibilities $\chi_{dc}(T)$ for $x = 0.03$ Ga and Co, including the pure Gd compound for comparison, as measured on cooling at $H = 1$ kOe. The inset shows the corresponding inverse susceptibility $1/\chi_{dc}(T)$.

For $RuSr_2Gd(Cu_{1-x}M_x)_2O_8$ (M= Ga, Co, Ni) the dc magnetic susceptibility was also measured at an external field 1 kOe. The pure sample is included for comparison. Fig. 5 shows $\chi_{dc}(T)$ for $x = 0.03$ Ga and Co. As can be observed from the figure, the magnetic susceptibility of the Ga-doped compound lies well below that of the pure and Co substituted compound as Ga is nonmagnetic whereas Co has a large magnetic moment. The magnetic ordering temperature T_M is not changed as it reflects the fact that the ferromagnetism in Ru-1212RE originates from the RuO_2 layers. The inset shows $\chi^{-1}(T)$ for the same concentration of Co and Ga. As in the pure compounds, the data were fitted in terms of two independent Curie-Weiss functions. As it is expected, the effective magnetic moment of the Ru sublattice for the $x = 0.03$ Co-doped sample, (μ_{eff}= 3.37 μ_B) is larger than for the 3% Ga-doped sample (μ_{eff}= 2.96 μ_B) due to the magnetic contribution of Co to the Ru moments. This is in contrast to other HTSCs, where the magnetism is enhanced for both, magnetic and non-magnetic copper substituted cations [28].

dc-Resistivity

The zero-field resistivity ρ of bulk Ru-1212$Gd_{0.5}Eu_{0.5}$ exhibits a maximum around T_c = 41 K and drops to zero slightly above 19 K. For pure Ru-1212Eu/Gd, ρ shows maxima around 34 and 51 K, respectively [29, 21]. It is clear that any small structural changes affect the superconducting and magnetic properties. The decrease of T_c with substitution of Gd by Eu can also be attributed to the change of the bond angle ϕ (Ru-O-Cu) which affects the transfer of charge carriers between RuO_2 and CuO_2 layers. From the Rietveld analysis, it was found that the value of ϕ (Ru-O-Cu) for Ru-1212$Gd_{0.5}Eu_{0.5}$ is slightly smaller than that of pure Ru-1212Gd compounds.

The dc resistivity of Na-doped samples was measured at zero external magnetic field, as shown in Fig. 7. The resistivity increases with increasing Na concentration but the sample with $x = 0.03$ is still superconducting with an onset temperature of $T_c = 45$ K.

The observed decrease of T_c by substituting Sr partially by Na was not expected. Since Na is a monovalent ion and Sr is divalent, Na substitution can be considered as a hole doping leading to an increase of the charge carrier concentration in the system. Therefore, other factors must prevent an increase of T_c with increasing Na concentration such as trapping of the charge carriers. Another reason of a decreasing T_c induced by La and Na doping may be due to the closeness of the SrO layers to the superconducting CuO_2 layers so that even minor changes on a local scale within the SrO_2 layers may have a strong effect. As shown in the inset of the Fig. 7, the Na-doped sample for x = 0.10 has a very similar semiconducting behavior in the whole temperature range as the 10% La-doped compound.

As reported earlier [31, 32, 33, 34, 35], impurities in the copper planes rapidly destroy superconductivity due to pair breaking. The concentration range of impurities and the rate of the T_c suppression depend on the type of the substituting cations, as well as on the HTSCs material itself. The resistivity of $RuSr_2Gd(Cu_{1-x}M_x)_2O_8$ (M=Ga, Co and Ni) was measured at zero external field. Fig. 8 shows the temperature dependence of the resistivity of Ni-doped samples as a typical example of copper substitution. It is noticed that the resistivity increases with increasing Ni and superconductivity is fully suppressed at $x = 0.03$. The upper inset in this figure shows the resistivity for $x = 0.03$ Ni which is non-superconducting and the resistivity increases for decreasing T in the

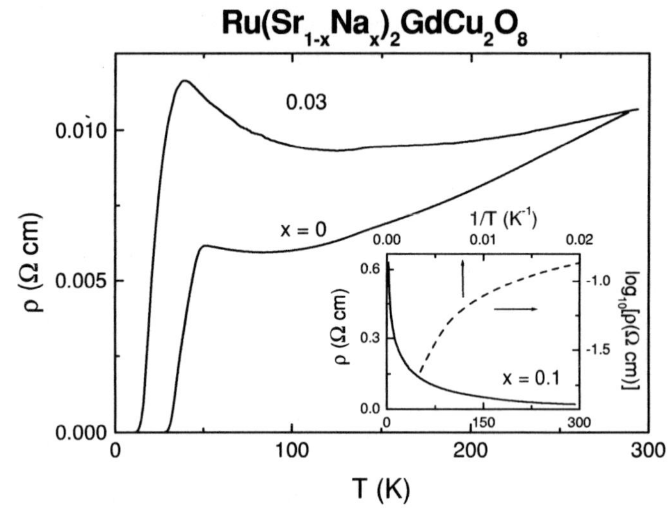

FIGURE 6. dc resistivity vs. temperature of $Ru(Sr_{1-x}Na_x)_2GdCu_2O_8$ for x = 0, 0.03 and 0.1. The resistivity for x = 0.10 is shown in the inset: ρ vs. T (solid line, left and lower scale) and log ρ vs. T^{-1} (dashed line, right and upper scale).

whole temperature range. Similarly, Bernhard et al.[36] reported that T_c in Ru-1212Gd was completely destroyed for 0.03 Zn-doping. The lower inset summarizes the change of T_c with different substituted cations. Evidently, Co suppresses the superconductivity faster than Ga and Ni. This may be attributed to the large magnetic moment of Co, leading corrspondingly to a strong pair breaking effect. Compared to other HTSCs, Xiao et al.[37] reported on the substitution of $3d$ elements at the Cu site in $YBa_2Cu_3O_{7-\delta}$ (YBCO). They observed that T_c is reduced to zero beyond a critical concentration which varies from element to element. For example, non-magnetic Zn ions have a stronger influence on T_c than magnetic Co ions. Moreover, Tarascon et al. [28] show that superconductivity in YBCO is preserved for a large doping range up to $x = 0.3$ for Fe or Co and up to $x = 0.5$ for Ni. On the other hand, superconductivity in Ru-1212 is completely suppressed at very low concentrations of copper substitution ($x \leq 0.03$). Similarly, 2.5% Zn impurities are sufficient to suppress the superconductivity in $La_{1.85}Sr_{0.15}Cu_{1-x}Zn_xO_4$ (Ref.[34]).

CONCLUSIONS

We have successfully synthesized two novel substitutional series $Ru(Sr_{1-x}Na_x)_2GdCu_2O_8$ ($0 \leq x \leq 0.10$) and $RuSr_2Gd(Cu_{1-x}M_x)_2O_8$ ($M = \mathrm{Co, Ni, Ga}$, $0 \leq x \leq 0.03$), as well as the mixed compound $RuSr_2Gd_{0.5}Eu_{0.5}Cu_2O_8$. Powder x-ray diffraction revealed single phase materials and only minor crystallographic changes as compared to pure Ru-1212Gd for all investigated compounds. Changes of the physical behavior of Ru-1212Gd are known to be sensitive to both, substituent and substitutional site. Na

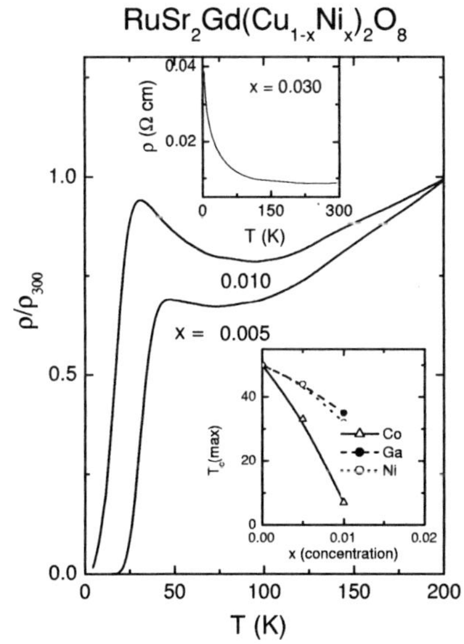

FIGURE 7. Normalized electrical resistivity (ρ/ρ_{300}) vs. temperature for Ni-doped Ru-1212. The upper inset shows the resistivity for $x = 0.03$ while the lower inset shows the change of T_c for different copper substituting elements.

doping has been chosen in order to study the effect of hole doping. Moreover, exchange of Cu ions for Co, Ni or Ga respectively probes the effect of impurities within the superconducting CuO-planes. The magnetic and superconducting properties of doped Ru-1212Gd were studies by means of resistivity and magnetic measurements.

Because the ferromagnetism in the rutheno-cuprate system is due to the RuO_2 layers (apart from the very low temperature region), there is no change of T_M for an equal mixture of Eu and Gd rare earth ions (Ru-1212$Gd_{0.5}Eu_{0.5}$). Similar to the effect of La-doping, T_M is enhanced and T_c suppressed by Na-doping but at a different rate. It can be considered that monovalent Na-doping increases the hole concentration in the system because it substitutes a divalent cation (Sr^{2+}). The decrease of T_c with increasing Na content may therefore be attributed to other effects like trapping of charge carriers. Superconductivity is suppressed rapidly by copper substitution due to pair-breaking as generally observed in HTSCs. The effect of Co substitution is stronger than the effect of magnetic Ni or non-magnetic Ga impurities because of its large magnetic moment. Compared with 3d-element doped YBCO, superconductivity is destroyed very rapidly in Ru-1212RE and implies that the major contribution to superconductivity originates from the copper-oxide planes. Moreover, the decrease of the magnetic moment at low temperatures for Ga substitution and its increase for Co or Ni suggests that there is a contribution of the CuO-planes to the total magnetic moment of the system.

ACKNOWLEDGMENTS

This work was partly supported by Bundesminterium für Bildung und Forschung (BMBF) via VDI/EKM 13 N 6917 and by the Deutsche Forschungsgemeinschaft via SFB484, Augsburg.

REFERENCES

1. V. L. Ginzburg, Sov. Phys. JETP, **4**, 153 (1957).
2. "Superconductivity in ternary compounds", edited by M. B. Maple and Ø. Fisher,(Berlin, **II**,1982).
3. L. N. Bulaevskii et al., Adv. Phys. **34**, 176 (1985).
4. P. Fulde and R. A. Ferrell, Phys. Rev. **135**, A550 (1964); A. I. Larkin and Yu N. Ovchinnikov, Zh. Eksp. teor. Fiz **47**, 1136 (1964).
5. L. Bauerfeind, W. Widder and H. F. Braun, Physica C 254, 151 (1995); J. Low Temp. Phys. **105**, 1605 (1996).
6. I. Felner et al., Physica C **311**, 163 (1999).
7. C. Artini et al., Physica C in press.
8. T. Papageorgiou et al., Physica C in press.
9. A. C. McLaughlin et al., Phys. Rev. B **60**,7512 (1999).
10. O. Chmaissem et al., Phys. Rev. B **61**, 6401 (2000).
11. J. W. Lynn et al., Phys. Rev. B **61**, 14964 (2000).
12. R. S. Liu et al., Phys. Rev. B **63**, 212507 (2001).
13. K. Kumagai, S. Takada and Y. Furukawa, Phys. Rev. B **63**, 180509 (2001).
14. K. Otzschi et al., J. Low. Temp. Phys, **117**, 855 (1999).
15. P. W. Klamut et al., Physica C **341-348**, 455 (2000).
16. B. Lorenz et al., Physica C **363**, 251 (2001).
17. J. Tallon et al., IEEE. Trans. App. Super, **9** , 1696 (1999).
18. A. C. McLaughlin and J.P.Attfield, Phys. Rev. B **60**, 14605 (1999); A.C. McLaughlin et al., J. Mater. Chem. **11**, 173 (2001).
19. P. W. Klamut et al., Phys. Rev. B **63**, 224512 (2001); P. W. Klamut et al., Physica C **350**, 24 (2001).
20. J. T. Rijssenbeek et al., Physica C **341-348**, 481 (2000); S. Malo et al., J. Inorg. Mater. **2**, 601 (2000).
21. P. Mandal et al., Phys. Rev. B **64**, 144506 (2002).
22. A. Hassen et al., Phys. Rev. B, submitted.
23. J. L. Tallon et al., Phys. Rev. B **61**, 4671 (2000).
24. M. Hrovat et al., J. Mater. Sci. Lett., **19**, 919 (2000); M. Hrovat et al., J. Mater. Sci. Lett., **19**, 1423 (2000).
25. D. P. Hai et al., Physica C **357-360**, 406 (2001).
26. C. Benhard et al., Phys. Rev. B **59**, 14099 (1999).
27. G. V. M. Williams and S. Krämer, Phys. Rev. B **62**, 4132 (2000).
28. J. M. Tarascon et al., Phys. Rev. B **36**, 8393 (1987); J. M. Tarascon et al., Phys. Rev. B **37**, 7458 (1988).
29. R. L. Meng et al., Physica C**353**, 195 (2001).
30. B. Lorenz et al., Phys. Rev. B **65**, 174503(2002); B. Lorenz et al., cond-mat/Nr., (unpublished).
31. G. V. M. Williams and J. L. Tallon, Phys. Rev. B **57**, 10984 (1998).
32. J. L. Tallon and G. V. M. Williams, Phys. Rev. B **61**, 982 (2000).
33. Ratan Lal et al., Phys. Rev. B **49**, 6382 (1994).
34. K. Westerholt et al., Phys. Rev. B **39**, 11 680 (1989).
35. G. V. M. Williams et al., Phys. Rev. B **54**, 9532 (1996).
36. C. Bernhard et al., Phys. Rev. B **61**, 14960 (2000).
37. G. Xiao et al., Phys. Rev. B**35**, 8782 (1987).

Space-Group Approach to the Wavefunction of a Cooper Pair and Its Application to High-T_c Superconductors

V.G.Yarzhemsky
Kurnakov Institute of General and Inorganic Chemistry
31 Leninsky, 119991 Moscow, Russia

Abstract

The general two-electron wavefunctions obeying the Pauli exclusion principle and Anderson criteria for Cooper pairs are constructed for D_{2h} and D_{4h} symmetries. It is shown that in axial symmetry groups the Blount theorem is violated and lines of nodes of triplet superconducting order parameter are required by the symmetry. Application of the results to high -T_c superconductors is discussed.

INTRODUCTION

The signature of unconventional superconductivity, both heavy fermion (HF) and high -T_c, is a breaking of different symmetries, including time reversal symmetry. In contrast with standard BCS superconductivity, where superconducting order parameter (SOP) is totally symmetric (belongs to a_{1g} irreducible representation (IR)), SOP of unconventional superconductors belongs to others IRs of point groups. We refer the reader to two reviews on high -T_c [1] and HF [2] superconductors. Since the underlying symmetries of unconventional superconductors are very significant in understanding of their pairing nature, there are three different group theoretical approaches to the description of the SOP symmetry: point group approach [1-14] , space-group approach [15-21] and unitary group approach [22-24]. The SOP symmetry of high -T_c superconductors was considered in a framework of all these approaches [11-14, 19, 22, 23].

The starting point of all approaches is the Anderson [3] description of the Cooper pair in ($\vec{k}$-$\vec{k}$) manifold. Anderson showed that the wavefunctions of electrons in a singlet pair are connected by the time-reversal. In $L-S$ coupling scheme singlet pairing operator is written as:

$$\left\langle C_k^\dagger C_{-k}^\dagger + C_{-k}^\dagger C_k^\dagger \right\rangle \left\langle C_\uparrow^\dagger C_\downarrow^\dagger - C_\downarrow^\dagger C_\uparrow^\dagger \right\rangle \qquad (1)$$

The wavefunctions of electrons in a triplet pair are connected by the time-reversal, space inversion and their product [3]. The Cooper pair's wavefunction

CP678, *Lectures on the Physics of Highly Correlated Electron Systems VII: Seventh Training Course*, edited by A. Avella and F. Mancini

is invariant with respect to lattice translations, hence it belongs to irreducible representation of point group and is characterized by parity [3]. Spatial part of a singlet pair is even and that of a triplet pair is odd.

In a point group approach the symmetry of the SOP is described in terms of algebraic functions belonging to the IRs of point group under consideration and hence its conclusions depend on the choice of basis functions and are ambiguous [10]. The results of point group approach in many cases correctly reflect symmetry properties of SOP, but they can't be used in quantum mechanical calculations, since they do not express Cooper pair wavefunction as two-electron wavefunction obeying the Pauli exclusion principle. The power-low temperature dependence of physical properties below T_c indicates that the superconducting gap vanishes at points and on the lines at Fermi surface [1,2]. On the other hand according to the Blount theorem, there are no group-theoretical requirement for the triplet SOP in a strong spin- orbit coupling case [5].

The basic idea of a space group approach is to construct general two electron wavefunction on the basis of the correct functions belonging to the space-group IRs and retain only those pairs for which the Anderson criteria are fulfilled. The space group approach was used to find all possible IRs for Cooper pair wavefunction but the general wavefunction of Cooper pairs were not constructed so far. It was also shown that different types of time-reversal symmetry violations result in violations of the Blount theorem [19-21].

In the present work the induced representation method is applied to obtain the nodal structure of SOP for D_{2h} and D_{4h} symmetries. It is also shown that the violations of the Blount theorem may be connected with symmetry breaking from highest point symmetry O_h to axial point group symmetry. Application of the theoretical results to high -T_c superconductors is discussed.

THEORY

The one-electron states in a solid are labelled by two quantum numbers: the wave vector $\vec{k}$ taken in the representation domain of a Brillouin zone and the index κ of small IR q_κ of the wave vector group H [25-28]. The structure of IR of the space group G depends on the left coset decomposition of G with respect to H:

$$G = \sum_i s_i H, \; i = 1 \ldots |G| / |H| \tag{2}$$

Where $|\;|$ stands for the number of elements in a group. The action of rotational elements of coset representatives s_i results in all prongs of the wave vector star $\left\{\vec{k}\right\}$.

The induced representation is defined by the following formula:

$$(q_\kappa \uparrow G)(g)_{i\mu,j\nu} = q_\kappa(s_i^{-1} g s_j)_{\mu\nu} \delta(s_i^{-1} g s_j, H) \tag{3}$$

where: $\delta(s_i^{-1} g s_j, H) = \begin{cases} 1, & if \; s_i^{-1} g s_j \in H \\ 0, & if \; s_i^{-1} g s_j \notin H \end{cases}$

If the basis set of IR q_κ $\{\psi_\mu\}$ is known , the basis set of induced representation is obtained by the action of left coset representatives on it:

$$\{\Psi_{i\mu}\} = \sum_{i} s_i\{\psi_\mu\},\ i = 1...|G|/|H| \tag{4}$$

The induced representation and its basis set defined by formula (4) may be or may not be irreducible. In the case of the space group G and its little group H the induced representation (3) is irreducible and is used throughout [25-28] in solid state physics.

Since the symmetry of one-electron states is defined, they will be used for constructing two-electron states, namely Cooper pairs, making used of general principles of quantum mechanics. According to the Pauli exclusion principle the spatial part of a singlet Cooper pair wavefunction belongs to the symmetrized Kronecker square of the IR $q_\kappa \uparrow G$ and the spatial part of a triplet pair belongs to antisymmetrized Kronecker square of the same IR. In the case of strong spin-orbit coupling Cooper pair's wavefunction pair belongs to the antisymmetrized Kronecker square of the double-valued IR.

The Kronecker square of induced representation may be decomposed making use of the Mackey theorem on symmetrized squares [26]. Its structure depends on the double coset decomposition of the space group G with respect to the wave vector group H:

$$G = \sum_{\sigma} H d_\sigma H \tag{5}$$

For every d_σ, including identity element, we consider an intersection subgroup $M_\sigma = H \cap d_\sigma H d_\sigma^{-1}$ and its representation defined by the formula:

$$P_\sigma = q_\kappa(m) \times q_\kappa(d_\sigma^{-1} m d_\sigma) \tag{6}$$

where $\times$ denotes direct (Kronecker) product of representations.

Corresponding wave vector $\overrightarrow{k}_\sigma$ is defined by the following formula:

$$\overrightarrow{k} + d_\sigma \overrightarrow{k} = \overrightarrow{k}_\sigma + \overrightarrow{b}_\sigma \tag{7}$$

where $\overrightarrow{b}_\sigma$ is a vector of a reciprocal lattice.

For $\overrightarrow{k}$ a general point in Brillouin zone $\overrightarrow{k}_\sigma$ equals zero if d_σ is a space inversion.

For any self-inverse double coset $\sigma = \alpha$ (that is $Hd_\alpha H = Hd_\alpha^{-1}H$) there are two extensions $P_\alpha^\pm$ of P_α into extended intersection subgroup $\tilde{M}_\alpha$:

$$\tilde{M}_\alpha = M_\alpha + aM_\alpha \tag{8}$$

(where $a = d_\alpha h_1 = h_2 d_\alpha$ and $h_1,\ h_2 \in H$). These two extensions correspond to symmetrized and antisymmetrized squares. Their characters on group M_α are the same, and for the elements of the left coset aM_α they are defined by two following formulae :

$$\chi(P_\alpha^+(am)) = +\chi(q_\kappa(amam)) \tag{9}$$

$$\chi(P_{\alpha}^{-}(am)) = -\chi(q_{\kappa}(amam)) \quad (10)$$

In the above notations symmetrized and antisymmetrized parts of the Kronecker square of induced representation are written as:

$$[q_{\kappa} \uparrow G \times q_{\kappa} \uparrow G] = [q_{\kappa} \times q_{\kappa}] \uparrow G + \sum_{\alpha} P_{\alpha}^{+} \uparrow G + \sum_{\beta} P_{\beta} \uparrow G \quad (11)$$

$$\{q_{\kappa} \uparrow G \times q_{\kappa} \uparrow G\} = \{q_{\kappa} \times q_{\kappa}\} \uparrow G + \sum_{\alpha} P_{\alpha}^{-} \uparrow G + \sum_{\beta} P_{\beta} \uparrow G \quad (12)$$

The first items on the right correspond to the double coset defined by the identity element for which the small representations q_{κ} are symmetrized (antisymmetrized) over the H group. The sums in the second items correspond to the self-inverse double cosets discussed above and the third items correspond to the not self-inverse double cosets for which $Hd_{\beta}H \neq Hd_{\beta}^{-1}H$. It follows from the Mackey theorem that types of quasi-particles which it is possible to construct from two equivalent electrons depend on the double coset decomposition of the space-group with respect to the wave vector group. Hence these types depend on the position of a wave vector in one-electron Brillouin zone. In the present work we consider pairs corresponding to the double coset defined by the space inversion. It follows from formula (7) that the wave vector of a pair equals zero in this case and its wavefunctions are classified according to IRs of the point group $\hat{G}$ which is a central extension of a space group G.

TABLE 1. Decomposition of Kronecker products for different planes of symmetry

product	character				decomposition
	E	σ_h	I	C_2	
P_{α}^{+}	1	1	1	1	A_g
P_{α}^{-}	1	1	-1	-1	B_u
plane (001)	h_1	h_{28}	h_{25}	h_4	
$P_{\alpha}^{-} \times E_g \downarrow T_{1g}$	3	-1	-3	1	$2A_u + B_u$
$P_{\alpha}^{-} \times E_g \downarrow C_{2h}$	2	-2	-2	2	$2A_u$
$P_{\alpha}^{-} \times A_{2g} \downarrow C_{2h}$	1	1	-1	-1	B_u
plane (100)	h_1	h_{26}	h_{25}	h_2	
$P_{\alpha}^{-} \times E_g \downarrow C_{2h}$	2	0	-2	0	$A_u + B_u$
$P_{\alpha}^{-} \times A_{2g} \downarrow C_{2h}$	1	-1	-1	1	A_u
plane (110)	h_1	h_{40}	h_{25}	h_{16}	
$P_{\alpha}^{-} \times E_g \downarrow C_{2h}$	2	0	-2	0	$A_u + B_u$
$P_{\alpha}^{-} \times A_{2g} \downarrow C_{2h}$	1	-1	-1	1	A_u

For $\vec{k}$ a general point in a Brillouin zone the extended intersection subgroup $\tilde{M}_{\alpha}$ is a group C_i consisting of the identity element E and the space inversion I. It follows from formulae (6), (9) and (10) that the spatial part of a singlet pair belongs to IR A_g and the spatial part of a triplet pair belongs to IR A_u of a group C_i. The induced representations $P_{\alpha}^{\pm} \uparrow G$ may be decomposed making use of Frobenius reciprocity theorem [28]: *the number of appearance of the IR* Γ *of the group* G *in the decomposition of the induced representation* $P_{\alpha}^{\pm} \uparrow G$ *equals to the number of appearance of the IR* $P_{\alpha}^{\pm}$ *in the decomposition of* Γ *when it subduced to* H. Hence for $\vec{k}$ a general point in a Brillouin zone we obtain that for the spatial part of a singlet pair all even representations are possible and for the spatial part of a triplet pair all odd representations are possible. The same conclusion is also valid for total wavefunctions of Copper pairs since multiplication by even spin singlet S^0 and spin triplet S^1 wavefunctions doesn't change the parity. It also follows from Frobenius reciprocity theorem that for $\vec{k}$ a general point in a Brillouin zone the number of appearance of any IR (even for singlet pairs and odd for triplet pairs) equals to its dimension. As the result, for two-dimensional IRs (E_g and E_u in D_{4h} symmetry) two linearly independent basis sets are possible. This is illustrated in the next section by direct construction of the basis set.

We adopt Kovalev's [25] notations for the point group operations. On the planes of symmetry in Brillouin zone the little group H contains a reflection operators and the extended intersection group $\tilde{M}_{\alpha}$ consists of four elements. The characters of symmetrized and antisymmetrized Kronecker squares belong to IRs A_g and A_u respectively. Since not all the IRs of the group C_{2h} are present, it follows from the Frobenius reciprocity theorem, that some of the IRs of the whole group are forbidden if the $\vec{k}$ vector is on this plane. The intersection of this plane with the Fermi surface results in a line of nodes of any particular IR. If, for example, this IR is responsible for superconductivity this corresponds to the line of nodes of superconducting order parameter. In a weak spin-orbit coupling case lines of nodes follows from the symmetry both for triplet and singlet pairs.

In a strong spin-orbit coupling case spins of electrons are coupled with their spatial parts in each $\vec{k}-\vec{k}$ manifold. To obtain the characters of the total wavefunction one should multiply the characters of the spatial part by the spin singlet and spin triplet wavefunctions. In the highest point symmetry O_h these spin parts belong to IRs A_{1g} and T_{1g} respectively. The character for the singlet pair is unchanged, but the decomposition of the character in a triplet case contains characters of all IRs of group C_{2h}. This means that there are no symmetry requirements for the lines of nodes for triplet pairs in a strong spin-orbital case. (Blount theorem [5]). When the symmetry is reduced from the highest point symmetry O_h to axial symmetry group D_{4h} the IR of spin triplet wavefunction is reduced to two A_{2g} and E_g. Decomposition $P_{\alpha}^{-} \times E_g \downarrow C_{2h}$ contains two IRs $2A_u$ of group C_{2h} and it follows that not all odd IR appear in the induced representation. Similar consideration is valid for the product $P_{\alpha}^{-} \times A_{2g} \downarrow C_{2h}$. If the interaction of electron spin with lattice is not negligible, the spin states E_g and A_{2g} are splitted and generally speaking only one of them corresponds to Cooper pairs. Hence it follows that group theoretical reasons for lines of nodes are possible for triplet states in strong spin-orbit coupling case. Similar decomposition is presented in Table 1. for all planes of symmetry of group D_{4h}. For all planes the spins of a electrons are in z direction. This approach corresponds to strong interaction of electron spin with lattice. Note that in Refs.[19-21] spin-lattice interactions were neglected and spin was quantized along the directions perpendicular to the planes. Making use of Table 1 and Frobenius reciprocity theorem possible IRs of Cooper pairs are constructed for groups D_{2h} and

D_{4h} and presented in Table 2.

TABLE 2. Possible IRs of spatial parts of Cooper pairs for $\vec{k}$ at the planes of symmetry for D_{2h} and D_{4h} point groups.

pair	D_{2h}	D_{4h}
plane (001)		
weak S-O coupling		
singlet	A_g+B_{1g}	$A_{1g}+A_{2g}+B_{1g}+B_{2g}$
triplet	$B_{2u}+B_{3u}$	$2E_u$
strong $S-O$ coupling		
triplet ($\hat{x}$,$\hat{y}$)	$2A_u+2B_{1u}$	$2(A_{1u}+A_{2u}+B_{1u}+B_{2u})$
triplet ($\hat{z}$)	$B_{2u}+B_{3u}$	$2E_u$
plane (100)		
weak $S-O$ coupling		
singlet	A_g+B_{3g}	$A_{1g}+B_{1g}+E_g$
triplet	A_u+B_{3u}	$A_{2u}+B_{2u}+E_u$
strong $S-O$ coupling		
triplet ($\hat{x}$,$\hat{y}$)	$A_u+B_{1u}+B_{2u}+B_{3u}$	$A_{1u}+A_{2u}+B_{1u}+B_{2u}+E_u$
triplet ($\hat{z}$)	A_u+B_{3u}	$A_{1u}+B_{1u}+E_u$
plane (010)		
weak $S-O$ coupling		
singlet	A_g+B_{2g}	$A_{1g}+B_{1g}+E_g$
triplet	A_u+B_{2u}	$A_{2u}+B_{2u}+E_u$
strong $S-O$ coupling		
triplet ($\hat{x}$,$\hat{y}$)	$A_u+B_{1u}+B_{2u}+B_{3u}$	$A_{1u}+A_{2u}+B_{1u}+B_{2u}+E_u$
triplet ($\hat{z}$)	A_u+B_{2u}	$A_{1u}+B_{1u}+E_u$
planes (110) and ($\bar{1}10$)		
weak $S-O$ coupling		
singlet		$A_{1g}+B_{2g}+E_g$
triplet		$A_{2u}+B_{1u}+E_u$
strong $S-O$ coupling		
triplet ($\hat{x}$,$\hat{y}$)		$A_{1u}+A_{2u}+B_{1u}+B_{2u}+E_u$
triplet ($\hat{z}$)		$A_{1u}+B_{2u}+E_u$

If IR responsible for the Cooper pairing is known, the absence of this IR in the decomposition for any plane indicated line of nodes of SOP. On the other hand, if experimental nodal structure of SOP is known, comparison with theoretical data may be useful to obtain index of IR responsible for Cooper pairing [20,21]. For $\vec{k}$ a general point in a Brillouin zone the basis functions belonging to the group $\tilde{M}_\alpha$ are the Anderson [3] functions for singlet and triplet cases. The basis functions for the whole group may be obtained by projection operators technique. Let us denote $\vec{k}_1$ the wave vector chosen in the representation domain of a Brillouin zone. Making use of Kovalev's notation h_{25} for the space inversion the spatial parts of Anderson singlet and triplet functions are written as:

$$\Phi_1^s = \psi_1^1\psi_{25}^2 + \psi_{25}^1\psi_1^2 \tag{13}$$

$$\Phi_1^t = \psi_1^1\psi_{25}^2 - \psi_{25}^1\psi_1^2 \tag{14}$$

Where the superscript of ψ denotes the number of electronic coordinate and subscript of ψ the prong of the $\vec{k}$-vector star.

Acting by h_2 (180° rotations around the axis X) on the functions (13) and (14) we obtain two other basis functions:

$$\Phi_2^s = \psi_2^1\psi_{26}^2 + \psi_{26}^1\psi_1^2 \tag{15}$$

$$\Phi_2^t = \psi_2^1\psi_{26}^2 - \psi_{26}^1\psi_2^2 \tag{16}$$

Note that in Kovalev's [25] notations for O_h group multiplication of pure rotation element by I corresponds to adding 24 to the element number. To construct full basis for D_{2h} group we need also functions $\Phi_3^{s(t)}$ and $\Phi_4^{s(t)}$, which are obtained from $\Phi_1^{s(t)}$ by the action of 180° rotations around the axes Y and Z respectively. In addition, for D_{4h} group the elements h_{13}, h_{16} (180° rotations around the axes $(\bar{1}10)$ and (110)) and h_{14}, h_{15} (90° and 270° counterclockwise rotation around Z axis) are required.

These functions span the space of Anderson functions under the action of all point group operations. Since the space inversion is already included in the basis function, their total number equals to the half of number of point group operations. The action of pure rotations on the initial vector $\vec{k}_1$ result in a star, whose number of prongs is half of the number of prongs in the the wave vector star, and we call it a half-star. The action of the space inversion on the basis vector corresponding to any prong of a half-star doesn't change a vector but introduces multiplies -1 for the triplet case. Making use of standard projection operator technique and functions $\Phi_{1-4}^{s,t}$ we easily obtain the basis functions for Cooper pairs belonging to all IRs of D_{2h} group. The results are presented in Table 3.

Before going to the projection for D_{4h} group it is useful to remind the followig correspondence of IRs in the subduction $D_{4h} \downarrow D_{2h}$: A_1 *and* $B_1 \longrightarrow A_1$, A_2 *and* $B_2 \longrightarrow B_1$, $E \longrightarrow B_2 + B_3$. The basis functions for one-dimensional IRs of D_{4h} group are immediately obtained by projection operator technique. Since each of IRs $E_{g(u)}$ appear twice in the Kronecker product decomposition, there are two independent basis sets labeled by additional quantum numbers . Bearing in mind the above reduction scheme of IRs, we begin with basis sets corresponding to IRs B_2 and B_3 of group D_{2h}, making use of formula (4) and choosing h_{13} as a left coset representative, we obtain the remainder results of Table 3. The results for triplet pairs for D_{4h} group pairs are not presented in the Table 3. To obtain wavefunctions of triplet pairs one should replacc subscripts g to u in the first column and all superscripts s to t in the second column without changing of the signs. For $\vec{k}$ a general point in a Brillouin zone all IRs are possible for Cooper pair. But when the $\vec{k}$-vector approaches any mirror plane, the mirror reflection image of $\vec{k}$ also approaches the

$\vec{k}$-vector. Total number of states decreases and lines of nodes are eventual. There are two possibilities. If two-electron function is unchanged under the action of the reflection, the function under consideration is nonvanishing on the mirror plane. On the other hand, if the function changes its sign, two mirror counterparts are cancelling on the plane. This corresponds to the line of nodes. Note, that the space inversion changes the sign of the spatial part of the triplet function. Making use of the above rules we can easily obtain nodal structure of basis functions of one-dimensional IRs of groups D_{2h} and D_{4h} presented in Table 3. The results are in agreement with the data of Table 2 obtained making use of Mackey theorem.

TABLE 3. Spatial parts of Cooper pair wavefunctions for point groups D_{2h} and D_{4h}

D_{2h}		D_{4h}	
IR	pairing function	IR	pairing function
A_{1g}	$\Phi_1^s+\Phi_2^s+\Phi_3^s+\Phi_4^s$	A_{1g}	$\Phi_1^s+\Phi_2^s+\Phi_3^s+\Phi_4^s+\Phi_{13}^s+\Phi_{14}^s+\Phi_{15}^s+\Phi_{16}^s$
B_{1g}	$\Phi_1^s-\Phi_2^s-\Phi_3^s+\Phi_4^s$	A_{2g}	$\Phi_1^s-\Phi_2^s-\Phi_3^s+\Phi_4^s-\Phi_{13}^s+\Phi_{14}^s+\Phi_{15}^s-\Phi_{16}^s$
B_{2g}	$\Phi_1^s-\Phi_2^s+\Phi_3^s-\Phi_4^s$	B_{1g}	$\Phi_1^s+\Phi_2^s+\Phi_3^s+\Phi_4^s-\Phi_{13}^s-\Phi_{14}^s-\Phi_{15}^s-\Phi_{16}^s$
B_{3g}	$\Phi_1^s+\Phi_2^s-\Phi_3^s-\Phi_4^s$	B_{2g}	$\Phi_1^s-\Phi_2^s-\Phi_3^s+\Phi_4^s+\Phi_{13}^s-\Phi_{14}^s-\Phi_{15}^s+\Phi_{16}^s$
A_{1u}	$\Phi_1^t+\Phi_2^t+\Phi_3^t+\Phi_4^t$	$E_g(B_{2g})$	$\Phi_{13}^s-\Phi_{15}^s+\Phi_{14}^s-\Phi_{16}^s$
B_{1u}	$\Phi_1^s-\Phi_2^s-\Phi_3^s+\Phi_4^s$		$\Phi_1^s-\Phi_2^s+\Phi_3^s-\Phi_4^s$
B_{2u}	$\Phi_1^s-\Phi_2^s+\Phi_3^s-\Phi_4^s$	$E_g(B_{3g})$	$\Phi_1^s+\Phi_2^s-\Phi_3^s-\Phi_4^s$
B_{3u}	$\Phi_1^s+\Phi_2^s-\Phi_3^s-\Phi_4^s$		$\Phi_{13}^s+\Phi_{15}^s-\Phi_{14}^s-\Phi_{16}^s$

Two dimensional IRs appear twice for $\vec{k}$ a general point in a Brillouin zone. It is seen from Table 2 for group D_{4h}, that in the basal plane, two-dimensional IRs are forbidden in some cases, and in vertical planes they appear only once. In this case a direct analysis of nodal structure of basis functions of Table 3 is required. The analysis shows that basis functions of IR $E_g(B_{2g})$ vanish in the planes(100) and (001) and that of IR $E_g(B_{3g})$ vanish in the planes (010) and (001). Linear combinations of these basis functions $E_g(B_{2g})\pm E_g(B_{3g})$ vanish in planes (110) and $(\bar{1}10)$ respectively and both vanish in plane (001). Hence it follows that only lines of nodes in basal plane follow unambiguously from the symmetry. It should be noted, that point group approach also results different nodal structure of different two-dimensional IRs [4].

DISCUSSION OF THE RESULTS

The analysis of broad set of experimental data on the of high-T_c superconductors [1] led the most of the authors to the conclusion of singlet pairing and A_g SOP symmetry in these compounds. Angular resolved photoelectron spectra of high-T_c superconductors [29] reveal a strong trough in the diagonal of xy plane indicating $d_{x^2-y^2}$ -pairing with line of nodes. On the other hand some experiments reveal also totally symmetric s -pairing without nodes. In many cases an interplay between these two types of pairing [1] both belonging to A_g IR exists. It is seen from tables 2 and 3, that A_g pairing function, obtained group theoretically is noddles and that other IRs have nodes in the coordinate planes only. Hence it follows that nodal structure of high-T_c superconductors is more complex then that which follows from the symmetry only. To explain this one can consider two wave vectors $\vec{k}_{-}$ and

$\vec{k}_\beta$ symmetrical with respect to diagonal of the deformed square. Note, that the orthorombicity [(b-a)/(b+a)] of YBCuO is about 2 % only [1]. Two types of basis functions of Cooper pairs belonging to A_g IR Φ^s_α and Φ^s_β are easily obtained from the Table 3 by introducing additional subscripts α and β. One can suppose that due to the interaction two self-vectors are linear combinations of these basis states :

$$\Phi^S_s = C_\alpha \left(\Phi^S_{1,\alpha} + \Phi^S_{2,\alpha} + \Phi^S_{3,\alpha} + \Phi^S_{4,\alpha}\right) + C_\beta \left(\Phi^S_{1,\beta} + \Phi^S_{2,\beta} + \Phi^S_{3,\beta} + \Phi^S_{4,\beta}\right) \tag{17}$$

$$\Phi^S_{x^2-y^2} = C_\beta \left(\Phi^S_{1,\alpha} + \Phi^S_{2,\alpha} + \Phi^S_{3,\alpha} + \Phi^S_{4,\alpha}\right) - C_\alpha \left(\Phi^S_{1,\beta} + \Phi^S_{2,\beta} + \Phi^S_{3,\beta} + \Phi^S_{4,\beta}\right) \tag{18}$$

Both combinations belong to IR A_g of group D_{2h}. First one corresponds to the noddles s-pairing and the second to the $d_{x^2-y^2}$−pairing with line of nodes in the diagonal of xy- plane. In the limit of zero orthorombic distortion the symmetry group is D_{4h} and $C_\alpha = C_\beta$, subscripts α and β are dropped and the sums in second brackets in right hand sides of (17) and (18) are written as $\Phi^S_{13} + \Phi^S_{14} + \Phi^S_{15} + \Phi^S_{16}$. In this case combination (17) belongs to IR A_{1g} and combination (18) belongs to IR B_{1g} of the symmetry group D_{4h}. Hence it follows that the nodal structure of SOP in high- T$_c$ superconductors is defined by hidden symmetry D_{4h}.

CONCLUSION

Space group approach to the wavefunction of a Cooper pair is developed by making use of projection operator technique and applied to construction of Cooper pair wavefunction in D_{2h} and D_{4h} symmetries relevant to high-T$_c$ superconductors. The nodal structure of SOP is obtained and discussed for D_{2h} and D_{4h} symmetries. It is shown that violations of Blount theorem accompany symmetry breaking to axial point groups. It results in appearance of lines of nodes for triplet SOP in strong spin-orbit coupling case. It is obtained that experimentally observed nodal structure of SOP in high-T$_c$ superconductors follow from the hidden symmetry D_{4h}.

ACKNOWLEDGEMENTS

Author is grateful to International Institute for Advanced Scientific Studies "Eduardo R.Caianiello" and Russian Foundation for Basic Researches for financial support of his participation in VII Training Course in the Physics of Strongly Correlated Electron Systems and High Temperature Superconductors. The research is supported by Russian Foundation for Basic Researches.

REFERENCES

1. Tsuel C.C. and Kirtley J.R., *Rev.Mod.Phys.* **72** , 969-1016 (2000).
2. Sigrist M. and Ueda K., *Rev. Mod.Phys.* **63**, 239-311 (1991).
3. Anderson P. W., *Phys.Rev. B* **30**, 4000-4002 (1984).
4. Volovik G. E.and Gor'kov L. P., *Sov. Phys. Jetp.* **61**, 843-859 (1985).
5. Blount E. I., *Phys. Rev.B* **32**, 2935-2944 (1985).
6. Ozaki M.-A.and Machida K. , *Phys.Rev. B* **39**, 4145-4163 (1989).
7. Ozaki M.-A., Machida K.and Ohmi T., *Prog.Theor.Phys.* **75**, 442-444 (1986).
8. Ueda K. and Rice T.M., *Phys. Rev. B* **31**, 7114-7119 (1985).

9. Sauls J.A., *Adv. Phys.* **43**, 113-141 (1994).
10. Yip S.and Garg A., *Phys. Rev. B* **48**, 3304-3308 (1993).
11. Sigrist M. and Rice T.M., *Z. Phys. B.* **68,** 9-14 (1987).
12. Annett J.F., Goldenberg N.D. and Renn S.R. *Phys. Rev. B* **43,** 2778-2782(1991).
13. Li Q.P., Koltenbah B.E. and Joint R., *Phys.Rev.B.* **48,** 437-455 (1993).
14. Wenger F. and Östlund S., *Phys.Rev.B.* **47**, 5977-5983 (1993).
15. Izyumov Yu. A., Laptev V. M. and Syromyatnikov V. N., *Int.J. Mod. Phys.* **3**, 1377-1402 (1989).
16. Yarzhemsky V.G., *Superconductivity:phys. chem. techn.* **3(11)**, 2504-2510 (1990) (in Russian).
17. Yarzhemsky V.G. and Murav'ev E.N., *Sov. Phys. Dokl.* **36(4)**, 323-326 (1991).
18. Yarzhemsky V. G.and . Murav'ev E. N, *J.Phys:Cond.Matter* **4**, 3525-3532 (1992).
19. Yarzhemsky V.G., *Z. Phys. B* **99**, 19-23 (1995).
20. Yarzhemsky V. G., *phys. stat. sol. (b)* **209**, 101-107 (1998).
21. Yarzhemsky V.G., *Int. J. Quant. Chem.* **80**, 133-140 (2000).
22. Zhang S.C., *Science* **275**, 1089-1096 (1997).
23. Rabello S., Kohno H., Dehmer E., Zhang S.C., *Phys. Rev.Let.* **80** 3586-3589 (1998).
24. Murakami S., Nagaoka N. and Sigrist M., *Phys.Rev.Lett.* **82**, 2939-2942 (1999).
25. Kovalev O. V., *Irreducible and Induced Representations and Corepresentations of Fedorov's Groups* (Moscow: Nauka, 1986, in Russian).
26. Bradley C. J. and Davies B. L., *J. Math. Phys.* **11**, 1536-1552 (1970).
27. Bradley C. J. and Cracknell A. P., *The Mathematical Theory of Symmetry in Solids. Representation Theory of Point Groups and Space Groups* (Oxford: Clarendon, 1972).
28 Altman S. L., *Induced Representations for Crystals and Molecules.* (New York Academic Press 1977).
29 Ding H.M., Norman M.R., Campuzano J.C., Randeria M., Bellman A.F., Yokoya T., Takahashi T., Mochiki T. and Kadowaki K., *Phys. Rev. B* **54**, R9678-R9681 (1996).

Two-Dimensional *t*-*J* Model in a Staggered Field

Yasuhiro Saiga* and Masaki Oshikawa*

Department of Physics, Tokyo Institute of Technology, Oh-okayama, Meguro-ku, Tokyo 152-8551, Japan

Abstract. The two-dimensional *t*-*J* model in a staggered field is studied by exact diagonalization up to 20 sites. For the low-hole-density region and a realistic value of J/t, it is found that the presence of a staggered field strengthens the attraction between two holes. With increasing field, the $d_{x^2-y^2}$-wave superconducting correlations are enhanced while the extended-s-wave ones hardly change. This implies that coexistence of the $d_{x^2-y^2}$-wave superconducting order and the commensurate antiferromagnetic order occurs in a staggered field.

INTRODUCTION

One of the contentious issues in high-T_c cuprates is the interplay between superconductivity and antiferromagnetism. Recent experiments have updated information about this issue [1, 2, 3, 4, 5, 6]. Elastic neutron scattering experiments in $YBa_2Cu_3O_y$ with $y = 6.5$ [1] and 6.6 [2, 5] show that magnetic intensity emerges near room temperature at the momentum (π, π) in units of the reciprocal lattice parameter. The intensity increases continuously with decreasing temperature. Remarkably an upturn of the intensity is observed at a superconducting transition temperature. This suggests coexistence of superconductivity and the commensurate antiferromagnetic (AF) order. Nuclear quadrupole resonance (NQR) measurements have revealed the presence of magnetic moments in the superconducting state on $Hg_{0.8}Cu_{0.2}Ba_2Ca_2Cu_3O_{8+\delta}$ [6]. The observed magnetic moments in these materials are often regarded as a consequence of the formation of the d-density wave (DDW) order [7]. However, the upturn of moments in the superconducting state may be due to different origin because the DDW and superconductivity compete against each other [8]. Another candidate for the origin of the moments is Cu spin magnetism. Thus it is highly required to consider theoretically whether the commensurate AF order induced by Cu spin coexists with superconductivity.

For the ordinary two-dimensional (2D) *t*-*J* model, the possible coexistence between antiferromagnetism (by spin) and $d_{x^2-y^2}$-wave superconductivity has been discussed at low hole doping, from variational approach [9] and quantum Monte Carlo simulations [10]. Commensurate AF correlations should be strengthened by a staggered field at arbitrary hole doping, and would attain the long-range order. If the commensurate AF order favors the pairing formation, superconducting correlations would be enhanced by a staggered field. In fact, for the 1D *t*-*J* model in a staggered field, the correlation exponent near half filling is larger than unity for a realistic value of J/t, which means that superconducting correlations are the most dominant [11, 12].

The purpose of this paper is to demonstrate that the presence of a staggered field

CP678, *Lectures on the Physics of Highly Correlated Electron Systems VII: Seventh Training Course*, edited by A. Avella and F. Mancini

strengthens the attraction between two holes and helps the $d_{x^2-y^2}$-wave superconductivity within the *t*-*J* model on a square lattice. To this end, exact diagonalization is employed for the 4×4, $\sqrt{18}\times\sqrt{18}$ and $\sqrt{20}\times\sqrt{20}$ clusters with periodic boundary conditions [13]. We also show that the hole-pair hopping process is produced from the second-order perturbation in a strong staggered field.

MODEL

We consider the following Hamiltonian given by

$$\begin{aligned} H &= -t\sum_{\langle\vec{i},\vec{j}\rangle\sigma}\left(\tilde{c}^{\dagger}_{\vec{i}\sigma}\tilde{c}_{\vec{j}\sigma}+\text{H.c.}\right)+J\sum_{\langle\vec{i},\vec{j}\rangle}\left(\vec{S}_{\vec{i}}\cdot\vec{S}_{\vec{j}}-\frac{1}{4}n_{\vec{i}}n_{\vec{j}}\right) \\ &\quad -h\sum_{\vec{i}\in\mathrm{A}}S^z_{\vec{i}}+h\sum_{\vec{j}\in\mathrm{B}}S^z_{\vec{j}}, \end{aligned} \tag{1}$$

where $\langle\vec{i},\vec{j}\rangle$ is nearest neighbors. The constrained fermion operator $\tilde{c}_{\vec{i}\sigma}$ is given by $\tilde{c}_{\vec{i}\sigma}=c_{\vec{i}\sigma}(1-n_{\vec{i},-\sigma})$, which means that double occupancy at each site is excluded. The last two terms are due to the presence of a staggered field whose magnitude is denoted by *h*; A and B represent the two sublattices on a square lattice. We refer to this model as the *t*-*J*-*h* model. In this work we fix $J/t=0.4$ which is considered as a realistic value, and vary h/t as a parameter.

NUMERICAL RESULTS

We first discuss the binding energy defined by [14]

$$E_{\mathrm{B}}=E_0(N_{\mathrm{h}}=2)+E_0(N_{\mathrm{h}}=0)-2E_0(N_{\mathrm{h}}=1), \tag{2}$$

where $E_0(N_{\mathrm{h}})$ denotes the ground-state energy with N_{h} holes in N sites. A negative value of E_{B} indicates the presence of hole binding. In Fig. 1 we show the dependence of the binding energy on the staggered field *h*. The binding energy is negative in the whole range of h/t and has a peak at $h/t\sim0.8, 1.2$, and 1.2 for $N=16, 18$, and 20, respectively. The obtained results apparently implies that hole pairing is suppressed by a small staggered field. However, being an intensive quantity, the binding energy is severely affected by finite-size effects [15]. In fact, although at $h=0$ (and $J/t=0.4$) the binding energy for $N\leq20$ takes negative values, an extrapolation from finite-size data up to $N=32$ rather indicates a positive value in the thermodynamic limit [16, 17]. Therefore, it is difficult to discuss the effect of a staggered field on hole pairing from the present data on the binding energy. Thus, in the following we will study other quantities which is less sensitive to finite-size effects. On the other hand, for a large h/t the binding energy is still negative and there is little size dependence. This suggests that hole pairing does occur at sufficiently large h/t.

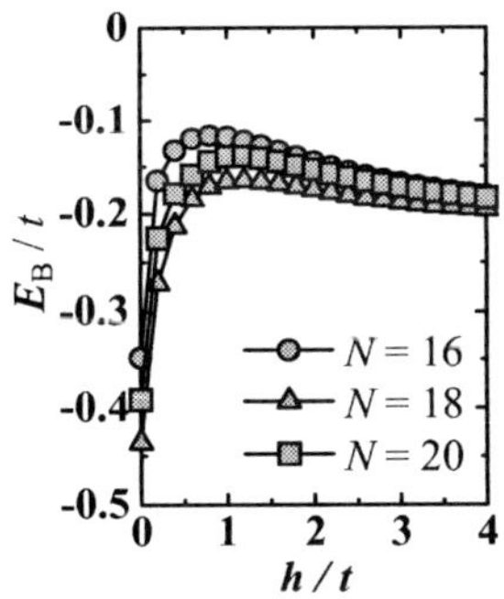

FIGURE 1. Binding energy as a function of h/t in the 2D t-J-h model with $J/t = 0.4$.

A tendency of hole binding can be obtained also from the hole correlation function, which is given by

$$C_{\text{hole}}(\vec{r}) = \frac{1}{N}\sum_{\vec{i}} \langle n_{\text{h}}(\vec{i})n_{\text{h}}(\vec{i}+\vec{r})\rangle, \tag{3}$$

where $n_{\text{h}}(\vec{i}) = 1 - n_{\vec{i}}$. In the left panel of Fig. 2 we show the distance dependence of $C_{\text{hole}}(r)$ with $r \equiv |\vec{r}|$ for the hole densities $N_{\text{h}}/N = 2/18 \simeq 0.111$ and $4/18 \simeq 0.222$. For two holes and $h = 0$, the most dominant correlations are at $r = \sqrt{2}$, namely, when the holes stay at the nearest distance on the same sublattice [18, 19]. As h/t increases, correlations at the nearest distance on the lattice ($r = 1$) become stronger than those at $r = \sqrt{2}$, and contribution at longer distances is suppressed. For four holes, correlations at $r = 1$ are enhanced while ones at the largest distance hardly change. This means that the presence of a staggered field makes the interaction between two holes attractive but the hole-pairs are well separated.

In order to analyze the obtained data in the two-hole case, we calculate the root-mean-square separation of the hole pair [15, 20] defined as $r_{\text{rms}} \equiv \sqrt{\langle r^2\rangle}$ where

$$\langle r^2\rangle = \sum_{\vec{r}(\neq\vec{0})} |\vec{r}'|^2 C_{\text{hole}}(\vec{r}) \Big/ \sum_{\vec{r}(\neq\vec{0})} C_{\text{hole}}(\vec{r}). \tag{4}$$

Here $|\vec{r}'|$ denotes the substantial distance between two sites under periodic boundary conditions. The right panel of Fig. 2 shows the h/t-dependence of r_{rms}. It is clear that the separation of the hole pair becomes smaller with increasing staggered field. For a large h/t the value of r_{rms} is less affected by finite-size effects, as the binding energy is. The present result suggests that the staggered field does help binding of two holes, and that hole binding survives in the thermodynamic limit at least for a sufficiently large h/t.

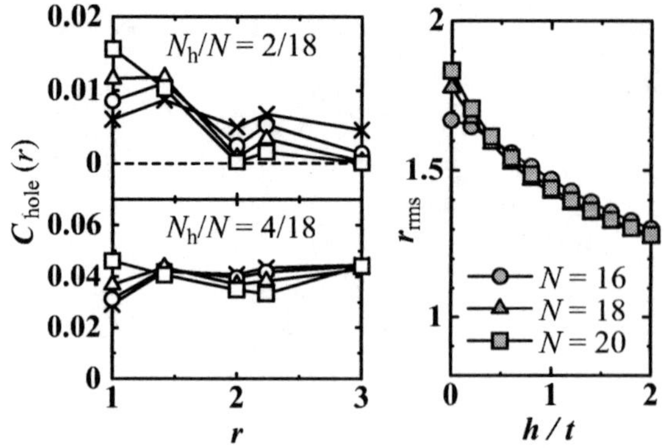

FIGURE 2. Left: Equal-time hole correlations as a function of distance in the 2D t-J-h model with $N = 18$, $J/t = 0.4$, and various values of h/t. Crosses, circles, triangles, and squares are the data for $h/t = 0.0, 0.4, 1.0$, and 2.0, respectively. Right: Root-mean-square separation of the hole pair as a function of h/t. $J/t = 0.4$.

Next we turn to exploration into superconductivity. We calculate the equal-time superconducting correlations given by [21, 22]

$$C_\alpha(\vec{r}) = \frac{1}{N}\sum_{\vec{i}}\langle\Delta_\alpha^\dagger(\vec{i})\Delta_\alpha(\vec{i}+\vec{r})\rangle. \tag{5}$$

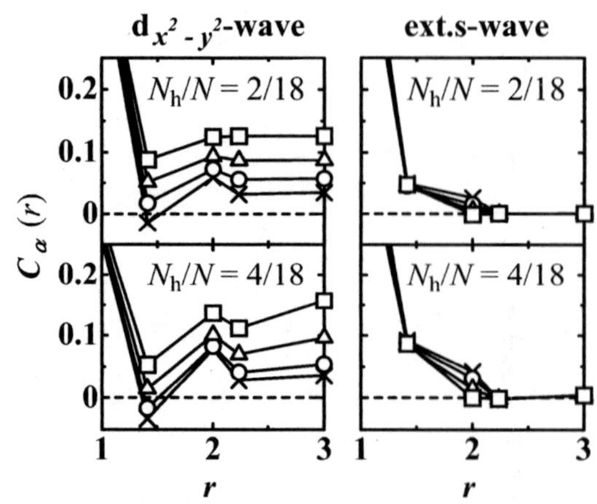

FIGURE 3. Equal-time superconducting correlations as a function of distance in the 2D t-J-h model with $N = 18$, $J/t = 0.4$, and $h/t = 0.0, 0.4, 1.0$, and 2.0. The symbols are the same as in the left panel of Fig. 2.

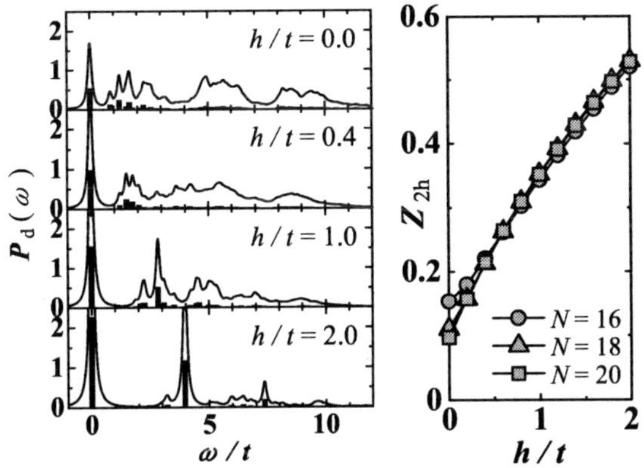

FIGURE 4. Left: Pair spectral function with $d_{x^2-y^2}$-wave symmetry in the 2D t-J-h model with 18 sites, $J/t = 0.4$, and various values of h/t. The delta functions (vertical bars) are broadened by a Lorentzian with a width of $0.1t$ (solid curves). Right: Spectral weight Z_{2h} as a function of h/t. $J/t = 0.4$.

The singlet pairing operator $\Delta_\alpha(\vec{i})$ is defined as

$$\Delta_\alpha(\vec{i}) = \frac{1}{\sqrt{2}} \sum_{\vec{\varepsilon}} f_\alpha(\vec{\varepsilon}) \left(c_{\vec{i}\uparrow} c_{\vec{i}+\vec{\varepsilon},\downarrow} - c_{\vec{i}\downarrow} c_{\vec{i}+\vec{\varepsilon},\uparrow} \right), \tag{6}$$

where $\vec{\varepsilon}$ is $(\pm 1, 0)$ and $(0, \pm 1)$. For the extended-s-wave pairing symmetry ($\alpha = \text{s}$), we put $f_s(\vec{\varepsilon}) = +1$ at all $\vec{\varepsilon}$. For the $d_{x^2-y^2}$ symmetry ($\alpha = \text{d}$), we put $f_d(\vec{\varepsilon}) = +1$ at $\vec{\varepsilon} = (\pm 1, 0)$ and $f_d(\vec{\varepsilon}) = -1$ at $\vec{\varepsilon} = (0, \pm 1)$. Figure 3 shows the distance dependence of $C_d(r)$ and $C_s(r)$ for the hole densities 0.111 and 0.222 in 18 sites. For $n_h \lesssim 0.2$, with increasing field, the $d_{x^2-y^2}$-wave superconducting correlations are enhanced at all distances with $r \geq 1$. In contrast, the extended-s-wave ones hardly change, especially at long distances. This implies that the presence of a staggered field helps the $d_{x^2-y^2}$-wave superconductivity.

Calculation of the pair spectral function [23, 24] should provide another evidence for the $d_{x^2-y^2}$-wave pairing enhanced by a staggered field. The pair spectral function is given by

$$\begin{aligned} P_\alpha(\omega) &= \sum_n |\langle \Psi_n(N_h = 2) | \Delta_\alpha^{\text{tot}} | \Psi_0(N_h = 0) \rangle|^2 \\ &\quad \times \delta(\omega - E_n(N_h = 2) + E_0(N_h = 0) + \mu), \end{aligned} \tag{7}$$

where $\Delta_\alpha^{\text{tot}} = \sum_{\vec{i}} \Delta_\alpha(\vec{i})/\sqrt{N}$, $\mu = E_0(N_h = 2) - E_0(N_h = 0)$, and $|\Psi_n(N_h)\rangle$ denotes an eigenstate with energy $E_n(N_h)$ in the N_h-hole system. The left panel of Fig. 4 shows the ω-dependence of $P_d(\omega)$ for various values of h/t. The overall feature approximated by a Lorentzian is insensitive to the system size at each h/t. The peak at $\omega = 0$ (i.e., the

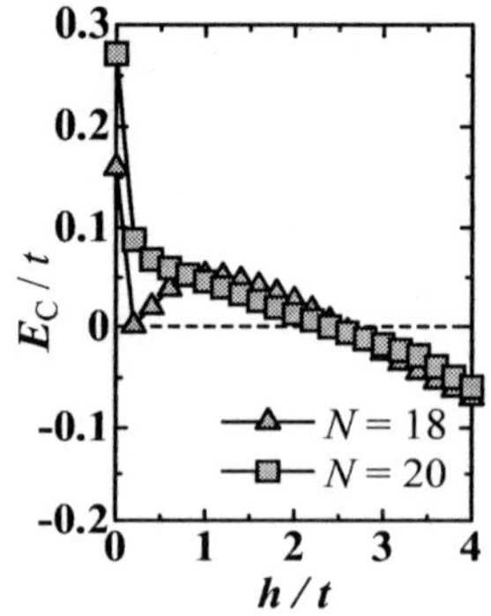

FIGURE 5. Clustering energy as a function of h/t in the 2D t-J-h model with $J/t = 0.4$.

coherent peak) grows with increasing h/t. The contribution for $\omega > 0$, which seems to be a continuum spectrum, is relatively suppressed, but a peak with secondary dominant intensity appears. Concerning this peak for $h/t = 2.0$, the values of energy ω and residue z are $(\omega/t, z) = (3.94, 0.268), (3.96, 0.273)$, and $(3.95, 0.265)$ for $N = 16, 18$, and 20, respectively. The weak size dependence of both values of ω and z indicates that the secondary peak may be a delta-function contribution rather than a part of continuum spectrum in the bulk limit.

We estimate the spectral weight defined as [24, 25]

$$Z_{2\mathrm{h}} = \frac{|\langle \Psi_0(N_\mathrm{h} = 2)|\Delta_\mathrm{d}^\mathrm{tot}|\Psi_0(N_\mathrm{h} = 0)\rangle|^2}{\langle \Psi_0(N_\mathrm{h} = 0)|(\Delta_\mathrm{d}^\mathrm{tot})^\dagger \Delta_\mathrm{d}^\mathrm{tot}|\Psi_0(N_\mathrm{h} = 0)\rangle}, \tag{8}$$

which corresponds to the coherent peak of $P_\mathrm{d}(\omega)$ at $\omega = 0$. Note that $Z_{2\mathrm{h}}$ is between 0 and 1 because the denominator of Eq. (8) is equal to integration of $P_\mathrm{d}(\omega)$ over ω. The right panel of Fig. 4 shows $Z_{2\mathrm{h}}$ as a function of h/t. The weight is monotonically increasing function of h/t. Again, the size dependence of $Z_{2\mathrm{h}}$ is weak for a large h/t. Therefore we expect that the coherent peak survives in the thermodynamic limit for a sufficiently large h/t.

While our results suggest the enhancement of hole pairing by a staggered field, it is possible that the attraction between holes leads to phase separation. We calculate the clustering energy defined by [14]

$$E_\mathrm{C} = E_0(N_\mathrm{h} = 4) + E_0(N_\mathrm{h} = 0) - 2E_0(N_\mathrm{h} = 2). \tag{9}$$

If this quantity is negative, phase separation is expected to occur. The results for $N = 18$[1] and 20 are shown in Fig. 5, which suggests that the region $0 \le h/t \lesssim 2$ is not interrupted by phase separation.

[1] The clustering energy for $N = 18$ has a dip at $h/t \sim 0.2$, but this is likely to be due to finite-size effects.

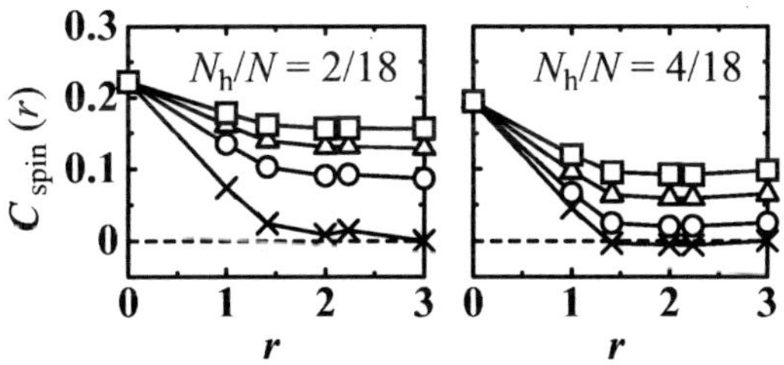

FIGURE 6. Equal-time staggered-spin correlations as a function of distance in the 2D t-J-h model with $N = 18$, $J/t = 0.4$, and $h/t = 0.0, 0.4, 1.0$, and 2.0. The symbols are the same as in the left panel of Fig. 2.

When the staggered field is present, the finite magnetic moment per site should be produced. Therefore the commensurate AF long-range order is expected to be realized. In Fig. 6 we show the result on the staggered-spin correlations given by

$$C_{\text{spin}}(\vec{r}) = \frac{1}{N}\sum_{\vec{i}} (-1)^{r_x+r_y} \langle S^z_{\vec{i}} S^z_{\vec{i}+\vec{r}} \rangle, \tag{10}$$

with $\vec{r} = (r_x, r_y)$. Actually the correlations are found to reach up to long distances for $h/t > 0$ as expected. An important point is that *both* the $d_{x^2-y^2}$-wave superconducting correlations and the staggered-spin ones are enhanced by a staggered field. Therefore we expect the presence of the $d_{x^2-y^2}$-wave superconducting order and the commensurate AF order in a certain range of h/t.

PERTURBATION FROM LARGE-H/T LIMIT

Why does the presence of a staggered field help the pairing formation? This can be understood from the large-h/t limit. In order to treat analytically, we consider the t-J_z-h model where the isotropic Heisenberg term (J) in Eq. (1) is replaced by the Ising term (J_z) [11, 12]. For $h/t \gg 1$ we can regard the single-particle hopping term (t) as a perturbative one. In the Hilbert space with all spins along the direction of the staggered field, the second-order perturbation leads to the effective Hamiltonian given by $H_{\text{eff}} = P\tilde{H}_{\text{eff}}P$ where

$$\begin{aligned}\tilde{H}_{\text{eff}} &= J_z \sum_{\langle\vec{i},\vec{j}\rangle} \left(S^z_{\vec{i}} S^z_{\vec{j}} - \frac{1}{4} n_{\vec{i}} n_{\vec{j}} \right) \\ &-\tilde{t} \sum_{\langle\vec{i},\vec{j},\vec{\ell}\rangle\sigma} \left[\tilde{c}^{\dagger}_{\vec{i}\sigma}(1-n_{\vec{j}})\tilde{c}_{\vec{\ell}\sigma} + (1-n_{\vec{i}})(1-n_{\vec{j}})n_{\vec{\ell}\sigma} + \text{H.c.} \right] \\ &+(\text{other terms}).\end{aligned} \tag{11}$$

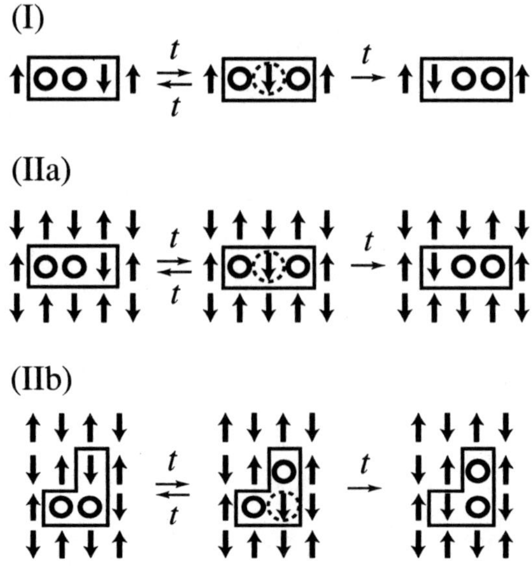

FIGURE 7. Typical processes of the second-order perturbation in the t-J_z-h model for $h/t \gg 1$. (I) is the 1D case, while (IIa) and (IIb) are the 2D case. The spin surrounded by a broken circle is against the direction of the staggered field. The three sites surrounded by a box correspond to $\langle \vec{i}, \vec{j}, \vec{\ell} \rangle$ in the second term of Eq. (11).

Here $\langle \vec{i}, \vec{j} \rangle$ and $\langle \vec{i}, \vec{j}, \vec{\ell} \rangle$ are nearest neighbors, and $P = \prod_{\vec{i} \in \mathrm{A}} (1 - n_{\vec{i}\downarrow}) \prod_{\vec{j} \in \mathrm{B}} (1 - n_{\vec{j}\uparrow})$. The second term in the right-hand side of Eq. (11) includes hopping of a hole which occurs only if there is another hole in the neighboring site. Namely, it gives hopping of a hole pair. This is generated in the second-order processes shown in Fig. 7. In the 2D case the pair-hopping integral $\tilde{t}$ is given by $t^2/(h + 3J_z/2)$ where the denominator indicates the energy difference between the initial state and the intermediate one [see Fig. 7(IIa) and 7(IIb)]. We note that $\tilde{t}$ is replaced by $t^2/(h + J_z/2)$ in the 1D case [see Fig. 7(I)] [11, 12].

The fact that the effective Hamiltonian includes the hole-pair hopping implies that hole binding and superconductivity may be naturally realized in the antiferromagnetic background forced by the staggered field. The effective pair-hopping integral $\tilde{t}$ decreases as h/t increases. This suggests that a too large staggered field makes the hole pairs less mobile, which may be related to the phase separation at large h/t discussed above. On the other hand, it does not mean that a small h/t is better for superconductivity, because the calculation is based on the perturbation theory in t/h.

SUMMARY

We have investigated the binding energy and various correlation functions for the 2D t-J model in a staggered field. For the low-hole-density region and a realistic value of J/t, the presence of a staggered field strengthens the attraction between two holes and helps the $d_{x^2-y^2}$-wave superconductivity. This implies that coexistence of the $d_{x^2-y^2}$-wave superconducting order and the commensurate antiferromagnetic order is likely to occur in a staggered field.

ACKNOWLEDGMENTS

We thank G. Misguich for calling our attention to recent experiments on high-T_c cuprates. Y.S. is supported by JSPS Research Fellowships for Young Scientists. The present work is supported in part by Grant-in-Aid for Scientific Research from MEXT of Japan.

REFERENCES

1. Y. Sidis, C. Ulrich, P. Bourges, C. Bernhard, C. Niedermayer, L.P. Regnault, N.H. Andersen, and B. Keimer, Phys. Rev. Lett. **86**, 4100 (2001).
2. H.A. Mook, P. Dai, and F. Doğan, Phys. Rev. B **64**, 012502 (2001).
3. J.A. Hodges, Y. Sidis, P. Bourges, I. Mirebeau, M. Hennion, and X. Chaud, Phys. Rev. B **66**, 020501 (2002).
4. C. Stock, W.J.L. Buyers, Z. Tun, R. Liang, D. Peets, D. Bonn, W.N. Hardy, and L. Taillefer, Phys. Rev. B **66**, 024505 (2002).
5. H.A. Mook, P. Dai, S.M. Hayden, A. Hiess, J.W. Lynn, S.-H. Lee, and F. Doğan, Phys. Rev. B **66**, 144513 (2002).
6. H. Breitzke, I. Eremin, D. Manske, E.V. Antipov, and K. Lüders, cond-mat/0210652.
7. S. Chakravarty, R.B. Laughlin, D.K. Morr, and C. Nayak, Phys. Rev. B **63**, 094503 (2001).
8. K. Hamada and D. Yoshioka, cond-mat/0211406.
9. A. Himeda and M. Ogata, Phys. Rev. B **60**, R9935 (1999).
10. S. Sorella, G.B. Martins, F. Becca, C. Gazza, L. Capriotti, A. Parola, and E. Dagotto, Phys. Rev. Lett. **88**, 117002 (2002).
11. J. Bonča, P. Prelovšek, I. Sega, H.Q. Lin, and D.K. Campbell, Phys. Rev. Lett. **69**, 526 (1992).
12. P. Prelovšek, I. Sega, J. Bonča, H.Q. Lin, and D.K. Campbell, Phys. Rev. B **47**, 12224 (1993).
13. E. Dagotto, Rev. Mod. Phys. **66**, 763 (1994).
14. J. Riera and A.P. Young, Phys. Rev. B **39**, 9697 (1989).
15. D. Poilblanc, J. Riera, and E. Dagotto, Phys. Rev. B **49**, 12318 (1994).
16. C.T. Shih, Y.C. Chen, H.Q. Lin, and T.K. Lee, Phys. Rev. Lett. **81**, 1294 (1998).
17. A.L. Chernyshev, P.W. Leung, and R.J. Gooding, Phys. Rev. B **58**, 13594 (1998).
18. D. Poilblanc, Phys. Rev. B **49**, 1477 (1994).
19. D. Poilblanc, Phys. Rev. B **52**, 9201 (1995).
20. P.W. Leung, Phys. Rev. B **65**, 205101 (2002).
21. E. Dagotto and J. Riera, Phys. Rev. Lett. **70**, 682 (1993).
22. E. Dagotto, J. Riera, Y.C. Chen, A. Moreo, A. Nazarenko, F. Alcaraz, and F. Ortolani, Phys. Rev. B **49**, 3548 (1994).
23. E. Dagotto, J. Riera, and A.P. Young, Phys. Rev. B **42**, 2347 (1990).
24. D. Poilblanc, Phys. Rev. B **48**, 3368 (1993).
25. E. Dagotto and J.R. Schrieffer, Phys. Rev. B **43**, 8705 (1991).

Author Index